Samuel Plumbe

A Practical Treatise on the Diseases of the Skin

Arranged with a View to their Constitutional Causes and Local Characters

Samuel Plumbe

A Practical Treatise on the Diseases of the Skin
Arranged with a View to their Constitutional Causes and Local Characters

ISBN/EAN: 9783337269913

Printed in Europe, USA, Canada, Australia, Japan

Cover: Foto ©berggeist007 / pixelio.de

More available books at **www.hansebooks.com**

SELECT MEDICAL LIBRARY
Extras.

SUBSCRIBERS to the *Library* and *Bulletin*, and the *Medical Faculty in general*, are respectfully informed that the Publishers will furnish the following Works as *Extras;* for which purpose they are stitched in thick paper covers, with strong elastic backs, similar to the *regular* numbers; they can be sent by mail at the Periodical charge for Postage, which is per sheet, if under 100 miles, $1\frac{1}{2}$ cents, exceeding that distance, $2\frac{1}{2}$ cents.

To the name of each work is stated its number of sheets and the selling price; so that any gentleman desirous of having one or more *Extras* will, by remitting us a note, (or order payable in Philadelphia,) be furnished, by return of mail, with whatever he may select, to the amount.

In order to encourage literature and the sciences generally, the Post Office regulations on Periodicals are such that remittances can be made with little or no expense to Subscribers or Publishers,—the Postmaster enjoying the *privilege* of *franking* all such letters; which, we are pleased to state, their courtesy has generally led them to exercise, and for which they will please accept our thanks.

BARRINGTON & HASWELL.

N.B. Those works comprised within brackets are bound in one volume, and must be ordered as one Extra.

LEE'S OBSERVATIONS on the PRINCIPAL MEDICAL INSTITUTIONS and PRACTICE of FRANCE, ITALY, and GERMANY, &c., with an Appendix on ANIMAL MAGNETISM and HOMŒOPATHY.
JOHNSTONE'S SYLLABUS of MATERIA MEDICA.
LATHAM'S LECTURES ON CLINICAL MEDICINE.
} 14 sheets { $1 35

A TREATISE ON TETANUS, by Thomas B. Curling.
BOUILLAUD ON ACUTE ARTICULAR RHEUMATISM in general. *Translated from the French,* by James Kitchen, M.D.
} 8 sheets { 0 80

PRACTICAL OBSERVATIONS on DISEASES of the HEART, LUNGS, STOMACH, LIVER, &c. By John Marshall, M.D., &c.
WEATHERHEAD on DISEASES of the LUNGS; considered especially in relation to the particular Tissue affected illustraitng the different kinds of Cough.
} 8 sheets { $0 80

†

PRICHARD on INSANITY and other DISEASES affecting the MIND.
14 sheets. 1 25

EPIDEMICS of the MIDDLE AGES, viz. *The Black Death* and *Dancing
Mania;* translated from the German of Hecker, by Dr. Babington, F.R.S.
7 sheets. 0 60

The ECLECTIC JOURNAL of MEDICINE, by John Bell, M.D., from
November, 1836, to October, 1837. 19 sheets. . . . 2 00

PLUMBE on DISEASES of the SKIN; with splendid *coloured Engravings.*
17 sheets. 2 25

TURNBULL'S TREATISE on the MEDICAL PROPERTIES of the ⎫
Natural order RANUNCULACEÆ, &c. &c. ⎪ 9 sheets
THE GUMS; their *Structure, Diseases, Sympathies,* &c. By George ⎬ 0 85
Waite. ⎪
An ESSAY on DEW, &c. By W. C. Wells, F.R.S. ⎭

COLLINS'S PRACTICAL TREATISE on MIDWIFERY. 11 sheets. . 1 25

EVANSON and MAUNSELL on the MANAGEMENT and DISEASES
of CHILDREN. 13 sheets. 1 50

The SURGEON'S PRACTICAL GUIDE in DRESSING, and in the
Methodic APPLICATION of BANDAGES. *Illustrated by 100 Engravings.*
By Thomas Cutler, M.D., &c. 4 sheets. . . . 0 60

EDWARDS on the INFLUENCE of PHYSICAL AGENTS on LIFE:
with observations on ELECTRICITY, &c. 10 sheets. . . 1 00

HORNER'S NECROLOGICAL NOTICE OF DR. P. S. PHYSICK. ⎫
IS MEDICAL SCIENCE FAVOURABLE TO SCEPTICISM? ⎪ 3 sheets
By Dr. Dale, of Newcastle, Delaware. ⎬ 0 30
ON DENGUE; its HISTORY, PATHOLOGY, and TREATMENT. ⎪
By Prof. Dickson of S. C. ⎭

THE FOLLOWING ESSAYS ON PHYSIOLOGY AND HY- ⎫
GIENE:— Reid's Experimental Investigation into the Functions ⎪
of the Eighth Pair of Nerves. ⎪
Ehrenberg's Microscopical Observations on the Brain and Nerves; ⎪
with *numerous Engravings.* ⎪
On the Combination of Motor and Sensitive Nervous Activity; by ⎪
Prof. Stromeyer, Hanover. ⎪ 13 sheets
Vegetable Physiology. ⎬ 1 25
Experiments on the Brain, Spinal Marrow, and Nerves. By Prof. ⎪
Mayer, of Bonn; with *wood cuts.* ⎪
Public Hygiene. ⎪
Progress of the Anatomy and Physiology of the Nervous System, ⎪
during 1836. By Pro. Muller. ⎪
Vital Statistics. ⎪
REID on the FUNCTIONS of the EIGHTH PAIR of NERVES. ⎭

FRECKLETON'S OUTLINES of GENERAL PATHOLOGY. 7 sheets. . 0 75

URINARY DISEASES, and their TREATMENT. By R. Willis, M.D., &c.
10 sheets. 1 00

MILLINGEN'S CURIOSITIES of MEDICAL EXPERIENCE. 15 sheets. 1 50

ANDRAL'S MEDICAL CLINIC: *Diseases of the Encephalon, Spinal Cord,*
&c. &c. 13 sheets. 1 60

The ECLECTIC JOURNAL of MEDICINE, by John Bell, M.D., &c., from November, 1837, to October, 1838. 21 sheets. . . . $2 00

LECTURES on the PHYSIOLOGY and DISEASES of the CHEST. By Prof. Williams. *With Engravings.* 15 sheets. 1 75

LECTURES on BLOOD-LETTING. By Dr. Clutterbuck. 5 sheets. . 0 65

MEDICAL and TOPOGRAPHICAL OBSERVATIONS upon the MEDITERRANEAN, and upon PORTUGAL, SPAIN, and other countries. By G. R. B. Horner, Surgeon U. S. N., &c. *Illustrated with Engravings.* 9 sheets. 1 00

MAGENDIE'S LECTURES on the BLOOD : *its Changes during Disease,* &c. 12 sheets. 1 25

The ECLECTIC JOURNAL of MEDICINE, by John Bell, M.D., &c., from November, 1838, to October, 1839. 20 sheets. . . 2 00

HOLLAND'S MEDICAL NOTES and REFLECTIONS. 16 sheets. . 1 60

ARMY METEOROLOGICAL REGISTER for the YEARS 1826, 1827, 1828, 1829, and 1830. HINTS on the MEDICAL EXAMINATION of RECRUITS for the ARMY; and on the Discharge of Soldiers from the Service on Surgeon's Certificate : Adapted to the Service of the United States. By Thomas Henderson, M.D., Assistant Surgeon U. S. Army, &c., &c. } 6 sheets 0 65

MACARTNEY on INFLAMMATION. 5 sheets. . . . 0 50

BURNE on HABITUAL CONSTIPATION—its Causes and Consequences. 7 sheets. 0 75

A PRACTICAL TREATISE on VENEREAL DISORDERS, &c. By P. Ricord of the Venereal Hospital, Paris. AMUSSAT'S LECTURES on the RETENTION of URINE, CAUSED by STRICTURES of the URETHRA, and on the Diseases of the Prostate. *Translated from the French* by James P. Jervey, M.D. } 7 sheets 0 75

ESQUIROL on MENTAL DISEASES. AN ESSAY ON HYSTERIA. *With numerous Illustrative and Curious Cases.* By Thomas Laycock. } 10 sheets 1 00

CLINICAL REMARKS on some Cases of LIVER ABSCESS presenting externally. By John G. Malcolmson, M.D., Surgeon Hon. E. I. C. Service, &c. THOMSON'S NOTICES of INFLAMMATORY AFFECTIONS of the INTERNAL ORGANS after EXTERNAL INJURIES and SURGICAL OPERATIONS. } 4 sheets 0 45

The ECLECTIC JOURNAL of MEDICINE, by John Bell, M.D., &c., from November, 1839, to October, 1840. 20 sheets. . . 2 00

STOKES'S LECTURES on the THEORY and PRACTICE of PHYSIC. *With numerous Notes and Twelve Additional Lectures,* by John Bell, M.D., Lecturer on the Institutes of Medicine and Medical Jurisprudence. 28 sheets. 2 75

" We feel great confidence, as well as pleasure, in recommending the present edition of Dr. Stokes's Lectures as a valuable addition to the systematic works on the Theory and Practice of Medicine now in general use—as well for the instruction of the student, as for the frequent reference of the medical practitioner."—*Amer. our. Med. Sciences.* Jan. 1841.
" Dr. Bell's additions to the labours of Dr. Stokes are highly valuable. No writer on Practical Medicine of the present day has higher claims to the confidence of the profession, than the distinguished teacher at Dublin; and Dr. Bell has long stood in the first rank of American medical writers. Besides his Lectures, twelve in number, Dr. Bell has added copious notes to the lectures of Dr. Stokes. The volume is handsomely printed, on good paper."—*Western Journ. of Med. and Surgery.*

GOOCH'S PRACTICAL COMPENDIUM of MIDWIFERY. 14 sheets. . 1 00

A

PRACTICAL TREATISE

ON THE

DISEASES OF THE SKIN,

ARRANGED WITH

𝕬 𝖁𝖎𝖊𝖜 𝖙𝖔 𝖙𝖍𝖊𝖎𝖗 𝕮𝖔𝖓𝖘𝖙𝖎𝖙𝖚𝖙𝖎𝖔𝖓𝖆𝖑 𝕮𝖆𝖚𝖘𝖊𝖘

AND

LOCAL CHARACTERS:

INCLUDING THE SUBSTANCE OF THE ESSAY TO WHICH THE ROYAL COLLEGE
OF SURGEONS AWARDED THE JACKSONIAN PRIZE,

AND

ALL SUCH VALUABLE FACTS AS HAVE BEEN RECORDED BY CONTINENTAL
AUTHORS ON THESE SUBJECTS TO THE PRESENT TIME.

BY SAMUEL PLUMBE,

LATE SENIOR SURGEON TO THE ROYAL METROPOLITAN INFIRMARY FOR CHILDREN,
AND ACTING SURGEON TO THE ST. GILES'S AND ST. GEORGE'S
PAROCHIAL INFIRMARY, ETC., ETC., ETC.

FROM THE LAST LONDON EDITION,

REVISED, CORRECTED, CONSIDERABLY ENLARGED, AND WITH ADDITIONAL
ENGRAVINGS.

PHILADELPHIA:

HASWELL, BARRINGTON, AND HASWELL.

1837.

HENRY HOARE, ESQ.

My Dear Sir,

It has been the custom from the very dawn of science and literature for an Author to seize the opportunity afforded by a new Publication to testify his respect or gratitude to a Friend or Patron in the form of a Dedication. A part of his Book, usually so called, lasts as long as the work itself, which the Author sometimes imagines will live after him.

That this Book will live after us both, is my sincere prayer; and THAT *in order that it should remain a testimonial of my humble sense of your uniform, conscientious, zealous, and independent conduct in the discharge of those public trusts which your station in society imposes on you, as also of my personal respect.*

Science and politics have, it is said, seldom much to do with each other; and it has been somewhere written down as an axiom, that they never ought. Still the real and true patrons of science are never, practically speaking, men who do not feel, like yourself, how essential the encouragement of science is to the prosperity of our country. This consideration must be the excuse for the liberty I have taken with you; I fear not that in doing an act of justice, I am likely to commit an offence, and therefore I omit the too common plea, "dedicated by permission."

I am, my Dear Sir,

Very truly,

Your obedient Servant,

S. PLUMBE.

Southampton Street,
Bloomsbury Square,
December 1836.

PREFACE.

HAVING in view the extension of the sphere of utility of the following work, I have endeavoured to reduce its expense to the student; first, by curtailing as far as possible, consistent with utility, all historical, pathological, and descriptive details; and, secondly, by avoiding numerous and expensive illustrations.

In a former edition I expressed an opinion that careful and accurate descriptions, attentively and carefully perused, possessed great advantages over engraved illustrations (unless executed with extraordinary skill and at enormous expense) in conveying practical knowledge to the student and practitioner in a majority of the cases of cutaneous disease which may come under their observation. I am confirmed in that opinion by the results of experience. There is great difficulty in obtaining, through draftsmen's and engravers' and printers' hands—each failing, perhaps, in one point or another of their parts—any thing like *vraisemblance* as regards the larger portion of such diseases as they occur in the British Islands.* Hence ·

---

* In using this expression, I would have it understood that there is commonly a wide difference between the delineations and descriptions of French and English authors on this class of diseases. As regards the delineations, either of Alibert or Rayer, if attentively examined, the utmost liberality must be exercised to enable us to admit them to be representations of individual diseases as occurring here and known by the same names. Doubtless both these authors have strongly, if not faithfully, performed their tasks in this respect; but the delineations often appear to an Englishman to amount to caricatures, and the descriptions to abound with exaggerations.

That both may be correct, as regards the cases which come under the French practitioner's eye, it is far from my intention to deny, or even to doubt. It must be considered that the individual portraits are, for the most part, drawn from the persons of the lowest and poorest of the French population,—a class of people so poor and so badly fed in the main as would lead us, *à priori*, to expect a greatly aggravated form of those diseases among them. A slight examination of the extensive and clever illustrations of these authors will convince the reader of the correctness of my observations on this point; and really the descriptions are still

I have given all the attention in my power to historical and descriptive details under every different head.

On referring to the works of French authors we are struck with the influences which diet, and regimen, and habits of life have in regulating the degrees of violence of cutaneous disease. As regards the middle and higher classes of the two countries, it would appear that the French have the advantage over the English: most certainly they are not so frequently the victims of those troublesome, annoying, and unsightly affections as ourselves. Nor is this difference more difficult of explanation than that which is manifested between such diseases of the poor of the two countries. If the aggravated sufferings of the poor of the continent depend on starvation or the alternative of using disgusting and unhealthy food, the more fortunate of their countrymen enjoy, perhaps, all that is necessary to health without attaining that degree of repletion which so often gives occasion to disease among their equals in this country. It must be highly satisfactory to the philosophic observer to perceive that the habits of all but the very lower classes in England have been, for many years, approximating to those of our continental

more widely separated from the reality, as the diseases are seen in England, than the illustrations themselves. "J'ai recueilli," says M. Alibert, "à l'hôpital Saint Louis l'observation suivante sur deux petites filles, Virginie et Julie Calandini, Italiennes d'origine. L'une avoit atteint sa septième année, l'autre sa cinquième. Elles jouissoient d'une constitution très-forte, ainsi que leurs parens. Lorqu'on nous apporta ces deux enfans, leur tête étoit couverte de tubercules faveux circulaires, d'une couleur trèsjaune, creusés en godet dans leur milieu, offrant des bords prééminens, tels en un mot, que nous venons de les décrire dans le cas précédent. On observoit une régularité parfaite dans les tubercules ; mais la plupart d'entre eux étoient cohérens et disposés par plaques sur le cuir chevelu. Au milieu de cette masse croûteuse, devenue informe dans quelques endroits de la tête, par les progrès de la maladie, on appercevoit des sillons et des crevasses d'une profondeur considérable. Le cuir chevelu y étoit sanguinolent et presque détruit par l'effet terrible de l'ulcération existante. Il y avoit des croûtes brisées qui tomboient en petits grains au travers des cheveux. Mais il est surtout impossible de dire quelle quantité de poux infestoit la tête de ces deux malheureux enfans. Ces animalcules dévorans étoient cachés en nombre infini sous les croûtes ; ils excitoient un pruit intolérable. Cette éruption dégoûtante exhaloit en outre une odeur de souris difficile à supporter. Au surplus, la Teigne dont il s'agit, n'avoit pas seulement son siège dans le cuir chevelu. Il est à remarquer que les deux sœurs avoient des plaques faveuses dans des parties absolument analogues du reste du corps, comme aux sourcils, aux tempes, au bas des épaules, aux lombes, sur la région du sacrum, à la partie externe et supérieure des cuisses, à la partie moyenne des jambes."

Hillocks of accumulated filthy secretions, rising an inch or two above the surface of the scalp, attached at their bases to the hair, while the more superficial part having become dried to the extent of resembling rotten mortar, split on the vertex and sides of the head into parts, forming huge chasms, in and round about which are seen myriads of crawling vermin, giving, by their movements, motion to the half-detached parts, are the predominant features of other cases described by these authors, the odour being as disgusting as the sight. This is not by any means an extreme case selected to confirm my opinion, there are other instances equally striking to be found elsewhere.

brethren. Those indubitable evidences of " good fellowship," blotched and bloated countenances and " Bardolph " noses, are becoming extinct among the English almost as fast as the seams and pits of another pest of the human race—that of the small-pox.

The portion of the field of medical science to which the following pages refer has been taken possession of, to a considerable extent, by the translations from French authors (these having benefited largely by the labours of Willan and Bateman, and others of our countrymen.) Now it is evident that I should have too much to do, consistent with the space I have allotted for this work, to place the pretensions of English practitioners and students of this class of diseases, in every individual instance, on their proper and merited footing, and yet it is a duty which I cannot altogether overlook. The fact is unquestionable, that our countryman Willan lead the way in opening this part of the science to the world. His recorded Facts and his published Illustrations being of origin and date antecedent to that of the French school by several years. There have been numerous English contributors too, to the fund of information on these subjects ; and it is not a day of triumph to English practitioners when, in the study of these diseases, they are expected to peruse the works of continental authors in preference to those of their countrymen, under an unfounded assumption that they are more valuable and instructive than those of the latter.

There are two curious points of evidence affecting this question.

First, there has been for years, in a work published in the French capital, purporting to be largely occupied by descriptions of its public institutions, an uncontradicted statement, that though the hospital of St. Louis is dedicated to the reception and treatment of cutaneous diseases, furnished with baths and " all appliances and means to boot," yet the treatment adopted there is not found more successful than that adopted elsewhere under less favourable circumstances. Secondly, in that very institution, the treatment of a very large portion is surrendered by the physicians and surgeons into the hands of secret *nostrum mongers.*

The study of cutaneous diseases, after having been nearly neglected for many years, appears suddenly to have become a favourite one, particularly in France.

The splendid Tableaux of Alibert in folio retained for some time full possession of the field, as the best representations of them. The expense of the experiment, however, seems to have deterred his cotemporaries from competing with him, except in the form of unadorned, detached essays and treatises, and very inferiorly executed engravings. Among the most distinguished of the latter was Rayer, who, in 1826. published his work in octavo, with badly executed engravings, manifestly copies to a considerable extent of those of certain English authors. This work having been noticed with praise by one or two teachers of medicine and surgery in their lectures in London was speedily translated by a young English

surgeon,* who certainly executed his task with as much faithfulness as talent. Unfortunately for him, however, the same dicta which gave the impulse to him, gave it also to M. Rayer and the French booksellers, both in Paris and London; and before Mr. Dickinson's translation had had its fair chance of sale and circulation, out comes simultaneously in Paris in the French, and in London in the English language, a new edition of Rayer, the letter press of which, amounting to upwards of *twelve hundred pages of most closely printed octavo*, accompanied by an atlas of twenty-two engravings of royal quarto size; the number of figures in the whole being about four hundred! In the mean time, however, the veteran Alibert had concocted not only a new arrangement, but a new and almost totally different nomenclature, and employing partly his old engravings, and partly some of more recent execution, he had fully supported his reputation for correct representation and colouring, but risked that part which depended on judcious nomenclature and arrangement. He had constructed an "Arbre de Dermatoses," a figure of which he inserted as a frontispiece to his last edition.

I must confess, that this design exhibits proofs of great zeal on the part of M. Alibert; and for the information of the reader, who may not have had an opportunity of seeing it, a few short words of description will not be misapplied. The trunk represents the head of the family "des Dermatoses," the branches emanating immediately from it, amounting to twelve in number, pass from it and divide into limbs and sprigs, each having its cognomen attached, according to the author's altered nomenclature. The subject was already so seriously encumbered with designations without accurate definitions—with divisions and subdivisions into species under the old nomenclature, that students of the subject had barely obtained a superficial knowledge of them; a few only perhaps arrived even at that, so that it is too much to expect that they will begin *de novo* with M. Alibert's new nosology.

It will be perceived, I hope and believe, that these expensive illustrations—this multipled nomenclature, these complicated treatises of twelve hundred pages octavo, &c. will have served me chiefly not as assistants in my task so much, as by affording me landmarks to warn me of my pledge, to make my work as cheap as useful to the reader. Despite the comparatively inexpensive labours of the engravers in France, I feel very strongly impressed with the opinion, that by the judicious employment of those of our countrymen on a more limited scale, we shall be able to convey far more accurate notions of the characters of these diseases as they occur among ourselves.

Even in our own country, the engravings of Willan and Bateman have sunk into utter inutility, because of their enormous expense : nor does the more recent attempts to revive them in an economical form, under the superintendence of one of the most industrious and

* Mr. Dickinson.

talented men of the age, promise much for their effective resuscitation.* If information be expected to be conveyed extensively it must be conveyed cheaply ; and neither the engravings of Willan and Bateman, those superintended by Thompson, or those of Rayer, can be brought within that moderate expense which is necessary to secure their extensive circulation.

I have not been unmindful, while I have steered as clear as possible of diffusiveness, of the really valuable matter of recent continental authors ; a proof of which will be seen in the illustration of the structure and anatomy of the skin, furnished in a following page, with descriptive details. The researches of MM. Breschet and Roussel de Vauzeme, have furnished me with a large portion of these. The figures No. 11 and 12, which I have added, will be found useful in illustrating my views of the pathology of the different forms of Acne, Sycosis, Porrigo, and some others.

Any notice of Measles, Scarlatina, Small-pox, or, indeed any portion of the exanthemata, in this work would be out of place. They are diseases, of which the condition of the surface forms a very small feature, and their treatment is for the most part fully understood.

Lest the profession and the public should expect from me an elaborate account of all diseases of the skin, "British and Foreign," it behoves me to say, that, being able to speak conscientiously and confidently by the results of practice and experience of those most frequently seen in the different classes of society in this country, I have not expended my own time, or drawn on that of my reader, by recording second and third hand accounts of authors of my own or other countries. I have, nevertheless, I hope, with some pretensions to diligence, availed myself of all sources which I could command, to make my work such as the English practitioner may place some confidence in. If a valuable fact or idea has appeared in print, no matter in what book, journal, or country, I have spared no pains to obtain possession of it, and if possible, to convert it to my object of making my present undertaking a register of things not theorized on, but known.

If the records of ancient authors have obtained but little attention from me, it is not because I have undervalued them, but because they have been so thoroughly sifted by Willan, Bateman, Alibert, and lastly Rayer, as to leave no chance of finding an unappropriated grain among the remaining chaff for my own or my reader's use.

Of the writings of modern French authors generally, it will be perceived, that I have in two senses of the word made a liberal use. " Palmam qui meruit ferat" has been my motto in dealing with them, and the " utile et dulce," my guide in handing the fruits of

* Mr. A. T. Thompson's Atlas, and Seventh Edition of Bateman's Synopsis.

their labours to my countrymen.*   I despair not, therefore, though this edition meet as another did with *concealed* enemies,† it may make for itself a few *open* friends.

S. P.

* The cases multifarious and long which are to be found in the French Authors whom I have so often quoted, I have not considered of sufficient value to obtain possession of more than a very few pages, and these only, where they seemed likely to be useful in solving the doubts of the student, and of an unquestionable illustrative character.

† " We have received some severe strictures on Mr. Plumbe's work from more than one or two ANONYMOUS sources. *Most of them bear the impress of the detestable odium medicum, the disgrace of the profession.*"—Editor of the Medico-chirurgical Review.

DESCRIPTION OF THE ENGRAVINGS.

PLATE I.

DIVISION I.

Fig. 1.—The upper part of this cluster of figures is intended to represent the uninflamed follicle; the lower the commencement and progress of inflammation, and its termination in the formation of matter.—2. The enlarged and indurated tubercles (A. indurata), with matter formed in their centre, which occur in bad constitutions.—3. Inflamed and suppurated follicles, forming sycosis on the beard.—4. The appearance of spots of Porrigo scutulata, where no fluid secretion or scab has been formed.—5. The partially denuded scalp of long-established cases of the latter, where scabs have been allowed to accumulate, and where great irritation prevails, the remaining hairs insulated by pustules.

DIVISION II.

Fig. 20.—The two inferior spots representing the first appearance of the spots of Lepra before the first scale separates. The superior, large, round, and scaly; the disease in a spreading state.—21. Psoriasis.—22. An enlarged representation of the morbid and discoloured cuticle forming Ichthyosis. The numberless fissures caused by the cracking of this hard, dry substance, and dividing it into thousands of pieces, are well represented. It is introduced here out of its proper place, it being the only subject of the 6th division worthy of representation.

DIVISION III.

Fig. 11.—Porrigo favosa.—12. P. larvalis, both from cases of considerable standing.—13. The pimples of infants, some of them surrounded by considerable inflammation; their representation in clusters connected by patches of inflamed skin (S. intertinctus, &c.) has been omitted.—14 and 15. The pimples of adults, termed Lichen; the first of these, as it sometimes occurs on the arms and other parts covered by the finer kind of hair, each hair occupying the centre of a pimple: the second as it appears on other parts.—16. The pimples of Prurigo, the tops of some of them scratched off, leaving a peculiar little, black, bloody scab on their apices.—17. Two of the commoner forms of Urticaria.—18. The vesicles of Herpes in an advanced and partly flaccid state.—19. The carbuncular Furuncle.

DIVISION IV.

Fig. 23.—The inferior portion exhibiting an enlarged view of the vesicles of Impetigo. The superior, the disease in an advanced stage, with the scab partially covering it.—24. The vesicles and enlarged pustules of the itch.—25. The appearance of the skin in Eczema mercuriale.—26. The tubercle of Erythema nodosum.

DIVISION V.

Fig. 6.—Petechiæ or Purpura simplex.—7. The enlarged spots of Purpura hæmorrhagica.—8. Different stages or degrees of the Ecthymatous eruption.—9. The conical scabs of Rupia. The similarity of character between the two latter is rendered very distinct.—10. Pompholyx. The superior vesicle discoloured by the admixture of blood from the vessels of the surface.

PLATE II.

For Description, *see* pp. 27, 28.

PLATE III.

The changes produced on the scalp in long-established cases of Porrigo have been attempted to be represented in this plate. The adipose structure, described as secreting the hair, undergoes a considerable diminution or wasting, as the effect of the ceaseless irritation and discharge from the surface of the scalp, while the hair is for the most part extirpated; small tufts here and there distributed over the diseased surface only remain, and these are observed to have pustules among them. The diseased skin covering the wasted adipose structure appears to the touch to have nothing between it and the pericranium; and the greater degree of thickness of the healthy skin along the superior line of margin of the disease presents the appearance of a sudden declivity or depression. This wasting, however, is not permanent; and as the irritation on the surface is subdued by the plan detailed in the preceding pages, and the remaining hair removed, the skin gradually, except where deep ulceration has occurred, reassumes its original solidity and thickness, and the new hair begins to make its appearance.

PLATE IV.

The intention of the little engraving here presented is to convey to the reader an idea of four occasionally changing forms of disease of the scalp which have been differently designated, according to their existing states, by French and English authors. Teigne furfuracée—Amiantacée.—Faveuse.—P. Favosa.—P. Larvalis.—P. Lupinosa, &c. &c. The scabby and inflamed state of the forehead fairly represents the Porrigo favosa. The scalp covering the parietal bones is occupied by the honey-comb, cupped or lupin-seed scab—the occiput by the furfuraceous form of the disease, or P. furfurans of Willan and Bateman. If not interfered with, which is often the case in France, the scurf accumulates, the exfoliations from the surface assume the appearance of silvery scales, which, as they mix with the hair and approach the surface, have certainly from time to time justified the comparison to asbestos. The back of the ear again gives an accurate notion of what would possibly be called P. Larvalis, showing, however, the occasional tumefaction of the glands adjacent.

CONTENTS.

DECEMBER, 1837.—*S*   2

SECTION II.

CHAPTER I.

CHAPTER II.

OF PSORIASIS,

CHAPTER III.

PITYRIASIS, SCURF OR DANDRIF.

SECTION III.

CHAPTER I.

PORRIGO FAVOSA.

CHAPTER II.

ON THE PAPULAR ERUPTIONS OF INFANTS AND ADULTS, DENOMINATED STROPHULUS, LICHEN, &c.

DIRECTIONS FOR PLACING THE ENGRAVINGS.

Plate 1.—Copper-plate (coloured,) to *face* the title-page.
Plate 2.—Lithograph (not coloured,) to *face* page 27.
Plate 3.—(Lithograph) exhibiting the destructive effects of long-standing Porrigo, to *face* page 88, with its description appended.
Plate 4.—(Lithograph coloured,) with its description, to be inserted with plate 3 at 88.

A

PRACTICAL TREATISE,

&c. &c.

PRELIMINARY REMARKS

*On the Anatomy and Physiology of the Skin and its append-
ages, as connected with its diseases.*

"MEN," says M. Alibert, "commonly see nothing in the skin of
the human race beyond the means of defence which it affords, against
the contact of external substances, capable of injuring more vital
parts; but the physiologist and clinical physician look at it as the
double instrument of exhalation and absorption—as the deposit or
reservior of an exquisite sensibility—as a means of conveyance of
salutary remedies—as the agent of most favourable crises in disease,
and as the seat of a multitude of diseases, the very nomenclature of
which excites terror."

The absence of due consideration of the anatomy and physiology
of the skin, which a large portion of those who have written on the
subject of cutaneous diseases have manifested, as well in their descriptions as methods of treatment, will, it is hoped, afford ample
excuse for introducing a work of this nature with the following remarks.

The most distinguished authors on dermoid pathology, have boldly
fixed the seat of important and obstinate diseases in parts of the
skin denied by equally distinguished anatomists to have any existence
in reality; while whole pages have been wasted in the description
of affections obviously consequent on derangement of a well known
function of this structure, without the slightest reference to such
function, or the part of the skin in which it exists. Hence, methods
of treatment have been adopted, destitute of sound principles, and
productive of nothing but mischief; while principles have been overlooked which would have pointed out methods of treatment as efficient as simple, and utterly incapable of any deleterious influence.
Always excepting due regard to those universal agents in the production and modification of local diseases, disordered states of the

organs of digestion,* there is nothing affords so fair promise of improvement in the management of cutaneous affections, as a due regard to the anatomy and different functions of the skin. A careful reference to these will be found the only road to a principle of treatment of many affections of this structure, and these not the least unpleasant and unsightly which come under our notice, while it is calculated to disclose important facts in the history and character of others which have been hitherto entirely overlooked.

The skin has been hitherto understood to consist of three layers of differently formed structure. It has been so described by anatomists since the days of Malpighi, each layer having its different offices assigned to it. The true cutis or skin, that substance which is so readily converted to the purposes of the mechanic in the various forms of leather, being the most deeply-seated of these strata ; the rete mucosum next superposed ; and, lastly, the epidermis, cuticle, or scarf skin ; designed by its peculiar properties to protect the more sensible and vascular structures beneath it from the irritation of extraneous substances or violence.†

The cuticle is manifestly endowed with properties fitting it for such offices. It is of a dry horny structure, and it is insensible ; but it is at the same time, so elastic and accurately fitted to the folds and flexures of the more vital and nervous structure beneath, as readily to be able to accommodate itself to all the flexions and extensions of the latter in the different parts of the body.

The inquiry of the chemist into the composition of the cuticle, just furnishes us with evidence that thin and filmy as it may be in the human or other class of animals, it is easily identified with that of the horns and hoofs of the larger animals. It is an insensible covering to parts most peculiarly endowed with sensibility, the protection of which is vitally necessary to the health and existence of the animal, of whatever species it may be. This appears to be its primary and chief office as regards the animal economy. It is evident, however, that such a structure, while exercising the office of a protector of the parts beneath it, must be endowed with power to administer to other wants and necessities of those parts. It must be pervious, to admit of exhalation ; it must be pervious, to admit of the passage to the surface of the contents of the sebaceous follicles ; it must be, in short, while deriving its own supply and means of

---

* That the skin should be susceptible of the slightest deviation from health in the action of internal organs, is not surprising. It is a most important emunctory, and the chief office of these is to keep up the balance between the ingesta and egesta: hence, the effects upon it of disordered stomach, impeded biliary and urinary secretion, &c.; if more blood is sent to it, or its usual supply abstracted in consequence of such disorders, it suffers.—*Dr. Jackson's Dermato-pathologia.*

† Gaultier, an enterprising French physician, seems to have established the existence of a distinct papillary structure. According to M. Breschet, he had industriously investigated these structures, and was only prevented from continuing to prosecute them by his duties in the field. "He died," says M. B., "in the disastrous campaign of Moscow."

existence from the cutis, available to all the services and objects of that structure.

Anatomists have differed very much as to the mechanical structure of the cuticle. That medicinal remedies for disease could be introduced through its substance, and by its agency, has been a long established axiom, but the manner in which this takes place or is effected, is still, notwithstanding the researches of British and French authors, a matter of extreme doubt. The subject is, and has been for a long time unsettled, and probably is still destined, in the opinions of many, to remain so.

Mr. Chevalier, the only English physiologist who had of late years directed his attention to this subject, thought that the cuticle was permeable by means of a "velaminous structure;" others have been content to ascribe this property to pores, or minute and direct orifices in it, leading to the subjacent structures. More recently, ideas somewhat different from either of these have been entertained, but they are all alike subject to the objection of having been founded on microscopic observation and experiment, or by comparison with structures supposed to be similar to that of man in the larger species of animals.

In former editions of this work I took occasion to observe, that evidence obtained by these means was very likely to be fallacious ; for that microscopic inquiry, in order to be efficient, must at least have the advantages of light posterior as well as anterior to the object examined. That this desideratum is not attainable by such means in the living or dead animal, is obvious enough, supposing the different layers of the skin to remain *in situ naturalis.* The forcible, or ever so ingeniously contrived separation of the cuticle from its parent membrane, the cutis, must destroy of necessity either a poriferous or velaminous apparatus. A familiar illustration of this fact is seen in the application of a blister ; if the latter be applied to a surface teeming with perspiration, (and it is now and then so applied) the moment the irritation is sufficient to produce an effusion of fluid from the vessels of the cutis, the elevated cuticle becomes a water-proof impervious bag. Its connexion, whether with the cutis or supposed rete mucosum, is at once destroyed ; and whether the connexion be originally of the one or the other of the two kinds mentioned, the apparatus is evidently broken up, and incapable of renewal. The dead corneous substance, of which the cuticle is formed, may be, and often is, carelessly allowed to rest on the inflamed cutis after the fluid is discharged, and it is never found to be possessed of vital powers sufficient to renew its connexion with that structure : on the contrary, it exercises all the deleterious influence of an extraneous body, and often produces, by irritating the abraded cutis, very unhealthy sores. The latter result is very common among the children of the unhealthy poor of the metropolis, not unfrequently endangering the life of the sufferer, and even proving fatal.

There is an evident dependence on each other, between the cutis and cuticle. If the latter by collision, by external injury of any kind, becomes separated from the former, the means of forming a fresh supply vested in the cutis are found in instant operation ; there is no appearance of attempt to resuscitate or amend the torn and old cuticle, but, like magic, a new structure is produced from the vessels of the cutis, as efficiently administering to the wants of the cutis as the original.

It has been truly observed by Mr. Chevalier, " that the skin is as dependent on the cuticle, as the cuticle is on the skin. Neither can subsist in a state of integrity without the other."

In vesicular eruptions, not artificially induced, as, for instance, the herpes zoster, &c., a similar result follows to that produced by cantharides ; the vesicle being once formed, however minute it may be, the vitality of the cuticle is taken away ; and if the latter be suffered to remain in contact with the abraded cutis, irritation on a proportionately smaller scale is found to be produced.

It is evident, therefore, that the separation of the cuticle from the cutis, and its consequent impermeability, is not the mere effect of any peculiar influence of external stimulants, but results from that separation only, whatever its cause may be.

Although the office of perspiration, so far as the cuticle is concerned, has never, in an obvious degree, been connected with the origin of cutaneous diseases, it is yet a common practice to refer many of them to the suppression of this excretion. That this occurrence may be an indirect cause of eruptions of different kinds is easily enough understood. Several forms arise as the evident results of reaction after a chill, becoming the agents by which mischief to internal organs is averted, but the influence of cold on the cuticle itself can be but little; it becomes shrivelled and contracted, because the vessels of the cutis have been emptied by the determination of its blood to internal organs; but its elasticity soon enables it to conform to, and re-establish its connexion with the cutis, when reaction is established. There are, however, occasionally occurring slight inflammations of the skin, and disturbances in the production and growth of cuticle, manifested by redness and scurfy exfoliation, in consequence of the internal use of cold drinks, when the individual has been in a state of perspiration ; and it appears to me, that the only rational mode of explaining this effect is this : the collapsed state of those vessels or structures, produced by the shrinking or half emptied condition of those of the cutis during the chill, offers an obstruction when sudden reaction takes place to the increased quantity of excretory fluid produced. These slight cutaneous disorders are commonly designated surfeits, but though generally evanescent, they sometimes lead to the establishment of chronic irritation of the cutis, of an obstinate character.

"How," inquires Mr. Chevalier, "is the office of perspiration carried on through the cuticle ? What and where are its pores which have been supposed to transmit this fine exhalation? Are there any in reality ?" and he very confidently, seeing in his own opinion that the separated cuticle ought to be just as porous as in its natural state with its natural connexions, infers that it cannot be by a porous structure that perspiration or inhalation is effected.

"Pressed by this difficulty," said he, " I began an examination of both surfaces of the cuticle, the exterior and the interior."

Duly considering the different duties of the two surfaces, the external being an insensible and hard defence against external injury, the other destined to lay in soft contact with the highly sensible surface of the cutis, he separated the cuticle of different parts of the body from the cutis, and laid it in the field of a microscope magnifying one hundred and forty times, hoping to detect the pores; and failing in that, he discovered "an infinite number of minute velamina, regularly arranged, of exquisite tenuity, presenting a follicular appearance, and separated from each other by bands of a thicker structure, crossing and intersecting them so as to render them distinct."

Now it is just as possible that the appearance of these said velamina was produced by the act of separation of the cuticle from the cutis, as that any original organization or formation of the kind existed ; and whether the perspiration passed through pores or through a structure like that described by Mr. Chevalier, it is quite clear that either of such structures must have been destroyed by the degree of force necessary to break the connexion existing between the cutis and cuticle, even under the most favourable circumstances and most careful management.

Nevertheless he expressed his conviction that the terminal vessels of the cutaneous apparatus are lodged in these velamina : that "so long as the vessels maintain a vital connexion with them, they transmit their secretion through them as through a bibulous and exquisitely hygrometrical covering of the finest delicacy and perfection, while through the same medium, and dependent on subjacent tubes taking a contrary course inwards, absorption is carried on to a great but less certain extent and continuity." If such an apparatus as this be really in existence, it amply accounts for the failure of other anatomists in discovering pores or holes for the transmission of the secretions of the perspiratory organ to the surface of the cuticle. It appears to me that so long as it is necessary to separate the supposed layers of the cutaneous covering from each other, and subject them to inquiries made through a microscope, we have little chance of adding to our anatomical or physiological knowledge of either.

MM. Breschet and Roussel de Vauzeme appear to have prosecuted very industriously and philosophically a lengthened inquiry into the structure and functions of the skin. These gentlemen agreed to designate by the name of epidermis the entire corneous substance which invested the cutis. "This epidermic matter," they say, "is applied on the cutis in the same manner as a mask of liquid plaster with which it is usual to cover the face of a person when modelled, and which becomes adapted to the inequalities of this surface." The comparison is well founded. The fœtus in utero has no sooner obtained shape and form, than it has obtained a cuticle modelled on

3*

the form and fitted to all the flexures of the cutis; and from this period the wear and tear of the cuticle, and supply by new formation begins and continues through life, the surface exfoliating in the form of scurf in a proportion equal to that of its re-production by the vessels of the cutis when the process is not interfered with by disease or disordered action.

Following the example of Breschet and Roussel de Vauzeme in reasoning from analogous structures, we find in the hoof of the horse an illustration of our subject. Experiments instituted some time since afforded the following results.

An aged horse, with the disease termed sandcrack, was selected, which, under the care of an experienced veterinarian, promised to afford satisfactory results as to the cure. This disease of, or rather mechanical injury to, the hoof, appears to have obtained its name from the crack or fissure in the hoof being usually found partially filled by sand or other extraneous matter. It is an injury under any circumstances producing lameness and much inflammatory action. The fissure emits, generally, an abundant discharge, in which is usually found flakes of a curdy consistence apparently half dissolved. The treatment found to be most successful consists of paring down the edges of the fissure to the sensible parts—of freely cutting away even to the extent of producing a discharge of blood. The flakes in question were evidently composed of ill-formed horn, generated by the sensible structure intervening between the hoof and coffin bone in a highly irritated state, and the cause of that irritation was as evidently the motion of the broken insensible edges of the fissure of the hoof. Fungous granulations were found to spring up notwithstanding the excision of the horny edges, if merely simple dressings were employed ; but where pressure was had recourse to by means of pledgets and bandages, the discharge ceased, and the bottom of the chasm became speedily covered by a thin film of horn, in all respects similar to that of the original substance of the hoof. The existence of a rete mucosum covering fungous granulations could not in such a case be believed: the new horny structure must have been produced by the direct agency of the sensible structure. The results obtained were found exactly to correspond in other cases in which the experiment was made.

In former editions of this work I had uniformly expressed my disbelief of the existence of a rete mucosum, separate and distinct from either the cutis or cuticle, or as the peculiar deposit of the colouring matter of the skin as attached to or forming a part of either. Such evidence as I could adduce, however, seemed to be rather of a negative than positive character. I could neither trace it in the human race or any other species of animal, of whatever complexion or colour of the hair. Its existence had been doubted by several anatomists of repute; but it was described, and still continues to be described, though not *demonstrated* by teachers, in our schools. It is, moreover, said to be shown in a variety of preparations in the museum of the College of Surgeons. Besides all this, there was the authority of Malpighi for its existence ; the question has however now, I think, been set at rest by the report of MM. Breschet and Roussel de Vauzeme.*

As these gentlemen's views cannot be out of place here, I may submit a brief summary of them.

Their inquiries seem to have been conducted under circumstances exceedingly favourable, and the results are likely to carry conviction to all who have confidence in the powers of the microscope, and belief in the existence of uniformly analogous structures in the organs of animals having similar offices with those similarly designated in man. To have been defeated ourselves in arriving at satisfactory conclusions by measures of the same nature as those employed by others, is no reason why we should withhold confidence from them, when, as I must confess is the case in these gentlemen's inquiries, they produce reasonable evidence of their success. They appear, moreover, to have entered the field with the utmost caution, distrusting the microscope, and confessing that they were often deceived by it,—a fact which entitles them to much more confidence than is due to the professed and sanguine microscopist, who is but too apt to deal with illusions as realities.

The skin of the whale and other cetacea appears to have been rendered largely instrumental, in this inquiry, to the establishment of the opinions now promulgated by MM. B. and R., and, doubtless, comparative anatomy generally, as regards the structures in question, has been scientifically and liberally drawn upon.

The results of these inquiries, they say, leads them to consider the skin as formed of six constituent parts.

1. The dermis. A cellular canvas, dense, fibrous, enveloping and protecting the capillary blood-vessels, the lymphatics, the nervous filaments, and the parenchyma of other organs contained in its substance.

2. The papillæ. The organ of touch, termination of the nervous system, developed under the form of nipples, slightly inclined, terminating in blunt points, concealed under several envelopes. In the whale, the summits, or points, are olive formed, while in man they are conical.

3. The perspiratory apparatus. The organs of secretion and excretion of perspiration. It is composed of a glandular parenchyma and the sodoriferous canals.

The parenchymous, or secretory organ, is seated in the dermis, and from it arises the excretory canals, spirally shaped, and taking their course obliquely, passing between the papillæ, and terminating on the surface of the cuticle.

4. The apparatus of inhalation, or absorbent canals. These canals resemble lymphatics, they are situated in the corneous substance or mucous body which forms the bed of the external dry cuticle, for the latter is dependent on the former for its production and support. These inhalent vessels do not appear to have open mouths on the

* Nouvelles Recherches sur la Structure de la Peau. Paris, 1835.

surface, but seem to originate in the form of a cul-de-sac, or small round protuberance ; all else is unsatisfactory as regards their origin on the surface. At their opposite extremity they communicate with and terminate in a net-work of vessels, composed of an intermixture of lymphatics and veins.

5. The apparatus producing the mucous matter.—Composed, 1stly, of a glandular parenchyma, or organ of secretion, situated in the substance of the dermis ; 2dly, of excretory canals originating in that structure, and depositing the mucous matter among the papillæ before described.

6. The apparatus producing the colouring matter, or " appareil chromatogène."—Composed of a glandulous parenchyma, situated a little more deeply than the papillæ. The excretory ducts passing from this structure terminate under the cuticle amidst the papillæ. The excretory vessels of the organ of the colouring matter distribute that secretion over the surface of the dermis and its papillæ, and this being mixed with the mucous secretion before described, seems to have led to the mistake of Malpighi. " De ce mélange resulte le prétendu corps reticulaire de Malpighi et l'épiderme ou cuticle."[*]

This complicated and minute apparatus, if it really be in existence, (and that it is, or at least something resembling it, is rendered manifest to the naked eye in a variety of ways by its productions and effects,) must necessarily be subject to an infinity of derangements. Disturbance in its movements, even in the slightest degree, would à priori be expected to be followed by some external signs or deviations from health on the cuticle, the joint production of, and necessary part of the whole.

The question as to the manner in which the cuticle is produced or secreted appears to be settled. It is formed by the gradual adhesion to, and union with its inner surface, of the mucous secretion described as poured out amidst the papillæ, which becoming gradually dry and hardened by the action of the absorbents in depriving it of its fluid parts, assisted probably by evaporation from the surface, assimilates itself to, and becomes identified with this structure, supplying it from time to time with new substance proportioned in quantity to the necessities of the cuticle in all parts of the body.

MM. Breschet and Roussel de Vauzeme selected the skin of the heel posterior to the plantar arch for particular inquiry, on account of the thickness of the skin and corneous tissue affording a development most favourable to the study of the subject, availing themselves of the advantages of comparing it with similar structures in other animals at the same time. It must remain for the reader to decide what degree of confidence may be placed in microscopic examinations of a part so differently situated from all other portions of the cutaneous structure. The heel, it should be remembered, in the natural erect position of the body, sustains by far the greatest degree of pressure, and nature may in its wisdom have provided to meet exigences arising out of this circumstance, an apparatus very different

* Breschet and Roussel.

from that of other parts. Perhaps, therefore, we ought not to take the structure of the skin of the heel as a fac simile of the whole?

The reception of the whole of the report in question, as composed of indisputable fact, is, I must confess, rather too much to be expected by our continental brethren from us at present. True it is that the reporters pursued their inquiries under the eyes of a large portion of their scientific cotemporaries, and secured the confidence of the latter ; but it is equally true that Mr. Chevalier, in a somewhat similar position, pursued his investigations here, and arrived at results so different, as to leave no doubt that either he or MM. Breschet and his *confrère* must have been deceived by their microscopes.

The object before me, however, being in as brief a manner as possible to notice the organization of the skin as connected with or influencing the character of its diseases, it remains for me to remind the reader that, in addition to the complicated and beautifully represented structures placed with the unpresuming character of copies before him, there is, First,—the apparatus secreting the hair. Secondly, the cutaneous or sebaceous follicles, each of which are too frequently connected with the more formidable diseases of the skin to deserve secondary consideration.

The representations in the engraving of these two structures have not been copied from any other author, they have been drawn with the assistance of the microscope with the utmost care. I believe they are correct representations of the mode in which the hair and follicles derive their support from the blood-vessels destined to supply them.

We have then to consider this vastly complicated structure as being the seat of or involved in the duties of no less than seven or eight different functions.

1. The sudatory pores, or hydrophorous canals.
2. The papillary structure, the seat of the senses of taste and touch.
3. The absorbent or lymphatic vessels.
4. The structure secreting the colouring matter.
5. The organs producing the germ of the cuticle.
6. The sebaceous follicles.
7. As perforated by, and lending its assistance in the growth and nourishment of the hair.

These are all things which we must deal with as realities, because we have the best evidence of their existence in their products, and the obvious necessity for them in the animal economy. The forms and figures of the different organs under the eye of the microscopist, is perhaps of little importance; but the pathologist must plainly perceive that here is a structure so complicated,—with so many duties to perform, and so dependent on the healthy performance of internal and vital organs, as to leave no doubt that neither local applications nor internal remedies are to be exclusively relied on in treating the diseases of the skin, whether they wear an active or chronic character. The absence of prominent features of general good health justifies dependence on internal remedies no more than their presence justifies dependence on merely local remedies ; comprehensive views must be taken by all professional men who hope to treat cutaneous diseases successfully ; and although the application of caustic in its different forms in some of the more formidable of these diseases in our hospitals, has obtained a reputation, and I may almost say established a practise by its success, the surgeon has been often deluded by neglecting to consider how differently his patient is situated in the wards of that hospital, and under control as to diet, &c. from that when he was first admitted to its advantages.

If there be any truth in the representations of the structure of the skin we have before us, there is obvious risk of doing harm by the undiscriminating use of any particular application, not even excepting the warm bath.   Simple as that is in its operation, and beneficial as it undoubtedly is in all those cutaneous diseases which wear an active character, it is still an instrument of great mischief where the energy of the circulation has been diminished.   On the other hand, looking to the complicated organic structure of the skin, what is to be expected from caustic and stimulating applications when the habit of the patient is full ?   The answer to this query is soon afforded to the practitioner who is tempted to rely on them.

I am of opinion that the variation of *form* of cutaneous disease is produced by the derangement, more or less, of certain portions of the cutaneous apparatus.   Now the application of stimulants and caustics deranges all these, and therefore generally lessens our chance of success, because, although we may be acting beneficially on one, we are probably exciting and disordering another.   Unless indeed we make up our minds to destroy the original structure entirely, and draw largely on the powers of nature to replace it by new ; stimulants and caustics are rarely justifiable, and even then comes again with redoubled force the consideration of those powers, or in other words, of the general health.   Is the latter manifestly such as to justify the destruction of a part of the surface, the seat of long-standing disease, with the expectation mentioned ?   I believe this question will seldom admit of being answered in the affirmative.

In any such cases as it has been customary to use these agents, such, for instance, as those denominated lupus, or the gnawing ulcer, a large portion of the cases of which are of scrofulous, or other hereditary origin, and few or none co-existent with an ordinarily healthy state of the constitution, *it is impossible to take measures of this kind, and call it less than a mere experiment ; and it is seldom that the skin remains long the seat of serious derangement where there is no deviation from health in what we are accustomed to consider more vital organs.*

I have availed myself largely of the labours of MM. Breschet and Roussel de Vauzeme, in the following illustrations of the anatomy of the skin and its functions : indeed, with the exception of the figures 11 and 12, it is but fair to repeat that I have selected them from the engravings of these authors ?*   An inquiry conducted as theirs

* Nouvelles Recherches sur la Structure de la Peau, par M. G. Breschet, Docteur en Medicine, &c. et M. Roussel de Vauzeme. Paris, 1835.   Extrait des Annales des Sciences Naturelles.

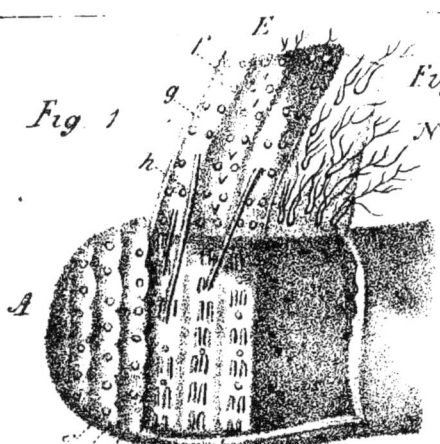

Fig 1

Fig 2

Fig 3

Fig. 4

Fig 5

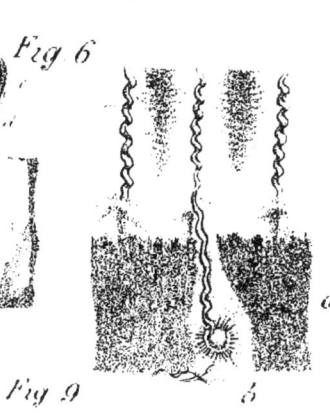

Fig 6

Fig 7

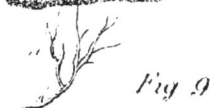

Fig 9

Fig. 8

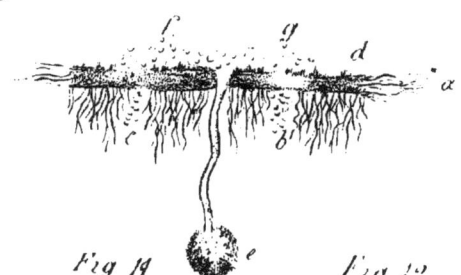

Fig. 10

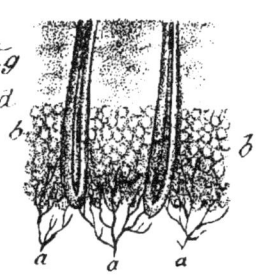

Fig 11

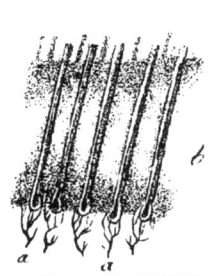

Fig 12

has been would have entitled its results to consideration here, even if it had not passed under the critical ordeal to which it has been subject in Paris. It was a great desideratum among English practitioners to have the true anatomy of the skin fully explained and demonstrated. More than half which has been written respecting it has been imaginary, and very seldom indeed admitting of proof; but M. Breschet and his *confrère* seem to have set the question at rest.

Looking at the complicated structures here represented, we should naturally expect that a very changeable and varying character of diseases of the surface must be generally expected where derangement of any one individual function took place.

EXPLANATION OF THE PLATE.

Figure 1.—*A*, the heel viewed from below, is the subject of this figure. The lines and intermediate dots represent the undisturbed surface of the skin *in situ naturalis*, (*b*) the projecting lines of papillæ which separate the transverse fissures, between which are seen (*c*) the sudatory pores, the orifices by which the perspiratory fluid approaches the surface; (*d*) furrows parallel with the projecting lines of papillæ. *E*, the first turned back lamella, the posterior margin of which is defined by the letters *f, g, h*, represents the internal surface of the cuticle separated from the dermis, and turned backwards: (*f*) the holes which receive the papillæ, (*g*) interpapillary spaces, (*h*) the extended thread-like appearance of the sudorific canals when the cuticle has been detached and everted.

The second turned back lamella *N*, represents the internal surface of the dermis separated from its attachments, the fimbriæ delineated consisting of vessels which passed through it backwards and forwards in its natural state; (*i*) the orifices by which the sudatory canals passed; (*k*) the papillæ for the most part found to be disposed in pairs in a lineal direction, between which are seen the passage for absorbent or inhalent vessels; (*m*) furrows of the dermis, in which the colouring matter is poured out by the excretory canals of that structure; I the outer margin of the foot; (*o*) the margin of the cut surface of the sole, showing the adipose structure beneath the cutis.

Figure 2.—A fragment of the inner surface of the epidermis in contact with the dermis, the same as depicted in figure 1st; but, examined with an instrument of great power, being dried, it exhibits the perforations here represented: it is the reticular membrane of Malpighi. (*a*) Projections fitting the furrows of the dermis, pierced by the lymphatic vessels; (*b*) interpapillary perforations for the sudorific canals; (*c*) the sheaths of the papillæ.

Figure 3.—A group of papillæ of the human skin, highly magnified; (*b*) the dermis.

Figure 4.—(*a a*) The papillæ in their sheaths; (*b*) the corneous matter of the epidermis; (*c*) the dermis.

Figure 5.—Represents the apparatus which constitutes the organ of touch and taste in the human subject; (*a*) the nerve entering the dermis, where it becomes capillary; (*b*) its entrance into the papillæ; (*c*) the neurilema furnished by the dermis; (*d*) the proper covering of the papillæ; (*e*) the bed, more or less thick, of corneous matter, the organ by which it is protected.

Fig. 6.—The sudorific organs; (*a*) the dermis; (*b*) the secretory organ, seen sometimes in the form of an oblong sac, surrounded by extremely fine fimbriæ; (*c*) the excretory canal, of a spiral form, which passes between the papillæ, traverses the epidermic corneous matter, and opens on the surface of the skin.

Fig. 7.—Fragments of the sudorific canals separated from the cutis and corneous matter.

Fig. 8. (*a*) The dermis (*b*) the papillæ, (*c*) the corneous matter raised at (*d*) to show its origin in the furrows of the dermis, between the papillæ, the torn prolongations corresponding to the excretory canals of the structure producing the colouring matter.

Fig. 9. shows this structure torn in two places. (*b & c*) Escape of the fragments, of which the torn filiform vessels were composed. (*d*) Small excretory canals which are torn when the corneous matter is raised from them. (*e*) The secretory organ of the mucus of the colouring apparatus. (*f*) The fluid state of the corneous matter, in which the colouring matter is floating in minute scales. (*g*) Beds of the corneous matter, which spread themselves right and left, and becoming condensed, form in succession their portion of the substance of the cuticle.

Fig. 10. A view of the skin cut lengthwise as to the furrows; (*a*) blood-vessels dividing themselves into capillary filaments in penetrating the dermis, (*b*) nerves similarly dividing, (*c*) mucous glands placed at unequal distance from the surface, the excretory canals penetrating the corneous matter; (*d*) the spiral-formed sudorific canals, (*g*) the chromatogenous organ covering the excretory canals, (*h*) the papillæ.

Fig. 11 is a representation of the sebaceous follicles originating in the substance of the cutis, and terminating with open mouths on the surface of the cuticle; (*a a a*) the vessels going to support the structure and supply its secretion; (*b b*) the trunk of the follicle containing the sebaceous matter free from obstruction and in a healthy state.

Fig. 12 represents the bulbs of the hair projecting below the cutis, and resting in the adipose structure, from the vessels of which they derive their support; (*a a*) the bulbs, with the vessels supplying them; (*b*) the hair passing through the skin and the cuticle to the surface.

Of the circumstances which influence the character, degree of activity, and progress of cutaneous disease.

I consider these to be six in number :

1. Hereditary conformation and irritability of the skin.
2. Climate and seasons of the year.
3. Uncleanly habits.
4. Coexisting internal malignant disease.
5. A scrofulous constitution.
6. Syphilitic or pseudo-syphilitic disease.

1. *Hereditary conformation of the skin.*

Hereditary conformation of the skin will often be found concerned in influencing the character of its diseases. In children, whose parents have enjoyed the advantages of a robust habit of body, and have been free from scrofula in any of its forms, and yet, in early life annoyed by the varieties of acne, the same temporary and almost intractable form of disease is sure to make its appearance about the age of puberty, continuing for a few years, disfiguring the countenance during that time with all their children, yielding to nothing in the shape either of internal medicines or external applications, and disappearing perhaps at the age of eighteen or twenty, leaving no marks of its former existence.   There is in such cases hereditary proneness to inflammation of the sebaceous follicles—that inflammation spreading itself over the forehead, face, and chest, and ending as it often does in suppuration, destroys in a few years the follicular apparatus altogether, and leaves a complexion according to the common estimate, far better than that enjoyed by those on whom such an affection has never taken place.   In such instances there is hereditary irritability of the skin also, and on the degree of this very much depends the duration of the disfigurement produced by the inflammatory and disfiguring redness.

On the other hand, parents with dingy, sallow, or dark complexions seldom find their children subject to this disease.   The follicles may the blocked up, the sebaceous matter occupying them—their orifices may be stained to utter blackness; innumerable black points thus formed, may give to the countenance the most unhealthy character, yet no inflammation of the follicle, no pimple or suppuration results.

Again, in the case of hereditary scrofulous disposition in the constitution, the disease of the surface, whatever the primary character may have been, when it *has involved the cutis for any length of time, is almost sure to put on the general characteristic features of scrofulous disease.*   The varieties of acne, sycosis, and some forms of porrigo, afford numbers of proofs of this fact, when they have been long enough in existence.

2. *The influence of climate, and of the seasons of the year, on diseases of the skin, as they occur in Europe.*

It is a fact, which admits of no controversy, that the heat of the climate influences the character of cutaneous diseases in all portions of the globe.   They prevail more extensively in warm and variable climates than in cold ; and when seen in the latter, seldom wear that character of active irritation, which almost uniformly attends them in the former.   This circumstance would appear to the physiologist to admit of ready explanation ; it may be taken as the simple result of increased determination to the vessels of the surface, but there are some features of certain diseases which are common here, not reconcileable entirely to that theory ; for impetigo and lepra, with some others, are occasionally found to become aggravated towards the approach of winter, after an absence, or state of quiescense, which had lasted during the summer months.   That class of eruptive diseases most influenced by the temperature of the atmosphere, consists of the varieties of lichen and strophulus, but the common itch is particularly seen in its most aggravated form in warm weather.

In tropical climates we have examples of the influence of heat in increasing the violence of irritation belonging to cutaneous disease, in what has been named lichen tropicus, or prickly heat, and also in the ¦Malabar itch, a disease, which there seems no reason to believe differs from the itch of our northern climes except in degree.

The lichen tropicus, during the time it continues to inflict its punishment on the newly arrived European, differs in no respect from summer affections of the skin common in Europe, except in the increased inflammatory redness, and comparatively insupportable itching, tingling, and sometimes smarting sensations of the sufferer.

The Malabar itch, as it is found affecting the native of India, answers to the description of that form of the itch of Europe, distinguished by the name of "pocky itch"—a state of the disease which is not now very frequently seen here. It is, in fact, common scabies, aggravated by neglect of cleanliness, as well as by heat of climate, each being alike favourable to the propagation of the insect producing it, and to that powerful determination to the vessels of the skin, which, in a few short hours, changes a pimple into a vesicle, and a vesicle into a pustule.

The determination to the cutaneous surface on the accession of the summer season, is familiar to us all. Nævi, almost imperceptible in winter, show themselves readily and prominently, as the weather becomes warmer ; and hence of course, the vulgar notion of their resemblance to the different kinds of fruit which happen to be in season. "Summer rash," the familiar appellation given to the varieties of lichen strophulus, &c. is seldom misapplied ; and the homely phraseology is, perhaps, far better than the classical, for, in truth, although it conveys little of pathological information, it conveys almost all that is correct as to etiology, and points out to the simplest mind a judicious plan of treatment.

Notwithstanding the notoriety of the fact, that the healthy performance of the functions of the skin is of vital importance to the whole of the internal organs, the preservation of the health of this structure continues to obtain but little attention.

When disfigurement of the face, eruptions on the hand, or any part exposed to observation take place, the patient often bends rather to his delicacy or pride than to his judgment. He will suffer eruptive disease, and continue to neglect or treat it as a trifle for years, if it does not show itself to others ; and the misfortune of the physician is, that when he is first consulted, he is called on to combat an enemy who has been for years established in the fortress, and well fed and supported during the whole period ! The habits of patients of this description forbid any experienced medical man to expect or to promise much ; for chronic disease of the cutis, when it has become the habit of the part, is almost as difficult to impede, check, or subdue by artificial means, as to suppress the normal functions of that part.

Patients so afflicted, give very different accounts of their impressions as to the influence of temperature on their disease. We generally find that they have suffered it to creep on, step by step, occasionally using palliatives,—merely deriving perhaps a little benefit, from time to time, from them,—then allowing the

disease to return with redoubled strength ; then again obtaining
a temporary benefit by similar measures ; and lastly, in but too
many instances we find them consoling themselves with the conviction that the disease will not shorten their lives, and compromising
the inconvenience to avoid trouble. This class of sufferers are generally advanced in life, and their diseases for the most part are of a
scaly external character, complicated with prurigo.   Indeed " pruriginous psoriasis" would not be a bad designation to apply to the
larger portion of such cases.

Exposure to the rays of the sun, however, invariably aggravates
chronic scaly diseases, particularly where a low and impoverishing
diet is submitted to by the sufferer.   Thus in the pellagra of the
peasants of Lombardy we find that, so long as these unfortunates are
compelled to live on food insufficient in its nutritive properties to
maintain a vigorous circulation, that disease gains strength, from
summer to summer, till it destroys life.

The degree of irritability of skin in some instances, is extraordin-
arily great, and is not unfrequently of hereditary origin.   Individuals so circumstanced, seldom pass a spring and summer without
experiencing some irritation on the surface, either of a scurfy or
papular character, unless their pursuits tend to promote free perspiration ; and in the latter case, the irritable skin is seldom in a state
consistent with comfortable feelings, without very frequent ablution
is practised.

Diseases of a scrofulous nature, affecting the skin in any part of
the body, become more aggravated in the winter.   In warm weather
they occasionally put on the character of healthy progress, and eventually disappear, although seldom without leaving marks of their
former existence on the part.

3. *On the influence of uncleanly habits in inducing diseases of*
the skin.

French authors say, that a multitude of chronic diseases of the
cutaneous structure are produced by uncleanly habits ; and M.
Rayer quotes the opinion of Willan in support of this assertion, as
regards those prevalent among the poor in England.

There is no doubt that irritable conditions of the skin, especially
during the first weeks of the summer season, are prevalent among
the poor as well as among the rich in England ; but their termina-
tions are far more frequently in the same light and inconsiderable
manner in which they first appeared, than in their assumption of the
character of a troublesome disease.   In the event of the sufferer
being so totally regardless of cleanliness, as to neglect the use of a
little warm water and soap, the irritation leading to scratching of the
parts, occasionally ends in superficial ulceration.   But I am warranted in saying, that in no case does the perseverance in uncleanly
habits give occasion to the establishment of *tedious* diseases of the

skin.   Strange as it may appear to a French dermologist, it is nevertheless true, that in England, with the exception of itch, and the different forms of scalled head, the poor are on the whole less subject than the affluent to disease of the skin.   I am not unprovided with sufficient data to establish this point, and am therefore by no means a subscriber to the notion of the omnipotent power of baths, warm or cold, medicated or simple in preventing them.   Common attention to cleanliness ; and those frequent changes of linen which every artisan commands in this happy country, are more than sufficient to ensure a healthy state of the skin, if no other agent of disease be at work.

I have spoken of scabies as a disease of the poor, or rather one which is commonly allowed to flourish among the poor.   I do not mean by this to say that it is confined to this class ; on the contrary it is a disease which ever has, and ever will, occasionally show itself among the better classes.   The common chances of masters and mistresses and servants as to the touch of the handle of a knife or napkin remain the same, the itch will, from time to time, certainly make its advances to the nursery and the drawing-room so long as it is necessary to touch an article which a fellow-creature has used.   I have been consulted in the most opulent and highest grades of society—families of children, with the parents, to the number of ten or twelve, having suffered from the torments of the disease for ten or twelve months, and never having heard the ungenteel name pronounced, yet were cured by three nights' use of the sulphur ointment, one and all !   So much for dirty habits, as the cause of all the cutaneous diseases of the poor of England !

French authors humanely condole with us, and with our poor, that we have not advantages as regards baths equal to them. " The active philanthropy of the administration of hospitals of Paris, have placed within reach of the poor, privileges, to the possession of which the rich alone aspire in other countries !"   In the hospitals of St. Louis only, in one year, the gratuitous baths were, to itch patients only, 130,000.   Yet if we read from time to time of the cases in the hospital of St. Louis, we find details that are not to be equalled in misery among the lowest class of paupers in England. The descriptions of French authors are in fact such as seldom apply to those diseases of the skin as are seen either among the poor or rich of happy England—there is a wide and clearly defined line of demarcation between them as regards either.

French authors, it will be observed, by those who study them attentively, favour us chiefly, if not solely, with the details of observations made on the poor in hospitals.   English medical men well know that it is not in such institutions that an accurate knowledge of cutaneous disease, such as it occurs in England, and particularly in respectable private practice, is to be obtained, and those who, in their earlier years of practice, are accustomed from day to day to visit hospitals and parish work-houses, will seldom see a case, which, with their utmost ingenuity they can bring to accord with a French description.

4. *On the influence of internal malignant diseases on those of the skin.*

It has happened to me to trace the conjoint progress of these diseases in a variety of instances, and so uniform have been the results, as to lead me to speak with some confidence as regards prognosis whenever they may be found coexistent.   The result is almost always fatal, for if the cutaneous affection happens to arrive at positive ulceration of the cutis, however superficial that may be, healthy cicatrization scarcely ever takes place.   On the contrary, as the internal disease advances, the superficial ulceration spreads ; not however after the mode of common sores, but showing from time to time little points of new cuticle, promising a satisfactory though slow cicatrization ; these will increase in extent for a day or two, and give rise to sanguine hopes ; but on the third morning, perhaps on examining the sore, you find all vestige of them gone, while the absorbents have been busy excavating the surrounding cutis, and the whole surface of the sore exhibits a glossy languid appearance and discharges a thin transparent sanies.   Females in the better walks of life, of light complexion, and very abstemious habits, constitute the majority of such cases as I have had the misfortune to see.

It has been customary to consider the superficial sore under these circumstances, as identical in character with the internal disease, and that the more malignant the latter may be, the stronger the presumed identity.   That this view of the case is however erroneous, I am led with confidence to believe, first, by the existence of the organic disease for a considerable period, antecedent to the cutaneous sore (which has been in two cases, which I have seen the result of a mere scratch); secondly, by a better explanation offering itself of its intractability and indisposition to heal in the coldness and pallor of the cutaneous surface on which it is situated ; and thirdly, by the absence of any signs of enlargement or disease in the absorbent glands adjacent to the ulcer on the skin, of however long standing the disease may be.

Abdominal tumours, of a malignant nature, have gone through their progress to the destruction of life, and an unhealthy sore of the kind described, in the course of years, only destroying the cheek, has never been accompanied by disease of the neighbouring glands.

I have ascertained, in other cases of similar nature to those described, that a little pimple on the skin of the face has been scratched, and thus given origin to the sore on the face, but the skin having been thus once broken, no other pimples appeared, the sore scabbed once or twice, and then the languid deceptive sore showed itself, and spread from time to time as the constitutional debility consequent on the internal disease increased.

Still there is a difference between the sore of mere constitutional debility and the preceding.   In cases of lepra, where, as I shall hereafter be able to show, great deficiency of tone in the vessels of the skin, as well as constitutional debility, exists as its sole cause, even though the skin may be half covered with the scales of the disease, an accidental injury, such as a severe scratch of a pin, a few inches in length, will be studded with small branny scales, such as make up the substance of the larger scales of the disease. In this case there are evident attempts at reparation, but in the other the whole powers of the system seem to be laid prostrate by the malignity of the internal disease. It may perhaps be alleged, that in the origin of such cases of sores on the face as I have mentioned, by the scratching off the head of a pimple, that pimple may have been of a tubercular character, and similar to that which appears more or less numerously on the unbroken integuments of a scirrhous breast; but in no instance which I have had an opportunity of noticing, or learning the history of, was there any resemblance to it. The absence of the power necessary to the effecting the healing process, and the incessant demand of the absorbents to supply the wants of the system in the absence of healthy action of the digestive organs, account for it, without driving us to the opinion, that it is a part of a malignant disease, and not one of its consequences.

The scrofulous, syphilitic, and pseudo-syphilitic diseases of the skin will be found in their proper places ; they do not require preliminary observations. As I have before said, when the cutis is once deeply involved in scrofulous ulceration, the features of that extraordinary disease are stamped on it for months and years. Even the glands go through the slow processes of suppuration and ulceration, and the sores appertaining to them will heal ; yet the skin remains broken—appearing to promise to heal from time to time, and, after many years, heals only with the greatest disfigurement of the parts.

Of the dangers of repulsion, or cure of cutaneous disease.

The prejudice is of old standing that cutaneous disease ought not to be suddenly removed, and many and long are the tales recorded of vital mischief following their sudden disappearance.

To combat that prejudice successfully is not to be expected by any author, but something may yet be alleged in the shape of argument and fact, calculated to diminish or modify it.

There can be no denial of the powerful influence of irritation, .inflammation, vesication, or ulceration of the cutaneous surface in warding off the mischief and counteracting the progress of internal disease. A practice, founded on this principle, is universally adopted by medical men, and very frequently with the greatest success, where manifest symptoms of internal disease exist. But the question naturally occurs, how, if there be no signs of such internal disease, can the practitioner be afraid of curing his patient, as speedily as possible, of the external affection ? The truth is, that few or no diseases of the surface require any but (to use a phrase not common in medicine) *off-hand* treatment, unless there be reason to apprehend that internal organic disease really exists or is threatened. It is seldom indeed, under such circumstances, that diseases of the skin are of more than secondary importance, and hence the general position I have endeavoured to lay down, of considering them in all their parts and characteristics in conjunction with the state of constitution of the patient.

When, therefore, there is no reason to suspect the integrity of any vital organ, I apprehend nothing, in the form of mischief may be expected if the skin disease be got rid of at once, whether by external applications or internal medicines. The doctrine of the humoral pathology has supported and still continues to support a contrary notion, but I believe it to be entirely fallacious ; and I have never found cause to regret having acted on that opinion.

That cases often occur of very extensive disease, and of a most active kind, where no internal disorder exists, I could afford abundant proofs of, and of the general results of such cases in perfect recovery. Common prudence would direct the practitioner to inquire as to the habits of life of the patient ; of the then existing state of the sanguineous system : he would reduce the diet, use reducing aperients, or even abstract blood if he had to deal with a case of impetigo of many months' duration, occupying the whole surface of the lower extremities in a man disposed to indulgence and corpulency ; but he need not be afraid of curing such disease after he had taken such precautions.

Diseases of the skin, accompanied by manifest marks of repletion of the system, particularly if of an active character, such as impetigo, can only be treated safely by preliminary steps of a depleting character, and the necessity of such steps are generally sufficiently manifest. It is necessary to make room in the sanguineous system for the increased quantity of blood which the suppression of the discharge from the surface throws upon it ; but that provision being made, and provided no vital organ has deviated from the healthy state, the cure of the external affection may be proceeded with in safety.

Baron Alibert has expressed strong opinions on this subject, and appears to have been almost an unreserved subscriber to the notion, that danger is often an attendant on the cure of cutaneous disease. Many other authors have insisted on it in the strongest terms ; and one of the most able and experienced physicians in London has only lately declared that the cure of common itch has been, in several instances, followed by a fatal result.

Nevertheless I am compelled to withhold my belief in the existence of any such danger, in any case where no vital organ has suffered from long-continued derangement or actual disease. It appears to me that a well grounded fear of removing superficial disease must be constituted only of positive knowledge, or at least reasonable suspicion of existing disease of vital organs, or of a state of repletion of the sanguineous system. In the first case it cannot be wise to suppress a long-standing inflammatory affection of the surface, without substituting some analogous counter-irritant for it; in the second, it is generally easy to make the necessary degree of preparation. There are many affections of the surface which will preserve their inflammatory character for years, while internal organic disease is advancing. When the latter approaches its destructive termination, these seem to disappear suddenly. The pathologist is able to explain this circumstance ; but the ignorant witnesses are but too apt to confound cause with effect. Thus, young persons who die victims to phthisis, who perhaps having been anxious about their complexion and personal appearance, have had recourse to some fancied detergent nostrums or other means, are very often considered as obstinate, though unconscious suicides.

There must be a weak part in some vital organ, to justify the medical practitioner in abstaining from the free exercise of his art in the cure of cutaneous disease ; for though the latter does not often *endanger*, it very commonly *embitters* the life of the patient, and becomes gradually the cause of bodily and mental decay.

The reputed influence of baths of certain kinds in the cure of diseases of the skin

Medicated Baths.—It is but fair to persons who have laboured to bring baths of this description into use in England, to say that they have succeeded in offering to the profession and the public, a promise likely to be redeemed, of their powerful agency in the cure of certain kinds of disease ; but there must be further time for their fair trial. With the exception of one particular disease of the skin, however, I have had many reasons for withholding my confidence in them as regards this class of affections; and, to be brief, the result of my experience only leads me to depend on the sulphur vapour bath in cases of lepra and psoriasis. In such cases again the instrument is sometimes worse than inert, if the state of constitution be not attended to with the utmost care. Lepra is certainly a disease of which debility of system is the most prominent feature. When combined with a scorbutic diathesis, wearing the character of the lepra nigricans of Willan, the discipline of the hot sulphur vapour bath is capable of doing very serious mischief, and it has often happened to me to witness the degeneration of the dark scaly spots into actual ulceration and extensive sores, by the injudicious and too frequent use of it. The weakened state of a patient unfitted him, in the case to which I shall refer more particularly hereafter, as an illustration of many others, to bear the drain of perspiration, and increased his general debility, and his life had been nearly sacrificed.

When, however, the chronic affections, designated lepra and psoriasis, do not yield to constitutional remedies directed towards the invigoration of the system, the bath in question is often a collateral agent of great power. It gives tone to the debilitated vessels of the cutis, while it diminishes their irritability.

Beyond this, none of these so much lauded and advertised baths
have proved of any more utility in diseases of the surface, than the
common warm water bath, if followed by a little friction with the
towel.

Of the influence of mineral waters exhibited internally over cutaneous diseases.

The reputation of these waters is of ancient origin, but there is
reason to believe that the statements of their miraculous power here
and there to be found, are for the most part entirely fabulous.   It is
easy to perceive that the anxious and ever-toiling man of business,
from the eastern parts of London and the city, will be in the course
of many months of exertion as liable to feel some depreciation of his
bodily, as well as mental powers, as persons of fashion who waste their
time in heated rooms, over the luxuries of the table ; or those who
spend half their nights engaged in their duties as senators.   All
alike in this busy scene, are for the larger part of the year drawing
largely on the powers of their constitutions, and returning little in
the shape of occasional relaxation to make up for the wear and tear
so constantly going on.   Hence ill health or simple debility, or other
deviations from the natural condition of the system supervene after
a long season of exertion and a hot summer.

Of those deviations from health which we find thus originating,
diseases of the skin form no inconsiderable portion.   Debility is the
characteristic of a very large portion of these, while repletion and
disorder of the digestive organs arising from irregular habits give
occasion to many others.   Now entire relaxation from the toils and
cares of business, hurried meals in a close atmosphere, abolished ;
substituting for them leisurely enjoyed meals in an open and healthy
air, and add to these, quiet and peace of mind !   Here is enough to
account for many of those changes which the invalid experiences in
going to our highly-extolled watering-places.   It is to the change of
habits and the change of air—the comparatively quiet mode of life
—the absence of the accustomed exhaustion of either the animal or
mental powers, that he is indebted, far more than to any medicinal
effect of the waters, for his recovery.

The history of almost all our watering-places, show that the
worldly philosophers who obtained for them the notice they have
enjoyed, had far clearer sighted notions as to their value than the
invalid who drank, in his faith, to obtain a cure for disease !   Far
be it from me, however, to say that *all* are inert as regards disease of
the surface.   On the contrary, I consider that as *collateral* agents
several of them are extremely useful.

I have made a special allusion to them individually, where I have
had reason to believe the disease under discussion has been beneficially influenced by their use, in different chapters of the following
work.   It is not necessary here to put forward more than one or
two general propositions.

Iron and sulphur are the only two agents capable of influencing cutaneous disease found in the mineral waters, beyond that which any simple aperient possesses in obviating constipation of the bowels.

There are now twelve of these springs recognised in different parts of England ; Buxton, Matlock, Tunbridge Wells, Harrogate, Bath, Bristol, Cheltenham, Leamington, Malvern, Isle of Wight, Brighton, and the Beulah Spa at Norwood.

Some of these are of little note. The chalybeate of Tunbridge Wells, as being the nearest to London, would, we should expect, experience the largest consumption of its water ; yet though there is nothing repugnant to the taste or smell, it appears to be taken when it is taken by visitors to that place, either reluctantly or with slight expectations of benefit. Still if the iron contained in the water be capable of acting as a tonic, active cases of impetigo, or other affections of the skin accompanied by repletion, or marked by local irritation to a great degree, would be more likely to become aggravated than cured by it ? In truth as regards the chalybeate waters we must become converts to the homœopathic system to believe that the small quantities of iron contained, even though the waters be taken liberally, can be of any great use as tonics.

As regards the waters containing sulphur, and with reference more particularly to the Harrogate, I have reason to feel convinced that in many chronic cutaneous diseases they have been singularly efficacious as adjuvants. The powers of sulphur, as administered in this form, are considerable in equalizing the circulation on the surface, and promoting perspiration. Partial chronic psoriasis has been often entirely cured by the liberal use of these waters, the general health being of course not made a secondary consideration.* Such of the foreign waters as have obtained reputation I shall take due notice of under the heads of the different diseases in which they have been found useful.

ON THE CLASSIFICATION OF DISEASES OF THE SKIN.

I have already disclaimed any intention for the present of interfering materially with the nomenclature of these diseases, and in the only instances in which I have done so at all, I flatter myself the reader will think me fully justified, after he has examined my reasons for it ; and though, as observed by Rayer, it is by itself a long and painful study, a majority of practitioners know something about it, and such attempts as those of M. Alibert to supplant the old, are not likely to meet with much support for a variety of reasons, the chief of these being the difficulty of understanding that which he proposes as the new.

---

* My late friend, Dr. Armstrong, assured me that he had seen patients in the afternoon who after taking the water plentifully in the morning, had gone to bed with entirely clean linen, and who emitted from their bodies a strong sulphureous vapour.

The nomenclature of Willan is entitled to be considered the standard, and was so treated in the first place by Alibert and more lately by Rayer. The bold attempt of Alibert to substitute for it such as he has lately published, cannot be supported by those best acquainted with its different subjects. To settle a nomenclature is a task which no one man can be equal to, the substitution of a new one for that which has been a long time established, is not to be accomplished by merely a few ; but when the attempt to do either comes from the pen of a single individual, and he ushers in his new or altered designations and divisions into genera and forms in number upwards of two hundred, it is evident that he must have failed in his attempt to make the knowledge of the subject more easily attainable. I entirely coincide with M. Rayer, who observes that "In the course of time it will no doubt become necessary completely to re-model the nomenclature of diseases of the skin on an uniform basis, and to dispense with all Greek, Latin, and Arabic words, &c. the etymological sense of which gives slight or inadequate ideas. At present a complete reform of the nomenclature would only swell our vocabularies." "Pathologists," says he, "must *agree on the number and external characters* of the objects that are to be denominated, that is to say on the number of morbid peculiarities which the skin may present."

Any attempt at present to alter the nomenclature of diseases of the skin, must be futile. The opportunities of studying it are not extensive to more than a few, and that few are prepared to consider the multiplicity of terms employed by authors a great evil ; and to look upon their reduction in number, and augmentation in sense, as the only desideratum.

I shall not trouble my readers on the subject of the classification of my predecessors on this subject very largely. Mercurialis,[*] and after him our countryman Turner, made the first attempts at classification, and these consisted of the simple separation of the diseases of the hairy scalp from those of other parts of the body. Here then, in the very commencement, we find authors aware of great differences between diseases of the skin of the head, and parts not covered by hair ; and we find, subsequently, that Turner and Lorry considered another, and equally sound division, as justifiable, namely as regard other parts than the scalp, and the influence of local or constitutional circumstances in producing them.

M. Rayer on this point remarks, that the etiology of diseases of the skin is too obscure to serve as the basis of their classification ; but he is mistaken ; the etiology of diseases of the skin is capable of being converted to far more extensive usefulness, both as regards classification and treatment, than he at present seems aware of. "Abandoning," says he, " so futile a distinction as the diseases of the scalp and other parts of the body, and taking no account of the *etiological* division above alluded to, Plenck adopted the arrangements of maculæ, pustulæ, vesiculæ, bullæ, papulæ, crustæ, &c. &c.

* De Morbis Cutaneis, 1633.

&c. and this classification was much superior to all that had been published previously." A classification of the external and ever-changing forms of the accumulated secretion of disease on the surface —one day a pimple, the next a vesicle, on the third a scab, or crust, the fourth a falling scale, the fifth a red spot! This might have served the purpose at the time for want of a better, but to pronounce it a better classification than one founded, whether with solid foundation or not, by its originators, on etiology, or the *causes* external and internal, of the cutaneous disease, would be manifestly absurd? There was a grand and fundamental error committed by Plenck, and imitated by Willan and Bateman, Alibert, Rayer, and others, in classing cutaneous disease in the manner mentioned. Their nomenclature may and must stand for a time, but the principle of their classification is manifestly deficient.

These remarks, or as some may call them, assertions, have been made advisedly, and to serve the purpose, not merely of introducing, but again advocating the cause of a classification which, if it may be considered fitted for nothing else, is fitted for the illustration of my own opinions and practice as to the treatment of disease, in the following pages. It is a classification founded on due consideration of the anatomy and physiological functions of the skin, and the influence of certain conditions of the constitution in giving origin to them, in determining their external character and progress, and distinctly declaring itself independent of the divisions of Willan and Bateman, founded merely on the external characters of the disease, or the notions of Rayer. The latter says, "that Willan confines himself to the consideration of the external appearances of cutaneous disease, while he himself has taken as his basis the conformation, the *structure, and phenomena* of the *alterations.*" Now if the results of practical inquiry shall arrive at the positive establishment of the etiology of disease, what is the claim of the " structure and phenomena of alterations," to form the basis of classification ? To know the cause is surely to know more towards its successful treatment, than the practice of the eye on the ever-varying forms of external morbid secretions can ever promise ?

It is impossible to fail to notice in the classification of Willan and Bateman, the adoration of precedent in their adoption of Plenck's arrangement as a starting point. Willan could have known little indeed, practically, of the subject, when he first began to direct his attention to it, even as an author ; and it is really surprising that he did not at the commencement see the great and fundamental deficiencies of that system. Many years since, when first directing my particular attention to this branch of medical science, it was apparent that no better guide could be found than afforded by their works. In point of fact, nothing else of value did really exist in print, either as regarded description, pathology, or methods of treatment ; and a very large portion of these diseases were surrendered to the management of empirics and old women, almost as a matter of course, before Willan's publication appeared. Indeed, to say the truth, in many parts of the country, antecedent to this period, the blind practice of the empiric was of so successful a character, compared to that of the medical practitioner, as to exclude the latter from any very extensive opportunities of studying such diseases.

Having, on the one hand, however at the period I have mentioned the only legitimate guides in Willan and Bateman, I had on the other abundance of illegitimates whose assistance I could obtain, and whose notions and practice in some of the more important of these diseases, I hesitate not to acknowledge that I derived advantage from. One of the largest parish workhouses of London, with its infirmary, fortunately came under my direction as surgeon, at the period I allude to, and here I found a considerable number of pauper females suffering from scalled head. Many of these had been a burthen to the parish for years; yet with the works of Willan and Bateman before me, I could not with any degree of certainty match their descriptions with such cases as I was called on to treat. Alibert had indeed described immense masses of dried and powdery scabs with chasms occupied by multitudes of pediculi, and there were several cases which bore out his description, such as Willan, it is fair to presume, had never seen. There were other cases answering to Willan's porrigo furfurans— others again to the tinea asbestina of Alibert; but it was apparent yet, that I had first to deal with these masses of scabs and scales, interwoven and matted with the hair of the parts, before I could know the exact condition of the surface of the cutis. So far as comb and brush were concerned, these scabs and scales were impenetrable. Ointments, the use of soap and warm water, and all other means, were persevered in for a long time before I could arrive at an opportunity of correctly examining the diseased cutis.

Having, at some personal risk, cleared away the masses of mortar-like scabs, in one or two cases, with the hoards of vermin they contained, and patiently watched the effect of frequent ablution and cutting, or rather clipping the hair very short, I found the scalp exhibiting all the characters of inflammation, denuded very extensively both of its cutis and hair; and wherever hair was growing, it was in unhealthy tufts, each individual hair growing through the centre of a pustule. I have endeavoured to give a representation of this fact in Plate No. 3.

Now this is the condition of the scalp, which will invariably be found to exist among the poor, whatever may have been the character of the disease at its first commencement. After it has been of long standing, large pustules may not show themselves where attention to frequent ablution has been adopted, but small ones certainly will; and where the scalp is cleared in the slightest case of scurf, or dandriff, the connexion of the cuticle with the hair, forming its sheath at its exit from the cutis will be found to be destroyed, and generally a minute portion of fluid will be perceived surrounding the hair.

In the institution to which I have alluded, that of which the very poorest of the poor in all London constitutes the chief inmates, I accidentally, at the very period at which I was investigating the cases mentioned, happened to have brought to me a poor barber, who had been a long time an inmate. He pointed out two or three of my then patients, as having been unruly ones of his own, and assured me he could have cured them months ago if they had been tractable ! I naturally inquired his treatment, and expected that he would produce a pitch cap. Not so however. He said, " I pull out the hair where there is matter in them, and then there is no more matter forms, and the new hairs grow up healthily." I inquired of him his means of doing this, and a child being brought forward, whose hair had been cut short, and on whose head the disease had been long established, and free pustulation going on at the roots of the hair, he took a common table knife, laid the back of, and pressing it, on the surface of the cutis, he took within the range of the point and bulb of the thumb, all the hairs he could, and, without much pain, divested the diseased cutis of its hair to a considerable extent. Pursuing this plan from time to time, the disease was cured.

This practice was, I found, common in the establishment, and very frequently successful : my own particular patients having been those declared intractable and incurable, by the ordinary means, brought themselves under my notice.

If the reader takes the foregoing facts into consideration, with a view to the causes of diseases of the skin, he will plainly perceive, first, that the description of masses of accumulated scabs and filth constitutes no description of a disease. and that the causes of a disease, or at least its protracted and obstinate character, are to be found *under those masses of filth,* whatever form they may assume ; in other words, in the actual condition of the cutis on which they are formed. Thus in the cases mentioned, we find an inflamed cutis, and growing through it unhealthy and diseased hair ; the disease of the cutis deprives the hair of its healthy cuticular sheath, and the hair thus situated becomes in every way a cause of increased irritation to the skin of the scalp, leading to pustulation, or the formation of matter around every individual hair, accumulating according to circumstances. Thus, I unhesitatingly say, of all those forms of chronic disease of the scalp, or other parts covered by hair, characterized by accumulation of unhealthy secretions, no matter what form these may assume, that we have their etiology clearly established, and a direct road to their cure plainly pointed out. Such is part of my justification for one of my divisions, or classes of cutaneous disease.

It is vain to allege that the diseases of the scalp are at their commencement very superficial ; as for instance, in slight cases of ringworm, commonly so called ; and that the formation of pustules does not take place, for what happens even in this slight form ? It is first discovered by the destruction of the hair, and if a microscopic examination be instituted, the margin, or ring of the disease, is found to consist of an inflamed circle of minute vesicles, every one of which has destroyed two or more hairs which previously occupied its site.

In this case, it is true, there is a specific poison acting as a cause, but it very often happens that, in order to cure it, repeated applications of blisters, caustics, &c. have been had recourse to, and then results a chronic inflammation of the scalp, which the hair, as long as it grows through it, will not allow to subside. French dermologists tells us,(and in so doing they show far more candour than is merely necessary to furnish a sound argument in support of my notions as to the causes of the obstinacy of scalled head in its different forms,) that the MM. Mahon are employed in the hospital of St. Louis to cure these diseases by a secret remedy, with the application of which the physicians and surgeons of the establishment have nothing to do, and of the composition of which they know nothing. Its *apparent simple results are the destruction of the hair of the diseased parts, and the consequent permanent cure of the disease.**

Sycosis, as it has been termed, affords another illustration of this part of my position ; but though situated on the chin and upper lip, and occupying the site of the beard, it obtains some additional features from the simultaneous disorder of another set of organs intermixed with those producing the hair. Nevertheless the extraction of the hair is almost invariably followed by the same results as in porrigo. We have then here again to look at the influence of the hair in supporting the disease. I have selected these instances, and so much of their history, as illustrative of the first division or section of an arrangement which I find every day more and more occasion to adhere to, namely, that of diseases which obtain their distinguishing characteristics from local peculiarities in the structure of the skin, and not dependant on, though occasionally influenced by, the state of the constitution. With the same view I now propose to my reader the consideration of those diseases of the skin "dependant on debilitated and deranged states of the constitution and consequent diminished tone of the vessels of the cutis," with the desire of urging still and again the necessity of departing from the old system of classification by scales, and scabs, and vesicles, and pustules, and with the hope that I shall be able to show reasons for so doing, or at least prove that the constitutional influence over another and more serious class of diseases, when duly attended to and considered, will be a far better guide in their treatment, a better pilot for the practitioner, than any arrangement dependant on the form the cutaneous eruption may assume.

Purpura or scurvy stands first among these. Its different states in different, and not unfrequently in the same persons, may, at uncertain periods, be found putting on the mask of pimple, vesicle, pustule, scab, or scale, blue flat discolorations, large vesicles half filled with bloody serum, and others smaller in dimensions, of a yellow transparent colour. Again, it may show itself in the form

* I was informed that it is a composition of crude potash and lime, but have found this not to possess the power of destroying the hair without great mischief to the scalp.

of what is commonly called thrush, either in the infant or adult. It may appear in the form of erythema nodosum, which is a scorbutic inflammation of the integuments ending in a partial effusion of serum and blood into the cellular structure. Ecthyma and rupia are the pustular forms of it, ending in scabs. The pathology of these will be hereafter more particularly explained : but the constitutional origin is the chief consideration, and constitutional measures form the only road to a cure.

The reader will I hope, duly appreciate my object in adducing these instances as illustrative of the claims of a classification founded as described ; it is simply to bring it into fair comparison with others of older date, and it leaves him to pronounce judgment between the two.

The basis of the classification or arrangement I have adopted is, in a few words, founded on the constitutional causes of the disease, and due consideration of the organic structure and physiology of the part of the skin on which it is seated. M. Rayer pays me the very equivocal compliment of calling it ingenious ; but there has been little demand on me for the exercise of ingenuity : *facts* have flown in upon me too quickly to render it necessary. This author may however very reasonably congratulate himself as far as regards the possession of ingenuity, for he takes the credit of improvement for " his basis on the conformation, phenomena, and structure of the diseased alterations, in preference to Willan's description of the external appearance of scab, scale, or whatever it may be." Rayer's classification had in his first edition consisted of four sections or divisions. The first of these comprehends all, and a vast deal more than Willan and Bateman ever thought of as diseases of the skin. The second " *alterations of the appendages of the skin.*" The third. " *the consideration of foreign bodies on the surface or in the substance of the skin.*"

This section he divided into "inanimate and animate species." The first of which is thus described —" Dirt ! dirt on the scalp of new born children ! inorganic matters, artificial colourations ! the second the vermin of the scalp and skin." *

The fourth section is thus headed ;—" *Diseases primarily foreign to the skin, but which sometimes produce peculiar alterations in this membrane.*" The only instance being the elephantiasis of the Arabs."

M. Alibert has recently invented an entirely new classification, with a partially new nomenclature. These he has illustrated by the assistance of the vegetable world, and selected the form of a tree, with numerous branches, sprigs, &c. which he calls the " Arbre des Dermatoses," the main branches of which, with their divisions and

* This third division is in the new edition reduced at the subject of " parasitic insects infesting the skin of man," the " inanimate" matter, the dirt, being wisely omitted. This is simply returning to the old arrangement of Plenck, who has his chapter on the " Insecta."

subdivisions, are two hundred and eleven in number! I shall freely select all such observations of M. Alibert as I may find out of the
reach of the *shade* of the branches of the said tree, as being most
likely to be valuable to the reader, and refer the latter to the work
itself, which is still of a valuable and practical character in its
details.

The reader will perceive that I have omitted, as well from the
classification as from the substance of the work, the consideration
of small-pox, measles, and scarlatina, the cutaneous eruptions appertaining to which are in all probability, entitled to no other estimation
than that of symptoms of a constitutional disease. Further they are
not to be interfered with with advantage; and lastly, they have
occupied so large a share of the attention of others, as to leave, I
believe, nothing new to be said on them. A work of the present
kind could not comprise a tenth part of what is necessary to be
learnt with respect to these diseases. Moreover, the first ten lines
of description even, would lead us into the precincts of that of specific constitutional disease.

For reasons of equal strength I shall omit erysipelas and the malignant pustule of the plague, which other dermologists have interwoven with the mass of their descriptions of cutaneous diseases.
Burns and the effects of frost, stand equally exempt from a work
professing only to speak of disease of a certain structure which is
less extensively involved¹ than others adjacent, from the effects of
frost or heat, or mechanical injury.

To assign a seat to many cutaneous affections in an individual and
integral portion of the cutaneous tissues, certainly appears impracticable. It is true that we may establish the connexion of the varieties of acne and sycosis with derangement of certain structures.
We may also confidently explain the origin, or at least the dependence of certain varieties of what is now denominated porrigo on
the peculiarities of the scalp or other parts covered by hair, but we
are soon met with the difficulty of accounting in such a manner for
the occurrence of the subcutaneous nettle-rash, a disease which
inflicts absolute torture for a time, and vanishes in a few hours,
leaving not a vestige of its existence—and also of the different characters of what is named prurigo or universal or partial, yet intolerable itching. Now there are two features common to prurigo and
urticaria; the slightest scratching of the skin, excited either by the
itching of prurigo or the tingling uneasiness of urticaria, invariably
produces an immense increase of irritation; manifested in the first
case by the multiplication of papulæ, and in the second, by the
peculiar elevations of the nettle-rash, in the forms of white lengthened wheals, or detached crops of round white spots, to which the
effect of a single sting of the nettle is the only resemblance. These
are each found compatible with a state of apparent good general
health, and inasmuch as they each occur on every part of the cutaneous system, a theory which is founded merely on local peculiarities of the part, will manifestly fail in its general application. Still

Willan's classification places urticaria among the exanthemata, and prurigo among the papulæ ; but neither the whitened wheals and and elevations of the one, or the pimples of the other, with a few exceptions as to the latter, are traceable before the itching and irritation has compelled the sufferer to rub or scratch, and thus irritate the part ; and the pimple of prurigo does not very essentially differ, under these circumstances, from the spots of urticaria ; but what is of still more importance, they are very often found to be co-existent with and detected as dependent on the same state of constitution. Here again I humbly conceive that an arrangement which shall interweave a due regard to constitutional causes with the local features of diseases of the surface must be better than one founded merely on their external characteristics.

I have little to say for my own arrangement beyond this ;* that I adopted it in the first place from a very firm conviction of its soundness and utility, and that subsequent experience has convinced me that I cannot amend it to any considerable extent. It does not seem to me to be necessary to proceed by circuitous arguments and a waste of time of the reader to obtain his concurrence in my view of the question. I have sought to simplify the study of the subject, and believe that the arrangement of the different forms of disease which I have adopted will tend to the accomplishment of that object. If, as is notoriously the fact, the road to the cure of these diseases is in the majority of instances through constitutional management, where can be the value of an arrangement in which this forms little or no part ? Nosology, classifications, and arrangements have no other objects than the *cure* of disease ?

SECTION I.

CHAPTER I.

On diseases which obtain their distinguishing characteristics from, or originate in, local peculiarities of the skin.

UNDER this denomination I purpose to consider the varieties of what have been termed acne, sycosis, the scrofulous form of disease of the follicles, lupus, or *noli me tangere*, and the different species of porrigo ; the four former depending on, or connected with, obstructed and diseased follicles, while the pustules of the more obsti-

---

* " Nosological classifications in medicine are generally a failure. *All that is wanted is an arrangement which shall connect things in themselves similar, and enable us to study with the least expense of time and trouble the particular facts and consolidated groups of facts that nature furnishes.*"—*Med. Chir. Review.*

nate forms of the latter, may be with propriety considered as the result of the local irritation of the hair.

ACNE,

OR OBSTRUCTION AND SIMPLE INFLAMMATION OF THE CUTANE- OUS FOLLICLES.

English authors have described four varieties of this affection under the titles of acne, simplex, punctata, indurata, and rosacea ;* while Alibert, under the head of dartres, describes the first and second of these as "dartre pustuleuse miliare," and "dartre pustuleuse disseminée." The acne indurata and rosacea are plainly his "dartre pustuleuse couperose," while the sycosis menti is well described and delineated in his works as "dartre pustuleuse mentagre."

These varied designations create difficulties to the student, and are not necessary or useful. The circumstances on which they are founded are for the most part accidental and unimportant, and dependent merely on ordinary deviations from the regular progress of inflammation, or the degree of activity or its reverse which such inflammation may assume in the structure concerned.

In its simple and most trifling form, the disease consists merely of obstruction of the sebaceous follicles, in consequence of their contents becoming too hard to pass readily to the surface. Inflammation of the follicle, and the production of what is called a pimple results, and is soon followed by the formation of matter; the follicle is destroyed by this process, the matter is discharged, a little redness remains for a day or two, and the part returns to the healthy state.

It has been customary among writers on cutaneous diseases to point out, in the discussion of each individual disease, the features or circumstances by which it may be distinguished from others. The necessity of this course will be rendered obvious, when it is considered that the external characteristics of the different affections are the only grounds on which we can proceed to make distinctions. Constitutional symptoms are seldom concerned in any great degree, and never, except in the exanthemata, to the extent which renders them of much service to us in diagnosis.

The situations in which the eruptions, designated as above, make their appearance, will generally be sufficient to enable us to determine their character. It has been observed in a preceding page, that the sebaceous follicles are chiefly distributed on the face, more particularly on the forehead, tip, and alæ of the nose, and the adjoining cutis, and less copiously on the chin. Next to these the chest, below the clavicle, to about the fifth or sixth rib, and the back to an equal extent, are most liberally furnished with them ; and as the disease consists in the derangement and inflammation of these

* Bateman's Delineations and Synopsis.

structures, it is these parts solely in which it makes its appearance. Hence, it has been described as " an eruption of distinct hard inflamed tubercles, which are sometimes permanent for a considerable length of time, and sometimes suppurate very slowly and partially; they usually appear on the face, especially on the forehead, temples, and chin, and sometimes also on the neck, shoulders, and upper part of the breast; but never descend to the lower parts of the trunk, or to the extremities. As the progress of each tubercle is slow, and they appear in succession, they are generally seen at the same time in the various stages of growth and decline ; and in the more violent cases are intermixed likewise with the marks or vestiges of those which have subsided."

In every case of this affection a very considerable number of black points will be observed, imbedded, in the cutis. These are nothing more than the blackened surfaces of the contents of the uninflamed follicles. The skins of different individuals differ greatly in the number, as well as size, of the sebaceous follicles : and hence, in a state of health, the complexions of some are said to be more clear than that of others ; the copious distribution of the black spots which have been described, giving a dirty and less healthy appearance to the part, while their minuteness in size and numbers leaves the agreeableness of the red and white unimpaired. It is evident, however, that the simple appearance of such spots ought not to be considered as a disease, or as in any respect a deviation from a state of health. The most desirable change which can be effected, therefore, where they exist to an unpleasant extent, is that which frequent ablution and moderate friction only can produce.

A constant attention to these latter points will usually prevent, where the skin is not very thickly furnished with follicles, any discoloration of the kind described : but the whole contents of the follicle, should this not be sufficient, may be easily squeezed out with a moderate degree of force, in a manner familiar to all. As a matter of precaution, this later step ought to be followed with respect to all such follicles as may exhibit the blackened surface described, where others are in a state of inflammation, as a preventive measure ; and the worm-like substance which the contents of the follicles produce by this operation is easily removed without the use of any kind of instrument.

It is under circumstances of actual and long continued obstruction only, that the treatment of this affection comes under the notice of the medical practitioner; the inflammation and redness of the follicle, with its accompanying tenderness on pressure, first directing the attention of the patient to it.

The immediate exciting cause of this inflammation is usually found to be some accidental disorder of the primæ viæ, the hardness of the contents of the follicle, and apparen tobstruction, frequently existing for some time without the excitement of mischief, till such obvious disorder occurs. Hence, instead of the solitary appearance of one of the tubercles here and there, on distant parts of the face, a considerable number, though not very thickly distributed, are observed simultaneously on the forehead, sides of the nose, chin, &c.*

* A writer in the Cyclopedia of Practical Medicine, doubts the uniform origin of this disease in the sebaceous follicles, but at the same time fairly enough quotes the following case from M. Gendrin (Histoire Anatomique des Inflammations).

" A man fifty-five years of age, who died of an organic disease of the stomach, was at the same time affected with acne rosacea which occupied the surface of the nose, and almost the whole of the cheeks. The day after his death the seat of the eruption was of a bluish red colour, and incisions showed the skin to be more dense and thicker than in the natural state. On the surfaces of the incisions a number of small round red-coloured bodies were observed, which examined by the lens appeared to be sebaceous crypts evidently double their natural size. On the surface of the skin were also to be observed the orifices of these crypts more dilated than in their natural state, from which a yellow caseous fluid might be squeezed. In the rete mucosum was seen a closely reticulated structure of blood-vessels of a violet red colour, and the corion generally was much increased in vascularity. Externally the skin was unequal, and elevated by small round grains which were felt more distinctly when a detached portion of the skin was pressed between the fingers. These bodies were produced by the development and induration of the sebaceous follicles. The general tenacity of the skin was considerably diminished." "Notwithstanding, (says the writer alluded to,) the proof apparently afforded by this dissection, we are convinced that the emphragma sebaceum is not the most common cause of the pustule of acne ; for although one sometimes finds, particularly on the face and chin, small pustules which are caused in this way, and from which, after the pus has been evacuated, a small oval body formed of indurated sebaceous matter may be expressed, and although a morbid accumulation of this matter is observed in several cases of acne, yet all this is far from proving that this state (i. e. the emphragma or enclosure of the sebaceous matter,) of the follicles is essential to this form of pustule. And therefore we very often see individuals whose follicles are in this state without presenting any trace of acne, whilst on the contrary we see acne unaccompanied by this state of the follicles, and when they do become complicated together, we never see the infarcted follicles become true pustules of acne, for by compressing the latter at the period of suppuration, pus is easily made to escape, and not indurated sebaceous matter."

Now as to the *apparent* proof alluded to, afforded by the case quoted, I think that if that case stood alone and unsupported in the results of dissection by any other single instance, it would be more than sufficiently conclusive to refute the objections in question ; but the fact is, that in every individual dissection I have made, and they have been not a few, the results have been on the whole so similar, that that description will apply with little variation to all. Indeed, on comparing it with one of my own notes, it corresponds almost verbatim. There is a scrofulous form of follicular inflammation it is true, and of that I shall more particularly speak hereafter, which differs from the simple and indurated forms, in certain circumstances being modified or altogether changed by its constitutional origin and dependence.

Then as to the remark, that " we very often see follicles in this state, (the state of emphragma,) without presenting any trace of acne ;* while on the contrary, we see *acne unaccompanied by this state of the follicles*; and when they do become complicated together, we never see the infarcted become true pustules of acne, for by compressing the latter at the period of suppuration, pus and not sebaceous matter escapes." I have not denied that follicles may be obstructed and enlarged without inflaming or suppurating even to the extent of forming

* I suppose the author means the inflamed tubercle of acne.

There is great difference in the period elapsing between the commencement of the inflammation of different follicles, and its termination in suppuration ; and consequently many of them are seen apparently only slightly inflamed, and presenting to the touch the resemblance of a small millet seed under the cuticle, while others have actually suppurated. It is not correct, however, as far as I have observed, that inflammation, as alleged by Willan and Bateman, when once begun, ever terminates in resolution ; for each tubercle thus formed, if punctured with a lancet when assuming an appearance of subsiding, is found to contain matter : the orifice of the follicle is in such cases closed at the commencement of the inflammatory action, and if the latter has been less violent than usual, the quantity of pus produced does not lead to the rupture of the superincumbent structure, and absorption, more or less speedily, takes place.

In the case of an extensive affection of this kind, many tubercles are found in the state described, gradually subsiding without the appearance of matter on their apices. Where, however, matter makes its way to the surface, a few hours only elapse before it is discharged, and the only vestige remaining a day or two after, is merely a bluish red speck, which rapidly disappears. Minute scabs probably for the first forty-eight hours may be found covering this : these are intermixed with a considerable number of the reddened tubercles before mentioned, and the blackened orifices of a larger portion of the uninflamed follicles.* In most instances, when inflammation in the follicle begins, its orifice is soon closed up by the attendant tumefaction, and the blackened speck disappears ;

the atheromatous tumour, but have expressly pointed out the fact. But because all which are obstructed do not inflame, that on the contrary the sebaceous matter being retained, many go on increasing to the extent of producing a tumour of considerable dimensions unaccompanied by inflammation, is it fair to assume that obstruction does not act as the cause of inflammation and suppuration, and in the formation of the pustule in others which do inflame ? Then in the inflamed tubercle of acne does the writer mean to state that the open mouth of the follicle is even visible wherever a tubercle exists ? that there is no obstruction ? If he does, he goes far towards proving that follicular obstruction has nothing to do with the disease. The fact however is, that the tumefaction of the spot destroys all vestige of an opening ; the latter is firmly shut up while the inflammation and suppurative process is going on within, and the whole apparatus is destroyed by the formation of a minute abscess—no part of the follicle remains. It is therefore impossible that acne can go through its course without the mouth of the follicle being closed. If obstruction did not *cause* inflammation, inflammation would, by the mere tumefaction of the mouth of the follicle, cause obstruction. How then can it be said that the tubercles of acne may exist without obstruction of the follicle ?

I shall leave the question in the hands of the reader, with the simple remark that the pathology of this form of cutaneous disease which I have ventured with increasing confidence for years to put forth, is now called in question for the first time on the grounds above mentioned, but that no other or different explanation of its phenomena has been attempted.

* A very good, though much enlarged view of the simple distended follicle with the discoloured contents, mixed with others containing matter, is given by Dr. Bateman, Delin. 62. of Acne Punctata Simplex.

but it sometimes happens, that the latter is observed in the centre of the pustule, in which case, the sebaceous matter is surrounded by pus, and eventually discharged with it ; still, however, it retains its solid form, though insulated in this manner, when it is easily detected by rubbing the contents of the pustule between the finger and thumb.

It has been constantly observed, that persons of a sanguine temperament and florid complexion have been most subject to follicular inflammation ; and among these, that young men between the ages of twenty and twenty-five have been the greatest sufferers from it. Females at the same age, are also subjects in whom its visitations are not unfrequently manifested ; but in the latter, it rarely proceeds with such rapidity to suppuration, or produces such unpleasant appearances as to extent. On the contrary, from the absence in a great measure of those exciting causes which difference in the habit of living produces in the male sex, females particularly of pale complexions and lax fibre, are more often seen with a really considerably enlarged state of the follices, without a vestige of inflammation.

Independent of the disposition to this affection which the habit of living may generate, some individuals are peculiarly disposed to it from original formation of skin ; and hence has arisen the idea of hereditary character. In such cases as I have had an opportunity of noticing, particularly where the parents of those affected had been formerly sufferers, I have entertained no doubt, that not only a free determination to the skin marked by florid complexion existed, and contributed largely to the materials of inflammation, but that the number and size of the follicles were considerably less than in subjects where the disease did not occur ; but it is not to be expected that we shall always find an obvious cause either in this formation of skin, or in disorder of the digestive organs, for every obstructed or suppurating follicle which may be noticed.

In the skins of some individuals, small pearl-like tubercles are observed, here and there distributed in parts where the follicles are most numerous. This appearance arises from the deficiency of any opening in the cuticle, corresponding to the orifice of the follicles in the cutis ; the contents of the follicle being thus retained at its orifice. The description of vitiligo, in the Synopsis of Bateman, is in some respects exceedingly applicable to this state of parts, and it is not improbable that the latter has given rise to the idea of such a disease as vitiligo. The tubercles I have described, however, seldom attain the size of a wart, as stated in the latter affection, nor are productive, if occurring on the scalp, of any serious mischief to the hair. They now and then suppurate, but sometimes remain stationary for a considerable period, and ultimately disappear by absorption, or are liberated by the exfoliation of the cuticle, and wiped away.

When disorder of digestive organs has been the immediate exciting cause of acne, the symptoms of such disorder appear to be materially alleviated ; and there is no doubt that more formidable mischief is sometimes prevented by its occurrence. In the treatment of the mild and transient kind described, therefore the state of these organs, and the general condition of the system, ought to occupy the chief attention ; not so much, however, from the fear of doing mischief by sedative applications in repelling the eruption. but as leading to the most direct remedial measure.

Repellents, as they have been termed, seldom have any effect in preventing suppuration, and this indeed ought to be considered the most desirable course it should take. Frequent bathing of the parts with warm water, and gentle friction with the mildest kind of soap constitutes by far the best kind of local application ; its effects being at once to allay irritation, and promote suppuration, and also remove any accumulation in the follicles of the part as yet uninflamed.

The use of stimulants, except in the old established and indurated state of this affection, (of which we shall have to speak hereafter,) as practised by the ancients, and approved by Dr. Bateman, is obviously improper. It is in truth opposed to the first principles of surgery. It requires no argumentative reasoning to prove, that such applications will not prevent the suppuration of the follicle, or that such suppuration is the most speedy and safe termination to which it can be conducted. Indeed the use of stimulants seems to have occurred to the writers alluded to, only as being preferable to sedatives, repellents, or such applications as, by checking the determination to the cutaneous surface, may endanger parts of more importance. They are capable, however of adding much to the irritation and extent of the disease, and consequently of protracting its duration and course, while the plan which conducts the tubercles to a safe and healthy suppuration is at once conformable to the obvious indications of nature, least painful to the patient, and free from danger of the supposed prejudicial effects of sedatives.

It may be not improper to notice here the recommendation of blistering, alluded to by Dr. Bateman, and originating, I believe, with Darwin. I entertain not the slightest doubt of its immediate effect in putting a stop to the disease ; the mouths of the obstructed follicles are at once cleared by the vesication, and their contents loosened and easily got rid of, while the inflammation in each particular follicle is checked by the counter-irritation produced on the surface in its neighbourhood ; but it is a remedy unnecessarily severe and exceedingly inconvenient.

The constitutional disposition to this affection, which is occasionally observed in young men, is usually found to depend on a plethoric state of system ; and as it is obvious that such condition would lead to a more solid and adhesive secretion of the sebaceous follicles, this fact is readily explained. The increased irritability attendant thereon will explain also the pain and tenderness of the tubercles, and the rapidity with which they run on to suppuration.

It has been observed that these affections are usually of too trifling a character, except in females, to induce persons affected to take professional advice : and Celsus has informed us, that for the sake of females only he has thought proper to notice it. Like all other diseases, however, where frequent attacks of inflammation of the affected structure take place, it unfortunately happens that a change is gradually wrought in the structure and appearance of the surrounding skin, and if due attention be not paid to it, great disfiguration and unpleasantness of appearance is likely to take place. The acne indurata and rosacea are instances of this occurrence, and no individual can be reasonably charged with too much anxiety regarding personal appearance, who would be desirous of avoiding the infliction of the characteristics of those on the face, nose, or in truth, any part of the body, whether or not exposed to observation.

*A. Indurata, A. Rosacea, Couperosa, Dartre Pustuleuse Cuperosa, &c.**

The foregoing remarks apply more particularly to the states and consequences of obstruction of the sebaceous follicles, denominated acne simplex and acne punctata. When it occurs more extensively, and habitual disorder of the digestive organs, or scrofulous diathesis

* " Cuperosa is a chronic inflammation of the skin, characterised by the successive eruption of small isolated pustules, the bases of which, more or less hard, are surrounded by an inflamed areola; they are scattered over the cheeks, nose, forehead, and sometimes even on the ears and superior part of the neck.

" 1°. In its more simple form, (*Acne simplex*, W.) cuperosa is announced by some red pustules on the face. Their successive development takes place without local heat, or any other sensation than a slight formication of the skin; each pustule is formed, suppurates, and dries, independent of the rest. The progress of suppuration is slow in the small pustules of cuperosa; it is not till towards the middle of the second week that their summits grow thin and become perforated, and then covered by a very small, thin light crust, produced by the desiccation of the seropurulent humour. These pustules are frequently intermixed with small black points, more or less prominent, formed by a thick, solid, unctuous humour, accidentally accumulated in the follicles of the skin. When these small points are numerous and close together, the skin of the nose has a fat, oily aspect, and that of the cheeks becomes rough and unequal (*Acne punctata*, W.)

" 2°. The pustules of cuperosa are sometimes larger, more numerous, and closer together; they are conoid in form, and have large hard bases; they are of a violet-red colour, are indolent, and suppuration takes place at their summits only after some weeks' duration. They are at times collected into groups, so close together as to appear to form a flattened tumour; the pustules are most inflamed in adults, particularly in those of a sanguine temperament. They are aggravated by the slightest excess of regimen, or by remaining long in an elevated temperature, &c. Under these different influences they run rapidly through their stages, but commonly succeed each other in great numbers. In this variety (*Acne indurata*, W.) the vascular tissue of the dermis is more deeply affected, and the subcutaneous cellular tissue itself sometimes participates in the tumefaction of the skin. On their disappearance, most of the pustules leave a livid tint on the skin, and a depression which is never effaced.

" 3°. Another variety of cuperosa belongs more especially to mature age (*Acne rosacea*, W.). Some red spots, developed on the nose and cheeks, become the seat of a disagreeable itching, after the ingestion of strong wines or spirituous

exist, a slow and unhealthy suppuration takes place, spreading from
the originally inflamed follicle, and involving to a considerable ex-

liquors. They soon enlarge, and become transformed into small pustules, not
very numerous at first, but which multiply and succeed each other without intermission. The skin on which they are developed remains habitually injected,
and retains a violaceous red tint, which is brighter around the pustules. The
points on which they have been renewed several times become tumefied and
indurated ; the small superficial veins are dilated, and give a bluish cast to the
skin ; the features are enlarged, and the expression of the physiognomy is altered,
and assumes a disagreeable aspect.

" Besides these fundamental differences in the form, number, and progress of
the pustules, cuperosa presents innumerable shades, according to the extent it
occupies, its duration, and the nature of the affections complicated with it. Sometimes the pustules, few in number, are isolated, and leave behind them only a slight
redness on the skin. At other times they are numerous, succeed each other
rapidly, cover the whole face, and extend even to the neck. When cuperosa
arrives at this degree of intensity, it is frequently followed by red or violaceous
tubercles, more or less voluminous; the conjunctiva are inflamed ; the gums
become painful and tumefied ; and the teeth loosen, after this chronic inflammation of the mouth. Lastly, in more rare cases, cuperosa does not extend beyond
the alæ of the nose, on which are elevated rugous livid tumours. All the
elementary tissues of this organ swell, so as to give it double or treble the dimensions which it usually has.

" Cuperosa is generally seen in men from thirty to forty years of age. Acne
punctata, however, is more particularly observed in youth ; men of mature age,
and old people, are seldom affected with it. The sanguine temperament of youth,
and bilious temperament of adult age, predispose to cuperosa. Its connexion
with chronic inflammation of the stomach and alimentary canal, is frequent, and
easily detected. Its dependence on an affection of the liver is more rare and
difficult to ascertain, notwithstanding the old and repeated assertions to the contrary. Women, more frequently the subjects of cuperosa than men, are usually
affected by it at the age of puberty, and on the cessation of the menses. This
eruption may also supervene on the suppression of the menstrual flux, and disappear on its return ; or it may coincide with simple dysmenorrhœa. Cuperosa
is seldom aggravated by pregnancy ; but often decreases or disappears during
gestation. This disease is hereditary, and may be transmitted to several generations successively. It has been supposed that cold, damp climates, have a
marked influence on the development of this eruption, as it is more frequent in
the north of Germany and in England, than in meridianal countries ; but this
may be explained by the abuse of spirituous liquors, indulged in by the people
of the north, which is a less equivocal cause than their climate.

" The excesses of the table, some vicious habits, more or less acute moral
affections, certain occupations which require a long continuance of the same atti-
tude, causing a determination of blood to the head, are the common cause of
cuperosa. Lastly, the contact of certain paints and astringent liquids, the use of
cosmetics, practised by women in the decline of life, are more direct and immediate
causes, the action of which is particularly evident when there is no predisposition
to the disease.

" The pustules of cuperosa are easily distinguished from those of other diseases.
They never have the dimensions, nor the adherent crusts, of the pustules of ecthyma. They are never covered by large thick crusts, like those of impetigo, or
the varieties of tinea. The pustules of cuperosa cannot be confounded with the
papulæ of lichen. The small light crusts formed on the summits of the pustules
of cuperosa, are very distinct from the thinner, and more extensive crusts of
chronic lichen, which may readily be mistaken for epidermic scales. The development of syphilitic pustules or tubercles is rarely confined to the face ; they
are most usually observed on all regions of the body at the same time, or, at
least, on a considerable portion of its surface. This circumstance sufficiently

tent those in the neighbourhood, producing, instead of a minute pustule, a considerable, though slowly formed collection of matter. The

distinguishes them from cuperosa. When syphilitic pustules occupy, exclusively, some parts of the face, they are commonly seated around the alæ of the nose, and commissures of the lips, and are almost always uneven and fissured, having the appearance of vegetations. They are distinguished too, by their shining surface and coppery colour. The tubercles of lupus, (*dartre rougeante scrophuleuse,*) at first superficial, and but slightly elevated, might be confounded with the tubercles which sometimes succeed to the pustules of cuperosa; but those of lupus enlarge more slowly, assume a livid tint, extend from the nose to the cheeks, and destroy, in ulcerating, all the subjacent tissues; thus rendering a mistake impossible.

" Cuperosa is curable when the subject of it is young, the eruption recent and slight, and the pustules not very numerous. On the contrary, when it shows itself in adult age, is connected with chronic affection of the digestive organs, or when it is hereditary, of long standing, and of large extent, the best treatment rarely succeeds in preventing the development of the pustules, or causing the resolution of the tubercles.

" In persons affected with cuperosa, the diet should consist of white meats, fresh vegetables, and ripe juicy fruits; they should carefully abstain from fatiguing exercise, excessive study, and from remaining in places of high temperature. If the patient is young and sanguineous, the pustules numerous and confluent; if the tubercles are inflamed and unite at their bases, bleeding from the foot, and the repeated application of leeches, behind the ears, to the temples, and alæ nasi, is generally useful. Ambrose Paré * advises copious bloodletting, as being efficacious. "A patient attacked with *rose-drop*," says he, "should be bled from the basilic vein, then from the forehead, then from the nose; and should have leeches applied on different parts of the face. He should also be cupped on the shoulders." If cuperosa depends on the suppression of the menses, or of a hæmorrhoidal flux, leeches should be applied to the vulva, or anus, at the period of the usual appearance of these evacuations. The employment of diluents, whey, cooling diet, clysters, tepid baths, bran lotions, and internally of tepid milk, bitter almond emulsion, decoction of quince-seed, assists the good effects of this treatment. It is seldom, however, that purely antiphlogistic treatment completely cures cuperosa, and we are obliged to resort to some *external means*, the efficiency of which has been proved by long experience. The ancients made frequent use of liniments, of which turpentine, vinegar, soap, myrrh, &c. were the base. At the outset of cuperosa, when the pustules are few, isolated, or not much inflamed, and in graver cases, after having bled freely, it is preferable to employ lotions of distilled rose-water, lavender, &c. mixed with about a sixth part of alcohol, according to the state of the pustules. Solutions of the deuto-chloruret of mercury † are also useful.

" The mineral and sulphureous waters of Barèges, Aix in Savoy, Cauterets, &c. administered as lotions, baths, and vapours, are very beneficial in old cuperosa.

" The nitrate of silver and hydrochloric acid have sometimes been employed, to cause chronic cuperosa to take on the acute form. These applications should, in general, be preceded by bloodletting, and made so as not to affect the skin too deeply, or they may be followed by erysipelas, ulceration, and indelible cicatrices.

" Aqueous vapour *en douche* is useful, after bleeding, to facilitate the absorption of the tubercles in this disease. Directed for twelve or fifteen minutes on the face, it produces a fluxility, rendering the skin softer and smoother to the touch. The action of vapour-baths is more general and less useful.

" The resolution of the tubercles may be attempted by repeated unctions

---

* Paré (Amb.) *de la Goutte Rose*, lib. xxvi. chap. xlv.
† Rose-water, Oj. Eau de Cologne ℥j. Corros. Sublim. gr. viij. for lotion.

latter, instead of finding its way quickly to the surface, accumulates and disorders the substance of the cutis to a great extent. Its course and extent are marked during the more active state of the inflammation by a florid-looking and very irregularly formed, rather prominent, tubercle, exceedingly tender to the touch, and after a time becoming in some one or two points soft : still the superincumbent skin is not ruptured, and it eventually assumes a dark blue colour. In this state it sometimes comes under the notice of the surgeon, when the existence of matter being ascertained, a lancet is thrust into it, and the contents discharged, leaving for some time an unhealthy livid edge to the orifice, which slowly heals, and, in most instances, leaves a mark of some duration on the part. If neglected beyond a certain extent, however, and this course is not taken, the matter is either absorbed in the most sluggish and protracted manner, or a small portion remains to find its way through the skin, leaving a minute sore, which soon dries up; a blue discolouration of the spot sometimes remaining for months.

I believe this state of the affection, as represented by Dr. Bateman in his plate of acne indurata, to be in all cases the result of established disorder of the stomach and bowels, or of scrofulous diathesis. Where the latter did not manifestly exist, the former seemed to have been very frequently aggravated by the greatest inattention either to the quantity or quality of the aliment received.

From the more extended character of the inflammation, doubts may at first sight be reasonably entertained as to the analogy of this species with the first mentioned ; in other words, as to its origin in the follicles of the skin. The identity of these affections are, however, sufficiently manifested where a minute examination is instituted in any case of extensive mischief ; for the incipient inflammation of single follicles is in such cases almost constantly found intermixed with the larger tubercles formed of inflamed clusters.

---

with the ointments of the proto-chloruret of ammonia, or proto-sulphate of mercury.

" Ambrose Paré and Darwin advise that obstinate cuperosa should be combated by the application of blisters over the face, but caution should be taken in using these means.

" To obtain a permanent cure, it is always necessary to continue the treatment for some time after the disappearance of cuperosa ; cold sulphureous applications, *en douche* and *en arrosoir*, are very efficacious in restoring the skin to its natural state.

" When cuperosa was looked upon as a depurative disease, the use of purgatives and vegetable juices, such as cochlearia, beccabunga, sage, &c. was recommended. It is now generally thought, at least, useless to load the digestive organs with these remedies. If it is complicated with gastro-enteritis, or chronic hepatitis, their action will certainly aggravate these internal inflammations, which ought to be combated by means adapted to their seat and nature.

" Ambrose Paré gives a case of obstinate cuperosa, treated successfully by the application of a blister to the face. The treatment employed by this celebrated surgeon has been too much extolled, and we possess other agents less painful, less dangerous, and equally efficacious."—*Dickinson's Translation of Rayer.*

No other eruption, moreover, ever assumes in the remotest manner an approximation in appearance to this affection, and is at the same time confined, as this is, to parts where the follicles are more particularly distributed.

As in the former cases, the first appearance, different stages of the progress, and final termination of the disease, are often seen coexistent on the face, neck, and breast of the same individual, and generally also a considerable number of enlarged and distended, though uninflamed, follicles of the acne punctata.

In the common treatment of this affection, it is lamentable to witness the complete deficiency of attention to common surgical principles ; and though it is noticed by different authors, that local treatment forms a great portion of what is necessary to be attended to, no notice is taken of any particular application beyond those of a generally stimulating character.

It is clear, from the most superficial observation, that a disease, consisting in its origin of active inflammation, and in its progress of those variations only which particular constitutions or states of system originate previous to final suppuration, can rarely, if ever, require local stimulants ; and in this particular affection, I have no doubt that such applications, before the matter has been discharged from the tubercle, are highly improper.

If any vestige of active inflammation remain, it should be soothed, and suppuration promoted by poultices and fomentations ; and if any tubercle should be found assuming the blue colour without signs of matter coming forward, it should be freely punctured.   As in the case to which I formerly alluded of simple healthy inflammation of the follicle, where the orifice had been obliterated by the inflammation, and no external appearance but that of a minute tubercle existed, suppuration will, in almost all cases, be found effected, and the cure will follow infinitely more rapidly than where stimulants only are relied on.

The contents of each tubercle being evacuated, and the temporary additional irritation which the puncture excites having subsided, the use of stimulants is plainly indicated.   The absorbent vessels are capable of producing a very considerable effect on the thickened and discoloured cutis of the part in a very short period, even unexcited ; but a very manifest advantage, notwithstanding, usually follows the employment of spirituous lotions.   The oxym. hydr. dissolved in proof spirit, in the proportion of five grains to eight ounces, with which the spots are lightly sponged, seems to be entitled to a preference over other formulæ as lotions ; but if made of greater strength, it frequently irritates the skin.   Were it not for the unpleasantness of the application, the mercurial ointment would answer the purpose by far better than any thing else. · In severe cases it may be safely depended on for doing all which any application is capable of.

With regard to the constitutional treatment necessary, in all such cases as have come under my notice, the tonic plan seemed to be

called for, and was followed, in conjunction with the local management detailed, by pretty uniformly successful results.   The history
of the patient usually disclosed a disposition to voracity, with little
anxiety at any time by what species of food gratification was to be
obtained.   A furred, yellow tongue, offensive breath, frequent heat
and feverishness, with considerable languor at intervals, and irregularity of bowels, seemed to have been the consequences, which,
however, with the exception of some debility, were speedily removed by a short alterative course of medicine, and a strict attention
to diet.   A plethoric state of system, with a generally healthy
condition, is, I believe, incompatible with this form of disease
altogether.

Sometimes the extent of the mischiefs of follicular obstruction
and inflammation is confined to the tip of the nose, producing a considerable enlargement of these parts ; and, perhaps, of all deviations
from health to which the human frame is liable, this is one which
obtains the least commiseration.   The general impression is, that it
is the offspring of what is commonly called good living, and in the
only case which I have met with, where hard drinking did not form
a part of the habits of the patient, an extraordinary appetite, not of
the most delicate kind, was well known to exist.

It would not appear to deserve a separate consideration from the
preceding, but from the circumstance of its limited extent and very
constant dependence on the habits alluded to.   From the latter fact,
it has been frequently considered an indication of diseased liver ;
affections of this organ, however, of a tuberculated character, the
known usual consequence of dram-drinking, are by far more frequently unaccompanied with the slightest disposition to this deformity; while persons in the possession of what they consider unimpaired
health, enjoying better appetites for food, and exhibiting fewer
indications of general disorder than falls to the lot of those who
labour under hepatic derangement, generally become the subjects
of its most aggravated forms.   It is a state of the disease, moreover,
rarely indeed seen in that class of society where indulgence in the
use of spirituous liquors to the most deleterious extent is quite
common, to the exclusion of necessary portions of good animal
food.

The chronic inflammatory redness and enlargement of the nose,
which constitutes the state of acne rosacea, the affection under consideration, is always the result of repeated attacks, or of long continuation of the obstructed state of the follicles, the direct exciting
cause being an overloaded state of the stomach, and frequent excessive and general excitement of the digestive organs.   There seems to
be a satisfactory explanation of the disease being first found to attack
the nose, in the fact, that the follicles are here very thickly distributed, while it is more exposed to the chilling effects of cold, and
consequently more frequently to the effects of reaction ; constant
checks being thereby afforded to the healthy progress of suppuration,
as well as means of increasing and extending the inflammation.

From the operation of these causes, collections of matter, equal in extent to the space occupied by three or four follicles, are frequently concealed under a smooth red tubercle for weeks; its existence under such tubercle not being suspected. As the disease advances, and the exciting causes arising out of the habits of the subject are still kept up, others are formed in the same way, which, now and then, as in the common indurated state, proceed to suppuration, leaving half the dimensions of the tubercle existing, which, perhaps, never is entirely removed by the absorbents.

From the protracted duration of the complaint, and the succession of the phenomena detailed, in every part of the tip and alæ nasi, the irregular tuberculated nose is at length fully formed, and every thing like follicular organization is completely destroyed. The usual secretions of the part are thus prevented, and, consequently, the turgescence is increased, and confirmed chronic inflammation takes place, which nothing but a plan of living diametrically opposed to that previously followed, affords any hope of removing.

A minute examination of the spot, when the disease first makes its appearance, will usually enable us to detect a state of parts where the puncture of a lancet is called for ; minute collections of matter may be in this way discharged from the centre of apparently incipient tubercles, and the thickening of the skin will often in this state rapidly disappear, if purgatives to a small extent and attention to diet are had recourse to at the same time.

Emollient applications may be employed with advantage for some time after ; and when the florid redness begins to be followed by marks of diminished energy in the vessels of the parts, the stimulant plan may be adopted with final success. Sometimes the alæ nasi, when in the above state of indurated enlargement, become suddenly covered by a vesico-pustular eruption, assuming a different appearance from the original disorder. This eruption does not last beyond a few days, and appears generally as the result of some temporary additional cause of excitement.

In my original essay submitted to the attention of the College, I alluded to the practice of a surgeon in the habit of advertising his success in affections of the kind under consideration, and I stated my belief that the practice in question consisted of the liberal use of friction by means of soft brushes to the part, assisted by soap and warm water. It is certain that this treatment is capable of producing a great diminution of the tumefaction and redness, and of bringing about a more healthy state of parts. In some cases in which I have been led to prescribe it, the cutis has become thinner, and enabled me to detect considerable collections of matter, which have been speedily discharged, and followed by importantly beneficial alterations in the appearance and size of the part. It seems that a more healthy stimulus is thus given to the vessels of the spot, and suppuration effectually promoted by the friction, while the attendant irritation is subdued by the use of warm water as a fomentation ;

the adjoining follicles being, as before stated, kept from accumulation and obstruction.   Having their secretion thus going on freely and copiously, they still further assist in relieving the turgescence of the part.

The most obvious symptom of disordered state of the stomach prevailing in this affection is acidity, and a great opinion has accordingly been entertained of the use of alkaline medicines ; but it must be sufficiently clear, that unless the patient can be prevailed on to correct the habits on which such acidity or other disorder depends, little ultimate benefit can be obtained.

It would be improper to pass over this part of my subject without noticing the occasional excessive dilatation and growth of obstructed follicles, where inflammatory action and suppuration do not take place.   The formation of atheromatous tumours, first, I believe, noticed by Sir A. Cooper, as dependent on follicular obstruction, is unquestionably the result of the absence of excitement in the follicle when first distention is effected.   If inflammatory action does not happen to occur at this period, an encysted tumour of small extent is formed and continues to grow, the follicle appearing, after distention to a certain extent, to lose its susceptibility.   I have extirpated a considerable number of these tumours, and have invariably found their contents to be precisely the same as the sebaceous matter of the follicles ; while the cyst, as it has been termed, which contains it, exhibits no greater difference from the delicate membrane lining the parietes of the follicle in a healthy state, than is common with other membranes lining cavities under circumstances of diseased enlargement of such cavities.

CHAPTER II.

On Scrofulous Inflammation and Ulceration of the Follicles.

The skin, particularly in those parts which are the seats of anything like glandular organization, exhibits very frequently all the distinguishing phenomena of scrofulous disease.   The cutaneous follicles of the face, chest, &c. are affected by it, and its rise, progress, and decay, are of corresponding tediousness with the disease as affecting other organs.   I have described the origin, progress, and termination of other forms of acne, but of the scrofulous form it is difficult to speak of its ending, for it very commonly continues unyielding to any medical means or measures with which I am acquainted for years, and yet ultimately disappears spontaneously. As far as I am aware of, no author has drawn a proper line of distinction between this and ordinary forms of follicular inflammation. Perhaps the peculiar features of scrofulous inflammation have not been suspected of fixing themselves in this structure, and hence writers on lupus or noli me tangere very frequently speak of the connexion of that disease with scrofula. In short, as far as opinions of men can go towards the proof of a problem, that destructive and terrific disease is always either scrofulous or malignant, or, in other words, intimately connected with cancerous ulceration and cancerous constitutional diathesis. I believe that mere scrofulous inflammation of the follicular apparatus, which I am about to describe, seldom runs into positive ulceration, and yet it is productive of waste of the skin and subjacent structure, and of irregular contractions of the diseased cutis, and often of great deformity. When ulceration does, however, take place, it is of a most intractable character.

There is scarcely any disease of the skin which may not become influenced by the scrofulous diathesis, so far as to deviate from its usual course. Impetigo is far more difficultly cured under such circumstances; and lepra, the next most common of diseases of the skin in England, becomes very often totally unmanageable.

Let me bring the reader's attention to the opinions of one of the most distinguished of English surgeons on the subject of scrofula, so far as regards the cutaneous structure. " The shortest, the most concentrated idea I can give of scrofula," says Sir A. Cooper, " is, that it is congenital or original debility, and this state of the body is marked by peculiar characters both in external formation and internal structure. The external characteristic of a scrofulous constitution is the state of the skin, which is peculiarly thin and delicate in its structure. It is usually of a light colour, but this is far from being uniformly the case, for it is sometimes dark ; but in either case, if it be gently pinched up, it will be found extremely thin when compared with that of a strong healthy child ; and, as this state of delicate fibre of the skin denotes a similar internal structure, it becomes an easy criterion of the general conformation of the body.

" This thinness and delicacy of the integuments are the reasons that the cheek often exhibits a fixed and florid colour, which is considered as a great beauty by the passing observer, but is regarded as a sign of weakness by the intelligent mind ; it arises from the blood of the arteries being seen through their delicately constructed coats and their thin cutaneous covering.

" From the same flimsy delicacy of the skin the veins are seen permeating the cellular tissue ; and the darkness under the eye, which is so common an attendant on this kind of conformation under slight indisposition, arises from congestion in the veins and difficulty in the free return of blood. From weakly vascular action also springs the thickness of the lip, as the blood is retained in this very vascular structure.

" Flaxen or delicately silken hair often attends this state of the skin, and in those whose hair is red, there is a strong natural propensity to scrofulous complaints.

" Black hair and a dark skin are, therefore, generally signs of a healthy formation, but if the skin be thin it is not a guarantee from a scrofulous disposition. The thinness and delicacy of the skin exist in each of its constituent parts : the cuticle, from a blast of cold air, chaps and desquamates, the sun's heat parches and cracks it."

Cutaneous inflammation or ulceration of any kind, affecting a skin corresponding with the above description, would, *à priori*, be expected to assume a character of obstinacy, the powers of restoration being so small ; and we find that, whether ulceration take place or not, the constitutional character of the disease is still preserved. It is seldom, indeed, that we find it unaccompanied by scrofulous disease of other structures.

I think I am justified in saying, scrofulous disease of the follicles occurs in two forms. In the first, an instance of which has been lately under my care, of which a drawing has been made by Mr. Stewart, the disease commences on the chin or forehead, involving the follicles and skin at first no more than to the extent of a few lines in circumference, one or more of these patches of inflammation are surrounded by minute pimples, which, on examination, are clearly constituted by inflamed follicles. In this case, there had been, for many years antecedent to the appearance of the eruption, extensive disease of the cervical glands, the sores of these had lately healed, and as the discharge from them diminished, the eruption spread till it occupied the whole space between the chin, and cheek, and nasal bones. A uniform thickened red condition of the integuments of the lips, cheeks, and alæ nasi existed. The surface was of a deep red, and irregularly tuberculated. True scrofulous ophthalmia had also existed, and been cured, a short time previously. This was the case of a pauper, but nevertheless, all the advantages which improved diet and appropriate medical treatment could command were put in requisition for a long time without the slightest benefit.

The second form is far more insidious, and approaches usually in the guise of the merest trifle of a pimple on the cheek. A slight degree of tingling or itching accompanies it ; if it is scratched it bleeds a little, and a bit of court-plaster is often put on it from day to day with the expectation of its healing without trouble. A minute gummy exudation, forming a pellicle of scab, is seen. This drops, or is picked off in a day or two, and is immediately succeeded by another of increased dimensions, a little blood oozes from the surface, and when the latter is cleared with a little violence it is found of a dark venous hue, and in only the smallest conceivable degree imbedded beneath the surface of the surrounding parts. This condition of things may go on for weeks, and yet the patient, from the little trouble it gives, does not obtain professional advice. The sore arrives at the dimension of a split pea, but even then it not unfrequently happens, that the surgeon considers it, as the patient has done, a trifle. There is little or no redness or apparent inflammation in the adjacent skin, nor is the latter in any degree thickened ; and at this period the worst which can be said of it is, that it is a languid-looking sore which will not heal.   Various local applications are tried with the hope of converting it into a healthy sore, they all fail : then, as a last resource, caustics are applied, a slough is produced, this separates, and still the same glassy, smooth, inert surface presents itself ; no attempt at healthy granulation is discovered, but, on the contrary, if the surface be examined with the microscope, the minute artery and vein may be seen wandering over the pale surface as tranquil as if no breach had been made, and no attempts at repairs were called for.   The "delicately-constructed coats of these vessels," to use the language of the distinguished surgeon above quoted, "are barely discernible, lying, as they do, almost on the surface of the glassy semitransparent sore."

The sore gradually, but generally very slowly, extends itself, it spreads in every direction, and as it spreads it deepens.   It involves the subcutaneous cellular tissue, but the process by which it extends itself cannot be called or considered an irritable one.   The surrounding skin, pale as marble, exhibits, on the edge of the sore, a regularity equal almost to that of an incision ; and if the sore be formed on the cheek, it preserves pretty nearly a circular form till it approaches the alæ nasi orbit or lip, in either of which cases, without ceasing to spread and destroy, it assumes the form which the motion and figure of the parts give.   Thus, in the case of a lady from Birmingham, who consulted me some time since, it involved the upper lip, and gradually wasted the cellular structure in the form of a fissure.   In another it spread, having commenced in the form I have described on the cheek, to the alæ nasi of that side, destroying the integuments and cartilage of the nose, taking the course of a central line drawn from the middle of the lip through the centre of the septum nasi along the dorsum of the nose, to within a quarter of an inch of the edge of the lower eyelid.   It continued its course from this point in a curved line, corresponding to the edge of the orbitar process of the superior maxillary bone to the outer angle of the eye ; and, before death took place, half the upper lip, half the nose, and half the integuments of the cheek, had been destroyed by the ravages of the disease.   While life remained it continued to spread in all directions but on the upper margin of the sore, and there, having approached the fold of the eyelid, when it was apprehended that this and the tarsus would give way and expose the eye-ball it stopped, and transfused its devastating powers to the lip and angle of the mouth.

This case was considered, by some of the many surgeons who were consulted, as a malignant cancerous sore, and the more confidently so, as there existed a tumour of an irregular form and great dimensions, in the epigastric region.   It did not, however, at any period of its course exhibit fungous, irregular granulations, or sloughs ; there were no tubercles on the surrounding skin, no enlargement of the cervical or other glands.   The surrounding skin pale as marble, flabby, and free from heat and irritation, exhibited no redness, while, to use a phrase of a late teacher of surgical science, the absorbents were nibbling it away by inches. On the other hand, there were present all those external characteristics denoting the scrofulous diathesis, as noticed by Sir A. Cooper. A peculiarly thin and delicate skin, a complexion, the fairest of the fair—a florid cheek—flaxen and delicately silken hair, to which may be added a blue eye, thin and delicate nails, and a light and delicate form. This lady too had always moved in the very highest classes of society, and sensible and prudent in her habits through life, had borne and brought up a large family, with perhaps less than the common lot of suffering and ill health falling to persons so situated. These facts submitted to the reader's consideration will enable him to judge of the nature of this case. I think it cannot be considered to have been a cancerous or malignant sore, for the reasons I have alleged ; but as no post-mortem examination was allowed, the nature of the abdominal tumour must remain undiscovered. I consider it to have been merely a scrofulous disease of the pancreas.

With reference to the treatment of either of these forms of scro-fula originating in the cutaneous follicles, I am hardly able to make a suggestion beyond those which have been before offered by other authors for a long time. These all relate to the best means of invigorating the constitution, and without attention to this, local measures are of little or no avail. The first, or, as I may desig-nate it, the chronic diffused scrofulous inflammation of the follicles is seldom benefited, either by local or constitutional measures. If leeches are applied, the redness and inflammation does not yield, and yet the wound made by the leech heals, leaving, after a few days, no vestige of its existence. It is more than probable, that if a leech were applied in the other case an almost incurable sore would be formed.

Neither of these forms are destitute of the power of "tantalizing" both the patient and surgeon, even from their very birth ; the first, where no ulceration or broken skin takes place, now and then be-comes pale and tranquil, and promises to disappear ; the other ex-hibits from time to time on different parts of its pale and glassy surface, little pellicles of new skin, which for two or three days spread and increase in size in every direction, promising the forma-tion of a bit of really sound skin ; when, on looking anxiously on the morning of the fourth you see the new skin perforated in the form of numerous little pin-holes—the absorbents have been nib-bling it away ! In short, a favourable termination or cure of either of these two forms is hardly to be expected. In the first, the con-stitutional vice stands in the way, although there appears on the part, evinced by redness, heat, &c., sufficient powers of repair. In the second, it is almost always evident on minute examination of the vessels on the surface of the sore, that their coats are so ex-tremely delicate as to be unable to bear the impetus of any thing beyond a very weak circulation, a circulation totally inadequate to the demand where repair is called for.

The foregoing descriptions are intended to illustrate scrofulous disease of the cutis, and do not apply to scrofulous inflammation and suppuration of the cellular membrane beneath it ; yet sores are formed now and then by the breach of the cutis, and the evacuation of matter taking place, the opening instead of healing becomes larger and larger, and the sore finally assumes a character nearly as intractable and obstinate as that last described.

CHAPTER III.

SYCOSIS.

Dartre Pustuleuse Mentagre, &c.

This disease is divided into two species by Bateman, according to the situation in which it appears. Its intimate analogy to the different forms of acne has been alluded to by that author, but the particular points of resemblance or dissimilitude have been passed over without notice. It is classed with acne in the order tuberculæ, with as much reason as that affection itself; but a correct pathological view of its character will lead to the opinion, that it really consists of a number of minute abscesses ; and has, therefore, little claim to the station allotted to it.

What has been termed sycosis is nothing more or less than acne or follicular obstruction and its consequences, occurring on parts covered by hair ; and though the necessity of distinction between it and the same state of parts not so covered be admitted, this circumstance cannot justify a subdivision like that adopted, of S. menti, S. capilitii, &c.

Like many of the formidable diseases described by Alibert, it is seldom seen in the better classes of society in a severe degree, though extremely common on a more limited scale ; and where cases occur answering to the descriptions of this author and Dr. Bateman, but little inquiry or observation will be necessary to show its identity with acne.

It "consists of an eruption of inflamed, but not very hard tubercles, occurring on the bearded portion of the face, and on the scalp in adults, and usually clustering together in irregular patches." On the chin "the tubercles arise first on the under lip, or on the prominent part of the chin, in an irregularly circular cluster ; but this is speedily followed by other clusters, and by *distinct tubercles,* which appear in succession along *the lower part of the cheeks up to the ears,* and under the jaw, *towards the neck, as far as the beard grows.* The tubercles are red and smooth, and of a conoidal form, and nearly equal to a pea in magnitude. Many of them continue in this condition three or four weeks, and even longer, having attained their full size in seven or eight days. But others suppurate very slowly, and partially discharging a small quantity of thick matter, by which the hairs of the beard are matted together, so that shaving becomes impracticable, from the tender and irregular surface of the skin. This condition of the face, rendered rugged by tubercles from both ears round to the point of the chin, together with partial ulceration and scabbing, and the matting together of the unshaven beard, occasions a considerable degree of deformity, and it is accompanied also with a troublesome itching."

The above description is that of the S. menti of Bateman—when it affects the scalp, " it is seated chiefly about the margin in the occiput round the forehead, temples, &c. and near the external ear, which is also liable to be included in the eruption. The tubercles rise in clusters, which affect the circular form ; they are softer and more acuminated than those on the chin, and they all pass into suppuration in the course of eight or ten days, becoming confluent, and producing an elevated unequal ulcerated surface, which often appears granulated, so as to afford some resemblance to the internal pulp of a fig."

The disease is shown in Alibert's plate as dartre pustuleuse, or Herpes pustulosus mentagre ; but it is quite clear that this author, like Bateman, was unaware of the real character of the disease, since there does not appear any allusion to the follicular apparatus, the derangement of which forms its most important feature. The descriptions of neither of these authors apply to the majority of cases of the affection, but rather belong to those where neglect of cleanliness has contributed much to aggravate and increase the irritation with which it commences. Few young men between the ages of twenty and thirty or thirty-five escape occasional attacks of the disorder on their beards ; and it almost always occurs simultaneously with more or less of irritation on the skin of the forehead. The resemblance to the pulp of a fig, whence it appears to have derived its name, only obtains in the worst and most neglected cases in the lower classes of society.

When affections of the cutis, no matter of what kind, occur on parts which are covered by hair, such affections almost invariably assume a more obstinate and formidable character ; indeed, the peculiarities of most of the common diseases of the scalp depend on the local irritation of the hair.

The part occupied by the beard is generally pretty well supplied with the sebaceous follicles, and these are of course equally liable to disorder with those in other parts of the skin from constitutional causes. When, however, any accidental circumstance brings on inflammation and disorder in them, the peculiarity of their situation abounds with impediments to its termination in the most desirable manner. The mouth of the inflamed follicle and the adjacent cutis is penetrated by hair, and the violence inflicted by frequent shaving, makes every individual hair a powerful mean of adding to the mischief. If a disposition in the inflammation of the follicle to subside may exist, the influence of the operation of shaving, for instance, is sufficiently great to prevent it, and to hurry on the inflammatory action to the suppurative process. If many tubercles are formed, it is capable of increasing their size, and extending the inflammation to the adjoining follicles. I believe these effects of the operation of shaving are manifested in every case of the disease, and it will be easily understood why it should be so, when the facts that the inflamed part is penetrated by a substance like the hair, and that the latter is violently stretched upon and through it by the use of the razor, are borne in mind.

From the local influence of the hair on the inflamed spot where it penetrates the cutis alluded to, which will be more fully elucidated in the following chapter on Porrigo, a secretion of pus is formed around many of them, the hair being situated in its centre, and these pustules intermixed with the tubercles formed by the inflamed follicles, some of which are also showing matter on their apices, make up the external characters of the disease. The same phenomena are also observed when it occurs on the scalp ; the extraction of single hairs here and there forming orifices by which the contents of a pustule are readily discharged. As in common acne, if a lancet be thrust into the tubercles of sycosis, (which is advisable where one or more hairs are not distinctly seen growing in it), matter will certainly be found, and a rapid subsidence of the inflammation follow : where hairs can be pulled out in the way mentioned, however, the evacuation of the matter is accomplished by gently squeezing the tubercle, without having recourse to the lancet.

From what has been said, it will appear that the existence of hair on the part, and its consequences in aggravating the inflammation, form the only difference between sycosis and acne ; and where the hair can be extracted without pain, this step should not be neglected in commencing the treatment of the former. After this, emollients and applications tending to promote suppuration and allay irritation are the best, no more good being to be expected from sedative washes or stimulants than in the treatment of the subjects of the preceding chapter.

The constitutional part of the treatment of sycosis has been already detailed ; the same general remarks on this point applying equally to it and to acne. The attention to the part should be constant and unremitting, and should consist of warm fomentations frequently repeated during the day, with poultices if they can be conveniently applied at night. Every little tubercle should be punctured at its first commencement, and every hair extracted from the part which may be got out without much pain. This practice, strictly followed up, is capable of removing the most protracted and troublesome cases without the use of any internal medicines beyond alterative aperients,* and any dependence on internal remedies, unassisted by it, will inevitably lead to disappointment.

* The internal remedies recommended by Willan, Bateman, and Alibert, consist of general alteratives and tonics.

It is very desirable, even in the most trifling cases, to find a comfortable and cleanly substitute for the razor, and, generally speaking, the razor-scissors, as they are called by the cutlers who manufacture them, will answer all the purposes of that instrument equally well, a little more time and care only being necessary than in shaving. They may be obtained at all the cutlers or surgical instrument makers in London.

Mentagra, the common designation of this disease used in the French school, has been by a translator of Rayer, before named, coupled with impetigo as its synonyme without the authority of his author, who does not appear to have thought of such alliance. I can confidently submit the translation of Rayer,* who owns his obliga-

* "Mentagra is a cutaneous inflammation characterized by the successive eruption of a number of small acuminated pustules, similar to those of cuperosa on the chin, submaxillary regions, and lateral parts of the face.

"The development of mentagra is usually preceded by a feeling of tension and heat on different points of the chin. The pustules, announced by slight smarting, are observed, at first, under the form of small red points, which gradually grow more prominent. Towards the second or third day of their formation, their summits become white, and enlarge ; but it is seldom they exceed a millet-seed in size. From the fifth to the seventh day, each pustule spontaneously breaks ; its walls shrivel ; but there is a slight exudation which produces a thin, slightly adherent crust. This unites at its circumference with the epidermic scales detached from the inflamed skin around the pustules. The inflammation does not extend beyond the reticular body, and cicatrices are never formed. In young persons the pustules are usually numerous and close together, sometimes the whole surface of the chin and lateral parts of the face are covered with them. The inflammatory circle which surrounds them is lost, they unite in groups, and the attendant inflammation being more acute, the painful tension of the parts they occupy is more marked, their progress more rapid, and the consecutive crusts more adherent. If several pustules unite and become confounded in their development, the inflammation may at once penetrate the whole dermis, gain the subcutaneous cellular tissue, and produce phlegmon. The skin is sometimes so deeply affected and tumefied as to assume the appearance of humid vegetations. *The bulbs of the hair of the beard frequently participate in the disease, and a more or less considerable portion of the chin is deprived of hair. Its destruction is temporary ; new hairs, at first light and feeble, reappear, and soon regain the colour and size of those which have fallen.*†

"In most cases, mentagra, like cuperosa, is composed of several eruptions, succeeding one another at greater or less intervals. When pustules are developed several times on the same place, the inflammation penetrates the dermis and subcutaneous cellular tissue, producing indurations, which are not long before they assume the form of tubercles. These are seen particularly in persons of delicate constitution, in whom pustulous inflammation is never followed by complete resolution. When the eruptions are numerous, intense, and close together, the tubercles multiply and extend over the whole chin. New pustules form on the tubercles, or in the interstices between them, and thus obscure the primary character of the disease. It is now that the confused mixture of tubercles, crusts, pustules, and scales, gives to mentagra its disgusting appearance. Arrived at this stage, it is always a serious disease, and difficult of cure.

"Mentagra is not contagious ; it more particularly attacks men, young and adult, of a sanguine or bilious temperament, and who have a great deal of beard. It is, above all, developed in those who are exposed to high temperatures ; as cooks,

† The author quoted observes, in the recent edition of his work, that almost all the pustules are *traversed by a hair.* I had stated and explained this fact long before.

tions to Biett, and contrasts it with the few lines I have myself penned on the subject as the result of my own practical observation. It is easy to say a *great deal more* on this as well as most others of

founders, sugar-bakers, &c. Excesses of diet, and the abuse of spirituous drinks, and spiced meats; want of cleanliness, some irritating applications, the use of a dirty or blunt razor, &c. seem to favour the development of this disease. It is rarely seen in women, and is complicated with cuperosa.

" It is important that mentagra should be distinguished from other inflammations which may appear on the chin, particularly from ecthyma, impetigo *figurata*, from pustulous or tuberculous syphiloid disease, and from furuncle. The pustules of ecthyma are larger and more inflamed than those of mentagra. The crusts are of greater extent, thicker, and more adherent. The psydraceous pustules of impetigo *figurata* are not acuminated like those of mentagra; they differ also by their more prompt or acute development, and by their being disposed in groups. In mentagra they are more frequently isolated and distinct; in impetigo *fig.*, they are grouped and more numerous. In the latter disease, they break about the third or fourth day, and the sero-sanguinolent humour which escapes is quickly transformed into extended yellow crusts, which increase in thickness in a few days. The pustules of mentagra do not break till about the fifth or seventh day, and the crusts are thin, slight, and insulated. These distinctive symptoms, however, are more obscure when the eruption is very considerable, its march more acute, and the pustules close and confluent. Syphilitic pustules are seldom manifested on the lower part of the face only; they are always seen on the alæ of the nose, on the forehead, and commissures of the lips. Those of mentagra, on the contrary, confined to the chin, and usually to the lower part of it, are acuminated and insulated on vivid red bases, showing more acute inflammation. Syphilitic pustules are flatter, and have tawny, copper coloured, and almost furfuraceous bases; they are not preceded by itching, nor by the painful tension which announces mentagra. Arrived at the tuberculous stage, mentagra may be easily confounded with syphilitic tubercles. Those of mentagra, however, are conoid, and their bases appear to penetrate the whole substance of the dermis, and to extend as far as the subcutaneous cellular tissue. Syphilitic tubercles are rounder at their summit, shining, and seem to push up the superficial layers of the dermis. The pustulous and tuberculous state of the skin in syphilis, is also more general, and is mostly accompanied by chronic inflammations of the throat and conjunctiva; and is almost always preceded by obstinate noctural cephalalgia, presenting a group of symptoms altogether different from those of mentagra. In furuncle, the inflammation spreads to the cellular tissue, and there is a core discharged by an aperture which always produces a cicatrix; in mentagra, the inflammation first attacks the skin, the pustules yield only a small quantity of pus, never a core, and by an aperture only investing the epidermis, and which is soon effaced.

" The prognosis of mentagra presents such difficulties, that it is frequently impossible for the most experienced practitioner to judge the termination of this disease. At times, when the decrease in number of the pustules, and of the violaceous tint of the inflamed skin, seems to announce an approaching cure, new eruptions, of greater or less extent, are observed to supervene without any known cause. At other times, when fear is entertained that a considerable eruption, which extends over the whole chin, may endure for many years, it is found to yield readily to antiphlogistic regimen and treatment. In general, the most obstinate cases are those in which the pustules and primary form of the disease continue in the chronic state. In this point of view, mentagra may be considered as one of the most intractable diseases of the skin.

" The first step on the development of mentagra, is to cut the beard quite close with a pair of curved scissors, the action of the razor always aggravating the inflammation. If the disease is recent, and appears in a healthy vigorous subject, and the pustules are so close as to cause a more acute inflammation, local bleed-

7*

our subjects, than I have said, but I leave to our continental brethren
that task.

M. R. has distinguished himself in no ordinary degree beyond
me in this respect ; and when, as in the case before us, he seriously
tells us that the disease is influenced prejudicially by excesses of
diet, abuse of spirituous drinks, the use of dirty and blunt razors ;
that when it is necessary to bleed, either locally or generally, the
strength of the patient should be considered ; that the disease does
not occur *on the chin of women, &c. !* we cannot help thinking he
tells his readers more than is requisite.

The fact is, that the simplest application of the first principles of
surgery is fully adequate to the removal of this disease ; nor does it
assume a character of obstinacy, except in scrofulous constitutions, if
the suggestions I have offered be duly attended to.   As in acne, the
practice of the application of stimulants is never necessary, and
generally exceedingly injurious.

ing must be effected and repeated, with the precaution of placing the leeches
beyond the limits of the eruption.   If mentagra reappears after remission, general
bloodletting must precede the renewed application of the leeches.   The employment of bloodletting, local and general, must depend, however, on the strength
of the patient, on the intensity of the inflammation, and on the extent and frequency of the pustulous eruptions.   Baths, topical emollients, mucilaginous and
acidulated drinks, seconded by spare diet, are particularly indicated.   When the
digestive organs present no sign of irritation, causing a slight revulsion towards
the gastro-intestinal mucous membrane by small doses of calomel, the sulphates
of soda, potash, or magnesia, may be advantageous.

" In mentagra of long standing, and when tuberculous induration has occurred,
local bleeding may still be useful ; but it must not be carried to the same extent,
any only practised in strong healthy subjects.   In weak individuals, and those
of lax fibre, it is never beneficial.   When the tubercles are softened under the
continued employment of emollient applications, resolvent ointments may be
used, of which those of the proto-nitrate, deutoxyde, and protochloruret of
mercury, are the best.   Sulphureous and alkaline ointments may be used for
the same purpose.   Their use must be suspended, should new pustules arise
during it.   Aqueous vapour *en douche* is often successfully employed, to soften
tubercles and favour resolution.   The sulphureous waters of Barèges, Cauterets,
and Aix, may be applied in the same way.   Lastly, obstinate mentagra of long
standing is sometimes ameliorated by cauterisation, either with the concentrated
acids, or a solution of caustic potash, which changes the action of the diseased
parts.   Laxatives are often beneficial in the treatment of chronic mentagra, developed in young robust persons.   Men in the decline of life, and those of weak
constitution, are more advantageously treated by bitter and ferruginous preparations.   In several cases the muriate of gold rubbed on the tongue and gums, has
caused or accelerated the cure of obstinate mentagra !   Mercury is very efficacious, even in persons who have not been previously affected with syphilis ! !"

CHAPTER IV.

Of Lupus, Noli me tangere, Herpes exedens, Dartre rongeante,
&c. &c.

THE reasons for which I have assigned this position to the above disease will be found in the following pages :—

It has been described by different authors under designations varying ad infinitum. In England the name by which it is best known is that of Noli me tangere, which certainly conveys an idea of no pleasing character as to its nature. It is in truth a very horrible disease, very tedious and uncertain in its progress, and very little influenced by remedies of ordinary use.

The pathology of Lupus is involved in the greatest obscurity.*

* The brief history I give of it justifies my printing the following extract. The formidable character of the disease makes it essential that all which can be taught should be learned. It is but justice to M. Rayer to say that he has made the best use of the practice and experience of the French schools, and that all which is known among his countrymen seems to be included in the following :—

"Lupus is a chronic cutaneous inflammation which usually appears in the shape of external tubercles of different sizes, singly or in clusters, of a livid colour and indolent character, followed either by ichorous and phagedenic ulcers, which become covered with brownish and usually very adherent scabs,—lupus *exedens ;* or by extensive changes in the structure of the skin, but without preliminary or consecutive ulceration,—lupus *non exedens.* This disease may be confined to the face, and even to one of its parts, or may attack at once or in succession several regions of the body. The two varieties which have been indicated, are very distinct in their external appearance, and to a certain extent in their mode of treatment.

" Lupus *exedens* or *noli me tangere*, is commonly developed on the alæ or tip of the nose. It makes its appearance as a small external tubercle of a dusky-red colour, and hard consistency, whose progress is usually tardy. It sometimes commences as a chronic inflammation of the mucous membrane of the nasal fossæ, with redness and swelling of the nose in general ; a thin scab or crust then forms at the opening of the nostrils; this is removed, and a second and thicker one succeeds it,—an ulcer has in fact been formed, and soon extends to the alæ of the nose. Under other circumstances a livid or purple tint and some swelling of the end of the nose are the first symptoms of the disease observed. The redness increases, a superficial sore is formed, which becomes covered with a scab, and the ulcer extends in depth. It often happens too that one of the alæ of the nose swells, grows painful and of a purplish hue ; a slight ulcer then forms and becomes covered with a little scab ; this the patient commonly picks off, when it is replaced with a thicker one, under which the ulcerative process continues to go on, the scab being found to increase in thickness every time it is renewed ; the patient scarcely makes any complaint ; the skin, and occasionally the cartilage, are silently destroyed, and an ulcer of a bad character, from which a fetid sero-purulent discharge is poured out, is at length discovered, as if by accident, established under the scab.

"The nose is occasionally red on the superficies only, and this in a very equal and regular manner ; sometimes instead of being of its ordinary size and shape the nose becomes pointed, sharp, and tapering, the nostrils tending continually to close. The cartilage at the angle which unites its two lateral halves

In public hospitals in this country cases are not unfrequently noticed in its middle and advanced stages, but private practice among the better classes of society furnish comparatively few instances of it.

superiorly, seems then to project, and presents a red tint that is even perceptible through the soft parts.

" The ravages committed by this disease vary extremely ; almost the whole of the nose disappears in one instance, and the point only suffers a little in another, in which case it often looks as if a piece had been removed with a cutting in- strument. When such ulcers have been arrested and healed up, new tubercles occasionally form on or near the cicatrices, and the parts which had been spared originally may be entirely destroyed by a renewal of the ulcerative process ; even the whole nose and septum may vanish before its destructive influence. The ulceration may prove rapid or slow in its progress : sometimes after existing for several years, a small portion only of the nose is lost : in other instances, happily of rarer occurrence, the whole of the member is destroyed in from fifteen to twenty days (lupus *vorax*). Sometimes if the disease is interfered with, it seems to acquire new energy : the point of the nose assumes a livid red colour, which, though it may seem to disappear, returns within two or three days ; in- crustations, which are attended with acute pain, and grow very thick in the course of a few days, form in the interior of the nasal fossæ, whence a puriform fluid distils, and the point of the nose is then rapidly destroyed. The disease seems every now and then to be advancing towards recovery, when the part that was almost cicatrized, turns to a vivid red, is attacked anew with painful ulcera- tion, and is covered with a thick scab, under which the destructive inflammation makes rapid progress.

" In lupus *exedens* of the skin of the nose, the mucous membrane of the nasal fossæ is almost always affected with chronic inflammation. In some rare cases, altogether independent of a syphilitic cause, the septum is even destroyed before the outer surface of the nose is implicated. When the destruction commences in the skin again, it extends to the pituitary membrane, spreads along the nasal fossæ, and is even occasionally reflected over the mucous membrane of the arch of the palate to the gums, which are then deeply furrowed. Besides the des- truction, which is discovered when the scabs that have long concealed the mis- chief wrought by this disease are removed, it is very common to find the openings of the nostrils *contracted* in a greater or less degree by the thickening of the affected parts, or by the indurations consequent on the formation of cicatrices.

" The tubercles of lupus *exedens* are occasionally evolved near the *commissures of the lips*. Thick incrustations cover the ulcers, and thé patient cannot open his mouth without pain. After having destroyed a considerable extent of parts about the commissures, the ulcers often extend to the moveable substance of the lips, in which case, by the shrinking of the cicatrices when the disease gets well, the opening of the mouth is apt to be considerably diminished.

" The *lower eye-lid* is occasionally attacked with lupus *exedens*, and the ulcera- tion generally spreads to the skin of the cheek and to the conjunctiva palpebralis. The eye-ball, in this case imperfectly protected, inflames, the conjunctiva thickens, the cornea loses its transparency, and by and by becomes so dim that total blindness follow. If the eye-lids are not entirely destroyed, the sores in heal- ing cause their eversion ; the eyes then appear of twice their usual size, a cir- cumstance which, added to the vivid red of the conjunctiva, produces hideous deformity.

" It sometimes happens that one or more tubercles are evolved on the face, which, after continuing long stationary, increase suddenly in number as well as in size. The skin in the spaces between them swells and becomes œdematous in appearance, the tubercles then unite by their bases, and the whole mass falls into *a state of irregular ulceration* of bad character. The sore is covered with a blackish looking and very adherent scab, and spreads by degrees to the neigh- bouring parts. Lupus *exedens*, in this way, frequently attacks almost the whole of the face. When the ulcerative process stops, cicatrization takes place in the

Hence it seldom comes under the eye of the pathologist at its commencement, the unfortunate occupants of our charitable institutions usually suffering it to make considerable progress before they apply for relief.

form of irregular white bands, which stretch from the parts where the mischief began to those in the vicinity, and are very similar in their appearance to the cicatrices that result from extensive burns.

" Lupus *exedens* occurs with symptoms of still greater severity. Whilst it is advancing among parts still untouched, it returns and attacks the cicatrices, whether of old or recent date, which have resulted from its previous existence. These disappear very rapidly, and fresh tubercles, developed like a hard, rough, and swollen band around them and the ulcers that exist, become open before long, and add to the havoc going on. In the course of a few months the disease may thus ravage not only the greater part of the face but a large extent of the surface of the neck. The nose is frequently implicated in the destruction, and when the scabs are detached, the alæ or a portion of its end is found gone. When by means, properly directed, the progress of the disease is arrested, the skin, beset with numbers of small, red, and sallow tubercles, becomes covered with minute squamæ, and white and firmer cicatrices ere long succeed upon the ulcerated points. When the ravages of lupus have been thus extensive, the face is seamed with irregular cicatrices, often of a dull white, but occasionally of a rosy red colour, tense, shining, pretty thick in some places, but so thin in others that they appear transparent and on the point of giving way. This latter character they present especially on those parts that have been oftener than once attacked. These cicatrices seem frequently to hold by their extremities to two different tubercles, between which they stretch like lines of communication. They often appear covered on various points of their circumference with blackish incrustations, which, however, are commonly soon detached.

" Lupus *exedens* rarely attacks the integuments of the chest, or of the extremities; these regions are more frequently the seat of the lupus *non exedens serpiginosus.*

" Lastly, in one variety of lupus *exedens* the ulcers are beset with small, soft, red and spongy-looking and very prominent *tumours*, which occasion much deformity. This variety is one of the most formidable, but it occurs rarely.

" Lupus *exedens* often continues for years, committing the most frightful ravages, without the general health appearing to suffer in any degree. Yet, when not only the skin but the cartilage of the nose is rapidly destroyed, some patients show unequivocal symptoms of chronic inflammation of the lining membrane of the stomach, intestines, or bronchi, and several even sink under a species of slow fever accompanied with colliquative diarrhœa ; such a termination of lupus, however, so rarely happens, that when it does occur, it ought probably to be ascribed to an accidental complication, rather than to the influence of the disease itself.

" Lupus *non exedens.* This variety occasionally presents itself as a single tubercle, of a yellowish red colour, developed in the substance of the skin,— lupus *non exedens simplex.* I have observed a solitary tubercle, possessing these characters, continue on one of the cheeks in several children for a number of years, and leave, at a subsequent period, a small cicatrice in the spot it had occupied. The disease, however, more commonly begins in the face, as an irregular cluster of little tubercles of a dingy red colour, and a flattened or lenticular form, scarcely rising above the level of the skin. In some cases, they implicate a great portion of one or of both *cheeks,* of the forehead, and even of the *face in general.* These tubercles do not ulcerate on their summits ; and the sores that are occasionally encountered in the circumference of the clusters, are so rare that they must be held entirely accidental. The disease spreads by the formation of fresh tubercles, which spring up near those that already exist, and thus by degrees increase the area of the diseased surface—lupus *non exedens serpiginosus.*

Some authors have described it as in irritable unhealthy ulcer, situated about the tip or alæ nasi, gradually extending itself, and destroying first the cutis and then the cartilages of these parts,

The skin and subjacent cellular substance often become affected with an indolent infiltration, and the diseased surfaces look puffy and enlarged. The yellowish-red colour of the tubercles disappears under the pressure of the finger. The patient does not experience any pain; but touching the parts causes uneasiness, and the diseased surfaces frequently become sensible after violent exercise, and indulgence in spirituous liquors. The tubercles begin to shrink in the centre of the group; the skin there grows red, shining, slightly furfuraceous, and subsequently assumes the look of such a cicatrice as is formed after a superficial burn; it is, however, beset with points of a yellow coppery colour, due to the tubercles which, from shrinking themselves; or from the tumefaction of the neighbouring subjacent parts, are brought to the level of the skin. The clusters appear mingled with white points and bands, which are evidently cicatrices, owing to the disappearance of tubercles of another date, and which in vanishing have caused this singular alteration of the integuments. The tubercles of lupus *non excedens* are habitually affected with an epidermic desquamation, which is usually most remarkable around the circumference of the clusters where they are best defined.

"The features occasionally become very much enlarged in this disease; the cheeks, soft and flabby, or pasty, preserve to a certain degree the print of the finger, and look as if they were affected with the Arabian elephantiasis. The forehead and eyelids are puffed, and the eyes, covered with hypertrophied masses, seem sunk and hidden in the bottoms of the orbits; the lips, too, are considerably swollen, and often form two enormous, flabby masses, which expose the mucous membrane of their inner surface turned outwards; the ears, in fine, occasionally participate in this general tumefaction of the face. The tubercles of lupus *non excedens* are rarely affected with ulceration; such as does occur is accidental, very superficial, and covered with thin, laminated, and but slightly adherent incrustations. This disease continues for an indefinite period; whether left to itself, or modified by the action of various therapeutic agents, the affected parts never regain their natural state completely; the tumefaction of the skin and subcutaneous cellular tissue diminishes, the tubercles shrink and disappear, but the skin continues thin, shining, smooth to the touch, and seems to have lost something of its proper thickness.

"Lupus *non excedens* occasionally appears on the *extremities* in one or more clusters of small flattened lenticular tubercles, of a yellow-red tint, changing subsequently into patches of an irregular circular shape, the areas of which are red, furfuraceous, often traversed by prominent bands, whose salient edges are evidently tuberculated, and covered with firmer and thicker squamæ. When this eruption is left to itself, fresh tubercles appear successively in the circumference of the primary clusters, and encroach more and more upon the healthy integument; in this way I have seen the disease extend over the whole of a limb, the arm, for instance, from the shoulder to the wrist. The extremity thus affected, became much larger than the one on the opposite side, and even acquired the dimensions which the same part presents in Arabian elephantiasis. The motions of the elbow-joint were executed with pain and difficulty; the extent of the disease was sharply defined towards the shoulder superiorly, and near the wrist inferiorly, by a tuberculated line covered with squamæ. The skin of the arm and forearm, changed into a kind of indurated tissue of a paler colour than the healthy integument, was puckered or ridged with numbers of bands similar to those that follow burns, and sprinkled over with lenticular spots of a brown and dirty yellow, owing to tubercles that had shrunk or were buried in the substance of this tumid skin. The subcutaneous cellular membrane was infiltrated, and, in several places, pitted under pressure. At intervals, variously remote from each other, sometimes under the influence of the deuto-ioduret of mercury exhibited internally sometimes to all appearance spontaneously, the skin and cel-

ultimately spreading to an uncertain extent on the cheeks ; while others more properly begin their accounts by describing one or more tubercles of a red or livid dull brown colour, which remain

lular substance under it were attacked with a low kind of inflammation, attended with some swelling, heat and pain, but without any distinct external redness. A serous exudation then took place from a number of fine pores or openings which were visible on the surface of almost the whole of the tubercles, and into which the point of a large pin might have been insinuated. This *intervening* inflammation was always followed by a diminution in the size of the extremity, and the dispersion of a certain number of tubercles. I have, however, seen this variety of lupus *non exedens*, disappearing about the shoulder, at the same time that the disease was making rapid advances around the elbow and forearm.

" Lupus *non exedens* is occasionally evolved *below the ear*, and on the *nucha*, whence it extends in one instance towards the throat and shoulders, in another towards the occipital region, which it despoils completely of hair.

" Both of these varieties of *lupus* appear at first sight to be diseases entirely of a local nature. Those who are attacked with them are commonly in the enjoyment of pretty good health at the time ; females, indeed, occasionally complain of something like derangement in the periodical discharge, especially when the constitution appears to have a scrofulous taint, and when the disease is of some extent ; yet I have known several women who suffered from old and inveterate lupus in whom the catemania had always been regular and pretty copious.

" Of all *intercurrent* diseases erysipelas is the one which is most frequently observed along with lupus. The occurrence of this inflammation is occasionally to be accounted a lucky accident, especially in the lupus *non exedens serpiginosus*, inasmuch as under its influence a certain number of tubercles are always dispersed, and the whole disease may even be brought to a fortunate conclusion. But it also happens now and then that an attack of erysipelas, even in effecting a desirable modification in the state of the diseased integuments, may be attended with nervous symptoms of such severity as to make its occurrence subject of deep regret. As to the disease of the skin and other affections which existed previously to the development of lupus, the whole, with the exception of scrofula, perhaps, appear to be strangers to its cause.

" *Causes.* Lupus is happily a disease rather of rare occurrence. It is most generally developed between the sixteenth and twenty-fifth year, and seldom shows itself after the age of forty. Scrofulous children are of all individuals those who are most obnoxious to its attacks ; yet it undoubtedly occurs among the robust who have lived in the habitual enjoyment of excellent health. The disease is frequently uninfluenced by the arrival of puberty, and may recur in those who have suffered from it in their youth. It seems to be more common in the country than in towns ; and, perhaps, also to attack women more frequently than men. The poor inhabitants of Haute-Auvergne who live on acrid food, such as old cheese, tainted meats, &c., and house with their cattle, are often attacked with it. The disease is not contagious, and is seldom seen among the better classes of society. Blows, falls, &c., under the influence of which it has seemed occasionally to be developed, can only be regarded as determining and occasional causes of the disease.

" *Diagnosis.* Lupus is easily distinguished from rosacea, Greek elephantiasis, syphilitic affections, and the other forms of cutaneous disease that present tubercles, or ulcers covered with incrustations of varying thickness. Scrofula is peculiar to individuals of a strumous constitution ; the tubercles, sores, and affections of the bones and lymphatic glands that accompany it have peculiar characters. Farther, scrofulous ulcers extend by the detachment of their edges from the subjacent tissues, and the formation of sinuses, in consequence of the softening and suppuration of lymphatic glands, of caries of the bones, &c. ; whilst the ulcers of lupus are the effect of a process that consumes the skin and neighbouring parts from without inwards—from the surface towards the deeper structures. The red colour, the erythematous areola that surrounds the circumscribed

apparently unchanged for a considerable length of time before ulceration commences.

Some have correctly stated it to originate, when on the nose, in

indurations which succeed the pustules of rosacea, and these pustules themselves, which are generally met with in the neighbourhood of such indurations, are so many characters distinguishing this affection from the discoloured and indolent tubercles of lupus. In Greek elephantiasis the general tawny colour of the skin, the form and arrangement of the tubercles, which are nearly of the same tint, and present themselves as small knotty and unequal tumours, as well as the partial augmentations in size, succeeded by swellings that deform the face, are so many symptoms foreign to lupus *non exedens*. Farther, the tubercles of this variety of lupus are commonly arranged in circular groups, the limits of which are strongly marked and covered with squamæ, which is not the case with Greek elephantiasis. With the slightest attention it seems impossible to confound the circular clusters of lupus *non exedens* though covered with squamæ, with the patches of lepra, the areas of which never show any thing like the cicatrice of a burn, nor the lenticular stains, of a tawny yellow hue, produced by such tubercles as have shrunk, or such as are arising in their circumference. § 674. The ulcers of Greek elephantiasis are always more superficial than those of lupus *exedens*, and show no tendency like them to attack the neighbouring healthy parts. Lastly, the tubercles of Greek elephantiasis are commonly disseminated over several points of the surface of the body, and are accompanied with many other symptoms that never occur in lupus.

"The crusts of impetigo, yellow, prominent, rugous, and often very slightly connected, especially on the face, are very different from the thick, brown and firmly adhering scabs of lupus *exedens*, which moreover terminates in ulceration and cicatrization, consequences never seen in impetigo. It is also of importance to distinguish lupus *exedens* from some cancerous and syphilitic affections of the skin. Cancerous tubercles, very hard and often painful, are evolved among subjects more or less advanced in life, on the nose, cheeks and lips, especially, and generally exist for some considerable time before they ulcerate ; lupus *exedens* on the contrary, almost never appears in persons past the prime of life, and its tubercles are unaccompanied with pain from the first. Phagedenic cancer frequently commences in a solitary tubercle ; in lupus *exedens* there are usually several, and in lupus *non exedens* there are almost uniformly a considerable number. Farther, cancerous ulceration is often attended with considerable swelling of the soft parts, and the neighbouring vessels are dilated and varicose ; these sores are also made worse by the application of escharotics ; when the nose is the part affected, the bones also suffer and exhibit alterations characteristic of the disease. Lastly, cancerous sores discharge plentifully and are painful ; they are not covered with thick dry scabs like lupus.

"Syphilitic tubercles un-ulcerated, are more *rounded*, larger, and prominent in a greater degree, they are also of a more coppery red colour, without epidermic exfoliation, and have less tendency to ulcerate than those of lupus *exedens*, whose tubercles flattened in their forms and accompanied with a slight puffiness of the skin, are almost always covered with a small epidermic lamella ; they are arranged in clusters, the circumference of which is *strongly defined*, while the centre is erythematous, furfuraceous and traversed by lines or bands of a dull white colour. To conclude, syphilitic tubercles, as consecutive symptoms of a venereal affection, are favourably modified by mercurial preparations, and commonly appear in individuals of a certain age, whilst lupus resists mercury in every form, and makes its attacks most generally among children and persons before the age of puberty. The ulcers that follow syphilitic tubercles are deep, and their edges are tumified, of a coppery red tint and sharply cut ; whilst those of lupus *exedens* differ in their causes and in their mode of destroying parts. In lupus *exedens* of the nose, the skin is commonly first ulcerated, the cartilages and bones being only attacked consecutively, and often after a very long interval has elapsed from the beginning of the disease ; in syphilis, on the contrary, the bones are the

unhealthy inflammation and ulceration of the cutaneous follicles of this part, but have omitted to explain its occurrence where the follicular apparatus is materially deficient or altogether absent. In a

structures that usually suffer first, and it is only after portions of them are affected with necrosis that the skin is perforated with ulcers. Lastly, tubercles and ulcers of syphilitic origin are almost always accompanied with symptoms that distinctly tell of their nature, such as nocturnal pains in the bones, exostoses, irritis, and often with tubercles or sores of the mouth, pharynx, velum palati, &c. The solitary tubercles of lupus *exedens* of the cheeks have frequently been mistaken during their stationary period for small sanguineous tumours or nævi; they differ from these, however, both in their structure and mode of formation, as well as in their tendency and termination.

" *Prognosis.* Lupus is always a very obstinate disease. Months and even years commonly elapse before it yields to any form of treatment. Lupus *exedens* is not generally subdued until a considerable extent of parts has been destroyed, and always leaves indelible and deformed cicatrices behind it. The disease proves so much the less troublesome and its effects less deplorable as we are called upon to treat it at an earlier period of its existence, and as its progress has been slow. So long as the cicatrices remain soft, blueish, and convey to the finger something like a feeling of fluctuation, and so long as they are surrounded with tubercles of different sizes, there are grounds to apprehend a renewed attack of erosive inflammation, in which case the tubercles ulcerate and the cicatrices already formed are not long again becoming open. Puberty, and the establishment of the menstrual flux, which among women produces favourable changes in the greater number of chronic skin complaints, appear to have little influence in modifying or mitigating the destructive character of lupus.

" *Treatment.* The first indication in commencing the treatment of lupus is to endeavour to modify the general constitution by appropriate remedies. The disease itself is at the same time to be combated by such external and internal medicines as appear to exert a salutary influence on the development and progress of tubercles and the ulcerative process. When individuals of a flabby or evidently scrofulous habit are attacked with lupus, they may every morning take with advantage a table-spoonful of a solution of the hydro-chlorate of lime, made by dissolving a dram of the salt in a pound of water; every eight days the dose may be increased by a spoonful, and the medicine may at last be carried the length of ten or even twelve spoonfuls in the course of the day with propriety. This is a preferable medicine to the hydro-chlorate of baryta, the activity of which sometimes gives grounds for apprehension. Chalybeate mineral waters and the artificial preparations of iron may also be tried. I myself make frequent use of a powder composed of carbonate of iron, cinchona and cinnamon. Others have recommended the carburet and sulphuret of iron. Sulphureous baths taken every day for one or two months, the patient continuing immersed for several hours each time, are also powerful means of modifying flabby and scrofulous constitutions. Bitter medicines, such as the infusion of gentian, &c., and above all, the preparations of iodine, are employed with the same views. Food of good quality, the moderate use of some generous wine, a residence in a dry and bracing air, are also powerful modifiers of constitutions of the above description.

" Each variety of lupus presents particular indications.

" 1st. It is seldom that we are called in time to attempt the resolution of the primary tubercles of lupus *exedens;* patients generally present themselves with ulcers more or less extensive. When any tubercles exist they appear disseminated in the neighbourhood of some ulcer, which it is then of the last consequence to check in its destructive progress. This is generally affected by means of caustic applications;—the animal oil of Dippel, the nitrate of silver, potassa fusa, butter of antimony, super-nitrate of mercury, arsenical powders and pastes, and the actual cautery, have each had, and still have, their supporters. When

majority of cases, however, a tubercular elevation of the cutis,
marked by the characteristics of languid circulation, and having a
soft and pappy feel, precedes and forms the basis of the ulcer.

the disease is very extensive, the cauterization ought to be at first confined to a
single part, and extended successively to the whole of the affected surfaces.
When the ulcers are covered with scabs, these must previously be got rid of by
means of softening poultices. In lupus *exedens nasi*, when this part is affected
with an indolent enlargement, and its surface is of a purplish hue and covered
with an epidermic exfoliation, it is often advisable to carry a hair pencil, dipped
in the animal oil of Dippel, gently but repeatedly over the whole extent of the
diseased skin. The nitrate of silver employed in solution, in a similar manner,
has, however, generally appeared to me a preferable application. In the more
serious cases, the nitrate of silver is very advantageously employed after the
parts affected have been once or twice touched with the acid nitrate of mercury
or the arsenical paste of Frère Côme: it is the most useful escharotic, indeed,
whenever the object proposed is to cauterize superficially and with little pain.
The acid nitrate of mercury may be applied not only to the ulcers of lupus
exedens but to the tubercles themselves, and to such of the cicatrices as remain
blueish and soft, and threaten again to break open. When the end proposed is
to produce an eschar of considerable depth, which is often necessary, the acid
nitrate of mercury has indisputable advantages over the other active caustics
that are often employed in the same intention, such as the potassa fusa and
butter of antimony. A small dossil of lint is dipped in the solution, and applied to a portion of the diseased surface twelve or fifteen lines in diameter ; a
little soft dry lint is then to be laid over the space cauterized ; this trifling operation is not performed without the infliction of severe pain. The surfaces destroyed in this way are at first of a greenish-white colour ; a yellowish and slightly
adhering slough or scab is gradually formed afterwards, and the eschar is finally
thrown off at the end of a week or fortnight. Arsenical pastes can only be employed with safety as escharotics by enforcing precautionary measures of the
greatest strictness against their entrance into the nasal fossæ. The use of the
acid nitrate of mercury as an escharotic has the same disadvantage as the
arsenical paste, in occasionally causing a considerable degree of erysipelatous
inflammation ; this consequence, however, is in general much less severe and
far more transient with the mercurial than with the arsenical preparation.

"In lupus of considerable extent, affecting children, women, and subjects of
irritable constitution, the ulcerated surfaces, freed from incrustations, may be
dusted over with a thin layer of powder composed of ninety-nine parts of calomel,
and one of arsenous acid. This application is much employed by M. Dupuytren,
according to whom it acts rather as a specific than as an escharotic. Should the affected parts be covered with an imperfect cicatrice, it is even advisable to destroy
this, and to apply the powder four-and-twenty hours afterwards. Should there
be any difficulty in getting the powder to adhere in sufficient quantity, it may be
mixed with a little gum-water or simple ointment, and applied ; in this case,
however, the dose of arsenous acid employed must be increased by one or two
hundredth parts. In every case we are to wait till the powder or unguent falls
off of itself, which usually happens after the lapse of eight or ten days, and then
renew the application until a complete cure takes place, which is occasionally
accomplished within eight or ten weeks, that is to say after five or six applications of the remedy. When the ulcerated surface is of considerable extent, it is
advisable to apply this compound powder to a space of about two inches square
at a time, and only to cover the whole extent of parts affected by degrees.
Lastly, when the ulcers of lupus are of very old standing, and extremely indolent,
great benefit is occasionally derived from covering them with a blister, before
beginning the application of the powder or salve above mentioned. The advantage possessed by this powder over the arsenical paste in common use, (*pâte
arsénicale de Frère Côme*) are those of not exciting erysipelas in the tissues
around the diseased structures to which it is applied, and in cauterizing much

The importance of the disease appears to warrant a description applicable to its characteristics in different parts. When it makes its appearance in situations where considerable thickness of cellular

less deeply, so that it may be used again and again without danger to the patient. The relative proportions of its component ingredients may be changed by the cautious practitioner, according to the extent and circumstances of the case he has in hand; but it seems important that neither be omitted, inasmuch as both appear necessary to its action, without our being able to determine precisely the part which each of them plays in the general effect.

"The arsenical powder of Frère Côme * is a more energetic external remedy, which seems more particularly available in cases of old and obstinate ulcerated lupus, the ravages of which have been found uncontrollable by other less active applications. It is well to have recourse to this preparation at the very outset in lupus *exedens* of the cheeks. To prepare it for use a small quantity must be made sufficiently thin upon a piece of broken earthenware plate, and by means of a spatula spread upon a surface that ought not to exceed eight or ten lines in diameter. This application is almost always followed by some degree of erysipelas, which, when slight, may be left to itself; but when it proves severe, when the face becomes very much swollen, and the patient complains of violent headach, whilst the pulse is full and frequent, blood-letting must be practised, leeches attached behind the ears, emollient or laxative lavements thrown up, stimulating pediluvia employed, &c.; the inflammation soon abates, and the only consequence of the application of the arsenical caustic that remains is a blackish, and thick eschar, which continues very long adherent. These are all much more effectual modes of producing eschars and arresting the current of diseased action in lupus than the use of the actual cautery, a remedy which in this complaint is nearly abandoned at the present day. In lupus *exedens* of the nose, indeed it frequently aggravates the disease, causing the cartilages to swell and be absorbed; it would appear to be rather more applicable in lupus of the cheeks. Whatever the form of cautery or caustic employed, when the eschars are detached, they seldom fail to expose an ulcerated surface of the best appearance beneath them; a single application of the caustic, however, is generally insufficient to accomplish a cure; ten, fifteen, twenty, thirty or more, may be necessary in the course of several years, when the diseased surfaces are very extensive and of ancient date; a cure, indeed, is seldom or never consummated but by perseverance and undeviating attention. Whilst this is going on, when the nose is the seat of the disease, it is necessary to guard against the contraction of the nostrils, which then frequently show a great tendency to close up. They must be enlarged, when they have shrunk in any considerable degree, with the knife or caustic, and kept from again shrinking during many months by means of tents of prepared sponge.

During the treatment, patients should avoid exposure to excessive heat or to rigorous cold. By want of attention to this simple precaution, cicatrices that appeared sound have frequently been seen to open out afresh. When the disease is accompanied with any evident functional disturbance, this must be remedied by appropriate means. When amenorrhœa complicates it, the return of the catamenia must be solicited, and the appearance of this discharge kept up by every means at our disposal; when it cannot be brought back, a few leeches should be vicariously applied to the external parts, when symptoms of general uneasiness, or pains in the head, &c. seem to indicate the periods of its natural flow.

"Lupus *non exedens* is a more obstinate disease, if possible, than the one we have just been engaged in describing; in the hope of effecting the resolution of the tubercles which characterize it, the most powerful medicines of every description have been pushed to the uttermost. The tisan of Feltz; † Pearson's

* White oxide of arsenic, cinnabar, and animal charcoal.
† A decoction of crude antimony, sarsaparilla, and fish glue.

membrane, or, other structure, intervenes between the cutis and
more solid parts, it presents the character of a soft venous coloured
tubercle.   A few of these are found usually on the fore-arm and

solution of arseniate of soda, in doses, increased from a scruple to a dram daily ;
Fowler's solution of the arseniate of potash, even in hazardous doses ; arsenic, in
a word, in every form, has appeared to exert but a very slight influence on the
evolution and progress of the tubercles of this variety of lupus.   The animal oil
of Dippel has been tried, but with results too diversified to enable us to have
any reliance on its powers.

"The deuto-ioduret of mercury, in doses gradually increased from one-four-
teenth or one-tenth to one-fifth of a grain daily, is the only one of all the internal
medicines that I have tried, which has appeared to me to exert an indubitable
influence upon the progress of the tubercles of lupus *non exedens*.   After continuing this medicine for a month or two, the parts covered with tubercles frequently
become painful, and after a kind of local or intestine inflammation, attended
with vague and irregular symptoms of general febrile disturbance, the tubercles
begin to shrink, and many of them disappear entirely.   After continuing the use
of the deuto-ioduret of mercury for two or three months, patients should leave it
off for some short time, and thus try its influence repeatedly.   Besides the un-
doubted action exerted by this preparation on such tubercles as already exist,
its influence on the general constitution in preventing completely the formation
of fresh tubercles, is not less remarkable.   Nevertheless, despite my utmost
care in the exhibition of the medicine, and all the watching I could give its
action, I have occasionally had to contend with symptoms of inflammation in the
large intestines which compelled me immediately to suspend its use, if, indeed,
I was not obliged to abandon it altogether.

" *External* applications have been administered with some success in this form
of lupus.   In the foremost rank of remedies of this class must be placed the inunction, over the districts occupied by tubercles, of the ioduret of sulphur ointment (R. Iodureti sulphur. gr. xviii. Adepis suil. ʒi.), of an ointment of the
proto-ioduret of mercury (R. Proto-iodur. hydrarg. gr. xxx. Adepis ʒi.), and of
the deuto-ioduret of mercury (R. Deuto-iodur. hydrarg. gr. xviii. Adepis ʒi.)
Under the influence of one or other of these unguents, rubbed on the affected
parts, the skin becomes hot and red, and the tubercles frequently shrink and
disappear entirely.   Yet cases do occur in which these applications produce
little or no amendment, if they do not perchance excite unpleasant erysipelas of
the face.

"The occurrence of this inflammation as a consequence of the internal use, or
external application of these preparations, however, frequently appears to have a
beneficial ulterior effect, not only on the part in which it is developed, if this
chance to be beset with tubercles, but, further, on the clusters at some distance
from the seat of this new affection.   This circumstance was presented to my
notice in a very recent instance in the Hôpital de la Charité :   A young woman,
labouring under lupus *non exedens* of the cheeks, forehead, and left arm, was
attacked with erysipelas of the face, and during the continuance of this kind of
exanthematous fever, the skin of the arm underwent as evident an improvement as
that of the cheeks.

" In the hope of preventing the extension of the tubercular clusters of lupus, I
have isolated them by an incision, as well as by a deep application of caustic ;
but in vain ; fresh tubercles appeared beyond the artificial boundary I had traced.
The destruction of the clusters by caustic, or their extirpation with the knife,
would be attended with no good effects in the greater number of cases ; to cure
this variety of lupus completely it were necessary to eradicate the unknown
cause under the influence of which its tubercles are evolved ; and it is probable
that the only means by which this can be in any way accomplished consists in
the exhibition of those medicines which pervade the whole system, and alter it
deeply.

about the elbow joint. In one or two instances, the largest was discovered to have an old cicatrix from venesection in its centre. The figure of the tubercles varies, and, where there are many, some will be found oval, and others round ; while others, more irregularly shaped, are evidently formed of two or three smaller tubercles joined together. They are most elevated and darkest coloured in their centre, and on being pressed and examined, feel as if they contained fluid. Their progress is extremely slow, and if cut into much before the skin becomes broken, their internal structure is found somewhat to resemble a stringy slough, similar to that to be described hereafter, as surrounding the basis of the conical scabs of Rupia, where pressure has been employed, though of a clearer white than the latter.

When ulceration takes place, it is generally in or about the centre of the tubercle, and the ulcerated spot is immediately covered by a shining gummy exudation, which grows to a hardened scab in the course of a few hours. The surface of the sore seems to sink with more or less rapidity below the surrounding parts, and gradually assumes the appearance of a deep excavation, the bottom, sides, and edges of which are lined with the scab described. In this manner the disease continues to extend till the original tubercle is wasted away ; when, if care has been taken to alter the state of the constitution, the surface gradually assumes a healthy condition. A sensation of heat and tingling is felt during the whole course of the disease till this change takes place.

Such has been the history of it in numberless cases which have come under my notice, in all of which a cachectic state of system was evident at the commencement.

In the treatment of these cases, the state of the stomach and bowels required for a considerable time the exhibition of mercurial alteratives. These were followed by tonics and a better regulated mode of living. The local applications consisted, while the heat and tingling continued, of poultices made with the nitrous acid lotion, and, subsequent to this, of the latter fluid by means of pledgets of lint.

When occurring on the nose, the tubercular prominence is considerably less extensive. As in the case of acne indurata, it may sometimes occupy the space of six, eight, or ten of the mouths of the cutaneous follicles. Like the A. indurata, it has a bluish inactive appearance, while the cutis around it wears a reddened and tumefied character. If the adjacent parts be minutely examined, the orifices of many of the follicles are found covered by a thin pellicle of scab of a yellow colour, which easily separates and gives exit to a sanious discharge very unlike the natural secretion of the follicle.

In the course of time the skin in the centre of the tubercle breaks, and a little unhealthy speck of ulceration is perceived, which gradually enlarges in circumference, and more or less rapidly deepens. The character of the ulcerated surface is however peculiar, and when the scab is cleared away it appears as if covered by a layer of dark-brown varnish, and little or no discharge takes place from it. As the sore extends, it takes within its boundaries some one or more of the adjacent already diseased follicles, and thus spreading over the surface, it converts the whole of the parts into a ragged brown scabby ulcer.

The process by which the destruction of parts is effected has a marked peculiarity in this disease ; a gummy exudation forming scabs, instead of a purulent secretion continues through its course, and yet the absorption of structure will be at some periods proceeding with the utmost rapidity, and the excavation and destruction of parts in proportion, while at others no change will be perceptible for days and weeks. While the inflammation and swelling is considerable, there is great tingling, itching, and a sense of smarting or burning pain. On exposure to cold, the patient experiences a very severe degree of pain ; while warmth, if suddenly applied, is followed by excessive heat and excitement, and a proportionate aggravation of the disease. I have never met with any case where the patient had not been aware of a change for the worse during the winter season, provided his pursuits compelled him to leave his house.

As the disease proceeds, some one or other of the affected parts assume for a time an appearance of improved and even healthy action, and tolerably healthy granulations are formed : and now and then, even cicatrization will be effected to a small extent. The diseased action in such a case seems only to leave one part, in order to propagate itself more extensively in another ; and it seldom happens that if a portion of the diseased surface heals, another does not exchange a passive for an actively mischievous character. Eventually the tip and alæ of the nose are destroyed, and the openings of the nostrils, in consequence of the thickening of the parts and the accumulation of the scabby secretion about them, are nearly closed. The redness and inflammatory action extends up the nose, and even spreads to the eyelids and eyes. The lips and cheeks sometimes suffer, and cases are on record where the bones of the nose have been destroyed.

Sometimes the original sore is not the only one concerned in the work of destruction ; one, two, or more of the inflamed and ulcerated follicles, described as at first covered with a scab, appearing to take on the same character, spreading and destroying the skin in the same way, till they all join and form one common sore. In some cases the tubercular formation is altogether absent from the beginning, when the disease appears to commence at the mouths of a few of the follicles. In such cases the little scab described is perhaps rubbed off, another and another forms, fresh irritation and mischief is produced, till the character of the disease is established, and it pursues its course.

The spreading destructive ulcer which answers to the description of Noli me tangere of some authors, has been observed in a few instances to occur as the apparent consequence of chronic inflammation of the cartilages of the nose. I have been acquainted with more than one instance of such origin, and in the inquiries as to the history of some others, have been led to the opinion that such cases are more numerous than have been supposed. The particulars of one instance, and that of a very distressing nature, have been furnished me by a medical friend.

The subject was a young and interesting female, the daughter of an artist of considerable eminence. The attention of the latter was first directed to it by the discovery of a very trifling enlargement of the right ala, which was unattended with pain or discolouration. As the enlargement increased, medical advice was obtained, but none of the applications had recourse to appeared to be productive of benefit, though combined with judicious constitutional management. After a time, a speck appeared on the most dependent part, which terminated in the kind of scab described as characteristic of the disease. From this periŏd the wasting of the integuments and cartilages went rapidly on, (the sore extending over the tip to the opposite side,) till the whole organ was destroyed.

On the subject of the causes of Lupus, much difference of opinion has been expressed by different authors. Alibert, Rayer, and other French writers, consider it of scrofulous origin, whilst most others who have noticed it appear to have had no suspicion of such a connexion. On the whole, it seems probable that the cases which, similar to that mentioned, have originated in inflammation of the cartilages, are really of this nature, while the more common forms seen in England, are the results of disorders which the habits of the individual have induced. In the case mentioned, a scrofulous diathesis was certainly manifest ; but I believe that for every such case, twenty others come under our notice, where the subjects are accustomed to indulgences in spirituous potations, and habitual violence to the digestive organs, whose histories disclose nothing of what M. Rayer has so particularly stated to have occurred in the earlier periods of life of the subjects of his observation.

At the period of its development, moreover, and even in its advanced ages, as it occurs in this country, where much distress of mind, and a considerable aggravation of constitutional disorder has been observed, the glandular system has been seldom affected, or any other mark of scrofula existing ; a fact which in itself may be considered almost conclusive against the opinion in question. It would seem, indeed, that no form of disease of a clearly scrofulous origin approaches in similarity to this in any other points than its tediousness and intractability.

It has been compared, and probably sometimes confounded, with scirrhous ulcerations of these parts ; but the features by which it may readily be distinguished from such affections are,—

1st. Its situation ; cancerous disease of these parts usually first occurring on the lower lip.

2d. The uneasiness belonging to it is in no case described to be worse than are comprehended under the general designations of heat, itching, tingling, or smarting, while scirrhous ulceration is accompanied by severe darting pains.

3d. Diseased enlargements of the contiguous glands do not often make their appearance in its train, even though the disease has existed for years, which is not the case with cancer.

4th. The surface of the sore is never occupied by fungous granulations, or has thickened and hardened or everted edges, but retains its peculiar character to the last.

The experienced eye will discover at first sight several other points by which it may be known, but unfortunately successful treatment is by no means a consequence of such knowledge ; and, as far as I am aware, even up to the present time, all which can be said on this point consists of a simple detail of such measures as have appeared to be beneficial in a few cases, and failed in perhaps a hundred others.

Referring to the different descriptions of this disease already alluded to, as to its commencement in the form of a tubercle, or of ulceration of the sebaceous follicles of the integuments of the nose, it would seem a matter of interest to inquire, first, whether it originates in an actual tubercular growth of a portion of these integuments, when it happens to wear such an appearance ? and second, whether when spreading in the manner described from the orifices of one or more of the follicles of the part, the correctness of the opinion be established as to its origin in unhealthy inflammation and ulceration of such follicles ?

Now, the description of acne indurata with the pathology of this affection given in a preceding page, will enable us to discover points of similarity of a most important nature between it and the tubercle of lupus, before the skin of the latter is ruptured, and ulceration begins, though after this period the resemblance is lost. A brownish or livid tubercle is formed in both, and formed in the same way i. e. by the accumulation of the contents of the diseased follicles, and consequent inflammation of the adjacent parts. In the former case, if punctured, it heals slowly, but in the end satisfactorily ; while in the latter the healing process may be courted in vain for months and even years, and the destruction of parts will still go on. If a follicle suppurates in a healthy way, and the contents of the minute abscess are discharged, as in acne simplex, it speedily heals; but if its orifice be covered by a thin scab, if heat, itching, and much irritation be present, it does not heal, but assumes the diseased action described as characteristic of the disease under consideration.

Reasoning on these facts, and bearing in mind the organization of the parts which are so frequently the seat of the two diseases, we must be compelled to consider Noli me tangere to originate frequently in an unhealthy inflammation and ulceration of the cutaneous follicles on and about the nose, spreading to and involving the adjacent cellular structure.

If this opinion be correct, it follows that the tubercular formations on other parts of the body, which are considered to be of the same nature, are really essentially different, so long as they wear a tubercular character, their interior being made up of a solid organized structure. The similarity of the mode of extension of the ulcerations, after the skin has been broken, has however so much of identity of character in it, as to render a wider separation than we have adopted unnecessary.

The disease has been already stated to occur chiefly among the lower classes of people, and its most severe and aggravated cases will be found among those of individuals who have been before subject to the worst forms of acne, which circumstance appears to form another point of similarity between the histories of the two diseases.

The published results of any inquiries as to the causes of the extreme obstinacy of lupus may be, at present, thought problematical. The constant exposure of the parts to extreme vicissitudes of temperature consequent on their situation would, it is evident, very much operate to prevent a healthy process of reparation where the disease is once established. A still more powerful obstacle will generally exist in the want of resolution on the part of the patient to abstain from habits which originated it : and it is evident, that the cellular structure and cutis covering the cartilages of the nose are so thin as to be inadequate in energy of circulation to overcome obstacles of this nature to the healing of the part.

Nevertheless it must be always clear to an observant surgeon that excessive heat and irritation of the margin of the broken surface prevails, and this fact should operate as a caution against the employment of stimulants and caustics, which constitute the practice too often resorted to, and which has become, as I have before said, more than half established in our hospitals.

This disease is in many cases occurring in the middle and lower classes of society considered by the patient to be of syphilitic origin, and much mischief and misery has arisen from such mistake, for the use of mercury to any extent seldom fails to aggravate it in the extremest degree ; and although it is commonly said that mild and soothing local applications do not agree with the sore, there are but too many instances coming from time to time under our notice, where the influence of internal medicines will fully explain the intractability of the external sore—the internal well intended medicines create and support a morbid irritability of the system, and if the error in question be committed, the good intentions of nature are impeded*—the original sore spreads, and new tubercles form, terminating like the first in ulceration.

* The feverish irritation of the system produced by the use of mercury, often converts the sluggish and gradually destructive sore into one of the utmost activity.

Treatment of Noli me tangere.

According to the statement of authors, of all the medicines, or medicinal applications, which have been tried in the treatment of lupus, arsenic has been found most frequently useful. " Ulcers of this kind differ exceedingly from one another in their degree of virulence, but they are all so far of the same nature, that arsenic, in general, agrees with them, and puts a stop to their progress, while they are aggravated by milder dressings."*

General alteratives, as Plummer's pill and the decoct. sarsæ comp. are in most cases advisable, and now and then of the greatest utility, any important derangement of the digestive organs, or febrile excitement, being first removed by adequate means. If no advantage is derived from these, in combination with common sedative applications, arsenic ought to be used, both internally and externally. This valuable medicine is usually exhibited in the form of Fowler's solution.† It is proper to increase the dose gradually, till some manifestation of tendency to disorder of the stomach and bowels occurs, when it should be entirely withheld, and purgatives, with opium substituted, till such symptoms have subsided. The form of lotion is that best adapted for external application, and I have usually tried, though only now and then with advantage, a preparation somewhat resembling that above. Ointments, or any other greasy applications, with one exception, are highly objectionable : they appear usually to increase the heat and irritation, instead of diminishing it. The spirituous arsenical solution is, in most respects, superior to others, possessing, as it does, the specific powers of the arsenic, with the sedative properties of spirituous evaporation.

Solutions of nitrate of silver of various strength are said occasionally to do good when the arsenic fails, as also, that touching the smaller sores with caustic is occasionally followed by healthy granulations : in situations, however, where the formation of a slough is not likely to add to deformity, the free use of caustic ought to be had recourse to, as if no great disorder of constitution exists, a healthy state of parts is often speedily produced at once. The dependence on the uncertain effects of milder applications is, in many cases, followed by disappointment in the end, and by a greater destruction of parts than the caustic ever effects. In two different cases within the last year, where I had reason to be confident of the attention of the patient to the prescribed rules of diet, &c. I have

---

* Sir E. Home on Ulcers, edit. 2d, p. 267.

† The following recipe has been preferred in some of our public institutions.
 R. Kali arsenicati, gr. iv.
   Aq. menth. sativ. ℥iv.
   Sp. vini tenuior, ℥j. Misce et cola.
The solution alluded to for internal exhibition is but of half the strength. It is employed at the same time internally in doses of two drams three times a day.

freely applied the nitric acid to the parts, and produced a healthy sore, which speedily healed.

From an attentive observation of several cases of lupus continued for a long time, I am satisfied that the want of success in its treatment arises, either out of the difficulties of keeping the parts free from the influence of atmospheric vicissitudes, or the patient from habits destructive to his general health ; in either of which cases medicinal treatment, or local applications, might *à priori* be expected to fail.

Rayer, indeed, in the work already quoted, observes on the increased difficulties attending its management in the winter season. The note I have introduced will show somewhat of the practice, and a few of the opinions of our French brethren on this very troublesome disease.

The preparations they employ as applications to the part are, on the whole, considerably more active than such as we are accustomed to.; but I am inclined to think, that where the habits of the patient can be directed and controlled, they will not prove to be more successful. A great error has been usually committed in the treatment of this disease, in the non-attention to the temperature of the atmosphere in which the patients live. Because the system is not seriously affected, they are allowed to pursue their avocations at all seasons of the year, and under all changes of this kind. The diseased parts are alternately chilled and excited, the healthy action of one day is changed for a different one on the next, and the operations of nature constantly interfered with, instead of encouraged. The only cases of recovery with which I have been acquainted as taking place under medical treatment, have been those where the patients were protected from these mischievous agents.

Another cause of disappointment and difficulty may be traced to the practice of directing the attention to the centre, or other broken parts of the surface of the disease, to the exclusion of proper notice of the surrounding diseased skin. Thus the adjacent parts, to the extent of several inches, studded with the diseased follicles, far advanced towards ulceration, are not interfered with in their progress in any manner ; and if perchance a healthy action should be brought on the ulcerated surface, the vexation of perceiving some of the former far advanced, and rapidly taking on the scabby ulceration of the latter, is often the result. If proper applications are made to the circumference of the diseased skin, and healthy action be first established there, the disease will often give way, till the skin resumes its natural appearance, even up to the margin of the ulceration ; and it is evident, that under these circumstances, the latter has a much better chance of being covered by healthy granulations. I am entitled to say, that this is a practice which has been much more frequently successful than any other, consisting of applications to the worst, or broken parts of the disease.

The application of sedative washes over the whole extent of inflamed skin during the night (the ulceration being covered by some simple application), and the rubbing in around its margin, every night and morning, a portion of the ointment prescribed below, constitute two of the most powerful measures which I have ever seen employed. Combined with proper constitutional treatment, I have found them successful in some of the most inveterate cases.*

CHAPTER V.

PORRIGO.

Les Teignes, Dermatoses Teigneuses of Alibert, Favus of Rayer, Ringworm, Scalled Head, &c. &c. of English Authors.

The degree of obstinacy evinced by the larger portion of cases of this disease, the interruption it frequently occasions to the education of children, (its known infectious nature preventing their admission into schools,) combine with other circumstances to give it a peculiar interest. As in many other cutaneous diseases, much of the trouble which has been taken by different authors in its investigation has been far from contributing effectually to the knowledge of its pathology or treatment. Much confusion and discouragement to the student, moreover, is attributable to the division into so many species which has been adopted ; and what is known is thus prevented from becoming so generally known as it might be.

From a patient and lengthened inquiry into the history and progress of porrigo, I am compelled to believe that the abolition of these distinctions will be eminently useful in advancing the knowledge of it ; and that rather than continue to uphold the original system of describing the form, consistence, or colour of the accumulated secretions, and designating them according to these accidental circumstances, as different species of the disease, it will be better to trace the latter at once from its first appearance and original character, up to the periods when the state of the parts in question is become such as described by authors who have preceded me. To notice also those circumstances or facts which influence the formation and consistence of the diseased secretions, and render the study of the subject more simple by pointing out a distinct line of connexion between the most simple forms and those which are considered most inveterate ; or, in other words, by showing that the latter are

* R. Hydrarg. subm.
  Plumbi superacet. a. $\bar{3}$ss.
  Ung. hydrarg. nitrat.
  —— cetacei a. $\bar{3}$ij. M.

DIFFERENT FORMS OF PORRIGO & TEIGNE

only the result of injudicious treatment or gross neglect of the former.

In pursuing this plan, it is necessary to show whether any, and if any, what kind of primary disease is capable of taking those varied forms of scab and filthy accumulated secretions, which have been described by Bateman and Alibert under a variety of designations, as Porrigo furfurans, P. lupuosa, and P. scutulata, Teigne faveuse, T. granulée, T. furfuracée, T. amiantacée, and it will not be difficult to ascertain, in most cases, that they have originated in one of two particular states of disease of the cutis about to be described. I have never met with any instance (when satisfactory evidence could be obtained of the form of the disease at its first commencement) leading to an opposite conclusion.

Nevertheless, as these diseases are sometimes seen here in those advanced and neglected forms which are described by Alibert as occurring under his notice in the hospital of St. Louis, it would be improper to omit altogether his very accurate descriptions. It will be observed, however, that they are more precise as regards the form and structure of the diseased secretions than the actual condition of the diseased surface.

The " Teigne faveuse" of Alibert is evidently the Porrigo lupinosa of English authors. It is described by the former as follows :

" It developes itself in the form of an eruption of minute pustules, which create an itching, more or less violent, on the scalp. The contents of these pustules dry up, and give place to the formation of small circular scabs, hollowed in their centres, enlarging gradually in their dimensions, but still preserving their circular form. As these scabs sometimes form in great numbers on different parts of the head, their edges approach each other, forming by their aggregation plates of considerable extent, in which the eye nevertheless distinguishes with facility the cup-like form of the individual scabs. This cup bears some resemblance to the cells of the honeycomb or the fructifications of certain species of lichen. When the disease has recently appeared, the scabs are either of a yellow or fawn colour ; *but as they get old and dry, they become white, wear off, break, and detach themselves from the scalp, and then you only perceive their remains, which cease afterwards to assume a regular form.*

" The scabs of this species have from the beginning their bases deeply enchased in the skin and strongly adhering to it ; so much so, that I have never been able to detach them without great pain and some discharge of blood. Sometimes the cutis is deeply involved in the irritation of the disease, and cracks to a considerable extent are formed, from which an ichorous or purulent matter is discharged; and now and then, though very rarely, the cutis and cellular membrane are destroyed to such an extent as to expose the bony substance of the skull.

" With some individuals the disease does not confine itself merely to the head ; I have seen it appear on the forehead, the shoulders, the temples, the lower part of the shoulder-blades, the elbows, and forearms ; I have also seen it extend from the top of the loins to the sacrum, the front of the knees, legs, &c.

" The attendant itching is in some cases almost intolerable, and the children who are the subjects of it are induced to scratch themselves severely ; they appear to experience a kind of voluptuous enjoyment in tearing the scalp with their nails. Pediculi, which multiply in great numbers under and between the scabs, add further to this irritation ; all the cavities are full of them, and the whole mass appears agitated by their movement. The smell emitted by the scalp is as disgusting as its aspect, and always preserves the same character. It resembles the urine of a cat in this particular, or chambers which have been infested by mice.

" When by the help of emollient poultices the favous crusts fall off, this smell changes its character, but is still offensive. Between the different clusters of the favous pustules the skin is covered with furfuraceous scales, which are produced by the general irritation of the dermoid system of the head.

" Proceeding to the examination of the scalp, after the separation of the crusts and scabs which have been softened by repeated lotions and poultices, you see the reticular structure, it has become red and erythematous. The epidermis has disappeared, and a yellowish viscous and fœtid fluid runs here and there from numerous ulcerations. You may likewise perceive small abscesses, more or less in number, dispersed over different parts of the head, taking a lenticular form, and appearing as so many centres of inflammation; but one of the most remarkable symptoms attending the disease when neglected and abandoned to its progress, is the *alopecia*, which in some cases has become almost universal.

" In the places where the hair has been rooted out, the skin remains smooth and shining ; you may perceive here and there notwithstanding some thin hairs altered in their structure and colour, and having a languinous appearance."

" *Teigne granulée.*"—This form of the disease, or rather the state of the accumulated secretions from which it has obtained its name, is not described by English authors, nor is it represented in Bateman's delineations. It is, however, not uncommon among the children of paupers when first admitted into the parochial establishments. The ordinary attention to cleanliness adopted in these institutions commonly leads to the speedy removal of the mass of filth which Alibert describes. It is only the sequence of an old standing utterly neglected case of T. faveuse or of P. favosa.

The Teigne granulée does not in general occupy so large a space of the scalp as the T. faveuse, and occurs most frequently on the upper and back part of the head. Its external characteristics are composed of small brown or dark grey crusts, which resemble fragments of mortar, or the plaster falling from old walls which have been discoloured by damp and dust. The dimensions of these granulations vary exceedingly, and their form is altogether irregular. They are often very hard, and have even a stony consistence, which poultices cannot soften. The patches of the disease are generally a little distance from each other, are not so deeply enchased in the dermoid system as those of the T. faveuse ; but sometimes, like those, they are surrounded by a considerable number of thin scales, dry and furfuraceous. It has a nauseous smell, which greatly resembles rancid butter, or milk which is beginning to turn. This smell is particularly perceptible when any moisture exists on the parts, but disappears when exsiccation is completely effected.

The itching attending the disease is very great. When the crusts are separated from the scalp, the places they occupied are red and erythematous, smooth, polished, and often swollen. Here and there small whitish abscesses are perceived, not elevated above the surface, from which issues a small quantity of viscous colourless fluid, which thickens and dries on the part, forming new crusts analogous to those preceding. The T. granulée scarcely ever attacks other parts of the body, but confines itself to the scalp and immediately adjacent parts.

The "Teigne furfuracée," or Porrigo furfurans of Willan and Bateman, when once established, so as to answer the following description, is a most tedious and unmanageable state of disease, and by far more frequently seen in England than those conditions constituting the T. faveuse and T. granulée. It does not form crusts of any thickness, but whitish furfuraceous scales more or less thick, sometimes damp and adhering to the hair by the help of a viscous and foetid discharge, and at others dry, and detaching themselves with the greatest facility.

The T. furfuracée begins by a slight desquamation of the cuticle of the scalp, and is often accompanied by a considerable itching ; an ichorous matter flows at the same time from the affected surface, which dries, and forms the scales of scurf from which the disease is named. As the disease increases by degrees it spreads in time over the greater part of the scalp ; the layers of scurf thicken, and at this period they resemble a coating of bran or coarse flour, the under surface of which is saturated with fluid.

If the scalp is carefully freed from this adhesive substance, it is found to be divested of its cuticle. It is usually of a pink colour, and offers a smooth, polished, shining surface resembling varnish.

Some authors have denied the existence of this form of the disease, because it is not often seen in hospitals. Others consider it only as the same diseases which have been designated T. faveuse or T. granulée, a degree less advanced ; but the scabs which characterise this form are of quite a different kind from those described under these denominations. The hair is in this case matted and glued together, and when the finger is pressed upon it, the whole mass yields softly in every part nearly alike.

The T. furfuracée is chiefly confined to the scalp, but sometimes extends a little beyond its margin on the forehead, forming crusts resembling quantities of bran cemented by some adhesive fluid ; the edges of these are sometimes dry and perfectly white. Much itching attends it, and great numbers of pediculi are observed ranging freely over the affected parts. Ulcerations to a small extent also occur here and there, by which the quantity of fluid secretion is increased, and is found generally to have the smell of sour milk.

I have never observed this disease attack adults, but it often happens to children who have passed their first septenary, though a contrary opinion on this point has been advanced.

The " Teigne amiantacée" of Alibert is evidently merely a variety of the P. furfurans, occurring perhaps on the scalp of persons with less irritability of skin. It is evident, that such a state of skin would be less likely to favour pustulation or the fluid exudation mentioned as belonging to the latter ; and in fact, that the inflammatory symptoms would generally assume a character more approaching to chronic than active disease. Hence, in the following description we find the phenomena of chronic inflammation of those vessels of the cutis whose office it is to produce the epidermis, and a constantly repeated exfoliation of this substance in minute scales unattended with fluid exudation. The sheath which the cuticle gives to the hair a little beyond its exit from the scalp, and which is in a healthy state almost transparent, and scarcely perceptible by the naked eye, becomes by this chronic inflammation of the vessels producing it more rapidly elongated upon the hair ; it grows dry and harsh, and gives to the hair near its root a shining silvery appearance resembling the fibres of asbestos ; and hence Alibert has given it its extraordinary designation.

" The Teigne amiantacée does not form crusts, but shining silvery scales, which, by their concretion, harden and unite the hairs nearly their whole length in parcels, and its silky and delicate appearance gives it a resemblance to asbestos.

" It generally occupies the upper and fore part of the head, and is particularly characterised by very small fine scales of a silvery or or mother-o'pearl appearance, which surrounding the hair does not a little resemble that thin transparent pellicle with which the feathers of young birds are surrounded when they are first hatched. When the hair, thus hardened with this scaly substance, is cut off with the scissors, the skin appears furrowed ; it is red and inflamed, but less so than in the T. furfuracée. *The itching sensation is inconsiderable*, and as the diseased parts are usually destitute of moisture, no unpleasant smell is emitted."

The " Teigne muqueuse," or Porrigo larvalis, completes Mons. Alibert's series of this class of diseases ; but the arrangement adopted in this publication compels me to place it in a distant part of the work. In his general observations on " les Teignes," he furnishes us with still more disgusting descriptions of the ravages he has seen committed by it, and cases are alluded to certainly without a parallel among those observed in England. The common history of the scalled head of long standing here, seldom discloses any connexion between the local affection and the constitution. Its history is comprised in a few words : the secretion from the pustules is allowed to accumulate, scabs are formed of it upon the surface, which being confined by their adhesion to the remaining hairs, become the means of further accumulation of this irritating, fluid. The inflammatory action of the part materially increased in this way, the quantity of fluid secreted, and the means of extension of the disease, receive likewise a proportionate increase. The itching and irritation is considerable, the scabs accumulate and increase in thickness, and emit an offensive odour, and myriads of pediculi are seen creeping over and about them.

I submit here a brief table, showing the designations of English and French authors, in speaking of diseases of the scalp, and have taken the liberty to reserve one of its columns for a very short name or definition, according to my own views. The old nomenclature of the French school is retained in the table in one column, and the more modern, though not the last, of the Baron in another. I fear the reader will hardly think the latter an improvement, and as to the variation in the descriptions of the different forms of the disease as agreed on by both English and French authors, the older must be allowed to bear away the palm. Hence I have retained it in the different pages of this work.

English Authors.	French Authors.	
P. Larvalis . . .	Teigne muqueuse	Impetigo capitis infantilis.
P. Furfurans . .	{ T. furfuracée and T. amiantacée	{ Chronic inflammation of scalp producing a scurfy covering.
P. Lupinosa.	T. faveuse	{ Lupin-seed scab of the head. Neglected scab of P. scutulata.
P. Scutulata . .	{ Not described by the French Authors, till Alibert mentioned it in his last edition, and described it as very rare	{ Common ring-worm of England, identified with P. lupinosa, where cleanliness is not observed.
P. Decalvans . .	Alopecia	Baldness—no disease.
P. Favosa. . . .	{ No correspondingdescription in French—has been confounded with T. faveuse	{ Very likely in a totally neglected state to answer the description of the worst forms of T. granulée or T. faveuse.
Not described by English Authors . .	T. granulée	{ Accumulated and dried scabs, and secretions of the diffused Porrigo.

9*

The " genre premier" of the " Dermatoses Teigneuses" of Alibert's last work referred to, is headed, " Achore," and of this there are two species, l'Achore muqueuse and l'Achore lacteumeux. The line of distinction he has drawn does not exist in England, and this part of the subject may be dismissed with the simple remark, that crusta lactea and P. larvalis are the names by which it is known here, and that the treatment is as well known as the names.

The second species is " Porrigine," or Porrigo, of which we have—

La porrigine furfuracée.
La porrigine amiantacée.
La porrigine granulée.
La porrigine tonsurante.

The first three our readers are well acquainted with—the last is the P. decalvans, baldness or alopecia, which I deny to be a disease in the ordinary acceptation of the term, or to admit of benefit from medicinal applications of any kind.

What, my readers will perhaps impatiently ask, do French authors tell us is the proper treatment of these diseases? to which I respond, with great regret, we are not permitted to know it. M. Alibert would have told us if he could ; so would MM. Rayer and Biett. The secret rested twenty years ago with MM. Mahon, frère et fils, and now is confined to MM. *les frères Mahon !* Pity it is that the spirit of the three days of July did not visit the hospital of St. Louis ! M. le Baron Alibert, physician to the king, refers to MM. Mahon, who have treated thirty-nine thousand cases of disease of the scalp by a secret method, for a statement of the proportion one form of the disease bears to another, and thinks it worth his while to give, as an authority, MM. Mahon's answer !

A book is published in Paris, which is read by ninety-nine out of every hundred of the visitors of that city, and it says, that though the hospital of St. Louis is dedicated exclusively to the treatment of diseases of the skin, the practice of the professional attendants does not appear to be more successful than in other places. How can it be otherwise, when looking at the immense proportion which diseases of the scalp bear to all other cutaneous diseases, (being considerably larger in number than all the rest put together.) They are, without investigation and without trouble of any kind, placed under the care of M. Mahon. It is impossible to deal with patience with this part of my subject any further.

The forms I am about to describe, each occasionally occur spontaneously, but are also very frequently the results of infection. They have also the power of producing each other, as I have ascertained, beyond doubt, in a variety of instances ; a fact which is, perhaps, by itself, sufficient to establish their identity. The description given by Bateman, as P. scutulata, applies tolerably correctly to one only, while the other is undescribed by this or any other author, except which perhaps may be the case, the incipient stage of T. faveuse or

P. lupinosa be meant to be applied to it; and it seldom indeed happens that the children of this country are so neglected as to allow the resemblance to proceed any farther than the first formation of the pustules. I by no means question the propriety or utility of tracing in a description the features of a disease at the period when it first comes under the notice of the observer, and I have therefore inserted those of Mons. A. They may be useful as references, when the practitioner finds a case presenting itself where the utmost possible neglect has contributed to its advancement. But as by far the greatest portion of the cases occurring here are not of this nature, I shall be content to describe them as they are most frequently met with. In doing this, I am not sorry to be compelled to re-sign almost entirely the nomenclature of the authors I have men-tioned.

The first of these is that noticed by Turner,* in which "the hair falls off not altogether from the root, but by piecemeal." It is sup-posed by this author, and by Sennertus, to be produced by some insect, but they have not noticed the state of the skin of the affected parts. The attention is first attracted to it by the *falling off* of the hair of the part ; there is little attendant itching, and no apparent fluid secretion on the spot. Sometimes, but not always, the patches are of a pretty regularly circular form, the margin being clearly defined, and exhibiting a line of scurf considerably thicker than that in the centre. In the centre of the spots the skin is scurfy, and the hair thinned, and easily extracted by the finger and thumb. What remains of it is unhealthy in appearance, some hairs being thin and delicate, others being the remains or stumps of those which have been broken or dropped off. There is a downy or towy looking sub-stance just rising above, and mixing with the scurf, evidently formed by feeble attempts at the production of new hair. Two, three, or more of these spots, varying in dimensions, are usually discovered on examining the head more particularly ; and when the hair has been removed by shaving, they exhibit a red and slightly inflamed appearance. Several others in an incipient state will be discovered in different parts. The latter may be known before the hair begins to fall off, when they exhibit nothing beyond the appearance of a small discolouration about the size of a spangle ; the hue is of a yellowish red, somewhat resembling the bran of the darker-coloured wheat. Others a little larger have decidedly assumed the ringed form.

With children of light complexion, with thin and delicate hair, with no constitutional disorder, or no great irritability of skin, this will be the state in which the disease will always be found, provided no interference by stimulating or other applications has occurred ; the margin of the spots not exhibiting any distinct appearance of pustulation. The diameter of these continue to enlarge rather slow-ly till they join each other, and a great part of the scalp is divested

* A Treatise on Diseases of the Skin, 1736, p. 200.

of its hair ; but if stimuli in the shape of ointments be applied, a more active condition often takes place, and minute achores form, not only on the margins, but on other parts.   Much irritation, heat, and itching arises ; the disease spreads with greater rapidity, and changes its chronic inactive character for one directly the reverse. The pustules discharge their contents, and unless the head be washed frequently during the day, layers of lightish straw-coloured scabs are formed, under which the cutis is sometimes found to be abraded to a considerable extent.

At the commencement of the disease, and for some time after, spots evidently of the same nature as the affection of the scalp may be seen on different parts of the body ; but the former being usually protracted for a considerable time, from causes hereafter to be more particularly mentioned, these spots generally disappear before much improvement is effected on the scalp.   To describe them more particularly would be to copy very nearly the account of the first form of Herpes circinatus of Bateman ;* but the case, or cases, from

* "It appears in small circular patches, in which the vesicles arise only around the circumference ; these are small, with moderately red bases, and contain a transparent fluid, which is discharged in three or four days, when little prominent dark scabs form over them.   The central area in each vesicular ring is at first free from any eruption, but the surface becomes somewhat rough, and of a dull red colour, and throws off an exfoliation, as the vesicular eruption declines, which terminates in about a week with the falling off of the scabs, leaving the cuticle red for a short time.

" The whole disease, however, does not conclude so soon ; for there is commonly a succession of the vesicular circles on the upper part of the body, as the face and neck, and the arms and shoulders, which have occasionally extended to the lower extremities, protracting the duration of the whole to the end of the second or third week.   No inconvenience, however, attends the eruption, except a disagreeable itching and tingling in the patches.

"The herpetic ringworm is most commonly seen in children, *and has been deemed contagious.   It has sometimes, indeed, been observed in several children, in one school or family, at the same time ;* but this was most probably to be attributed to the season, or some other common cause ; since none of the other species of herpes are communicable by contact.   It is scarcely necessary to point out here the difference between this vesicular ringworm, and the contagious pustular eruption of the scalp and forehead, which bears a similar popular appellation.

" The itching and tingling are considerably alleviated by the use of astringent and slightly stimulant applications, and the vesicles are somewhat repressed by the same expedients.   It is a popular practice to besmear them with ink ; but solutions of the salts of iron, copper, or zinc, or of borax, alum, &c. in a less dirty form, answer the same end."

On the subject of the disbelief here expressed of the connexion between this affection and the disease of the scalp, I shall take leave to observe, 1st. That when the disease of the scalp is noticed at it commencement, some of the spots in question are almost always to be found on other parts of the body.   2d. That the original form and progress of the disease is nearly the same.   3d. That pustulation attends that on the scalp most frequently, only from the circumstance of the hair growing through the diseased part, and exercising the properties of extraneous substances.   4th. That the diseased secretion of the scalp affection is capable of producing by inoculation the ringworm of the skin on other parts, and *vice versâ.*

Several years ago, in a publication on this subject, I noticed the case of a lady, who from two of the circular spots on her arm, inoculated one of her children

which his plate has been taken, appear to have been a different disease altogether. The vesicles of the little patches, to which I allude as connected with the ringworm of the scalp (if vesicles they may be termed) are, when unbroken, scarcely discernible to the naked eye, and are ruptured in a very few hours after their formation. They are, indeed, rarely seen unbroken ; and when the attention is first directed to the spot, it exhibits the appearance of a small ring of scab, of a brownish colour ; and in this state is well known under the name of ringworm, and quickly yields to the application of any mild escharotic application. The frequent occurrence of these spots simultaneously with the disease of the scalp first led me to the suspicion of connexion between them, and for the reasons I have detailed below, I now entertain no doubt of their identity, the apparent difference of character being solely the result of the *mischievous influence of the hair of the scalp.*

The foregoing description and remarks apply, as already observed, to the cutaneous affection, as it occurs in children whose general health is unimpaired, and where the skin is not remarkably irritable. It would be superfluous to observe here, that the skins of different individuals, and even that of the same person, at different periods, exhibit widely different degrees of irritability ; and the influence of this property would *à priori* be supposed to govern the character of any cutaneous affection to which the individual may happen to be subject. What happens, therefore, not unfrequently in this affection, where much irritability of skin exists, is that instead of minute vesicles, followed by a delicate circular scab, distinct though small achores are perceived, having the same circular form, but drying and forming in a few hours a small yellow scab, which is firmly attached to the cutis. The ring of pustules enlarges precisely in the same manner as that of the vesicles, but the contents of the pustules still drying, become attached to the margin of the scab already formed, and from day to day increase its bulk and diameter. The circular scab, dry and unyielding, becomes a source of increased irritation ; and the pustules under its margin become enlarged and more elevated, they now cause the edge of the scab to be raised, and the cupped figure of the P. lupinosa and Teigne faveuse is rendered distinct.

That the contents of the pustules under the smaller scabs possess the power of inoculating with ringworm any part of the skin of other individuals, has been proved in several instances by experiment ; and I have also seen repeatedly in the same family, different children showing the different conditions of the ringed incrustation,

with the genuine ringworm of the scalp, the disease afterwards affecting several other children of the same family. I would further direct the reader's attention to Bateman's delineation, No. 39. He will there see the two forms as clearly exhibited as could be wished, *i. e.* the simple ringed incrustation on the skin of the temple, where no hair grows, and several patches on the scalp, marked by the destruction of the hair.

and the circular cupped scab, as well on other parts of the body as on the scalp.

The other form of the disease to which I have alluded never assumes the circular figure just described ; on the scalp it is pustular from the beginning, and marked during every stage of its progress by a much greater degree of irritation and itching. It is so generally diffused over a considerable space, even on its first appearance, as to warrant a designation founded on this feature, in contradistinction from the foregoing, which is so constantly circumscribed.

Like the latter it appears to be readily identified with an affection of the skin of other parts, which is in part vesicular, but chiefly consisting of papulæ of different sizes. These not being of much importance are usually little noticed ; but as soon as the disease appears on the scalp, alarm is immediately communicated to the parents, or others connected with the children who are the subjects of it.

At this period much itching and irritation are found to exist. The pustules are very thickly dispersed over certain parts of the head, *every individual pustule having a hair growing through its centre,* and the scalp in the insterlices being excessively red and inflamed. The child is feverish and irritable, the digestive organs evidently disordered, and a generally bad state of health will be found to have existed for a considerable length of time.

The absorbent glands at the back of the head and those of the neck are enlarged and tender, and in some neglected cases have proceeded on to suppuration ; but this is by no means common. Small abscesses form here and there from the inflammation of the cellular membrane under the scalp, which, in a few days, discharge their contents, and heal ; the spots which they occupied remain in some cases ever after completely bald, the adipose structure secreting the hair having been destroyed.

As the pustules become ruptured, and their contents distributed over the adjacent parts of the scalp, these parts become inoculated, the disease spreads, and yellowish scabs are formed, of an unpleasant odour and aspect, which, unless frequent ablution be had recourse to, rapidly accumulate.

This diffused or pustular form of porrigo, is chiefly found among children of dark hair and unhealthy constitutions. It is not attended with immediate loss of the hair, like the former ; but this event takes place if the parts be frequently washed in a short time.

From the greater quantity of fluid secretion occurring in this form, the power of infection, and rapidity of extension, is considerably greater than in the circumscribed species. The manifest increased susceptibility to irritation of the skin of the individual also favours this difference : not unfrequently, however, under an improved state of the general health and frequent ablutions of the part, assisted by sedative applications, the disease subsides into a condition much resembling that first described ; the hair exhibiting the same characters, and a very small number of pustules appearing at intervals amongst it.

The foregoing descriptions comprehend every thing essential to the history of Porrigo (except as regards the P. favosa and P. larvalis) as it chiefly occurs in the better classes of society in England, when not of long standing, where cleanliness is particularly attended to, where no stimulating improper applications have been made instrumental in changing or aggravating its character, and where the general health of the patient is not too bad to admit of ready correction : and they will frequently, where strict attention to ablution be persevered in, terminate spontaneously. Under other circumstances, both the above forms are liable to terminate in that most obstinate and intractable one designated by Bateman P. furfurans, and admirably described by Alibert as Teigne furfuracée and amiantacée. The pathological changes necessary to produce this condition are seldom brought about very speedily, but seem rather to be the result of long-continued irritation. Its principal distinguishing feature is the copious production, and rapid exfoliation, of morbid cuticle. The branny form of this substance greatly favours its entanglement by the adhesive fluid of the pustules, and they speedily unite to form the cement described by M. Alibert.

It will be observed, that in the commencement of the forms first mentioned, very little scurf is produced. A pustule forms, which breaks, and leaves a slight scab, easily separable; but the vessels which secrete the cuticle of the surrounding parts partake but little of the inflammatory action. In long standing cases, however, this inflammation ceases to be confined to the solid structure of the cutis, and to have all its energies concentrated in the formation of a pustule. One of its most important functions, the formation of its natural defence, the cuticle, becomes deranged : the latter substance is produced in immense quantities, losing in strength and utility what it obtains in quantity, and contributing largely to that character from which this condition takes its name. The exfoliations of cuticle uniting with the contents of the pustules, become the means of increased mischief, by matting the hair firmly together and impeding the measures likely to be of use. The scalp cannot be cleared without great pain and difficulty. It is useless to attempt shaving, for the razor passes through this covering no more readily than through half dissolved glue, and the scalp is liable to be sliced and cut in every direction.

When by the use of warm water applied for a considerable length of time, and at the expense of some pain, the scalp is cleared and examined, it exhibits not only the erythematous redness and utter privation of cuticle described by Alibert, in the interstices of the hair, but at a considerable enlargement of the passages, by which the latter find their way to its surface. The covering which the hairs receive from the cuticle is also destroyed, and in its place, partly filling up the enlarged orifice in question, a glutinous fluid may be seen exuding, and surrounding each individual hair. The quantity of this fluid which may be secreted depends much on accidental circumstances, and is greater at some periods than others. The proportion which it bears to that of the exfoliations of cuticle determines the consistence and adhesiveness of the covering of the diseased part ; and hence, when small in quantity, the latter is more dry, harsh, and shining ; a difference which, it will be seen, constitutes Alibert's T. amiantacée.

In this state of the disease, and also under all other circumstances where the accumulated secretions are considerable in quantity, the term " Scalled head" is generally applied. It is also sometimes used with respect to the porrigo favosa, which will be more particularly described in another place. The extreme rapidity with which the yellow scabs of the latter accumulate, may appear indeed to give it an equal claim to this designation ; but it is much more easily removed, and usually, if cleanliness be attended to, terminates spontaneously. With the assistance of a little constitutional management, even in the worst cases, it is but of short duration. The idea attached to the term, in its vulgar acceptation, moreover, is that of a disease of a tedious intractable nature, which rarely, if ever, terminates of its own accord. Its application, therefore, to the P. favosa is obviously improper.

On the subject of the causes of the different forms of porrigo but little requires to be said. With reference to the species first described as connected with the small herpetic ringworm of the skin of parts not covered with hair, I am able to state, that little, if any, constitutional derangement is discoverable at its commencement. In a great majority of cases, the infection can be very satisfactorily traced from one child to another, whether occurring in families or in schools ; and it frequently spreads in the latter to a considerable extent, without any symptom of other disorder accompanying it. That it originates spontaneously in some cases is unquestionable, and the same remark may be made of the pustular form.

When the disease appears on the skin of an individual in the form of pustules, the constitution is importantly concerned ; and, as has been already stated, the chief remedial measures will be such as are directed to the improvement of the general health.

The pathology of the disease has been very little understood, and in its treatment recourse has been had to an infinite variety of medicinal applications, not one of which appears ever to have been employed on known scientific principles.* As might naturally be

---

* It would be useless to trouble the reader with the detail of the particular effects of these supposed remedies, which, from time to time, have been earnestly recommended ; for the majority of the best informed practitioners of the present day will be disposed to attach equal importance to them all ; experience having taught them that none have uniformly succeeded in producing a cure, or even temporary alleviation. As a matter of curiosity, however, it may be as well to enumerate them ; and it may happen that some future cases, under particular circumstances, may be benefited by their use.

The ung. flor. zinci—ointment of the cocculus indicus, in the proportion of two drachms of the powdered berry to an ounce of lard—equal parts of sulphur oint-

expected, the patience of the practitioner, and that of the friends of the patient, are in the end mutually exhausted ; and the disease will often continue for years, under these circumstances, alternately improving and getting worse.

From the earliest periods to which its history can be traced, the alopecia, or falling off of the hair, has been constantly noticed ; and it has been as uniformly observed, that where the part has been entirely divested of hair, the disease seemed to have disappeared. In the more advanced stages, such as those of Alibert, whose descriptions we have quoted, it invariably happens that where by care and management the diseased secretions are cleared away, almost every individual hair will be found insulated by a pustule. That the irritation excited by this operation produces a more active inflammation of the skin, there is no doubt ; and perhaps for a time it increases the number of the pustules, but the latter preserve their peculiar character, and are never seen beyond the margin of the hairy scalp.*

ment and soft soap intimately mixed, and used as soap in washing—ung. hydr.—ung. hydr. nit.—ung. hydr. nitrico oxyd.—tar and sulphur ointments—ung. acidi nitrosi—lotions of solution of potash and of muriatic acid—ointment of calomel, acetate of lead and opium, of hellebore, turpentine, mustard, stavesacre, black pepper, capsicum, galls, rue, &c. Lotions of sulphate of zinc and copper, of equal parts of vinegar and oil, of oxymuriate of mercury, and of argenti nitras. The application of tinct. ferri mur.—blisters.

The whole of the foregoing have been noticed by Willan and Bateman ; and Alibert seems to have a decided predilection for sulphur, both externally and internally, in this, as well as in all other cutaneous diseases.

In the fourth volume of the Med. Repository, the use of common adhesive plaster to cover the spots of ringworm is strongly recommended by Dr. Clanny, as being capable of itself of effecting a cure. Messrs. Simmons and Bell, in vol. xiii. of the Med. and Phys., give a similar account. In the fifth volume of the Repository, Mr. Bidwell speaks highly of stimulating applications, such as strong savin ointment, ung. hydr. nitr. with the addition of nitrous acid, and recommends occasionally varying these applications where any one in particular should fail in doing good.

In the thirteenth volume of the Med. and Phys. Journ. a communication from Mr. Low rests chiefly on the efficacy of internal alterative medicines ; but in the cases which he has detailed, it is evident that more than ordinary attention to ablution of the seat of the disease was had recourse to. In the succeeding volume lotions of kali. sulph. are recommended by Mr. Barlow ; and a subsequent page of the same volume contains a terrific record of the effects of tobacco in the form of strong infusion to the scalp to cure a case of long established scalled head, which will certainly act as an admonition to all who read it.

More recently, the periodical works, on general as well as medical subjects, have circulated a statement in favour of tan water, and water obtained from the gas manufactories, as lotions. The result of my own experience in such applications as the foregoing enables me to see as little prospect of doing any good with them as hope of eliciting a sound principle of treatment from the chemical analysis of the matter of Tinea of the French chemists.

* The observation of this fact appears to have led Dr. Underwood and others to the opinion, that a morbid state of the roots of the hair was the cause of the mischief, and hence we find them recommending the use of the pitch cap. Turner and Alibert both entertain a favourable opinion of it also, and in old standing cases there is no doubt that it has been of signal advantage.

Alibert, in his fourth article on " Les Teignes," after noticing the opinions

It is evident from these facts, that the hair is materially concerned in the production of the phenomena of the disease as it occurs on the scalp. In the first described form, we see it on other parts of the skin, of mere ephemeral existence, either disappearing spontaneously, or removed by the application of any simple astringent. In the second, any sedative application, if combined with a light aperient, will be adequate to its removal ; but if in either case the scalp happens to be affected, the disease evinces a considerable degree of obstinacy : and, as respects its management, becomes entirely altered.

The rapid formation of pustules, as they occur in old standing cases, may be satisfactorily traced to the influence of the hair on the part. The latter, in passing through this structure, irritated and inflamed as it is, become in effect so many extraneous bodies ; and in this character each hair may be supposed to be very capable of producing suppurative action sufficient for the formation of a pustule. Their influence in the withered and blighted state first described, where no distinct pustules are perceived, is still of the same nature ; thus continuing to aggravate, though not so extensively, the cutaneous inflammation constituting the original disease. When pustules are formed, they seldom penetrate the cutis, and never extend to the adipose structure secreting the hair ; they cannot consequently be the chief cause of the destruction of the latter. Furthermore, the nourishment of the hair appears to be in all cases cut off, or materially interrupted, soon after the inflammation of the cutis is established. The manner in which the latter operates in producing this effect, would seem to be best explained by reference to the known effects of scarlatina on the hair. In scarlatina the cutis is for a considerable length of time the seat of great inflammation and congestion ; the adipose structure, situated immediately under it, whose office it is to nourish the hair, is inadequately supplied with blood for this purpose ; and hence the hair drops off to a considerable extent, in all cases where the attack has been severe.

That this view of the pathology of the disease is correct, is proved by a variety of circumstances which occur in the history of almost every individual case. It would indeed appear that after a time its specific character and power of infection is entirely lost, and that it subsides into the mere effects of the irritation of extraneous substances on diseased parts.

alluded to, of the origin of the disease in the roots of the hair, attempts to prove that the part most importantly affected is the rete mucosum. He, singularly enough, selects his arguments from the history of his T. muqueuse (*the P. larvalis of Willan*, &c.) and remarks, that the occurrence of this affection and T. faveuse (*P. lupinosa*) on other parts of the skin than those covered by hair, prove that the phenomena of all the forms of what has been termed Porrigo, or Teigne, depend on derangement of this structure. To say nothing of the doubts entertained by some of the best anatomists as to the existence of the rete mucosum, it will still be observed, that in a classification of the crusta lactea (or, as I shall designate it, Impetigo larvalis) with such a disease as that described in the preceding pages, shows but little attention to the ordinary rules of nosology. As far as I have been able to observe, no important analogy could be traced between them.

The well known effects of the use of the pitch cap in eradicating it may be adduced as another fact supporting this opinion ; for it is obvious that the extraction of the hair by the roots is the only effect which such an application is capable of bringing about, the direct result of this being the subsidence of inflammation and the disappearance of pustules.

I have had some opportunities of the post mortem examination of the scalp of subjects who have died from acute attacks of vital organs, while suffering from old standing porrigo, but have not been able to discover any thing like organic derangement about the adipose structure, or the bulbs of the hair beneath the cutis. Where the hair has dropped off, or appeared to have been extirpated by the disease, the adipose substance was evidently wasted away in consequence of the long prevalent excessive irritation in the superincumbent cutis. In other portions of the scalp, where the surface had remained unaffected and the hair grew strong, no deviation from health in the parts beneath could be distinguished.

Treatment of the first or common circumscribed form.

On this part of the subject under consideration, it may not be amiss to revert to the opinions of Alibert, though the observations we shall have to offer will be somewhat at variance with them. He observes, that the different eruptions which appear on the scalp have a manifest tendency to the preservation of the animal economy, and that it becomes a matter of important consideration whether proceedings directed to the removal of the disease under consideration can be adopted with safety to the constitution. He states also, that cases have occurred where the disease has been suddenly suppressed, of the instantaneous appearance of more formidable affections of internal organs. In truth, his opinions appear to coincide with those of the older authors (some of whom he quotes as to the suppression of long continued discharges) most completely.

When it is considered, however, that these observations are applied under a general head to all the forms of " Les Teignes," that is to say, to the crusta lactea, or T. muqueuse, a disease attended with excessive discharge and irritation alike with the dry and inactive forms which have been mentioned, and which spread chiefly by infection, it is obvious that they should be received with some modification. Undoubtedly the sudden suppression of the former would, if it could be accomplished, be followed by mischiefs to some vital organ ; but, as regards all those conditions which have formed the subjects of the preceding pages under the head of Porrigo, I am able to state that slight attention to the general health is quite sufficient to avoid any bad consequences from the suppression of the local disease.

It not unfrequently happens that the conformation of the child who is the subject of it exhibits marks of great delicacy ; that there may be glandular disease partly developed in different portions of the absorbent system ; that habitual disorder of the digestive organs exists, occasionally becoming aggravated and leading to febrile symptoms, and great increase of pustulation and discharge from the scalp.   If such a state of things exist, common sense would point out the propriety of considering the amendment of the constitution previous to the adoption of decided local treatment.   If any thing like hereditary disposition to phthisis can be traced, I believe there is danger in the removal of *any cutaneous affection of an active character ;* and as regards the scalled head, or that state of porrigo where much discharge and scabbing have existed, I have observed many instances, when it has been suddenly removed, where consumption has appeared to follow as a consequence.

On the contrary, if no indications of a weak part in the system are to be traced,—if the child's health is good, or can be made so by a little constitutional management, no mischief is to be apprehended, however speedily the removal of the local affection may be effected.

If no constitutional disorder be obviously present in the case of the form first described, that is to say, where the characteristics of the well-known ringworm of the scalp exist, whether accompanied or not by spots on other parts of the skin, the best and most effectual application is that of one of the undiluted mineral acids.   Of the three, perhaps the preference may be given to the sulphuric : and I have been accustomed, whenever a case has been brought to me in an incipient state, to direct the removal of the whole of the hair of the scalp by shearing, or in other words, cutting the hair as close as can be done with a pair of scissors, and then to apply this fluid lightly on every spot which could be discovered by means of a feather.   It should be suffered to remain on a few minutes, and should be carefully spread over the whole of the diseased spot, and a few lines perhaps beyond its utmost boundaries.   It usually produces a good deal of smarting, and a slight blush of redness on the surrounding cutis in a very short space of time ; and when this is observed, a sponge dipped in warm water should be made use of to clear away what remains of the acid, otherwise it is apt to affect the scalp too deeply.   When applied in the above manner, the cuticle is evidently destroyed by it, and the vessels of the diseased cutis are much excited.   In a few hours a tolerably copious exudation of coagulable lymph is produced on the spot, which forms a scab more or less thick.   No appearances are now observed of the circular spreading irritation of the original disease, and in a few days the scab dries and separates, and brings with it the remaining unhealthy hairs of the part.   A bald spot somewhat reddened is now observed, the colour of which soon disappears, and new healthy hairs begin to spring up and speedily cover the part.

This application may be resorted to with a certainty of eradicating the disease, in almost all cases where the health is good, and where the disorder preserves the chronic circular form.   Even if there be a few pustules distinguishable within the areola, it may still be safely relied on.

It frequently happens, however, that when an attentive examination of the scalp has been instituted in the manner described, and the acid applied to every individual spot then appearing, the lapse of two or three days discovers others in distant or contiguous parts, which are apparently the results of the application of the infectious virus from the originally diseased parts before the acid was applied. No discouragement should be felt from this circumstance, for the application of the acid puts an immediate and effectual stop to their spreading. Frequent recourse should be had to ablution during the day, and on all convenient occasions, in order to clear away the contents of any minute pustules which may be formed, otherwise the extent of the mischief and difficulty of subduing it will be materially increased.

The manner in which the acid acts in the removal of the disease, may appear to be explained either by its excitement of a new and different action in the vessels of the surrounding part, by its destruction of those superficial vessels which were the original seat of the disease ; or, by producing an adhesive scab which entangles, and when it drops off completely brings with it the hair of the part, the influence of which in keeping up the disease has been already described.

Treatment of the Pustular or diffused species.

With respect to the pustular and more diffused form of the disease next described, it may not be amiss to premise, that any attempts to eradicate it by means like the foregoing, would invariably be abortive and mischievous. The state of the general health will, in most cases, require the first attention, and the local applications will be such as are most calculated to subdue irritation and promote cleanliness ; to remove as speedily as possible the contents of the ruptured pustules, and prevent their dying on the part and matting the hair together, and thus increasing the local mischief. These measures constitute nearly all which can be done in the shape of local applications.

On the subject of the constitutional disorder, little need be said here ; the cutaneous disease is one of the Protean results which follow various deviations from health in the actions of internal organs. In cases where infection from the other species could be traced, it has either derived its aggravated character from a general unhealthy disposition of the constitution, combined with scrofulous affection of the absorbent glands, or a preternatural disposition to irritability of skin. In many cases it has appeared spontaneously, where the child has been ill fed and otherwise badly managed ; but it is also sometimes observed where these circumstances have not occurred. It is, however, by no means seen under such circumstances as would lead to the opinion of its having any salutary influence to exert, or office to perform in the animal economy, like what has been called the Porrigo favosa.

If the pustular or vesicular patches of eruption, mentioned as often occurring in conjunction with the pustules on the scalp, be numerous and diffused over distant parts of the body, the warm bath should be used every evening, and the determination to the skin which is induced by the latter should be afterwards encouraged by putting the patient into a warm bed, and exhibiting some warm diluent drink. Such measures will speedily, when assisted by proper internal medicines, subdue the irritation of the skin on which they depend, and diminish the chance of the application of the virus to other parts.

If the slightest accumulation of the diseased secretions be perceived on the scalp, it is to be carefully removed, together with the matted hair growing on the spot, by means of finely-pointed scissors. Indeed the complete removal of the hair over the whole scalp is almost always requisite : for while it remains, the chances of the appearance of fresh pustules in other parts are very great. The more expeditious manner of removing the hair, recommended in the dry circular disease by shaving, however, is not here to be thought of, as it excites great inflammation and produces so much pain as to render the child incapable of submitting to have it afterwards properly managed. The careful employment of finely-pointed scissors is free from these objections, and the hair may be removed quite as close to the scalp by them as is necessary for all useful purposes.

When this has been accomplished, the best applications for the first few days will consist of ablutions by water, heated to the highest degree which the patient can bear with comfort. These are not to be employed merely as conducive to cleanliness, but as the means of subduing the excessive irritation of the scalp also ; they should, therefore, be repeated several times in the course of the day, so as to keep the diseased parts constantly under their influence, and at the same time to clear away the irritating contents of the pustules as fast as they are discharged.

While this process is regularly attended to on the scalp, the internal management will require to be conducted according to the constitutional condition of the patient. It has been before observed, that in the larger portion of cases of this kind a debilitated state of system, and very often mesenteric disease will be found to exist. To enlarge here on the medicines requisite to remove such a condition would be unnecessary ; they are very well known, and require only to be judiciously selected. The best kinds of diet and regimen are also equally well understood, and laid down in books of authority.

As a tonic remedy, where mesenteric disease is suspected, I have been in the habit of administering the hydr. oxym. in doses of from one-eighth to one-quarter of a grain in tincture of bark three or four times a day, according to the age of the child. A few grains of hydr. c. creta, with an adequate quantity of rhubarb, being used about twice a week.

It will, however, now and then happen that a full habit of body and apparently robust health accompanies the eruption, in which case the method of treatment will be as obvious as in the foregoing. In truth, existing circumstances, and the history of each individual case, are the only correct guides in the constitutional management.

Treatment of confirmed or old-standing Cases of Porrigo or Scalled Head.

Cases of this condition of the disease are much less common than formerly in England. This may be considered simply the result of the progressively increased means of comfort and cleanliness which the lower classes of society have experienced. With the exception of the Porrigo furfurans, or Teigne furfuracée, none of the forms described by the older authors, or by M. Alibert, are often seen in the more educated classes, and it seems proper to leave this for subsequent consideration by itself. As has been already observed, it is the form which those which have occupied the foregoing pages terminate in when of long duration, when subject to constant stimulating treatment, and where frequently repeated ablution has not been thought of sufficient consideration during its progress.

The "scalled head," or that state of disease which it will be proper to consider as best entitled to this term, I should be disposed to describe as follows :—Diffused clusters of pustules, appearing in some cases on one part of the scalp only at a time ; sometimes in the form of patches on different parts, and always accompanied by much inflammation and itching. When the pustules of one part have broken and discharged their contents, the inflammation subsides and some of the hairs appear to drop off. Soon after, other parts exhibit the same appearances, and the pustules follow the same course of increase, maturation, and decline ; always, however, leaving behind them some inflammation and a little scurfy exfoliation. In the course of time the disease traverses every part of the scalp over and over again, and the consequence of these repeated attacks of inflammation is in the end a fixed inflammatory state of the cutis, which continues to produce at intervals over the whole scalp repeated returns of pustulation. During the whole period of its continuance, every individual hair appears situated in the centre of a pustule ; while, if the hair on any particular spot happens to have been extirpated, the pustules cease to appear. The vexatious disappointments, experienced both by the patient and practitioner from time to time, ultimately brings on despair of recovery from legitimate means, and the assistance of quacks is called in, often to render the mischief more serious and obstinate, it is true, but sometimes to the complete removal of the disease. The anathemas of Willan and Bateman seem to have completely exploded the use of the pitch cap and every thing else acting as a depilatory : and Alibert, while he admits that nearly the whole of the cases he has

seen recover have been cured by similar means, is also very energetic in its condemnation.

The first-mentioned authors are of opinion that depilatories do
more mischief to the scalp than the disease, if left to itself, will
effect in years; but they have not made it appear how such mischief is
produced by remedies of this sort.   Is it supposed to be by retarding the growth of healthy hair on the part?   It is plain that such is
not the fact ; for the adipose structure beneath the cutis, from which
the hair derives its origin and support, sustains no mischief even if
every hair in the scalp should be dragged out by the roots : consequently the growth of new hair is certain.   Is it because the disease
is aggravated, and a greater number of pustules produced ?   No
person, who will take the trouble of making the experiment, can
fail to observe that the reverse of this is the case ; for when the hairs
are removed, pustulation ceases, and the inflammation gradually and
sometimes rapidly subsides.

In this tedious and troublesome form of disease we have not a
choice of remedies ; fortunate, indeed, may it be considered that
one should be known capable, if properly directed, of subduing it,
or one principle of treatment found to be generally beneficial.   The
offensiveness of its external appearance, its seat being a part constantly exposed to observation ; the total destruction of the chief
ornament of the countenance, the hair, which it occasions ; the
horror entertained of contact by every person, together with the
exclusion from places where education can be best obtained, make
the majority of cases truly pitiable.   It is for those who know not
the distress and misery which tedious cases excite in the minds of
sensitive parents, to speak unheedingly of the *horrible torture of
extracting the hair from the diseased parts* as a means of cure.
In truth, they must know very little of the mode in which this ought
to be effected, and of the slight degree of uneasiness attending it, to
be able to think it improper or inefficient.

It is not the pitch cap, however, but a more discriminating application of the principle upon which this apparatus acts, which I
am here desirous of advocating ; and I am fully warranted in stating,
that where the disease has been long established, and that state of
chronic inflammation of the skin before described has taken place,
the removal of the hair affords the only hope of recovery.   In such
a case, even if no cause existed to keep up the irritation, it should
be recollected that the latter, by its length of duration, has become
almost identified with the natural action of the part, and would
require much time to subdue it.   What then would be the difficulty
when the additional excitement of extraneous substances is allowed
to remain in full operation?

In the hospital of St. Louis cases remain for years without permanent improvement, and seem ultimately to get well without any
obvious reason, unless the thinning or destruction of the hair be
considered such.*   Not unfrequently, however, they have recourse

* Alibert himself affords exceedingly good evidence in support of these observations.   He says, "I knew a man who was ignorant of our art, and who em-

to the calotte,* as it is termed, of their own accord, and succeed in curing the disease. They are perfectly aware of the manner in which the calotte operates, and persevere in its employment with a great deal of patience and resolution till they have extirpated the hair.

The inconveniences of the calotte, however, are of a serious nature. In the first place, it is applied to the greatest part of the scalp, where perhaps no pustules may exist nor any inflammation have taken place. Under such circumstances the tearing it off must be horribly painful, as the sound hairs are adhering to it quite as firmly as those connected with the pustules of the disease.

The most judicious plan of proceeding is as follows.—The hair over the whole scalp should be carefully removed by scissors, the razor not being admissible on account of the irritation which follows its use. When pustules are discovered, poultices should be kept applied during the night, not so much for the purpose of subduing irritation, as promoting free suppuration round the roots of the hair, and thus loosening their connexion with the cutis.

They will in this state be found to come away very readily, and with little or no pain to the patient, on the application of a pair of forceps. The latter may be broad at the point, if the pustules are very numerous, and the operation will be found neither one of pain to the patient or much tediousness to the operator. Any person may be instructed how to perform it, and a short space of time occupied in it every morning will soon enable us to discover an important improvement. If followed up with attention and perseverance, it will rarely fail to subdue the disease, and to be followed by new and healthy hair.

In conjunction with the above management, the frequent application of spirituous evaporating lotion to the scalp, where the hairs have been removed, will be found of signal advantage. During the day it ought not to be omitted for any length of time, and is well calculated to subdue any little increased irritation which the extraction of the hair may have produced.

The constitutional treatment, as in other cases, must depend on the apparent state of health of the patient. Good general health ought to be secured by appropriate remedies before the above plan

ployed a topical application, which he persisted in keeping secret, but which I nevertheless discovered had lime in it. He succeeded in curing seventy individuals in six months. The means he employed were very simple; he confined himself to rubbing the effected part with it (the scabs or scales having been cleared away by poultices). What happened? The *fallen hair* was replaced by more, first of a pale, then of a deeper colour. The disease was cured." He mentions a variety of other applications which have often been had recourse to here without effect. The person here alluded to was, most probably, M. Mahon.

* A plaster made of strong vinegar, rye, meal, and pitch, which is spread while hot, and applied in the form of a cap to the head. After remaining on three days it is torn off with violence, bringing with it great numbers of hairs by the roots.

of local treatment be instituted, otherwise impediments will be likely to occur to its continuance. A robust or full habit of body is however as unfavourable as an opposite state, if the latter approach to any thing like debility. Stimulating diet and wine should at all times be avoided, as being liable to bring on increased itching and pustulation.

The disease, as I have observed in repeated instances, becomes suddenly aggravated by attacks of fevers, particularly those of such kinds as children are most subject to, as scarlatina, measles, mumps, &c.; during the continuance of any of which, it will of course be impossible to pursue the treatment above described.

Treatment of the Porrigo Furfurans.

The description of this form of the disease given in a preceding note,* is sufficiently correct to render it unnecessary to solicit the reader's attention to any thing further relative to it than its treatment; for although Alibert, Willan, and Bateman, have described it as always originating in the form of pustules mixed with a superabundance of scurf; the most unmanageable cases, and these are by far the most numerous, are those which are preceded by the bad management or neglect of one or the other of those already described, or of perhaps a tedious case of pityriasis. The accidental occurrence of disturbance of the system, or of some eruptive disease, while a scurfy state of the scalp has existed, has been indeed in several instances which I have observed, the apparent foundation of the disease in its worst form.

A very superficial view of the part on which it is situated will enable us to see that local applications afford little chance of benefit, the compactness with which the hair is matted together by the mixture of scurf and adhesive fluid secretion prevents a possibility of the cutis being affected by them, while the apparent state of system leads us to no indication which may be of service through the medium of the constitution.

There are various confident statements on record as to the effects of particular applications (one instance is that of Dr. Hamilton, brought forward by Willan respecting the cocculus indicus used as an ointment), which a diligent trial of has been uniformly followed by disappointment.

In private practice, now for some years tolerably extensive in this class of diseases, I have seldom found the parents or friends of the patients sufficiently persevering to preserve the scalp free from the accumulation of scurf and adhesive secretions, to allow of the fair and full influence of local applications; but where this desirable point can be attained, considerable benefit will arise from sedative cooling lotions, provided these are employed diligently, and assisted

* See p. 91. Teigne furfuracée, or P. furfurans.

by the judicious use of mild alteratives and aperients. I must in candour, however, confess, that except where the operation of plucking the hair from the diseased part has been diligently followed up, and every other possible means taken to check inflammatory action, a perfect restoration to healthy action has seldom been produced.

In a few cases these measures have been followed by complete success ; but the majority have, from the circumstances mentioned, been ultimately resigned in despair. It is a curious, though very consoling fact, however, that a spontaneous disappearance of the disease occurs in nineteen cases out of twenty, before the age of puberty arrives.

My experience of the last fact has hitherto deterred me from the use of the lime ointment, and some other severe applications, mentioned by Alibert and others, with the view of eradicating the hair, the possible mischief which may arise under their employment being more than sufficient to justify us in waiting for such a result. I have been, moreover, frequently consulted by the friends of children, who have been sufferers for many years, and who, finding little relief in this country, have been actually taken to Paris for the purpose of being submitted to the care and treatment of the French professors. Among a great number of these, I have not yet heard of one who derived any advantage from their treatment.

SECTION II.

On diseases chiefly marked by chronic inflammation of the vessels secreting the cuticle, producing morbid growth of this structure, and generally dependent on debility of system.

CHAPTER I.

LEPRA.

THE most important and most common of the diseases coming under this head, is Lepra.

Many preceding authors on the subject of cutaneous diseases have experienced considerable inconvenience in speaking of Leprosy, from the confusion of the older writers, who seem to have applied the term to endless varieties of disease of the skin, which do not appear to have the slightest to that which is properly so named.

In England, Turner, Willan, Bateman, and other authors, have laboured successfully in clearing up the difficulty ; and Alibert, though still retaining the term in connexion with his description of Elephantiasis, is not unaware of the material difference between the latter and the true Lepra.

An attentive perusal and consideration of the descriptions of diseases of the skin, which are to be found in the works of ancient authors, leads us, as in most other subjects of scientific research, to the conclusion that but little really valuable practical information had been obtained by them, either as regards their diagnosis, pathology, or treatment. In the case before us, the details of the worst forms of all diseases of the surface appear to have been constantly selected as the foundation for a description of Lepra, as if the latter comprehended only what was disgusting or terrific in appearance, or incontrollable by any known scientific means. Hence it is, that Heberden and Cullen have spoken so strangely of it, each of whom declare their disbelief of its frequent occurrence. Alibert also observes, that it is not often seen within the circle of his observation, though a more careful perusal of his works would certainly lead us to an opposite conclusion. In truth, this author, as will appear on further consideration, has in some measure followed the example of the ancients ; for he applies the term to severe cases of other diseases, while the true Lepra is found buried beneath a mass of incongruous matter under the head of Dartres.

By a strange misunderstanding, probably arising from the circumstances above mentioned, Leprosy has been confounded with Elephantiasis : no two diseases of the skin however can possibly be less alike. The disease of the cutis which exists in the latter, moreover, is of infinitely less importance than that of other and more deeply-seated parts, in making up the characters of this malady, and partakes but little in the diseased actions which give it its distinguishing feature, viz. enormity of growth. Notwithstanding what we have above remarked, a sufficiency of evidence appears to have been extracted from the confused mass of materials alluded to, to justify the opinion, that by the term Leprosy, a disease of the cutis only was originally meant, terminating sometimes, perhaps, unfavourably, in unhealthy sores, or spreading sloughy ulcerations, marasmus, and decay of mental and bodily strength ; but in the majority of cases, where cleanliness may have been attended to, in a slow and gradual return to health.

The abhorrence in which the unfortunate subjects of Leprosy were held by their fellow-creatures in the earlier periods of history affords no argument against this comparatively light view of the disease. It should be recollected, that in those unenlightened times, the power of infection was supposed to be vested in any person, however slightly affected ; and that, besides, inasmuch as it prevailed chiefly among the lower classes, it would necessarily be considered as bringing with it as much of disgrace as of misfortune. In addition to these considerations, the connexion which the sacred writings seem to have established between the general prevalence of the disorder, and the incurrence of the divine wrath by the people, would have led the latter to the exclusion of lepers from their society, even though the disease, pathologically speaking, may have been exceedingly trivial.

In describing the disease as it is now seen in this part of the world, it is certain that we shall be obliged to employ language materially different even from some modern authors to whom we have alluded ; and indeed there seems to be little doubt that almost all diseases of the skin are less frequently occurring, and less formidable and disgusting in their external characteristics here, than those observed among our continental brethren.* Even in this country, however, and under its mildest form, Lepra is troublesome and tedious in its course, and attended with much unsightliness and irritation. Changing, when for the better, in a manner the most slow and imperceptible, and not unfrequently accompanied with a proportionate degree of despondency and mental distress.

In the description of Lepra it has been customary to divide it into three species ; Willan and Alibert have both adopted this plan. The former employs the terms L. vulgaris, L. alphoide, and L. nigricans. The latter has la Lepre blanche, L. noire, and L. Tyrienne, as varieties of his Lepre squameuse. His descriptions of Lepre crustacée and tuberculeuse, apply to what we better know as varieties of psoriasis, ecthyma, rupia, &c. and Elephantiasis ; and will obtain attention in another page. With reference to Willan's terms, it may be necessary to observe that his two first species are merely different degrees of the affection, or different stages of its progress, while the L. nigricans, that designated by Alibert as L. noire, is the result of a scorbutic state of the system operating in combination with the ordinary causes of the cutaneous disease.

It appears to me, that to describe Lepra in the simplest and most correct form, these terms are not at all necessary ; and that they may therefore be dispensed with with advantage, inasmuch as they have had their share in creating the confusion elsewhere alluded to, and discouraging the student in the prosecution of his inquiries.

It is to little purpose that preceding authors have expended so much time in investigating the confused records of ancient times, to determine what was meant by the term, and to ascertain the correct history of the disease, if new difficulties in its study are to be invented by encumbering it with useless and multiplied names.

The understanding in which the term is at present received is that of a disease exhibiting, on superficial notice, red inflamed patches, from which extensive and rapid exfoliations of scales of morbid cuticle are constantly taking place, but on which no appearance of vesicular or pustular formation ever occurs. Dr. Willan thinks it is sometimes caused by indurated papulæ originally springing from the true skin, and which, by their elevation and extension of the cuticle,

* Alibert, in alluding to Dr. Willan's account of Leprosy, observes, "M. Willan dit avoir observé plusieurs espèces de Lèpres en Angleterre, mais ces lèpres ne sont autre chose que des dartres, auxquelles cet auteur a imposé des nommes qui ne leur conviennent pas," a mistake which the above observation explains.

produce some injury to this structure. This is, however, probably mere matter of conjecture, as the eruption of Lepra is rarely, if ever, preceded by diseases of a papular kind.

It is impossible to confound this disease with the scabs formed by the exsiccation or drying up of pustules or vesicles, if any attention be paid to it at its commencement. It is not meant by this assertion to deny that such mistakes have occurred, as stated by Dr. Willan ; for in truth it may be asserted of almost all cutaneous diseases, that the eye of the careful pathologist has not been directed to them till very lately ; and we are not therefore to be surprised at their having been so little understood by ancient writers.

The appearance which Lepra assumes at its commencement is that of round red spots, elevated above the surrounding skin, and generally not larger than a split pea. If the finger be drawn over it, a degree of stiffness, or absence of the natural flexibility of the skin, is distinctly observed. In a day or two, if a minute examination be instituted, the spot is found to have a glossy hard surface, a perfect semi-transparent scale having been formed, the surface of which is smooth and polished. In a short time this scale separates ; when the site which it occupied exhibits no variation of colour, but a considerable degree of roughness and irregularity. The examination of the inner surface of the separated scale enables us generally to discover in or about its centre a minute protuberance, considerably softer than its bulk, and which has evidently occupied a corresponding hollow or excavation in the denuded surface. If force has been used to separate the scale a speck of blood usually occupies this excavation, and the point described as the centre of the inner surface of the scale is also similarly discoloured. As the diseased spot enlarges, fresh scales are produced, but these are entirely dry, and do not exhibit the appearances on their inner surfaces which belong to the original scale of the incipient disease above described. Neither are they found uniformly to extend over the whole diseased spot, in the form of one continued scale ; as in this state, they separate partially and in patches, their dried edges giving a whitened, scurfy appearance, in some situations, where they are partially detached.

As the number and size of the spots increase, the limb, of course, becomes gradually increased in a shield of scale ; and those in the neighbourhood of joints consequently become very troublesome, producing a considerable restraint on the motion of the parts.

There is one circumstance peculiar to this disease, which has been particularly mentioned to me by the patient, namely, a sensation of pricking, most frequently noticed a little before the separation of the first scale, when perhaps it has scarcely attained the size of a spangle. I am inclined to think this sensation the consequence of the raising up of the edges of the scale produced by the tumefaction and elevation of the inflamed margin, and fresh growth of scale ; the centre which was attached to the cutis being thus forcibly torn from such attachment. This conjecture is rendered more probable by the fact, that when the disease is obviously subsiding, when no new scales rise up, and thrust those which were before formed from their attachments, the pricking in question js no longer felt.

In cases of long standing, and which have been much neglected, the glands secreting the nail are sometimes affected by the diseased action ; in consequence of which the formation of these appendages of the cutis is incomplete, and they exhibit appearances at their roots resembling the deposit of specks of matter in their substance. Sometimes, also, they act as extraneous bodies on the parts to which they are appended, producing much aggravation of the irritation, and a fluid discharge. I believe these appearances, however, are not very common in cases originating in this country. I have, however, given the best descriptions I can of them when they do occur elsewhere.

The elevated character of the spots, both at the commencement of the disease, and during its progress, is partly given to them by the morbid layer of cuticle forming the scale, such elevation being evidently less striking when the surface is cleared from this incumbrance. I think the remarks of Willan and Bateman, as to the non-accordance of the cutaneous lines on the diseased surface with those on the adjoining healthy skin, are incorrect in the majority of instances.

The parts of the body stated to be the most frequent seats of Lepra at its commencement are the arms, fore arms, and legs, from whence it extends to the trunk of the body, and sometimes to the head. When the latter circumstance occurs, like all other affections of the cutis extending to, or occurring in this part, it undergoes some change of character, dependent on the irritation of the hair, and is attended with a fluid discharge. The scabs cease to assume the circular form, and the substance of which they are formed falls off in detached portions, having something of the appearance of dirty mortar. The fluid exudation becomes the cause of a more ready separation from the skin of the diseased secretions, and if the scalp be examined, it will be found of a red and shining appearance ; the inflammation is, however, found to exhibit the same characteristics of a chronic form as are perceived on the skin of other parts where the scales are detached.

In tracing the progress of the disease on parts not covered by hair, we perceive that in a day or two after the first described minute scale has become detached, a dull red areola of inflammation is formed, which extends a line or two beyond the site which the scale occupied. In two or three more days the diameter of this areola is increased, and other scales are formed in the centre, which either gradually dry, turn up at their edges and drop off, or increase in size, and still preserve the circular form. When the spot arrives at the dimensions of a silver penny, the red margin of the cutis surrounding the edge of the scales is observed to be distinctly elevated ; and by the closeness of its adhesion to the latter, sometimes presents the appearance of a continuous surface. This appearance continues as the disease spreads and obtains strength, and the scale manifests no disposition to detach itself till it has covered a great portion of the limb.

If the scales in the centre drop off freely, which usually happens if the general health is improved, this part of the diseased cutis manifests a disposition to healthy changes : though still of a darkish red hue, it ceases to produce the hard dry scale as at first, but in the course of every three or four days a slight transparent film is formed, which is easily separated. Under these circumstances also the rapidity of extension from the margin gradually diminishes, the redness becomes of a more florid hue, and the scales are lighter and more speedily detached. This state of the parts is usually the harbinger of a rapid recovery from the disease, and it is evidently the result of an invigorated action of the cutaneous vessels. The lighter hue of the diseased part is usually followed by the disappearance of the elevation of the cutis at the margin, and the scales soon become of the same character here as in the centre. Eventually, they assume the appearance of scurf, and the skin regains its natural colour.

If, on the other hand, the habits of the patient be not such as are conducive to health, if a depleting system of treatment be had recourse to, instead of the reverse, or if the constitution should be marked by original debility, the disease may be expected to take one or other of the following unfavourable courses.

The incrustations continue to increase in circumference, or to spread irregularly over the adjacent parts. In the course of time many of them become so dry and hardened as to irritate the cutis beneath, when much pain and tenderness is experienced and matter is formed, and perhaps confined for a considerable length of time ; the margin of the scale is imbued with a dark-coloured sanious fluid, similar to what is described elsewhere as surrounding the bases of the scabs of Rupia. The disease loses its chief characteristic of a dry incrustation, and an unhealthy ulcerative process is commenced by which the scale is as it were undermined and detached. In this state of things, the appearance presented is that of an excavated unhealthy sore with a white sloughy surface and elevated edges, the latter preserving the dark-red colour it obtained from the first appearance of the disease. From day to day other spots make their appearance, which spread in the same manner as those preceding, till a large portion of the surface is occupied by the disease, and the case becomes truly deplorable.

In the greater number of cases, however, the rapidly increasing marks of constitutional debility, as the disease proceeds, establishes the conviction of the necessity of tonics and better living, both in the minds of the patient and medical attendant ; and hence this ulcerative stage of the disease is seldom seen. It nevertheless frequently occurs in different parts of the continent, where poverty and inattention to cleanliness are more prevalent than in England ; and I have seen it repeatedly in the impoverished class of labourers who have been compelled to apply for parochial relief from ill health or want of employment.

In the better classes of society, where great debility has been brought on by sedentary and studious pursuits, by continued anxiety of mind, or by great bodily exertion and privation of rest, the scales are not followed by ulceration or fluid discharge, but they spread to a great extent, then separate piecemeal, and are followed by others. When one spot has returned to health, the disease appears and pursues the same tedious course in other parts, thus continuing for years, unless the pursuits or circumstances of the patient are materially altered.

I have already alluded to the black leprosy, or in other words, to the disease as occurring in scorbutic habits. I have known the dark hue from which this variation takes its name arise in the common Lepra of long standing in one or two cases, but it usually accompanies the eruption at its commencement. It has not, to my knowledge, been traced through its course by any author who has noticed it, and I hope to be able to supply this deficiency.

I believe this state of Lepra is never seen without strong marks of want of energy of system, accompanied by torpor of bowels and sallow unhealthy complexion ; there has been a manifest connexion between these symptoms and the cutaneous affection, at least in all cases which I have observed, and the termination of such cases has clearly showed such connexion to be that of cause and effect.

The spots of the black leprosy are the same in figure as that already described. They originate in the same manner by a small scale, which falls off, leaving a blue or blackish spot, which slowly increases, preserving its circular form, and is covered by scales considerably less thick than the former. As the disease traverses the surface, its course is marked by the dark hue which it leaves behind ; but on this dark-coloured skin, after one or two layers of crusts have been detached, the production of this substance ceases, the surface becomes shining, and feels soft and swollen. In some parts the skin is more raised than in others, and there is a feeling communicated to the finger resembling that which a small quantity of fluid would give if confined beneath the cutis. In a few weeks the most prominent of these soft discoloured parts give way, and a little orifice is formed, through which a sanious thin fluid escapes ; and if this orifice be examined by a probe, the cutis will be found undermined and completely separated from the cellular membrane beneath to a considerable extent, sometimes indeed as far as the discolouration extends. The orifice rapidly increases in size till the discoloured cutis is pretty nearly all removed, when the disease presents the appearance of a common unhealthy sore with dark-coloured inactive edges. This stage or condition of the disease is not very often seen, and is chiefly confined to the lower extremities.

The causes of leprosy, as it occurs in England, have been considered very obscure. To some authors it has appeared to be produced by the application of dry, irritating, or caustic substances to the skin, and by want of cleanliness ; while others have noticed this opinion only to oppose it, and to assign others of an equally local influence.   Local causes, indeed, appear to have been sought after rather than such as may depend on the state of the general health of the patient ; but it is evident that this view of the subject must be, in the great majority of cases, incorrect, since the disease is observed to occur most frequently in this country on the skins of individuals not subject to the imputation of neglect of cleanliness.

There is no part of the civilized globe, to which historical research on medical subjects has been directed, where evidence of the extensive existence of Lepra has not been obtained ;* but a comparison of the statements of authors, as regards the degree in which it is found to prevail in different parts of the world and in different ages, clearly demonstrates that in proportion as civilization has advanced, and as commerce has extended its benefits, its frequency and severity have become diminished.   The connexion of cause and effect, which is here manifested, admits of ready explanation ; civilization gives that excitement to the mental powers on which the healthy performance of the animal functions are known to depend, while the facility with which linen may be every where obtained, prevents that neglect of cleanliness which favours the progress of every form of cutaneous disease.

The hereditary origin of leprosy, as well as of some other scaly diseases, has been frequently spoken of ; but I am compelled to withhold my belief, that any thing beyond a certain dryness of the cuticle, dependent on the original formation of the cutis, can be communicated in this way, though such dryness and disposition to crack and form a species of Psoriasis may be much aggravated by accidental circumstances, and produce scales similar both to those of this disease and of Lepra.   In parts of the world where cretinism prevails, leprosy is almost constantly observed ; and owing to the other disgusting appearances which these outcasts of the human race exhibit, it is vulgarly supposed, like this degeneration of the species itself, to be hereditary.   Under such circumstances, however, its causes may be much more satisfactorily referred to the hardships of its miserable subjects ; for there can be little doubt that filth and idle habits are at all times capable of acting as powerfully exciting causes.   "Hereditary diseases do not naturally and necessarily attend the human race ; leprosy, madness, gout, scrofula, &c. spring out of certain practices : they were all acquired, and probably will be eradicated.   Leprosy, originating in want of personal cleanliness, has already given way to improvements which have taken place in that respect.   Linen is now substituted for woollen in many articles of dress, and other regulations equally friendly to cleanliness have caused leprosy almost to disappear."†

* Vide Précis Theorique et Pratique sur les Maladies de la Peau, par M. Alibert, page 5, tom. ii., 1818.
† Dr. Jarrold.

Diseases which somewhat resemble Lepra and Psoriasis are, as I have before observed, sometimes noticed in young people, particularly females, in respectable circumstances in life, whose parents at the same age had been subject to similar affections. A light complexion and peculiar harshness of the skin exist in such cases where the latter is sound, and the scaliness and disposition to crack and form fissures is usually most troublesome in parts exposed to the drying effect of the atmosphere, as on the arms, hands, &c. In other parts of the body, as the edges of the arm-pits, &c., if the cuticle be minutely examined, it will be found extremely delicate, and easily irritated by the friction of the clothes, while the palms of the hands are horny and dry, and presenting none of that moisture and smoothness which belong to the part in the healthy state. The causes of these affections seem to consist in the original formation of the skin, and there is little or no ground to expect benefit from medicinal treatment : the history of such cases, however, is consolatory ; for as the age of the patient attains maturity, and the constitution assumes the energy and stamina of the adult, the powers of the circulation become more adequate to the regular and healthy nourishment of every part of the body, and a sounder and stronger cuticle is formed.

As regards the precise state of system under which Lepra first makes its appearance, it may be observed, that medical men are in the majority of cases precluded from obtaining positive information. The disease has usually made some progress before it comes under their observation, and the state of system, existing at this period, may be materially different from that formerly existing ; it should, not, therefore, be set down as that in which liability to the disease naturally exists. Diseased actions in other parts of the body will go on even to the destruction of life, without the continuance of the cause which originally excited them, and why should they not do so in cases of Lepra ?

Some time since I had an opportunity of observing it in a stout corpulent man of forty, who had suffered from it at intervals from his boyhood. In early age he had been of delicate health and subjected to hard labour and much privation ; as he approached manhood, the additional exertions which a wife and increasing family exacted from him, with the anxiety of mind consequent thereon, produced a considerable aggravation of the disease ; and he has observed, that as his circumstances varied and his mind was more or less at ease, his disease has been more or less troublesome.

On the subject of its communicability by contagion it is not necessary to say much. It is clearly a mistake originating with the older writers, who confounded it with other affections of a materially different character. In some instances, it is stated to have been produced by particular kinds of food and drink, " which operate through the idiosyncrasy of individuals," but such cases must be of rare occurrence. None which have hitherto come under my observation have been of this kind ; indeed it is obvious that the

time usually occupied in the production of the first and most minute
leprous scale, is much greater than can be supposed compatible with
the transient operation of substances passing the stomach and
bowels in the form of aliment.

The results of long and carefully conducted inquiry into the
etiology of Lepra have completely convinced me of its connexion
with, if not entire dependence on, mental and constitutional causes.
Of twenty cases of the disease which have come under my observation within the last few months, no more than four have been
individuals who could be subjected to suspicions of inattention to
cleanliness, and these even were persons whose history was that of
misfortune and disappointment, accustomed up to a recent period
to better and easier circumstances. Unaccustomed privations and
mental distress were the immediate precursors of the cutaneous
disease in such cases ; and during the time these were experienced
by the unfortunate individuals alluded to, no change appeared to be
brought about by any medicinal remedies. The result of a system
of treatment calculated to ameliorate their condition, viz. a liberal
allowance of animal food, porter, &c. and means of active healthful
employment, was a speedy return to health and vigour of constitution, and the extinction of the cutaneous disease.

In this metropolis Lepra is certainly very prevalent, it is even
more common than Impetigo. It is not seen among persons in
affluent or independent circumstances to any considerable extent ;
and when a case does occur in this class of society, it is only as a
sequence of some other disease, where depleting remedies to a great
extent have been had recourse to, or where studious and sedentary
pursuits have been followed to the exclusion of proper exercise and
nourishment.

The class of persons who appear to be most subject to it are those
whose minds are anxiously occupied by the cares of business or
study, or who are accustomed to bodily exertion beyond what their
strength enables them to bear.

A young medical man had succeeded to the business of his pre-
ceptor, in a populous district of the country : though every way
competent, by several years' attentive study to discharge his duties
in this situation, he found himself obliged to compete with others
perhaps equally able to practise their profession, and that his hopes
of success must depend for their realization on no ordinary efforts
of his own. A wife and increasing family added not a little to his
anxiety, and being of an originally delicate constitution, his health
and strength became impaired by the extraordinary efforts he was
called upon to make. Under these circumstances the disease first
appeared. I give this case in as abbreviated a manner as possible ;
it is, however, a mere recital of many others which have been
brought under my notice, the history of the subjects of which does
not differ from it in any material degree.*

* Notwithstanding the examples placed before me by other writers, more particularly by Alibert and Rayer, I forbear to be more diffuse in the recital of any

A gentleman lately consulted me, whose habits in many essential respects were materially different from those of the individual last alluded to. He was not accustomed to hard and laborious exercise, as riding thirty or forty miles a day on business, nor was he anxious about future prospects. His present means and income were more than equal to all his wants, and derived from a source not liable to fluctuation and uncertainty. But his inclination for study and sedentary pursuits led him to neglect ordinary precautions against disorder of his general health. He spent too much time over his books during the day to enable him to take proper exercise, while he encroached very largely on the proper hours of rest during the night, in his favourite pursuits. An inactive state of the bowels, sallow complexion, loss of appetite, and other symptoms denoting great disorder of the general health, which were immediately attributed to neglect of exercise, led him to the opposite extreme ; and he determined incautiously on hard riding, and other violent exertion, which he followed with much resolution for a few days, till the leprous spots made their appearance. Shortly after this, I had an opportunity of seeing him ; and I was given to understand, that in the course of ten days more than forty spots had made their appearance, some of which had arrived at the size of a shilling.

The proper treatment of this case, and that previously detailed, and in fact of all others, is nearly the same : attention to the general health, and the invigoration of the constitution, seemed to be the chief means by which the cure was brought about.

I have never heard of an instance of the disease making its first appearance on the skin of a stout, muscular, healthy man, though, as I have before observed, it has sometimes continued beyond the adult period, where difficulties and uneasiness of mind have existed on the part of the patient, having first appeared at the age of from sixteen to twenty, under depreciated health, and impoverished circumstances.

The pathology of lepra is by no means obscure. Its first appearance on the skin is accompanied by amply sufficient proofs, that the seat of the inflammatory action of which it is constituted is chiefly, perhaps entirely, situated in those vessels of the cutis, whose office it is to secrete the cuticle. The nature of this action too is sufficiently clear, and is evidently of a chronic kind, as is proved by the dark red hue of the spots, compared to most other eruptions ; and by the absence of tenderness or smarting, which is usually observed.

Unless the state of the general health of the patient be greatly

particular case ; for though the high opinion of the value of such details entertained by some may be correct as regards a few internal diseases, they are worse than useless in those of the surface ; where every thing is distinguishable by the naked eye, and the characteristic symptoms are in no case very equivocal. In the works of the authors alluded to, the expense to the reader is increased, and the value perhaps diminished.

impaired, it will be difficult to discover any more important deviation from the natural action of the vessels of the skin than that in question ; but where a scorbutic diathesis has been induced by privations of nourishment, and other hardships ; and ulceration has followed around the margins of the scaly spots, (a case not uncommon among paupers who have known better circumstances,) the disease loses the characters of Lepra, and becomes what may be more properly termed scorbutic ulcer.   In this state of things the cutis is speedily perforated, and the cellular structure beneath suffers to a considerable extent from ulcerative absorption.   When cicatrization afterwards takes place, it is slowly accomplished, and leaves behind it ugly scars, which seldom disappear ; an effect which does not follow the leprous scales in ordinary cases of the disease.

The principles on which the treatment of leprosy is to be conducted, will be easily perceived from the foregoing observations.   Dr. Bateman observes, " that there is no one remedy, nor any invariable plan of treatment, which will succeed, under all the circumstances of its appearance, in different instances ;" and he adds, that the different degrees of cutaneous excitement, or inflammatory action, which accompany the disease in different habits, afford the most important guide to the successful application of remedies.   There is, however, but little variation in these respects in any case, for the cutaneous disease invariably shows the languid, inactive state of the cutaneous vessels, which has been described.   It is at no period attended by an active state of inflammation, or great excitement ; and always admits of the use of certain local stimulating applications, hereafter to be noticed, with advantage.

Dr. Willan, after noticing the practice of the Greek physicians, as respects the use of strong purgatives and bleeding, and the local application of a variety of remedies, some powerfully astringent, others corrosive or vesicatory, and others again remarkable only for their utter inertness, proceeds to the detail of the opinions of other authors, on the Bath, Harrogate, Croft, and Moffatt waters, applied externally or internally, which appear in some cases to have been of service.   As external applications, he notices the ung. hydr. nitrat. a solution of hydr. oxymur. in alcohol, and tar ointment. He informs us further, 1st. That antimonials, sulphur, and nitre, have not alone any considerable efficacy.   2nd. That decoctions of emollient herbs, of guaiacum wood, sarsaparilla, mezereon, or elm bark, by no means deserve the character of specifics.   3rd. That salivation produced by mercury in any form does no good, but *that a spirituous solution of the oxymuriate of mercury in small doses, if continued for a length of time, is very useful.*   4th. That the nitrous and muriatic acids have been given for months without advantage in many cases, while others appeared to do better by the use of the aq. kali puri, in doses of twenty or thirty drops exhibited thrice a day.

Fowler's solution also, according to this author, is capable of curing the disease ; and I have been accustomed to exhibit it till within the last few months, in conjunction with bark and other tonics, as I believed with considerable success, but more recent observation has convinced me that the vehicle in which it was exhibited combined with the effects of a more nutritive diet, has had more to do with the cure than the arsenical solution. I do not, however, mean to deny its powers, even when unassisted by tonics of more established character, of inducing a more energetic action in the cutaneous vessels than exists in any form of Lepra which I have observed ; and it is entitled, therefore, to consideration, where other tonics and better living fail in the desired effect.

The solanum dulcamare, or bitter sweet, has been reported to possess great power also in the management of the disease. It has been used in the form of decoction externally as a lotion, and has been exhibited internally at the same time,* the strength of the decoction being one ounce of the vegetable to a pint. On the effects of this remedy I am able to speak very fully. Having tried it without the least success, I vainly sought for encouragement in the reports of others, and I am in candour compelled to say that I have never heard of a well authenticated case where it has been of service. A lady, who had suffered from Lepra twenty years since, underwent a course of this medicine by the direction of the late Dr. Bourne of Oxford, from which no benefit was obtained. She came to London a few years ago, and completely recovered by the judicious use of good animal food and wine.

Alibert, on the subject of the treatment of this disease, has summed up all his remarks, and arranged them under three heads, comprehending, 1st. Vues† générales sur le traitement des lèpres. 2nd. Un traitement interne employé pour la guérison des lèpres. 3rd. Du traitement externe employé pour la guérison des lèpres. When it is considered, however, that what he has here said is intended to apply alike to the leprosy properly speaking, and to the varying forms of elephantiasis, but little confidence will probably be conceded to it. He mentions the waters of Barege and Tivoli as sometimes efficacious, and notices the observation of Dr. Willan, Dr. Girdlestone, and others, on the use of the tincture of cantharides, arsenical solution, &c., also those of Dr. Crichton on dulcamara, but without adding to or lessening the importance of their observations, by adducing the results of any experience of his own in the remedies in question. The remainder of this part of his subject is occupied by an irregular mention of other medicines, or local applications, which have been already alluded to, and tried here in numberless instances without success.

It is evident, from the perusal of these different authors, that the dependence of Lepra on constitutional causes has been much overlooked ; and hence local applications have been considered as the chief remedial measures within our power. This oversight probably

* See Dr. Crichton's Communication to Dr. Willan on Cutaneous Diseases.
† " Sur les Maladies de la Peau," tom. ii. p. 97.

admits of explanation.   I have already stated that the majority of
cases which come under our observation in this metropolis are in
subjects of the middle or better classes in life—a class in which
debility, arising from common causes, would not be expected ; but
it is nevertheless true, that the studious and sedentary, and the
anxious and laborious, more particularly if of originally delicate
constitution, breathing as they do, from week to week, and month
to month, an atmosphere like that of London, even though they
should take with an appetite an ordinary quantity of food, will lose
in the course of time that vigour of health, and power of circulation,
necessary to the perfect performance of all the animal functions,
and this even without experiencing any distressing feeling of
debility.   Under these circumstances, it is not a matter of surprise
that the disorder should be considered local, the change having been
so gradual, as not to be perceptible to the patient: it is a minute
inquiry only on the part of the physician, and that extending to the
history of the former for months or years, which can lead to satisfactory results.

The practice I am now accustomed to adopt is as simple as the
foregoing observations would lead the reader to expect.   To restore
the strength of the patient to its original standard, not simply
before the cutaneous disease appeared, but even before those habits
or pursuits were adopted, which may have for years preceded it.
If having been brought up to the age of puberty, or manhood, in
the country, the patient experiences his first attack after confinement
for a year or two to the laborious pursuits of trade, a case of which
I have seen a great many instances ; return to the country and
relaxation from care and labour is soon followed by the disappearance of the disease.   If privation of rest, as in the case of studious
persons, has contributed its effect as a cause, it need not be remarked,
that books and late hours should be immediately avoided, and give
place to healthy and amusing exercise by day, and lengthened repose
by night.   A liberal diet of animal food, and wine and porter in
proper quantities, ought to be adopted, unless the state of the
digestive organs should forbid it.   In all cases a change in the mode
of living, and a temporary suspension of exertion, either mental or
bodily, is of the utmost importance.

As regards local applications, and their effects, I am unable to
offer any evidence of value, except as regards one individual preparation, which I have already alluded to elsewhere.   In a former
edition of this work, I took occasion to observe, on the disappointments we have been accustomed to experience in the local application of various powerful agents to different diseases of the skin, and
it then appeared to me that such failures were mainly attributable
to the neglect of appropriating in the different formulæ of lotions or
ointments, the individual component parts to the manifest conditions of the cutaneous vessels.   Thus, for instance, in the case of
impetigo, where much irritation, itching, and a copious fluid secretion
exist, sedatives of any kind produce only temporary effect.   The

disease, after appearing to subside, soon returns with even increased activity, the vessels of the part have become so debilitated by the excessive excitement of the disease, as to be incapable of performing their healthy functions. It would appear, therefore, a great desideratum, to combine in one local application what shall produce a sedative effect on the vessels of the inflamed part, and restore to them that tone which they have lost during the stage of excitement. It is on this principle, that the sulphur vapour bath has always appeared to me to act in various diseases of the skin ; the warm vapour, by the copious perspiration it produces, unloading the cutaneous vessels, while the acid deposited upon them in this state acts as a direct astringent and tonic.

I know not whether the ointment prescribed under the head of Lupus, owes its good effects to the combined properties in question, or otherwise ; but it does that in Lepra, as well as many other cutaneous diseases less connected with the constitution, which no other application will do, viz. it quickly subdues the inflammation of the cutis, and produces a healthy cuticle. Its composition may not bear the test of chemical criticism, perhaps, but as remedies are estimated by their useful effects, this is of no importance.

Notwithstanding the value I have been compelled to attach to constitutional treatment, this ointment is by no means unnecessary. In all cases where it has been had recourse to, in conjunction with the tonic plan of treatment of lepra, the dark hue of the surrounding inflammation changes more rapidly to a brighter red, and the scales are more readily detached.

In the management of the scorbutic form, very little deviation from the above plan will be found necessary. The ointment should of course be omitted where actual ulceration has taken place, and the local treatment should be that of a common sore.*

---

* The word *lepra*, which has been applied to almost all chronic diseases, when highly developed, is used in a more restricted, and better determined sense, to designate a chronic squamous phlegmasia of the skin. The principal characters of this disease are, scaly plates of different dimensions, almost always circular, or orbiculated, surrounded by a reddish and prominent circle, and depressed at their centres, scattered over the surface of the integuments, and the development of which is not preceded by vesicles or pustules.

The alteration of the skin constituting lepra (*L. vulgaris*, W.) *dartre furfuracée arrondie*, Alibert,) is announced by solid elevations, around which small patches are served, a line in diameter, reddish, shining, *and circular*, and which are prominent. On passing the pulp of the fingers over these elevations, they feel firm and solid. Some resemble a hard voluminous papule, and from this circumstance, no doubt, Willan supposed Lepra might be the result of induration of the papillæ of the skin.

The summits of these elevations (which are usually successive in their eruption) present, a few days after their formation, small whitish epidermic scales, which are semi-transparent, smooth, and shining. This small scale, *resembling a spangle*, soon separates, and its fall is followed by a pricking or itching sensation. The small surface of the skin, which is left uncovered, appears scarcely changed ; but is unequal to the touth. On the under surface of the scale a small eminence is observed, of less consistence than its other parts. *This eminence,*

The designation "Syphilitic Lepra" is very often employed by authors, but I have many doubts whether it is with the prudence requisite. It has often occurred to me to be a witness of much mis-

which is slightly bloody when the scale has been forcibly torn off, appears to have lodged in a slight excavation of the skin.

The surfaces of these small scaly points, after being once denuded, progressively and pretty rapidly enlarge, till they attain about an inch diameter, but *always preserve the circular form.* They become again covered with scales, which are dry, thin, firm, and resistant, of a pearly grey or yellowish shade ; they are surrounded by red or purple and *slightly elevated* edges, so that their centres appear to be depressed. The squamæ formed on them become thicker and thicker, forming prominent layers. The scales, almost always adherent to the skin, are not uniformly extended over the leprous patches, which are not invariably covered by one entire scale. The external surface often assumes a whitish tint. They separate partially and irregularly. When they are detached, the small orbiculated surfaces which they have covered, appear red and shining, and do not rise beyond the level of the surrounding healthy skin. If the leprous patches are of recent formation, they do not represent the lines of the dermis; but these are seen on the older patches. When ancient, they sometimes exhibit even a sort of wrinkle or furrow, corresponding to that remarked on the inferior surface of the scale.

The scales are quickly reproduced after their removal ; the parts thus undergoing, in the course of some months, a series of successive desquamations.

The healing of these orbiculated patches, when spontaneous or procured by art, begins at the centre, and thence extends to the circumference. This is indicated by the fall of the squamæ and their non-reproduction.

It is generally supposed that some modifications of this disease, alluded to by the Greek physicians, formerly received particular names, (alphos, leuce, melas.) These distinctions have been revived with more accuracy by Willan and Bateman. The size of the squamous patches is sometimes not large ; they increase slowly, and are slightly prominent (*L. alphoides*, W.), their diameter not being more than a few lines ; they are seldom very close together, and are developed almost exclusively on the limbs ; they differ from the patches of *L. vulgaris* by their whiteness and the small size of the scales. This variety is more common in children than in adults and old persons.

The leprous spots may present a brown or livid tint, (*L. nigricans*, W.) more deeply marked towards their edges, which are of a dirty violaceous red, and this is observed through the whole thickness of the scales. In this form of Lepra the squamæ are more easily detached than in the other varieties, and the surface of the affected skin continues for a long time red and shining. Excoriation, however, may take place and cause the issue of a sanguinolent serosity, so that a new lamellous concretion may be formed. The blackish tint remarked on the site of the scales results from a change in the reticular body of the skin.

Lepra, at times, covers the whole surface of the body ; at others, it affects only the knees and elbows. The orbiculated spots are, ordinarily, first observed on the limbs, and most frequently at the bend of the elbow or knee. In most cases they affect the limbs of both sides. From these parts they may progressively extend along the arms and thighs, over the chest, shoulders, loins, and lateral parts of the abdomen. The patches are sometimes more numerous and prominent around the lower part of the belly. They are seldom seen on the hands, head, or scalp. When observed on the head they are of small size. They occasionally appear near the external orbit of the angle, and spread over the eyebrows, forehead, and temples. Several confluent, or, as it were, agglomerated patches, may join at their edges ; but even then the *orbiculated* form of each is still indicated by the arcs of circles which may be distinguished at their edges.

If Lepra is long neglected, and the leprous patches cover the fingers, the disease may be propagated to the reticular body beneath the nails. These then become thick, rugous, opaque, of a dirty yellow colour, and curved at the ex-

ery, arising from hastily formed opinions on this point. The usual medical attendant of a family meets with a cutaneous disease which he does not find it easy to cure, or even control—it may be a case

tremities. The surface is unequal and irregular, and the thickened root seems formed of a collection of distinct superposited layers. More rarely the dermis which secretes the nail is inflamed, and furnishes a more or less abundant sanies.*

" When the patches are few in number, and not much inflamed, Lepra is not accompanied by any morbid sensation, except slight itching, when the temperature of the body is raised by exercise, or the heat of the bed. This sensation is occasioned, *Mr. Plumbe says, by the rising of the edges of the scales, which causes tumefaction of the surrounding areolæ. Whether this explanation be the true one or not, it is certain that, when lepra is cured, and new scales are no longer formed, to raise up and replace those already developed, this feeling of pricking and itching is not experienced by the patient.*

" When the spots, on the contrary, are numerous, inflamed, and scattered over the whole surface, they may be attended by excessive pain, anxiety, and tension of the limbs ; they may then pour out a serous fluid, and present a surface analogous to that of ulcerated chronic eczema. I have seen this inflammation carried to such a degree as to render the motions of the joints difficult, obliging the patient to keep his bed, every movement being hindered by the stiffness of the epidermic scales, which produce a very remarkable crepitation.

" However, lepra does not ordinarily exert any influence beyond the parts which it occupies. This affection seems to be essentially local. If P. Frank, Alibert, and some others, have mentioned, in the symptomatic description of this disease, morbid phenomena developed in other organs, and, in particular, *an alteration in the voice,* it is in consequence of their having confounded lepra with elephantiasis of the Greeks, and regarded two diseases so dissimilar, as varieties of the same.

" All the elementary tissues, which enter into the organization of the skin, do not appear to be equally affected in lepra. Mr. Plumbe supposes that the vessels which secrete the epidermis are attacked by chronic inflammation, rendering this production more abundant, and causing the fall of the scales. If all the vessels of the skin are equally affected and in the same manner, it is difficult to explain how the result of this inflammation can be confined to the morbid secretion of the epidermis, without causing the development of vesicles or pustules. However, this hypothesis, like many others formed on the same subject, does not at all account for the *orbicular* form always assumed by the leprous spots. Some pathologists have supposed that the superficial vessels of the skin are disposed in small concentric circles, while others have imagined this disposition of the spots to be the natural consequence of their primary formation, being that of a solid elevation, around which the inflammation irradiates circularly.

" Lepra is common to both sexes, and all ages. I have never seen it in infants at the breast; but frequently after the second period of dentition ; it is most common in women.† Mr. J. Wilson asserts that it is more frequent in England now than formerly ; but it is possible it may have been for a long time mistaken, or incompletely noticed under some other name. Heberden, in particular, was deceived when he affirmed that lepra was very rare in England, ' *De vero scorbuto et leprâ nihil habeo quod dicam, nam alter rarissimus est in urbibus, altera in Angliâ pene ignota.*' This suspicion seems the better founded, from the fact, that several French physicians, having but confused ideas of lepra, and those different from what it really is, have asserted that this disease is seen only in some of the meri-

---

* There are few or no cases occurring in England, of any degree of severity, in which the gland secreting the nail does not participate in the general debility of circulation. Hence unhealthy and unsightly nails are formed. The nail becomes a dead substance a few lines from the matrix, and therefore a cause of irritation,—this is the explanation of the fact, that a sanies is sometimes discharged.

† The reverse of this is the fact as regards English society.

of common Lepra in the female, a disease of which debility constitutes the first and most important features, there may be some soreness of throat, a symptom not at all uncommon in cases of simple

dional provinces, while every year, in Paris, there are admitted into the hospital
of St. Louis, and that of Enfans Malades, a number of individuals, labouring
under this squamous affection of the skin, which I have also myself observed in
other classes of society. It may be here remarked, that all that has been written
lately in France on the origin, propagation, and disappearance of lepra, in different parts of the world, contains a multitude of errors ; this is the consequence
of borrowing from authors who have confounded lepra with elephantiasis of the
Greeks, elephantiasis of the Arabs, and other diseases not less distinct from one
another.
    "The etiology of lepra is, for the most part, very obscure. This disease is not
propagated by contact mediate, or immediate. Husband and wife may cohabit
without communicating it the one to the other. All that has been said concerning
the pretended contagion of lepra is erroneous ; and in this respect, the most false
inductions have been drawn from the existence of leprosies during the eighth,
ninth, and tenth centuries. Again, no reliance is to be placed on a case cited by
Niebuhr, of a leprous subject, who, by having connexion with a linen-woman of
the lazaretto, communicated the disease to her, and procured her admission into
the hospital.
    "Like some other diseases of the skin, of shorter duration, (erythema, urti-
caria. &c.), lepra seems to be caused by the abuse, or even use of stimulating
food and spirituous liquors. Bateman knew a person, in whom the ingestion of
spiced aliment, or a small quantity of alcoholic liquors, never failed to produce
it ; it has been known also to supervene soon after the ingestion of poisonous
substances. the salts of copper, for example; and to follow the abuse of acids.
It has also been attributed to the habitual use of game, salted and spiced meats,
fish, and mushrooms ; yet the disease is not more frequent on the sea-coast than
inland. It has been attributed to the effects of grief and poverty; but rich individuals, and those given to luxury, are also subject to its attacks.
    "Willan supposes that the development of lepra is owing principally to the
effect of cold and moisture, and to the action of certain dry and pulverulent substances on the skin. Bateman has seen examples from similar causes; and adds.
with truth, that bakers, and those who work in laboratories, &c., are rarely affected with this disease : while it is often observed in young women in classes of
society where cleanliness is an object of particular attention.
    "In some cases lepra is manifested after violent and long-continued exercise.
Hereditary predisposition to it has been several times noted. It must be acknow-
ledged that there still remains great uncertainty and obscurity as to the number
and nature of the causes which produce lepra ; and this is mainly to be attributed
to the fact of practitioners being seldom consulted on the first appearance of the
disease.
    "The diagnosis of lepra is one of the most important points in its history ; and
yet it must be confessed that it is one of the diseases the characters of which are
the best marked. It differs. in many respects, from other chronic inflammations
of the skin, and even from those developed under the *squamous* form. In psoriasis,
as in lepra, the epidermis is rough, scaly, and red on its inferior surface ; but the
form of the squamous patches is *irregular*, in psoriasis, but regularly orbicular in
lepra. In the former disease, the edges of the patches are neither elevated nor
inflamed ; their shape neither oval nor circular ; the surface of the skin beneath
the scales, often deeply fissured, is generally much more sensible and irritable
than in lepra. There is, however, a variety of psoriasis (*P. guttata.* W.) which
somewhat resembles lepra, and constitutes, as it were, an intermediate form between this disease and the other kinds of psoriasis. Indeed, the squamous spots
of *P. guttata* are distinct and isolated, like those of lepra ; but they are smaller,
rarely exceeding one or two lines in diameter, and their circumferences are not
so regular. It is more especially when lepra begins to heal that it most resembles

Lepra, and this together with the cutaneous affection has often laid the foundation of imputations, involving the moral character and domestic comfort of families moving in the most respectable grades

psoriasis *guttata.* However, in some inveterate cases of lepra, when the orbiculated patches are confluent and confounded at their edges, it is difficult to distinguish them from certain cases of psoriasis. Syphilitic *squamous* spots, which resemble lepra in their seat, approach it still nearer by their *circular* shape. They resemble also *black* leprosy, (*L. nigricans,* W.) both in their size and their copper or violaceous colour. The edges of these spots are sometimes elevated like those of lepra, and their central parts are flat, and covered by very thin scales; but they are seldom more than six or eight lines in diameter. But the dryness and roughness of the skin, so remarkable in lepra, are not observed in syphiloid disease; and when the latter is of long standing, the spots are almost always as soft to the touch, and as supple as the other parts of the skin. Besides, in syphiloid disease the circles are livid, violaceous, devoid of scales, and rarely complete. Lastly, syphilitic spots supervene after venereal infection, grow pale, and disappear under the influence of mercury; and the healing presents this peculiarity, that it commences generally at the *circumference,* while that of lepra begins at the *centre* of the spot. It would seem difficult to confound the *scales* of lepra with the *crusts* of pustulous and vesiculous diseases; yet Willan observes that these mistakes have often arisen. Thus, some have confounded lepra with impetigo *figurata,* or, rather, with the scaly state of the skin consecutive to the fall of the crusts. In fact, this pustulous phlegmasia appears under the form of circumscribed patches of various size, commonly small and circular upon the upper, large, oval, and irregular on the lower limbs; and on these patches small pustules may be traced, the desiccation of the humour of which forms yellow, brownish, or greenish crusts, very distinct from the exfoliations of the epidermis observed in lepra.

" When lepra is developed on the scalp, it may be distinguished from the tineæ, by bearing in mind the progress of these diseases.

" Lepra is neither preceded nor attended by pustules; it causes no oozing or exudation from the skin; and it does not commonly alter the hair, notwithstanding the assertions of a crowd of authors to the contrary. In psoriasis of the scalp, there is furfuraceous desquamation, but no scales; if the hair is plucked out, it will easily be seen that there is usually alteration of the bulbs. Tinea annulare might be confounded with lepra, only the former commences with the development of psydraceous pustules, and the latter, by that of solid elevations.

" Lepra has been also confounded with icthyosis by Plenck and Chiarugi. The name of *lepra* too has been given to two diseases, than which nothing can be more distinct: elephantiasis of the Greeks, and elephantiasis of the Arabs. It may be here observed, that not only are the descriptions of lepra hitherto published inaccurate, false, and unintelligible, but also those of these three diseases (*lepra elephantiasis Græc.* and *elephantiasis Arab.*). Although differing from one another in their seat and external characters, they have been regarded as simple varieties of the same affection, and confounded in the same symptomatic description.

" The duration of lepra is indeterminate. In old people it is nearly always incurable; it seldom heals spontaneously, and at times resists the most rational treatment: when it affects the limbs alone, it is not dangerous; if it occupies the whole surface of the body, cutaneous transpiration is diminished or suspended, and is generally equalized by the increase of that of the pulmonary and urinary organs. Leprous patches are often developed on different regions, disappearing on some, and showing themselves on other points, for several years successively. Lepra never degenerates into cancer, as some writers have affirmed. When approaching to a cure, the scales are detached, and the spots grow indistinct at their centres; their edges dry, the skin ceases to become squamous, and the

of society, the consequences have been in many.instances with which I have been acquainted, most deplorable. It behoves, therefore, every medical man to be cautious as to the character he may assign

redness disappears : but the healing *always proceeds from the centre towards the circumference of the spot.*

"Many plans, some irritating, others of an opposite nature, have, in turn, been recommended in the treatment of lepra. All may be futile ; but it is of importance that the treatment should be *adapted to the degree of inflammation of the skin.* It may be as well to remark, that the salutary action of remedies is more evident during summer than at any other time of the year.

"If lepra is recent and extensive; if the skin is highly inflamed, thick, and much injected ; the itching very inconvenient ; the motions of the joints difficult, the disease will certainly be aggravated by sea-water baths, frictions, sulphureous lotions, &c., which are too generally and injuriously recommended in the treatment of diseases of the skin. Bleeding, unctions with cream, milk, fresh butter, or well-washed lard, procure prompt relief, when the leprous spots are much inflamed. If the patches are large and few in number, leeches may be applied near their edges, and repeated if necessary. Vapour, emollient, or gelatinous baths, are useful, as principal, or auxiliary means. The simple vapour-bath will sometimes alone cure lepra, when recent.

"When the squamous spots are but *slightly inflamed*, and of long-standing, recourse is generally had to applications which cause more or less irritation; but the skin should be previously well cleansed by lotions, tepid-baths, and light friction. When the scales are very adherent, or disposed in thick layers, stimulating lotions, containing alcohol, sulphuret of potass, &c., favour the fall of the squamæ, and may advantageously modify the march of the disease. After the scales are thus detached, a light layer of the black pitch ointment, or of tar, or the ung. hydrarg. nitrat. diluted or mixed with saturnine ointment, may be used. These should be applied at bed-time, and washed off in the morning with warm soap and water. The use of these topical measures, continued for some months, has been known to render the skin of its natural flexibility, even when the disease has been treated internally without success.

"Under similar circumstances, sulphureous lotions and baths have been attended with success. In France, the waters of Barèges, Cauterèts, Bagnères, &c.; and in England, those of Harrowgate, Leamington, Crofton, &c. are often recommended.

"The employment of sulphureous vapour-baths has sometimes been followed by a complete cure. Though so much extolled in Germany, they frequently fail ; their principal advantage appears to be the high temperature at which they may be administered. Several observations have proved that acid vapour-baths, natural or artificial salt-water-baths, and alkaline-baths, such as those of Plombières, the hot-wells of Mount d'Or, Vichy, &c. may be also usefully prescribed.

"Tepid-baths cause the fall of the scales, and are very useful in keeping the skin clean. Sea-baths are very much recommended in England, and indeed in France ; but they occasionally produce such an excitement of the skin, that they are obliged to be substituted by simple tepid-baths.

"Vapour-baths accelerate the circulation, and may be employed for the detachment of the scales.

"When the spots are few, and very ancient, the cure is sometimes attained by covering them successively with small blisters, or by cauterizing them superficially with a solution of chlorine, or the nitrate of mercury diluted.

"Stimulating applications are generally proper when it is of advantage to excite the skin, and this is often the indication in lepra ; but it is oftentimes not until after several trials, that it can be determined which application may be best suited to any individual case.

"The same principles should direct the therapeutist in the administration of *external* remedies. Those whose action is most marked in ancient lepra are in

to chronic affections of the skin in the cases of persons who are married. There are many disagreaable circumstances in the active discharge of our duties to society and to ourselves, which we must

general very energetic, and it is desirable to have recourse to them as seldom as need be.

" The decoction of dulcamara ($\mathfrak{Z}$ij. to a pint) has been recommended, in the dose of from half, to an ounce and a half daily, by Dr. Crichton. In larger doses it produces vertigo, without having any increased action on the skin. The extract is less powerful than the decoction, and may be added to it in the cases of young robust subjects. Purgatives were much employed formerly, but are not so in the present day. These remedies, assisted by the use of the tepid and vapour bath, have cured, in a month or six weeks, lepra which had resisted all other measures. The tinct. cantharid. in the dose of from five to thirty drops, has caused the disappearance of lepra, when it has not been very ancient or extensive. The dose has been carried to sixty or eighty drops, watching always the effects it may produce. This has, of all the energetic remedies employed in lepra, the most marked effect on the skin; but it is liable to cause, insidiously, chronic inflammation of the digestive organs and urinary passages.

" When the leprous spots are neither painful nor much inflamed, the use of arsenical preparations has been advised, which, according to Willan, Bateman, and others, tend to stimulate the skin. Fowler's solution is the form most commonly used, in the dose of four or five drops a day. This may be gradually increased to fifteen, divided into four doses, and should be persevered in for some months. Some practitioners have carried the dose to fifty or sixty drops, but such hazardous practice usually causes acute or chronic inflammation of the digestive or respiratory organs. The beneficial effect of this preparation has been several times proved by Willan, Bateman, Mr. Plumbe, &c., and I have myself also witnessed it; but must say I consider external applications far preferable. It may be as well to repeat, that the administration of such active medicines requires much circumspection, joined with constant vigilance. If, during the administration of this remedy, the patient should complain of tension, stiffness, or swelling of the face, heat or shootings in the œsophagus or mouth, these symptoms, even when there does not exist any appreciable derangement of the functions of the stomach, indicate not only that the dose has been carried far enough, but that it ought to be decreased. If the tongue becomes red at the point and edges, if thirst and slight erythema of the face supervene, and the secretion of saliva becomes abundant, the medicine should be suspended. Lastly, it should always be discontinued whenever it produces nausea, vomiting, vertigo accompanied by cough and epigastralgia. These symptoms usually disappear on the intermission of the arsenic, without requiring recourse to the lancet. The arsenical solution of Dr. Valagin, that of Dr. Pearson, and of Dr. Lefebvre, and the arsenical pills of the Edinburgh Pharmacopœia, possess the same advantages and inconveniences, and require the same vigilance and reserve in their employment.

" Pitch, in the dose of ten, twelve, or more grains; turpentine, to that of fifteen, twenty-four, or thirty-six, have been employed under the same circumstances as above. But, like them, they may aggravate the eruption when attended by much irritability of the skin, and produce new disorders internally.

" The deplorable inefficiency of most remedies against Lepra, and the hope of substituting some more certain and less dangerous means for those already known, has given birth to a crowd of essays and experiments, for the most part empirical, of which it is only necessary to give a summary of the principal results.

" Antimony, and its sulphuret, sometimes produce amelioration, but never the cure of this disease.

" The utility of mercurials has been exaggerated by Mr. J. Wilson. Small doses of an aqueous, or alcoholic solution of the corrosive sublimate, are the best of all these preparations. Calomel, as a laxative, is beneficially employed in

perforce encounter, and ·there are many troubles arising therefrom. It is well known that anxiety of mind is not only capable of disturbing the functions of the stomach, but those of every organ of the body ; that it prevents a due supply of nourishment to the cutaneous structure therefore is sufficiently manifest.   Hence have occurred many cases where the simple effect of domestic unhappiness, operating on the constitution of the individual in producing disease of the skin, have for ever destroyed the comfort and character both of husband and wife.

The journalist and the critic may possibly think that I draw an ideal picture, but such is not the fact.   I have in several cases had the happiness, in consequence of a favourable result, of seeing domestic happiness restored, together with health and confidence. I encroach perhaps a little on the field of the moralist in these remarks, but they cannot be much out of place.

If any fact were necessary, in addition to those I have adduced, that debility of circulation and system were the causes of Lepra, I might allege that in young men, as confessed to me often, the practice of masturbation has been followed, and that to a very great extent, in the very worst cases which have come under my notice.

L. vulgaris, but causes salivation so rapidly when it becomes absorbed, that this must be looked upon as one of its disadvantages.

" The decoction of *daphne mezereum*, employed by Pearson in several cases of lepra, has procured temporary relief, but not a permanent cure. Its effects are more marked, however, than those of guaiacum, or sarsaparilla. Mezereon may occasion vomiting, hypercatharsis, and inflammation of the stomach and larynx; it causes heat and violent pain in the throat. This drug is less active administered in the form of syrup; some give it as an adjunct to arsenic. The liq. potassæ of the London Pharmacopœia, in the dose of twenty or thirty drops; the aqueous extract of the white hellebore, in doses of from two to four grains; different preparations of the ranunculæ, of the *rhus radicans, toxicodendron, &c.*, have sometimes produced amendment in the leprous spots, when numerous and much inflamed, and without causing any particular derangement of the digestive organs.   Nevertheless, the immoderate and inconsiderate employment of these kind of medicines easily converts them into true poisons.

" Experiments, subsequent to those of Dr. Lettsom, go far towards confirming what he said of the advantages obtained from the use of the decoction of the bark of the pyramidal elm ; but it is seldom thought of now in the treatment of lepra.

" Further researches, then, must be prosecuted in the treatment of lepra. They should more particularly be directed to extending the domain of external medication, and restraining the employment of internal remedies, which are without efficacy unless violent, and, on the other hand, are the more dangerous as they are the more active.

" Lastly, it may be remarked, that a sober, regular mode of life, habitual use of white meats and fresh vegetables, ripe and juicy fruits, and milk, tend to favour the action of the different remedies enumerated ; and which must, in turn, be had recourse to in so obstinate a disease, when it is not judged prudent to abandon the case to itself.

" A great number of cases have been published under the name of *lepra*, but most of them differ from the squamous inflammation, the subject of this article. On the opposite side, some cases of well-characterised lepra have been described under other names. Alibert has given two examples, as *dartre furfuracée arrondie*, and M. Marcolini has detailed a case under the designation of *maladie impetigineuse*."

The Lepers of whom we read, therefore, as excommunicated from their species, and remarkable for their disgusting salaciousness, have been probably the victims of the vice alluded to—the effect has been mistaken for the cause.

There are only a few remedial measures, either of a constitutional or local character, which I have found to be successful in the treatment of Lepra.

1st. As regards the Lepra of young persons, who have brought on the disease by habits of the kind I have alluded to, the treatment is often a difficult one to decide on, *i. e.* whether it should be moral or medical, or judiciously composed of both. If there be candour on the part of the patient, which is not common, advice may be given, and medicines may be prescribed with a fair chance of success; if, on the contrary, confidence be withheld from the practitioner, the utmost he can do is to direct such measures to be adopted as will meet, and make up, for the wear and tear of the constitution arising out of such habits.

It not unfrequently happens, nevertheless, that intelligent young men freely confess the folly of which they have been guilty, and readily submit to the measures which are necessary. Involuntary seminal emissions during sleep, they will say, have become habitual, and they truly refer this habit to the former voluntary practice.

The practitioner, in such a case as this has a moral as well as a medical duty to discharge. It is evident that the one should go hand-in-hand with the other. Without the total abolition of the practice, it is plain that all exertions on the part of the surgeon must be thrown away.

Nocturnal seminal emissions in the cases of young men affected with Lepra have been so commonly connected with that disease, that it has been customary with me for a long time to endeavour to arrive directly or indirectly at the knowledge of the fact—when it commenced, and of how long standing? I do not say that it is always the result of *bad habits;* on the contrary, it has sometimes appeared to be the result of, as many persons would call it, *good habits!* It is a doubtful and difficult point to deal with, and one which I need not further allude to. On the whole there can be no question that the Lepra of this country is like the Pellagra of Lombardy, a disease *induced by* debility.

Debility *comes in many forms as regards the diseases of the surface,* and it does not fail under any circumstances or climes to show its distinctive features, such as I have described hitherto. I shall give a further illustration, or rather proof, of the correctness of my opinion, when I speak of Purpura.

The treatment of cases of this description is rather difficult; our way, for the reasons mentioned, is seldom clear. We are obliged to sound the patient's *mind* to get the requisite information, and *that* he is seldom inclined to open to us. The *methodus medendi* consists of a few articles of very familiar acquaintance. 1st. Preparations of iron, as for instance the Mist. Ferri comp. 2. Quinine.

3. Preparations of arsenic.   4. Of Cantharides in the form of tincture, the latter being particularly efficacious where the practice before alluded to is known to have had existence for any lengthened period.

Arsenic, quinine, iron, in the form of the sulphate, and the judicious admixture of the tincture of cantharides with these, have produced a perfect cure in a variety of instances in which hope had been entirely given up.

———

CHAPTER II.

OF PSORIASIS,

OR, DRY SCALY TETTER—CHRONIC INFLAMMATION OF THE CUTIS, DARTRE, ITCH, SCALE, &c. *

The meaning allotted to the term Psoriasis, as invented by Dr. Willan, is that of a disease as clearly dependent on disordered action

* " Psoriasis is a chronic inflammation of the skin, limited to a single region of the body, or occupying almost its entire surface, appearing primarily under the form of solid elevations which change into squamous patches of different sizes, not depressed in the centre, and of which the edges are irregular and but slightly raised.

" Four principal varieties of psoriasis are reckoned ; these are : 1st. Psoriasis *discreta (guttata* Willan). In this variety numbers of small distinct elevations, and squamous patches, occur, from two to four lines in diameter, irregularly circumscribed, and of a form and appearance very analogous to that which results when the body is sprinkled with water and the fluid lies in large drops upon its surface ; such, indeed, appears to have been the origin of the epithet employed by Willan, to distinguish this variety.

" Each of the squamous patches is announced by a small solid, red elevation, the size of a pin's head, the summit of which soon gets covered with a minute dry scale of a dull white colour.   These patches are irregularly rounded, slightly prominent, especially towards their centre, and separated from each other by considerable intervals of healthy skin.   When the patches are freed from their investing squamæ, the corion appears red and irritable, and if the whole have been thus cleansed, the disease appears in the shape of rounded spots, from two to four lines across, slightly prominent and of a brownish-red colour.   These patches occasionally get well, like those of lepra, from the centre towards the circumference : in this case the middle of the patches presents an accidental depression, and acquires a slightly yellow tint ; in proportion as the cure advances, these patches become transformed into segments or small arcs of circles, and after it is accomplished, the skin still presents small stains of a greyish brown, or yellowish cast, in those points that had been possessed by the eruption.

" Psoriasis *discreta* is seldom accompanied with much pruritus, even when the body is heated by exercise or any other accidental cause.

" The patches of this variety of psoriasis may be confined to the hairy scalp, face, trunk, or extremities, or be disseminated over the whole of these regions, appearing either at once upon all of them, or upon each in succession.   Almost always very irregularly disseminated, the patches appear crowded in one situa-

of the vessels forming the cuticle, as that which has occupied the foregoing pages. " It may be briefly described as a rough and scaly state of the cuticle, sometimes continuous, sometimes in separate

tion, and very thinly sown in another; on the extremities they are always observed to be more numerous in the line of extension than in that of flexion. Psoriasis *discreta* makes its attacks most commonly in the spring and autumn, and occasionally disappears spontaneously during summer; it has been known to appear and disappear in this manner for several years in succession. In children it is more quickly evolved than among adults.

"2nd. Psoriasis *confluens* (*diffusa*, Willan.) Instead of being separate and distinct, it much more frequently happens that the primary papulæ of psoriasis are evolved so close together that the squamous patches which succeed them meet and blend by their corresponding edges. As may be imagined, the patches in this instance are of very various sizes and forms. As in psoriasis *discreta*, each of the smaller patches which goes to the composition of the larger clusters, begins in the form of a solid papular-looking eminence, on the top of which a dry scale of a dull white colour is soon formed. The patches increase in size, become confluent, and compose at length an irregular squamous surface, upon which, however, the original spots may often be distinguished. These large patches are sometimes irregularly intersected by red lines, and here and there show angular portions of integument which are free from squamæ.

" Psoriasis is more constantly *confluent* on the limbs than on the trunk; and the patches characteristic of this form of the disease frequently disappear on one district at the same time that the eruption which gives rise to them makes its appearance on another.

" The arrangement of the eruption of psoriasis in small distinct circular spots, or in broad confluent patches, is no evidence of diversity in the nature of the disease; it is frequently seen in the shape of psoriasis *discreta* on the body, whilst it has the character of psoriasis *confluens* on the limbs.

" Patients labouring under confluent psoriasis of recent date, complain of a considerable degree of pain and pruritus in the affected parts, which is always increased by the warmth of bed, the vicinity of a fire, and any other cause that tends to stimulate, or raise the temperature of the surface.

" Psoriasis *confluens* of the fore-arms and legs is occasionally seen forming a kind of irregular band, and sometimes, but more rarely, a sort of case which includes the affected limb, through its entire length. In this instance, instead of the usual micaceous squamæ of psoriasis we occasionally only distinguish an agglomeration of minute furfuraceous scales, the colour of which approaches that of the flour of mustard. When the squamæ, in such cases, have been got rid of by baths, vapour douches, &c., the surface they covered appears smooth, shining, and highly inflamed.

"3d. Whether this squamous inflammation have appeared under the form of small *distinct* patches (psoriasis *discreta*), or of confluent masses (psoriasis *confluens*), when it has existed during many months, or several years, especially when it can be traced to a hereditary taint, or attacks individuals of shattered constitution, the disease gets worse and worse, the skin becomes hard, thickened, tense, and inelastic, yielding uneasily to the motions of the limbs, and appearing to undergo a kind of hypertrophy; the primary patches of the disease are no longer distinguishable, but the integument is covered with hard, dry, and thick white scales; numerous chaps of various depths soon follow, furrowing the surface in all directions, but especially in those of the natural folds of the skin (psoriasis *inveterata* Willan, *agria* of the ancient writers); and in those rare cases in which trunk and extremities are involved in one common incrustation, the disease assumes a hideous appearance, and the surface of the body has been compared by some pathologists to the rugged bark of an aged tree. This circumstance has even led M. Alibert to designate this last and inveterate stage of the disease, under the name of *dartre squameuse lichenoide*. The squamæ at this stage frequently rise in strong relief from the skin, exceeding by from a

patches of various sizes, but of an irregular figure, and for the most part accompanied with rhagades or fissures of the skin.'' Most frequently it makes its first appearance in the spring or summer,

quarter to half a line the level of the neighbouring healthy parts. They are also then produced in such abundance, that quantities may always be gathered from the beds, and shaken from the clothes of patients. These squamæ are occasionally a full line in thickness. It is in the vicinity of the articulations, that chaps or cracks occur most commonly : these get deeper and deeper, bleed when motion is attempted, and often pour out a glutinous fluid that dries up into linear incrustations. Farther, the parts affected, are frequently the seat of a burning pruritus, especially during the night. To conclude, considerable superficial excoriations have been seen to form upon the back, buttocks, and lower limbs, when these parts were the seat of this disease, which caused the patients much and extreme suffering.

" 4th. Under the title of psoriasis *gyrata* Willan has described a squamous vermiform eruption, characterised by squamous bands, spirally twisted or arranged longitudinally, and traversed by numerous superficial lines, corresponding evidently to the natural folds or wrinkles of the skin. These bands, however disposed, are affected with a very slight furfuraceous desquamation. I have met with but two cases of this variety, in neither of which could I observe anything like papulæ, or round squamous patches, analogous to those of psoriasis *discreta :* this eruption is seldom attended with pruritus, even when the temperature of the surface is increased by exercise or any other cause.

" Psoriasis may continue from a few months to several years ; it is always of long duration when the patches are numerous and it can be traced hereditarily.

" Independently of the remarkable varieties presented by this disease, which have now been described, it still offers several additional peculiarities according to the regions of the body it attacks.

" 1st. Psoriasis seldom appears primarily upon the *hairy scalp.* It there usually occurs in the *distinct* form; the squamæ are always yellower and more pulverulent than when they are produced from the trunk. The *confluent* form of the disease is still rarer here. I have, however, seen it covering almost the whole surface of the hairy scalp, and extending to the forehead, in a line parallel with that of the implantation of the hair, under the form of a prominent band, an inch in breadth, whose surface was covered with rough squamæ of a dull white colour, and whose lower edge was red and much raised above the level of the healthy skin. The inflammation of psoriasis frequently attacks the bulbs of the hair, which is then detached from the points affected.

" 2nd. Psoriasis frequently attacks the face, at the same time that it appears on other regions of the body ; the eruption, however, may be entirely confined to the countenance. The patches that characterize the disease in this situation are red and furfuraceous, and the squamæ are usually very light and thin. The subcutaneous cellular tissue becomes tumified when the disease is of long standing and has assumed the characters of the variety designated *inveterata.* On the eyelids it is announced, as indeed it is everywhere else, by the formation of papulæ, which usually appear about the angles of the eyes, the eyelids soon become stiff, tense, and chapped ; and in children the eruption is occasionally followed by the loss of the ciliæ and hairs of the eyebrows. Several other scaly affections of the eyelids, as well as the disease which, following Willan, I formerly described under the name of *psoriasis of the lips,* now appear to me to belong to the group of the pityriases, among which they will be found discussed.

" 3rd. Psoriasis of the *trunk* rarely occurs alone, the disease almost invariably affects the extremities at the same time. When inveterate, its squamæ are commonly thinner and smaller than those observed in psoriasis of the extremities, especially when the knees and elbows are the parts affected.

" 4th. Psoriasis of the *scrotum* is a rare disease ;* when it does occur it ge-

* Psoriasis, both of the scrotum and prepuce (particularly the latter) is exceedingly common in England.—S. P.

when the determination to the skin is strongest, and when exposure to heat or exertion favours the latter. Some heat, itching, and irritation, is first perceived, a scale forms, which speedily separates,

nerally becomes *inveterate*, and is attended with excessive pruritus, and most painful chaps. I have met with a case of psoriasis *discreta*, in which the patches were disposed in a line parallel with the raphe. The circular spots of psoriasis *discreta* occurring on the scrotum and verge of the anus in children, have been known to be mistaken for syphilitic tubercles, from which they differ in their mode of formation, which is that of squamous papulæ, in the more decidedly scaly look of their surface, and generally by the almost uniform existence at the same time of scaly patches upon the belly and thighs.

"5th. Psoriasis of the *prepuce* is also of very rare occurrence, almost always becomes inveterate, and is attended with thickening of the skin, and painful and bleeding chaps, which are occasionally accompanied by some degree of swelling in the lymphatic glands of the groin. This form of psoriasis, usually very obstinate among adults, has occasionally required the operation for phymosis for its removal. It is of great consequence not to confound the squamous patches by which psoriasis of the prepuce begins with a syphilitic eruption occurring on the same part.

"I have, lastly, to make particular mention of three varieties of psoriasis of the hands :—

"1st. Psoriasis *palmaris* may be distinct or confluent; in either case the disease begins with broader elevations than those of the other varieties : they are reddish in colour, and the seat of a considerable degree of heat and itchiness; pressed upon strongly, or pinched, they become painful; and if rather numerous, the patient is compelled to give up every kind of manual labour. In the confluent species, the palm of the hand swells generally, and becomes of a uniform violet-red colour. The feeling of heat which was complained of at first, becomes gradually less distressing, and ceases at length in a great measure, whilst the painful pruritus that accompanied it, becomes much less troublesome; during this time the cuticle that surrounds the elevations grows considerably thicker, acquires a yellowish colour like the skin of the heal, dries up and becomes friable, and at last of a dead white on the summits of the patches. At this period the altered and cracked epidermis covering the elevations is detached, either spontaneously or by the nails of the patient, and leaves a new epidermis through which the corion shines. The epidermis in the vicinity of the diseased parts also undergoes modifications; it becomes considerably thicker than usual, and of a dirty yellow hue, it dries and looks mealy on the surface, and finally exfoliates irregularly, at first in the neighbourhood of the older patches, and then in that of the joints and natural folds of the skin of the hands. The desquamation in this case is always *irregular* and is very different in appearance from that which happens in the variety next to be described,—the psoriasis *palmaris centrifuga;* like it, however, and perhaps even more constantly, it is attended with linear fissures, which penetrate to the quick in the course of the lines of the palm, and with a still greater number of clefts that extend less deeply and do not even reach the corion.

"2d. Psoriasis *palmaris centrifuga.* This variety begins in the palm of the hand by a solid elevation, whose summit is covered with a small white and dry epidermic scale; this spot is then surrounded by a reddish ring upon which the epidermis dries and is thrown off circularly. Around this first circle a second is before long formed, upon which a similar process of desquamation takes place, and these circles may appear one after the other, becoming more and more eccentric, until the whole palm of the hand is implicated, and squamous patches even appear on the palmar aspects of the fingers. The diseased parts are affected with a very troublesome pruritus, which is increased whenever the hand is exposed to any elevation of temperature, or even when the fingers are moved for any length of time. When patients yield to the impulse to scratch the parts affected, the skin assumes a violet-hue, and at a later stage presents numerous

leaving a red inflamed skin. Another and another follow, each succeeding being thicker than the last, till some moisture, or even pustulation takes place ; the scales at length crack and separate piece

cracks that pursue the course of the lines habitually observed in the palm. The surfaces between two of these crevices are covered with very hard and thick squamæ, and the whole palm is stiff and dry. This form of psoriasis has been principally observed among washerwomen, and others whose hands are habitually exposed to the contact of alkaline leys, or among coppersmiths, silversmiths, and tin smiths, the palms of whose hands are irritated by repeated pressure, combined with the contact of different metallic substances. Psoriasis *palmaris* usually grows worse in winter and often gets well in summer. After recovery, the skin remains for some time smooth and of a dusky-red colour. It seldom happens that the disease does not recur several times after getting well, when those affected do not give up the craft they may be exercising at the time, and which has been recognized as its occasional cause at least.

"3d. There is a variety of psoriasis *diffusa* which is developed on the backs of the hands, and is known under the name of *grocers' itch*, because frequently seen among persons exercising this trade, although it also attacks bakers, laundresses, and even individuals of the better classes of society. The eruption begins in the shape of two or three squamous elevations, which spread to such a degree as at length to cover the whole dorsum of the hand. The integuments are before long seamed with numbers of dry and painful chaps ; these occur especially over the wrist joint and the articulations between the metacarpal bones and first phalanges of the fingers. This variety of psoriasis is distinguished from the confluent and chronic lichen of the backs of the hands by the circumstance of the latter being always preceded by a considerable eruption of small papulæ. It is also important to distinguish this true psoriasis from the squamous inflammations artificially excited, to which individuals exercising certain callings are subject. When one of these varieties, or any other form of psoriasis extends to the whole back of the hand, the matrix of the nails is occasionally attacked with chronic inflammation, in which case the nails themselves become thickened, bent, cracked, and at length detached ; they are in due time succeeded by others, but these are liable to be affected in the same way. Psoriasis *plantaris* is a rarer disease than psoriasis *palmaris*, and less uniformly attended with chapping of the skin.

"4th. With regard to psoriasis of the *inferior extremities*, I have only to remark, that the disease is very apt to become *inveterate*. The legs then appear furnished with a general adventitious squamous envelope, which does, indeed, bear some resemblance to the lichenous covering of trees with which it has been compared.

"Psoriasis rarely appears complicated with other inflammatory affections of the skin, if we except lepra. It has, however, been seen, especially among children, occurring with characters of the greatest inveteracy along with eczema *impetiginodes*. Amidst the thin squamæ that cover the diseased patches, vesicles, and particularly purulent points, are then perceived. At a later period, these surfaces may become excoriated, and form thin, lamellar, yellowish-coloured scabs like those of the eczema. It occasionally happens, especially when children are attacked with the disease during the period of teething, that on the first invasion, or during the course of psoriasis, a certain degree of derangement is apparent in the functions of the digestive organs. It was this circumstance, undoubtedly, that led Willan and Bateman to speak of epigastric pains, lassitude, headach, and various other symptoms, as precursors of psoriasis, which are, in fact, rarely observed in connexion with the disease, unless when developed under the circumstances indicated.

"*Causes.* Next to eczema and lichen, psoriasis is one of the most common of the chronic affections of the skin. Of all the variety of forms assumed by the disease, that which I have designated psoriasis *discreta* is the most common ; in a given number of cases this will be found in the proportion of three-fifths

by piece, and the fissures in the different scales extend beneath the surface of the cutis. If the part affected be situated in the neighbourhood of joints, or otherwise influenced by flexion and extension,

to the whole. Psoriasis shows itself principally among adults, between the age of twenty-five and thirty years; women of a nervous and sanguine temperament are particularly obnoxious to its attacks. Of all the chronic and noncontagious affections of the skin, psoriasis is that, the hereditary nature of which is most satisfactorily ascertained. The seasons have a very marked influence on the production of this disease; it is usually developed in the beginning of the spring or autumn. The influence of different trades or professions appears limited to a few local varieties. To conclude, irritation of every kind, direct or indirect, applied to the skin,.may prove the occasional cause of psoriasis; the disease has, in fact, been seen succeeding repeated attacks of lichen, prurigo, different other diseases of the skin, and the application of a common blister.

"*Diagnosis.* Psoriasis can only be confounded with three diseases, which like it affect the squamous form, namely lepra, pityriasis, and scaly syphilis. There exists, indeed, between lepra and psoriasis a very great similarity. The resemblance of psoriasis *discreta* to lepra is more particularly striking. Both of these affections of the skin commence as solid papular-looking elevations, both soon assume the shape of circular scaly patches, and these in the same patient frequently present the appearances characteristic of psoriasis,*discreta* on the trunk, and those distinctive of lepra on the knees and elbows.

"Some recent writers have consequently maintained that lepra and psoriasis were nothing more than two varieties of the same disease. There are, however, certain characters which distinguish these two eruptions, or if it must be so, these two degrees of the same form of inflammation from each other. The patches of psoriasis *discreta*, for instance, are never so broad as those of lepra, neither are their edges raised, nor their centres depressed like those of the latter; in psoriasis, too, the squamæ adhere more firmly and are less abundantly produced than those of lepra. The differences between lepra and psoriasis *diffusa* are still more marked. The patches of the latter are irregular and not depressed in their centres, those of the former are exactly circular, and even when several leprous spots are blended together, their originally circular shapes continue to be proclaimed by the arcs of circles presented on their circumference.

"Psoriasis is distinguished from syphilitic scaly patches (psoriasis *syphilitica*, Willan), by the coppery or livid hue of the latter, which is deeper towards their centre than their circumference; they are also without true squamæ, being covered by a transparent epidermic layer, which is usually detached circularly, exposing a smooth and shining surface of a coppery-yellow, bounded by an epidermic edge. Syphilitic blotches, farther, are not accompanied with any pruritus, and are generally complicated with evident venereal affections of the pharynx or conjunctiva, with nodes, nocturnal pains, &c., and get speedily well under the influence of mercurial medicines carefully administered. Lastly, these blotches disappear from their circumference towards their centre, a character which of itself distinguishes them sufficiently from the squamous patches of lepra and some of the varieties of psoriasis.

"Psoriasis *discreta* of the hairy-scalp differs from pityriasis of the same part, by appearing under the form of patches which, beneath the squamæ, always present a red central point that passes the level of the skin generally.

"The varieties of psoriasis designated by Willan *diffusa* and *inveterata* commonly prove very obstinate diseases, more so even than lepra. Psoriasis *discreta* vel *guttata*, in a general way, is less rebellious than the *confluens* vel *diffusa*, which in its turn is less intractable than the *inveterata ;* this last variety, indeed, is often absolutely incurable. When psoriasis *discreta* begins to get well, the amendment is announced by the sinking of the patches, and takes place at first in one or several points of a particular district, from which it spreads to the other regions affected. When the varieties *diffusa* and *inveterata* end in recovery, the chaps of the skin disappear, the inflammation of the corion diminishes gradually,

blood frequently issues from their bottoms, accompanied by much pain and smarting.

The term Psora, as adopted by the Greek physicians, was applied

the altered epidermis is replaced by one of less thickness, less dry and inclined to crack, and after many successive desquamations, the dermis in the points affected becomes covered with a cuticle similar to that of the healthy integument.

"The medical treatment, and dietetic plan pursued in psoriasis ought to be based on the same principles as those that guide us in lepra; the curative means only require to be modified according to the more or less highly inflamed condition of the integuments. When psoriasis *discreta* is recent, and an adult is its subject, the disease must be attacked by one or two venesections. I am in possession of a sufficient number of facts to show that this measure is constantly serviceable, and the same conclusion has been come to by many other practitioners. Plain baths, or better still, narcotic and emollient temperate baths, which lessen the irritation of the skin and the pruritus with which it is constantly accompanied, are to be employed at the same time and subsequently. The douche and vapour-bath are also frequently found of service in the cases of adults. By using these alternately with sulphureous water-baths, confluent psoriasis, accompanied with no great degree of inflammation, are frequently got rid of in the course of three or four months. When confluent psoriasis is of old standing, a modification of the skin may often be advantageously attained by anointing it with an ointment of the tartrate of antimony; the same practice has even been found serviceable in some cases of the *inveterate* disease, although in these the alteration of the skin is so deep that the affection is generally incurable, at least among the aged. *Inveterate* psoriasis is always improved by the use of the emollient and narcotic-bath, as well as of the vapour-bath and douche. Among the aged, attacked with inveterate psoriasis, whose skin is thickened and indurated in different parts of the body, the treatment must be limited to such palliative measures. The same plan also appears to me better than any other when the subject of the disease is a member of the labouring class of the community, who would certainly have a relapse as soon as he was thrown upon his old occupations and habits. Such patients, labouring under inveterate psoriasis, have been seen, who derived no benefit whatever from taking the vapour-douche and bath more than one hundred and fifty times, or having undergone the most energetic internal treatment, from which, indeed, unpleasant symptoms, of different degrees of severity, resulted.

"In the treatment of psoriasis *diffusa* and *discreta* the exhibition every day for some months, of half an ounce of Epsom salts, or two drams of sub-carbonate of potash, or a few grains of calomel in combination with the resin of jalap, so as to procure several alvine evacuations, has been favourably recommended, patients at the same time making use of the tepid-bath; the purgative medicine, however, is immediately to be suspended should unequivocal symptoms of gastro-intestinal disturbance at any time make their appearance. This plan seems more particularly available in psoriasis of the face and hairy scalp.

"The deuto-chloride of mercury, in the dose of one-fourth of a grain, and the sulphurated sulphite of soda in the quantity of a scruple, daily, have also appeared to accomplish some cures of psoriasis. The tincture of cantharides is another medicine which has been tried in the different species of psoriasis, particularly the *inveterata*, with some success; the dose has been carried gradually from five to sixty drops every day, when no appreciable disturbance in the digestive, respiratory, or urinary organs supervened. The medicine is often continued for three or four months before any favourable change takes place in the state of the skin. One or other of the common arsenical solutions has also been strongly recommended in this obstinate form of disease. Arsenic of course requires great care in its exhibition; its effects must be closely watched, and it is advisable to give up its use for a few days every now and then. It is undeniable that by means of these active medicines several of the varieties of psoriasis,

to the disease now termed Scabies; the characters of which it is scarcely necessary to say are totally unlike any dry scaly disease, in any of their stages; the term Psoriasis, therefore, as conveying an idea of analogy, is obviously improper.

From attentive observation of the varieties in figure, (for in point of fact, there are no other essential differences between this and the disease termed Lepra) I am fully convinced that all the purposes of useful discussion would have been effectually consulted in including them under one head with the latter. The information we have at present acquired in the modern study of cutaneous diseases does not enable us to find a better reason for their separation, than that afforded by the circumstances of its having originated with the ancients.

Dr. Willan says, that "From the Lepra it may be distinguished not only by the different form and distribution of the patches, but also by its cessation and recurrence at certain seasons of the year, and by the disorder of the constitution with which it is usually attended." Spots of genuine Lepra, however, fully answering the description of this author, as regards their local characters, not only appear now and then by themselves accompanied by symptoms of constitutional disorder, but very frequently mixed with the irregular patches of Psoriasis, without such attendant symptoms.

Medical men are often puzzled regarding the designation they shall give to an extensive scaly disease, because they are unable to determine to which of the two affections termed Lepra and Psoriasis it most nearly approximates. Its less tedious duration, and liability to recur, certainly constitutes a difference; but such difference depends simply on its extent, or the degree of irritability of skin of the patient. The latter circumstance also explains the superior

even the most inveterate, have been cured; but it is no less certain that the majority of the cures, thus accomplishéd, have been but temporary, relapses having occurred the following spring or autumn; that such relapses are more especially frequent among the labouring classes of the community, and, lastly, that the greater number of cases of psoriasis *inveterata*, treated by such means, have been in nowise amended, although the medicines were continued for five or six months. I am therefore of opinion that it is in general inexpedient to put patients affected with psoriasis *inveterata* upon an arsenical course, in the faint hope of deriving a mere temporary improvement, with the fear before our eyes of inducing some obstinate derangement of the digestive organs, or of permanently injuring the general constitution.

" In a word, then, psoriasis *discreta* and *diffusa* can be successfully attacked by measures less dangerous than those just mentioned, by the vapour-bath, ointment of white precipitate. &c. &c.; and to me a palliative plan of treatment appears the only one available in cases of inveterate psoriasis, especially when its subjects are individuals belonging to the labouring classes of society.

" The *local* varieties of psoriasis offer the same curative indications as the general disease. Fomentations, baths, cataplasms, emollient and narcotic unguents, are all useful when the skin is red and painful. In psoriasis *palmaris* simple baths and fomentations, the vapour douche, and calomel ointment are habitually employed. When the disease has arisen from, or seems to be kept up by, any evident outward cause, the first indication of course is to make this unavailing."

degree of tenderness of the abraded cutis now and then manifest
in Psoriasis ; but with respect to the rhagades or fissures spoken of
as another point of difference, it is proper to observe, that they are
sometimes seen in round leprous patches, situated in the neighbourhood of joints.

The irregularly formed patches of this disease are, according to
Dr. Falconer, (and it is an observation confirmed by the evidence
of others) frequently following checks to the perspiration, by copiously drinking cold water when heated by exercise : the eruption
of Lepra often occurs under the same circumstances.   A variety of
other occurrences, supposed in different cases to have given rise to
it, are detailed also in the list of supposed causes of Lepra.

Besides the local varieties of Psoriasis which Dr. Willan has
spoken of, he has enumerated four of a general character occurring
on different parts of the body, appearing to be produced from constitutional causes only.   These are the P. guttata, diffusa, gyrata,
and inveterata.*

* For the convenience of the reader we transcribe the descriptions alluded to.
   1. *Psoriasis guttata.* This complaint appears in small, distinct but irregular
patches of laminated scales, with little or no inflammation round them.   The
patches very seldom extend to the size of a sixpence.   They have neither an
elevated border, nor the oval or circular form by which all the varieties of lepra are
distinguished ; but their circumference is sometimes angular, and sometimes
goes into small serpentine processes. The scale formed upon each of them is thin
and may be easily detached, leaving a red shining base. The patches are often distributed over the greatest part of the body, but more particularly on the back part
of the neck, the breasts, arms, loins, thighs, and legs. They appear also upon the
face, which rarely happens in lepra.   In that situation they are red and more
rough than the adjoining cuticle, but not covered with scales. The Psoriasis
guttata often appears on children in a sudden eruption, attended with a slight
disorder of the constitution, and spreads over the body within two or three days.
In adults it commences with a few scaly patches on the extremities, proceeds
very gradually, and has a longer duration than in children.   Its first occurrence
is usually in the spring season, after violent pains in the head, stomach, and
limbs. During the summer it disappears spontaneously, or may be soon removed
by proper applications, but it is apt to return again early in the ensuing spring,
and continues so to do for several successive years.   When the scales have been
removed, and the disease is about to go off, the patches have a shining appearance, and they retain a dark red, intermixed with somewhat of a bluish colour,
for many days, or even weeks, before the skin is restored to its usual state.
   2. The *Psoriasis diffusa* spreads into large patches irregularly circumscribed,
reddish, rough, and chappy, with scales interspersed. It commences, in general,
with numerous minute asperities, or elevations of the cuticle, more perceptible
by the touch than by sight.   Upon these small distinct scales are soon after
formed, adhering by a dark central point, while their edges may be seen white
and detached.   In the course of two or three weeks all the intervening cuticle
becomes rough and chappy, appears red, and raised and wrinkled, the lines of
the skin sinking into deep furrows. The scales which form among them are
often slight, and repeatedly exfoliate. Sometimes without any previous eruption
of papulæ, a large portion of the skin becomes dry, harsh, cracked, reddish, and
scaly, as above described. In other cases, the disorder commences with separate
patches of an uncertain form and size, some of them being small, like those in
the psoriasis guttata, some much larger.   The patches gradually expand till
they become confluent, and nearly cover the part or limb affected. Both the psoriasis guttata and diffusa likewise occur as a sequel of the lichen simplex. This

The first of these, at its commencement, is formed of distinct and *small* patches, with *irregular* circumferences; they appear on almost every part of the body, and even on the face.   The second

transition takes place more certainly after frequent returns of the lichen. The parts most affected by psoriasis diffusa are the cheeks, chin, upper eyelids and corners of the eyes, the temples, the external ear, the neck, the fleshy parts of the lower extremities, and the fore-arm, from the elbow to the back of the hand, along the supinator muscle of the radius. The fingers are sometimes nearly surrounded with a loose scaly incrustation; the nails crack and exfoliate superficially. The scaly patches likewise appear, though less frequently, on the forehead and scalp, on the shoulders, back, and loins, on the abdomen, and instep. This disease occasionally extends to all the parts above-mentioned at the same time; but, in general, it affects them successively, leaving one place free, and appearing in others; sometimes again returning to its first situation. The psoriasis diffusa is attended with a sensation of heat, and with a very troublesome itching, especially at night. It exhibits small slight, distinct scales, having less disposition than the lepra to form thick crusts. The chaps or fissures of the skin, which usually make a part of this complaint, are very sore and painful, but seldom discharge any fluid. When the scales are removed by frequent washing, or by the application of unguents, the surface, though raised and uneven, appears smooth and shining; and the deep furrows of the cuticle are lined by a slight scaliness. Should any portion of the diseased surface be forcibly excoriated, there issues out a thin lymph, mixed with some drops of blood, which slightly stains and stiffens the linen, but soon concretes into a thin dry scab; this is again succeeded by a white scaliness, gradually increasing and spreading in various directions. As the complaint declines, the roughness, chaps, scales, &c. disappear, and a new cuticle is formed; at first red, dry, and shrivelled, but which, in two or three weeks, acquires the proper texture. The duration of the psoriasis diffusa is from one to four months. If, in some constitutions, it does not then disappear, but becomes, to a certain degree, permanent, there is, at least, an aggravation or extension of it, about the usual periods of its return. In other cases, the disease, at the vernal returns, differs much as to its extent, and also with respect to the violence of the preceding symptoms. The eruption is, indeed, often confined to a single scaly patch, red, itching, and chapped, of a moderate size, but irregularly circumscribed. This solitary patch is sometimes situated on the temple, or upper part of the cheek, frequently on the breast, the calf of the leg, about the wrist, or within and a little below the elbow-joint, but especially at the lower part of the thigh, behind. It continues in any of these situations several months, without much observable alteration. The complaint, denominated with us the baker's itch, is an appearance of psoriasis diffusa on the back of the hand, commencing with one or two small, rough, scaly patches, and finally extending from the knuckles to the wrist. The rhagades, or chaps and fissures of the skin, are numerous about the knuckles and ball of the thumb, and where the back of the hand joins the wrist. They are often highly inflamed, and painful, but have no discharge of fluid from them. The back of the hand is a little raised or tumefied, and at an advanced period of the disorder exhibits a reddish, glossy surface, without crusts or numerous scales. However, the deep furrows of the cuticle are, for the most part, whitened by a slight scaliness.   This complaint is not general among bakers; that it is only aggravated by their business, and affects those who are otherwise disposed to it, may be collected from the following circumstances:—1. It disappears about midsummer, and returns in the cold weather ar the beginning of the year; 2. Persons constantly engaged in the business, after having been once affected with the eruption, sometimes enjoy a respite from it for two or three years; 3. When the business is discontinued, the complaint does not immediately cease. The grocer's itch has some affinity with the baker's itch, or tetter; but, being usually a pustular disease at its commencement, it properly belongs to another genus. Washerwomen, probably from the irritation of soap, are liable to be affected with a similar scaly disease on the

consists of *large* patches, also *irregularly* circumscribed.   The
third, as its name implies, is of a tortuous or serpentine character ;
and the fourth begins in separate irregular patches, which extend,
and become confluent, until at length they cover the whole surface
of the body, except a part of the face, or sometimes the palms of the
hands and soles of the feet, with an universal scaliness, interspersed
with deep furrows, and a harsh, stiff, and thickened state of the
skin.   The production of scales is so rapid, that large quantities are
found every morning in the patient's bed.   The nails become con-
vex, thickened, and opake, and are frequently renewed ; and at an
advanced period, especially in old people, extensive excoriations
sometimes occur, with a discharge of lymph, followed by a hard,
dry cuticle, which separates in large pieces.   " *In this extreme degree it approaches very closely to the inveterate degree of Lepra
vulgaris in all respects, the* only *difference being in the form of
the patches before they coalesce.*" *

The constitutional treatment of these which has been found most
successful, consists chiefly of the exhibition of those remedies with
the mention of which the consideration of Lepra was concluded.
The state of the constitution and chylopoietic viscera should always
obtain the first attention ; and if any feelings of irritation in the dis-

hands and arms, sometimes on the face and neck, which, in particular constitutions, proves very troublesome and of long duration.

3. The *Psoriasis gyrata* is distributed in narrow patches or stripes, variously
figured ; some of them are nearly longitudinal ; some circular, or semicircular,
with vermiform appendages; some are tortuous, or serpentine ; others like earth-
worms or leeches ; the furrows of the cuticle being deeper than usual, make the
resemblance more striking, by giving to them an annulated appearance.   There
is a separation of slight scales from the diseased surface, but no thick incrustations are formed.   The uniform disposition of these patches is singular.   I have
seen a large circular one situated on each breast above the papillæ; and two or
three others of a serpentine form, in analogous situations along the sides of the
chest.   The back is often variegated in like manner, with convoluted tetters,
similarly arranged on each side of the spine.   They likewise appear, in some
cases, on the arms and thighs, intersecting each other in various directions.   A
slighter kind of this complaint affects delicate young women and children in
small scaly circles or rings, little discoloured ; they appear on the cheeks, neck,
or upper part of the breast, and are mostly confounded with the herpetic, or pustular ring-worm.   The psoriasis gyrata has its remissions and returns, like the
psoriasis diffusa ; it also exhibits, in some cases, patches of the latter disorder
on the face, scalp, or extremities, while the trunk of the body is chequered with
the singular figures above described.

4. The *Psoriasis inveterata* is characterised by an almost universal scaliness,
with a harsh, dry, and thickened state of the skin.   It commences from a few
irregular, though distinct patches on the extremities.   Others appear afterwards
on different parts, and, becoming confluent, spread at length over all the surface
of the body, except a part of the face, or sometimes the palms of the hands and
soles of the feet.   The skin is red, deeply furrowed, or wrinkled, stiff, and rigid,
so as somewhat to impede the motion of the muscles and of the joints.   So
quick, likewise, is the production and separation of scales, that large quantities
of them are found in the bed on which a person affected with the disease has
slept.   They fall off in the same proportion by day, and being confined within
the linen, excite a troublesome and perpetual itching.

* Bateman's Synopsis, p. 42.

eased parts should be experienced, saline aperients may be employed with advantage. The warm bath, as an application generally calculated to allay irritation, should be often had recourse to, in conjunction with gentle mercurial alteratives : and these together will be found adequate to the effecting the cure in the slighter cases ; but if the disease assumes an obstinate form, the exhibition of the arsenical solution, or the use of the common or sulphur vapour bath, may be individually or conjointly had recourse to, with a tolerable certainty that, by their judicious management, a healthy state of skin may be brought about.

The manner in which the return to a healthy state of skin, both after Lepra and Psoriasis, is effected, is evidently by the diminution of the morbid excitement of the vessels of the part ; the reddened margins of the diseased spots first losing their florid redness, and ultimately changing to a blue colour. When the scales fall off, a little scurf remains for a few days, and the natural colour of the part is resumed. This similarity forms another ground for the opinion of the analogy of the two diseases.

In the treatment of Lepra and Psoriasis, if no obvious benefit is derived by a patient exhibition of general alterative remedies, the frequent use of the warm bath, the common or sulphur vapour bath, the decoction of dulcamara, or such medicines as may be indicated by a cachectic state of constitution, the use of the liquor arsenicalis may be tried with the best hopes of success.

The *modus operandi* of this medicine is involved in much obscurity ; there is, however, some reason to suppose that it is not unlike that attributed by Mr. Hunter to mercury, in the cure of Syphilis. It is certain, that in the diseases under consideration, if employed in small doses, and not followed up and gradually increased to a much larger extent than is commonly supposed prudent, it loses much of the estimation which ought properly to belong to it. The extreme care and circumspection necessary in its exhibition is itself a restraining consideration in its employment ; but if an inveterate disease can be cured by medicinal remedies, such care ought to be readily given.

I am in possession of some notes of a case answering to the description of Psoriasis inveterata, which have been handed to me by Mr. Gaskoin, of Clarges-street, whose various and well-executed drawings of cutaneous and other diseases have obtained for him no ordinary degree of credit. The case in question was treated by Mr. G. in the hospital of St. Louis in Paris. The arsenical solution was commenced in doses of two drops exhibited twice a day. The case was attentively watched, and the dose gradually increased, without any untoward symptom occurring, for the space of two months; at the end of which period, no less than thirty-eight drops were exhibited for each dose. At this time nausea and sickness being complained of, the medicine was suspended ; a very important and rapid change in the state of the disease having occurred within the last few days. Three days after, the severe colicky pains commonly following the exhibition of the medicine, even in much smaller doses, if less gradually administered, came on, followed by cold perspirations and great constitutional disorder. Opiates and cordials soon relieved these symptoms; and so satisfied was the patient of the good effects of the treatment, that though a trifling degree only of the disease remained, he was anxious to return to the use of the medicine. The case was ultimately, though of years' standing, completely cured by adhering to the same plan.

Similar cases have occurred under my own observation, from an extensive use of this mineral, though more recently, in consequence of the comparatively trifling inconvenience of the sulphur vapour bath, I have had recourse to the latter at once, and thus saved the patient much pain and inconvenience.

Of the utility and effect of the sulphur vapour bath in Psoriasis, the following case is by no means an extraordinary instance.

" The subject was a gentleman about forty years of age, who from his boyhood has been afflicted with a most inveterate cutaneous disease, but from which he is sometimes quite free. His parents were not known to have had any skin disease, and he has no clue to enable him to account for its origin. When this gentleman was first subjected to the employment of the bath, he was unable to walk : his legs, from the toes to the middle of the thighs, were incased in a thick scab, with long deep fissures, from which issued an abundant ichorous discharge. The thighs above this were studded with impetiginous pustules, the sacrum, gluteus muscles, and part of the lumbar region, were likewise covered with scab containing deep cracks. The arms and axilla on one side were covered with large patches of scab, all discharging the same kind of fluid. He was obliged to move with great caution, lest he should extend these cracks, or occasion fresh ones, which were always attended with much pain and inflammation. He had been in the habit of taking and persevering in the use of medicine, but was not aware that he derived much benefit from it.

He was directed to take three emollient vapour baths, and afterwards to persevere daily in the use of the sulphureous fumigating bath, and to take occasionally opening medicine. After the third fumigation, the amendment was evident, and he expressed himself much more comfortable in his feelings ; he continued them for a fortnight with regular progressive improvement. At the expiration of a month, the arms and axilla were well, and the scab on the legs was now only in patches, the left leg being much the best. There now came on a fresh accession of disease, showing itself in pustules with yellow heads a little above each inner ancle ; higher up there were numerous small vesicles containing clear lymph : some other parts of the legs still, however, went on improving, and the fumigations were continued : this accession was not of long continuance, but scabbed over, healing underneath. He now complained of heat and itching in the face, particularly about the chin ; he had slight feverish symptoms, for which he was directed a dose of the submur.

hyd. and some saline medicine : he became relieved : achores
made their appearance and gradually spread over the lower jaw,
containing yellowish matter, forming scabs, with surrounding inflammation. To this part he pretty constantly applied flannels wrung
out of hot water, and covered it with a bread and water poultice
at night : there was no cracking of these parts, but much watery
discharge.

He had used the fumigations daily six weeks; he now complained
of stiffness, itching, and redness in the left groin. On examination,
the inguinal glands were enlarged, hard, and painful to the touch ;
as the swelling of the glands subsided, the redness increased, and
he had intertrigo, which became very troublesome to him. A space
the size of a large dollar next made its appearance on the right
clavicle, near the humerus, of a bright threatening aspect, and a
smaller spot of the same appearance showed itself on the left clavicle,
both itching very much, and with moist exudation ; these spots in
forty-eight hours became joined by an inflamed line about an inch
broad, hanging, as it were, like a necklace : round the larger spots
there came many small-sized pustules with yellowish matter in them,
these did not extend ; the large spot gradually became whiter in
the middle, extending to the edges, disappearing slowly, leaving a
shining, smooth red surface, which, with the connecting line, gradually went away : of the intertrigo he was not quite well when
business called him into the country. The legs, back, and arms
were quite well : the gentleman used the bath seven weeks.

" I frequently observe pustules and vesicles at the same time in
impetiginous patients. The affections of the face seemed to be the
Porrigo favosa : the intertrigo again was of a distinct order ; the
affection of the chest seemed to have been communicated from the
chin, perhaps conveyed by frequent looking at the intertrigo of the
groin, as was his custom when naked going into the bath, and,
perhaps, when going to bed at night."

The predominance of the scaly state over that in which pustules
existed in this case fully justifies its consideration as a case of Psoriasis, though at the period when the eruptions of pustules occurred,
the parts where the latter were situated very closely resembled the
more aggravated cases of Impetigo.

The arsenical solution, and the sulphur vapour-bath, promise, if
used scientifically and carefully, to become instruments of pretty
uniform power in the cure of the diseases under consideration. It
seems not improbable that the due and efficient administration of
the former has been often interrupted by the consequences of want of
circumspection, and inattention to the gradual increase of the doses,
and that the extent to which it has been hitherto administered
may be very considerably increased with advantage.

The irritation of skin attending scaly diseases generally may be
considerably diminished by frequent ablution with warm water ;
and, if the tenderness be not too great, gentle friction may be had
recourse to. By such means, that accumulation of scales, which so

encourages the lodgment of irritating secretions, will be prevented, and the latter removed almost as soon as formed. There are opinions on record, indeed, that these measures deserve the consideration of certain remedies in many cutaneous diseases.

In the 16th vol. of the Edinburgh Medical and Surgical Journal, Dr. Morrison, in a paper on this subject, describes a *modus operandi* of this kind which he has found extremely successful. " I dip," says he, " a sponge in lukewarm water, and, after squeezing it hard, so that only dampness remains, I cover it with oatmeal ; with this the parts are rubbed for some length of time, the sponge being frequently dipped in oatmeal, and this operation is repeated two or three times a day, according to the urgency of the itching, and other symptoms. After the parts have been sufficiently rubbed, they are washed, and gently dried. Oil is then applied by means of a varnish brush, and the parts covered up with slips of linen."

A case of Psoriasis inveterata, of twenty months' standing, is particularly mentioned as having been cured by this plan in the space of seven weeks. If bleeding should occur during the process, it is not permitted to interrupt the operation, and is rather productive of beneficial effects than the reverse. Should pustular formations occur, they are to be freely broken down, the secretion cleared away in the manner described, and their sites subjected to the friction in common with other parts. There is every reason to believe this practice capable of producing the best effects in such diseases, particularly in conjunction with the use of the sulphur vapour bath. Dr. M. seems to have been so convinced of its value, as to think its application justified in cases of small pox, to the extent of breaking down every pustule as soon as formed ; and according to his opinion it would be found capable of preventing the secondary fever.*

Previous to dismissing the subject of Lepra and general Psoriasis, it may be as well to direct the reader's attention to some observations of importance recorded in the pages of periodical works, from which additional information of importance, or at least suggestions worth attending to, may be extracted, under circumstances of difficulty, in treating these diseases or their sequelæ.

---

* However extravagant this proposition may appear at first sight, a due consideration of the question will, perhaps, lead to a different view, and give it rather the character of bold and energetic practice. It is very true, that the breaking down and rubbing away the contents of the fully-maturated vesicle in a confluent state would be attended with much pain and irritation, and, perhaps, dangerous from the extent of surface thus abraded ; but it is not so certain that positive mischief would arise, if such step was taken at an earlier period of the disease, and as soon as the vesicle was completed. Whoever has seen the contents of the fully-maturated vesicle resting on the cutis for so many days, its contents partly absorbed, and partly undergoing the process of exsiccation on the skin; the powers of the constitution having been exhausted in its original production, and being now evidently sinking from the prejudicial influence of the demand upon it to repair the extensive mischief which the disease has produced, will think it a matter worthy of serious reflection. The practicability of the plan is another question.

Mr. Cuming, in the 12th volume of the Medical and Physical Journal, records a case in which considerable difficulty was experienced in removing the cracks or fissures of the disease in the hands, after it had been removed from other parts of the body. The use of blisters, under these circumstances, produced the desired effect, and was followed by a healthy state of the skin.

On the subject of the internal use of arsenic, a communication by Dr. Girdlestone, of Yarmouth, in the 15th volume of the same work, contains some observations which are highly valuable, as inculcating the necessity of a due degree of caution in the exhibition of this remedy. In one of the cases which he has detailed, three doses of eight drops each were followed in twenty-four hours by erysipelas, leading to the temporary suspension of the remedy, which was subsequently renewed, and produced a speedy recovery. Another case, detailed in the same place, was cured by the same means, but was attended previous to its final disappearance by boils on the affected parts. Dr. G. further observes, that when the disease was removed by this medicine, a commonly previous effect of it was a temporary aggravation of the irritation which characterised it with the appearance of boils or fissures.

In the 17th volume another case is recorded, in which it appears Dr. Batty was induced to try the arsenic from Dr. Girdlestone's recommendation of it. In this case almost every species of local application had been tried without success. Ointments of calomel, of white and red precipitate, ung. hydr. nitrat., a solution of lunar caustic, ointments of tar and sulphur, and a variety of internal medicines, had all been subjected to a fair trial without important benefit, before this medicine was had recourse to : the effect, however, was a speedy recovery.

The use of kali sulph. is, I think, in the same volume, stated by Mr. Earnest to have been equal to the cure of a case of Lepra. It was exhibited twice a day in doses of two scruples, a solution of it being at the same time employed as a lotion. I have given this medicine a fair trial since, without any apparent good effect.

The local situations of the more limited forms of Psoriasis, as in the consideration of Herpes, have formed the foundation of separate designations in the arrangement of Dr. Willan. The P. labialis, P. palmaria, P. ophthalmia, P. preputii, &c. are instances of this. The remarks, however, which I have thought proper to make on this multiplication of terms in Herpes, and elsewhere, apply with equal propriety here, particularly as the same plan of treatment in such local varieties is found equally successful in all, the only local application generally necessary, after the exciting cause is removed, being that of a little of the ointment of nitrate of mercury in a diluted state.

This latter observation particularly applies to the disease as occurring about the eyes, on the scrotum, on the arms of washerwomen, and the backs of the hand. On the prepuce and palms of the hands greater difficulties are experienced, because the motion of the part is frequently tearing the bottom of the cracks open, and causing blood to flow from them, and producing great inflammation and increase of heat and dryness ; the same mischiefs occur in fissures of the lips, as most persons, at some period or other, have opportunities of experiencing.

The disease, as it has been found occurring on the back of the hand, a form stated to be peculiar to bakers, that also spoken of as peculiar to washerwomen, have their origin, generally, in entirely local causes. Inattention to cleanliness, and the constant application of dry, heated, powdery substances, are uniformly found to have acted as causes in the first ; while actual solution of the cuticle, and partial abrasion of the tender surface of the cutis, constitute the mischief which characterises the latter. It is well known that strong solutions of potash speedily act on the cuticle in any part of the body, and that in the labours of this class of persons, this alkali is used in great abundance. Its operation is also further increased by the addition of friction. The part which has been chosen for representation in Dr. Bateman's plate is about the middle of the fore-arm, to which part they are accustomed to immerse the arms, and where it is obvious the fluid lodges, dries, and becomes more concentrated. Sedative applications are most successful in these cases : the warm-bath and diluted liq. plumb. acet. alternately for the first few days, and afterwards the application of the nitrated mercurial ointment before mentioned, form the best plan of treatment ; but should any degree of obstinacy be manifested, the application of sulphur vapour will soon remove the difficulty.*

The P. palmaria is sometimes difficult to remove ; it is also the most common of the limited forms of the disease. As its designation implies, it chiefly originates in the palms of the hands, and does not disclose in its history and progress any important connexion with the constitution. It is, nevertheless, frequently obstinate and tedious, because, as already stated, the parts affected are incessantly in motion.

It usually makes its appearance with some irritation and itching along the course of the lines of the palm : a little scab is formed, which is rubbed off, and a slight discharge takes place. The new skin which is formed is delicate, and around its margin, the original cutis, much thickened, appears to detach itself. Its harsh and unyielding edge seems to give it the power of irritating the central points where the new skin begins to form, and this process is interrupted ; minute scales, instead of healthy cuticle, are formed, which, from time to time, become detached. The constant opening and shutting of the hand increases the mischief, and fissures, more or less extensive, which frequently bleed, and become excessively tender and painful, follow. The original hard and dry skin of the part separates in a circular form, around the originally diseased

* In these local affections it is desirable to have the vapour only applied locally. Mr. Green has so modified the apparatus of one of his baths, as to admit of this without the unnecessary trouble attendant on completely undressing, &c.

spot, and the course of another of the palmar lines or flexures, coming within the reach of the mischief, begins to assume the same features as that first affected. Eventually, the course of the greater portion of these flexures are occupied by fissures, more or less deep, and the hand cannot be extended without the greatest difficulty and pain.

In the treatment of this affection it will be obvious, that the chance of doing good will depend very much on our ability to keep the part in a quiescent state. It is impossible, without complete restraint, to calculate with certainty on the influence of any application, whether in the form of lotion or ointment ; but when this is attained, there are several which appear to be capable of expediting the cure. Among these the ung. hydr. nitr., diluted with an equal part of hog's lard, is frequently useful. Common saturnine lotions also are sometimes sufficient. Indeed it would appear that, in the majority of cases, a healthy state of parts is only prevented by the irritation consequent on motion.

When such measures fail, there still remains in the hands of the practitioner a weapon, which may he had recourse to with confidence and safety, though apparently severe in its direct effects. It consists in the application of the undiluted nitric acid to the diseased surface, and more particularly along the course of the fissures. The inflammation it excites renders a quiescent state of the parts a matter of compulsion on the part of the patient, and by the time it subsides, the site of the disease will have changed its appearance. A thickened scale or scab separates, and leaves a healthy-looking, though delicate cuticle covering the part, which gradually becomes hard and strong.

Due regard should of course always be had to the state of the stomach and bowels, for even in the most trifling cutaneous affections, disorder of these organs materially impedes recovery. As regards the Psoriasis of the lips and prepuce, some such disorder will be generally found to an extent capable of acting as a direct cause in the production of the disease, and therefore requiring the first attention.

On the subject of the disease situated on the latter of these parts, it may be necessary to notice, as we have done in a former instance, the description given by Mr. Evans. "This disease," says he, "appears in the form of deep cracks or chasms around the margin of the prepuce, which are extremely irritable, and apt to bleed whenever any attempt is made at retraction, and which are generally, from the loose cellular structure of the part, much deeper than when the disease occurs on the lips. The discharge is of a gluinous nature till the morbid action ceases, when it becomes purulent, and then the healing process begins, which is often very tedious.

"In the treatment of this complaint I have found the ung. hydr. nitrat. diluted to half its strength, the best application."

The same application may be considered entitled to equal consideration, in the case of fissures of the lips.

Psoriasis Infantilis.

The Psoriasis of infants is entitled to a separate consideration from the foregoing, on a very important account. There is no question, that it has often been mistaken for the consequence of venereal taint on the part of the parents of the child who is the subject of it, and has been, therefore, the means of inflicting much injury and undeserved distress. Dr. Willan, after alluding to one or two unimportant points, in which he states the venereal affection of the skin of infants to differ from that under consideration, advises "practitioners not to be too hasty in judging from mere inspection, and never to decide till they are *justified by collateral* circumstances." It is seen in infants from birth up to three years of age.

From a very minute and lengthened inquiry into the history of many cases of this disease, I am irresistibly led to the conclusion that it has no connexion with any form of venereal disease, except indirectly, *i. e.* when syphilitic affection may have combined with other causes in reducing the strength of the mother, or otherwise depriving the child of healthy sustenance. A generally healthy performance of the different animal functions, which is often found compatible with constitutional syphilitic disease, is not seen in the cases of infants affected with Psoriasis. It is found to occur, moreover, where the parents on neither side can be liable to suspicion.

Like the Psoriasis and Lepra of adults, it may continue long after its direct exciting cause has ceased to exist; and when taking place on the skin of unhealthy and poorly nourished children, it frequently continues when a better constitutional condition has been produced. From this circumstance, and the unyielding nature of the disease, it has no doubt sometimes happened that the suspicion alluded to has arisen, where, at the commencement of the eruption, it would not have been entertained ; and the treatment which has been adopted under these circumstances has been improper and injurious. In an institution to which I have the honour of belonging, cases without number have come under my notice, where poverty of circumstances existed on the part of the parents in a most extreme degree, evidently operating as the cause of the disease, when from a change to better living, consequent on the parents obtaining employment, the disease has speedily disappeared. Many such cases have been considered and treated as venereal without benefit; but, on the contrary, with aggravation of all the symptoms.

With respect to the class of society among which infants are chiefly affected with this disease, it may be safely asserted, that it is almost unknown among the rich or affluent, or even those whose circumstances bear them uniformly above the reach of want. It may sometimes occur from bad feeding or neglect, but it is generally the concomitant of poverty, and for the most part only seen in the cellars and confined apartments of the poor and unhappy, where privation of nourishing food and impurity of atmosphere unite their depressing powers.

When the disease first comes under our notice among the poor, it is characterised by red patches of inflammation on the cheeks, chin, forehead, nates, abdomen, &c. On the face some of the patches are small and irregular, others large, and assuming somewhat of a circular form. Some are covered with a horny scale, which is glossy and smooth to the touch, but which in a few days cracks and exhibits fissures of greater or less length and depth. Some are occupied by portions of scales of morbid cuticle rapidly detaching themselves, while others are forming beneath, which take the same course. The eyes sometimes partake of the inflammation, and generally the mucous membrane of the nose becomes inflamed and thickened, leading to snuffling, difficulty of respiration, and some fluid discharge.

In the more aggravated cases, the bottoms of the fissures emit a a bloody discharge, if situated in the neighbourhood of joints, but not otherwise ; and if no great emaciation may have existed at first, the irritation and pain belonging to the disease makes it soon apparent, and it generally increases with rapidity till death takes place.

The most irritable and painful spot which the disease is found to occupy is about the nates or between the legs; the perineum, labiæ, scrotum, &c. also, are not unfrequently the seats of much abrasion and tenderness, apparently produced by the stimulating properties of the urine.

It seldom happens that if the excretions be examined, a very disordered state of the alimentary canal with the secretions of the liver be not discovered, which may have been either concerned as an original cause, or impeding the restoration of health and strength. Hence small doses of the hydr. c. creta, with the occasional use of a brisker purgative, are necessary. Tepid bathing, to the extent of liberating all the hard scales which may have been formed, and allaying the general irritation of the skin ; frequent ablutions with warm water of the more irritable parts, change to a purer atmosphere, and more efficient nourishment, have been in most cases adequate to the cure, when assisted by internal remedies of this nature.

In closing the subject of Psoriasis, it may not be amiss to direct the reader's attention to the following note,* on a similar disease,

* The pellagra, as an endemic disease, prevails chiefly in the plains of Lombardy, which are fertile even to luxuriance. The principal objects of cultivation, besides the produce of vineyards extensively spread over their surface, are maize, rice, and millet. In some districts, and particularly between the rivers Adda and Ticino, the pastures are extensive, and yield a considerable produce of milk, from which the Parmesan cheeses are made. Latterly the disease has been confined to no particular situation, varying in the degree of fertility of soil, or purity of atmosphere, but has been recognised as increasing in every part of Lombardy, as well on the plains as among the hills which rise on their northern border towards the Alps. It is also seen in the province of Friuli, the district which

which is spoken of by authors as the pest of many parts of Italy, under the name of Pellagra.* It will probably be thought not assuming too much, to say that the Pellagra is nothing more than Psoriasis,

intervenes between the foot of the Carinthian Alps and the northern shore of the Adriatic, and in the districts of Milan and Padua, where it is peculiarly prevalent, it is computed to attack five inhabitants out of every hundred.

The pellagra is confined almost exclusively to the lower classes of people, and chiefly to the peasants and those occupied in the labours of agriculture. It appears first in a local cutaneous form, sometimes preceded by languor and indications of a general cachectic state of body. The local symptoms very generally show themselves in the first instance *early in the spring at the period when the mid-day heat is rapidly increasing, and when the peasants are most actively engaged in the labours of the field.* The patient perceives on the back of his hands, on the feet, and sometimes, but more rarely, on other *parts of the body exposed to the sun,* certain red spots or blotches, which gradually extend themselves with a slight elevation of the cuticle, and a shining surface, not unlike that of Lepra in its early stage. The colour of this eruption is somewhat more obscure and dusky red than that of erysipelas; it is attended with no other sensation than that of slight pricking or itching, and some tension in the part. After a short continuation in this state, small tubercles are frequently observed to arise on the inflamed surface, the skin almost always becomes dry and scaly, forming rough patches, which are excoriated, and divided by furrows and rhagades. Desquamation gradually takes place, which, though it leaves behind it a shining unhealthy surface in the parts affected, yet, in the first year of the disease, is rarely followed by a repetition of the appearances just described. Towards the close of the summer, or occasionally still earlier, the skin has resumed nearly its natural appearance; and but that the further progress of the disease is familiar to every inhabitant of the country, the patient might be led to flatter himself that the evil was gone by, and that there was no particular reason to dread its recurrence. According to Jansen, except the mere local affection, the health is not in the least impaired, the patient enjoys a good appetite, eats heartily, and digests well; all the excretions are healthy, and in females the menses are regular; while others assert, that even in the first period of the disease, certain general symptoms occur, which are important, inasmuch as they indicate the constitutional nature of the malady. Debility of the whole body; vague and irregular pains of the trunk and limbs, but especially following the track of the spine and dorsal muscles; headache, with occasional vertigo; irregular appetite, and general depression of spirits: these are the more ordinary symptoms which attend the early part of its progress. The bowels are, for the most part, relaxed, and usually continue so in the further course of the disease. There are no febrile symptoms; and in females the menstruation is generally continued without irregularity.

"The remission which the patient obtains during the autumn and winter of the first year is almost universally followed by a recurrence of his symptoms in the ensuing spring, under a more severe form, and with much greater disorder of the constitution. The cutaneous disorder is renewed, and spreads itself more extensively; though still affecting chiefly the hands, neck, feet, and other exposed parts of the body. The skin becomes callous and deeply furrowed; and large rhagades show themselves, especially among the articulations of the fingers. From the cases I have seen of the cutaneous affection in this stage, I should speak of it as most resembling the inveterate degree of Psoriasis, or of the Lepra vulgaris, with some affinity certainly to the Ichthyosis, under which, as I have already mentioned, Alibert has classed it. The debility is greatly increased in the second year of the complaint, frequently depriving the patient of all power of pursuing his active labours, and rendering him peculiarly susceptible of all changes of temperature. Partial sweats frequently break out without

* See Jansen's Treatise on Pellagra. Holland, Med. Chir. Transactions. Alibert, Icthyosis Pellagré, &c. &c.

aggravated in an intense degree by the peculiar force, in these parts of the world, of its common and general causes.

any obvious cause. All the nervous symptoms of the first year are renewed in a more severe degree ; there is a general tendency to cramp and spasmodic affections ; the mind begins to suffer under the disorder, and the feeling of anxiety and despondence is very strongly marked. The symptoms already noticed make progress as the heat of summer advances ; and with greatest rapidity in those patients who are much exposed to the sun. As in the preceding year, they begin to decline towards the middle or end of autumn ; but the remission, as well of the local affection as of the general disorder, is much less complete than before, and the patient continues to suffer during the winter from the debility, and other effects consequent upon the disease.

"In the third year every symptom is renewed at an earlier period, and in an aggravated degree. The constitutional malady shows itself under a variety of forms ; some of the symptoms having considerable analogy to those of scorbutus, all of them indicating a general cachexy of habit, and more particularly a lesion of all the voluntary functions. The debility now becomes extreme ; the patient is scarcely able to support himself, and the limbs, besides their feebleness, are affected with pains, which still further impede the power of motion. The diarrhœa continues, and tends of course to augment this debility. Frequently a dysenteric state of the bowels comes on in the latter stages of the disease. The breath is generally fetid ; and the odour of the matter perspired often extremely offensive. The appetite and digestion are irregular, yet, on the whole, perhaps less affected than most of the other functions. Dropsical effusions now frequently come on : occasionally ascites, but more commonly anasarca. Vertigo, tinnitus aurium, and double vision, are almost universally concomitants of this stage of the disorder, and all the senses become exceedingly impaired. Some spasmodic affections are general, and these not unfrequently take a very decided epileptic form.

"Connected with these latter symptoms is the effect which the pellagra produces upon the minds of those suffering under the malady, which effect forms one of the most striking circumstances in the history of the disease. The anxiety, watchfulness, and moral depression of the patient are rapidly augmented. In the hospitals appropriated to the reception of such cases, the pellagrosi afford a melancholy spectacle of physical and moral suffering, such as I have rarely had occasion to witness elsewhere. These unhappy objects seem under the influence of an invincible despondency ; they seek to be alone, scarcely answer the questions put to them, and often shed tears without any obvious cause. Their faculties and senses become alike impaired, and the progress of the disease, where it does not carry them off from debility and exhaustion of the vital powers, generally leaves them incurable idiots, or produces occasionally maniacal affections, which terminate eventually in the same state. As a striking proof of this tendency of the disease, I may mention the fact, that at the time I visited the lunatic hospital at Milan, there were nearly 500 patients of both sexes confined there, of which number more than one-third were pellagrosi ; people brought thither by the termination of their disorder, either in idiotcy or mania. Even this statement gives little adequate idea of the ravages of the disease in this mode of its termination. The public hospitals of the country are wholly incompetent to receive the vast number of persons affected with the pellagra : and the greater proportion of these unfortunate people perish in their own habitations, or linger there a wretched spectacle of fatuity and decay. Where debility, as generally happens, is the cause of death, it manifests itself in the latter stage, with the usual concomitants of colliquative diarrhœa, spasmodic affections, and coma ; and produces a degree of emaciation, scarcely to be surpassed in any other disease."

The mania consequent on pellagra is often of a violent kind : when it makes its appearance, the disease is in some degree retarded, and the strength less rapidly declines. Though the period which the disease lasts has, for the convenience of description, been spoken of as three years in the preceding account, it is generally of much longer duration, being renewed every spring, and disappearing

CHAPTER III.

PITYRIASIS.

SCURFY exfoliations of the cuticle in different parts of the body, usually unaccompanied by much irritation or fluid secretion, constitute the above form of cutaneous disease.

It has occupied a great portion of the attention of Alibert, and in its different situations and variations in appearance, obtains a large space in the pages of his account of " Les Dartres."

The most common situation of scurfy accumulation is the scalp ; and occurring in infants more frequently than at any after-period of

again in the autumn. Change of situation and mode of life is found useful ; but the class of society to which the patients belong renders this step generally impracticable. If no remedial measures are adopted before the third or fourth year of the disease, little benefit arises afterwards from the adoption of any plan of treatment.

Dissection discovers no organic affection of uniform occurrence, though visceral disease is not uncommon. Dr. Holland, however, considers these affections, when discovered, as more probably the consequence than the cause of the pellagra. Hereditary origin has been generally traced by different authorities in a satisfactory manner, both sexes being equally liable to it.

The disease has gradually become more and more prevalent within the last fifty years : from which consideration chiefly, Dr. Holland supposes that " it depends on some present peculiarities in the mode of life of the peasants of this country ;" an idea which historical facts completely support, as it is evident that a multitude of powerful and concurring circumstances have, during the period specified, continued to operate in breaking down the spirit, and impoverishing the resources of this unfortunate class of people. The wars which have so often devastated the country, frequent changes of political state, variable systems of government, heavy taxes and imposts, and last of all, a general heartbreaking conviction that patient submission to such grievances, or the most industrious efforts to bear up against them, were alike unrewarded by an ultimately beneficial change, are circumstances calculated to break down the proudest spirits, and poison, at their very sources, those incentives to contented and industrious exertion on which the health and comfort of the peasantry of all countries is well known chiefly to depend. To these causes may be added a decaying state of commerce, and a faulty system of arrangement between the landlords and the cultivators of the soil, all tending to depress agriculture, and to reduce the peasantry at large to much misery and privation.

There can be no question, that the observations of Dr. Holland, above alluded to, are, to their full extent, correct as regards the causes producing this disease : thus the terrific and heart-rending picture is presented to us of an extensive population, inhabiting a country rich in the gifts of a beneficent Providence, endowed in an extraordinary degree with all those properties of the soil which are necessary to the utmost luxuriance in the production of the requisites of life, given up to the ravages of a demoralizing and disgusting disease, from the simple influence of bad government, and the desolating effects of war.

Pellagra I do not hesitate to define as a speces of Psoriasis, aggravated by inattention to cleanliness, low living, and *extreme poverty ; the local cutaneous affection obtaining its only variations in character from this disease from the irritation attending continued exposure to the heat of the sun.*

life, it has obtained the name of dandriff, from nurses and others, under whose notice at this age it most frequently comes.

Though very common in infants, however, it is now and then extremely troublesome in persons of maturer age. Adult individuals of dark complexion, and those of native Indian extraction, are particularly liable to it ; and in such cases it is not unfrequently obstinate and intractable.

The P. capitis (as it has been designated by Willan and Bateman) or the dandriff of infants, is usually not of a very important nature. If the child who is the subject of it be not very much neglected in point of cleanliness, it usually disappears in a few weeks ; but now and then, under different circumstances, it is followed by considerable irritation, fluid secretion, and scabbing, or a state much resembling the Porrigo furfurans. Under circumstances of peculiar aggravation, indeed, there is no real difference between the two affections.

With respect to the causes of this form of disease, it seems clear, that it is partly, if not wholly, dependant on the disposition to determination to the head, so generally seen in infants ; and it is by no means improbable, that in those cases where much irritation and fluid secretion takes place, the vessels of the brain are relieved by it. The state of the child's habit and general health, therefore, requires serious consideration in its treatment ; sedative applications are rarely advisable, or even safe ; nor does the affection require indeed any notice beyond what is comprehended in attention to the general health, and keeping the scalp as free from accumulation as can be done with convenience, and without pain.

In the adult subject Pityriasis now and then is seen, wearing a character materially conducing to the discomfort of the individual. Different parts of the body become the seats of much itching and tenderness, and when friction is employed, scabs of considerable thickness drop off, some soft and moist, others dry and thin ; the whole of the parts from which they have been separated, exhibiting a red, shining, glossy, and sometimes slightly moist surface. The skin of the chest and back are the common seats of this form of the affection. The hairy scalp and its margin also partake of it, and the exfoliations from the latter surface exhibit an increased thickness, and moisture, the result of the increased irritation produced by the hair. The colour of the affected parts when covered by the diseased cuticle is of a lightish yellow or copper hue ; when this has been washed or rubbed off, it approaches more to red ; but the cutis at no time appears as the colour consequent on common abrasion. This is a marked peculiarity of the affection, and clearly proves the fact elsewhere insisted on,—that one set of vessels forming part of the dermoid system may be affected by disordered or inflammatory action without the participation of others. Those vessels only, whose office it is to secrete the cuticle, are affected in this disease; and hence, the lightness of colour alluded to, and the immense production of scurfy exfoliation, or imperfect cuticle.

The figure of the affected patches is exceedingly irregular : here and there are spots from half an inch or less, to two or three inches in diameter, more or less approaching to a circular form. These will perhaps be found around the margin of a larger patch, the outlines of which are as irregular as the outlines of a map of an island. The colour of these larger patches also varies from time to time in different parts, from a light straw to a reddish colour ; hence the term P. versicolor, P. rubra, &c.

Another form of cuticular exfoliation which I have seen in individuals who have been resident in warm climates has not, I believe, been described elsewhere. The chest and back are its usual seats. There is at times considerable itching attending it. When the body is exposed, the outlines of the affected parts are easily traced by the eye, and are of a darker brown hue than the adjacent skin. If the finger be drawn over them, no inequality or elevation is perceptible ; in fact, there is nothing but the colour which enables us to trace its boundaries. If a dry cloth be rubbed forcibly on it, large films of thin delicate cuticle are rubbed off; and where these thin pellicles are once broken, they may be gently raised and detached throughout the whole diseased surface. Where such detachment has been effected, the cutis is tender, and on being touched or rubbed, it smarts, and becomes more inflamed. Nothing like dry or moist scurf is found separating itself from the cutis in this form in any of its stages.

Another form has been found occurring, where hereditary dryness of skin, and a disposition to crack and form fissures, have been known to exist. In such cases the individuals have been able to detach many square inches at a time, without pain or difficulty, a sound cuticle of ordinary strength and thickness still remaining to defend the cutis. The exfoliated portions of cuticle exhibit the cutaneous lines and perforations very beautifully, and are nearly transparent.

On the subject of the pathology of this affection, and more particularly the form first mentioned, as occurring in the adult, we may remark that its pretty uniform occurrence in individuals of delicate health, and diminished energy of circulation, will lead us to some correct conclusions. In such states of system the cutaneous vessels partake of the general debility, and have the disadvantages of their locality as furthest from the centre of circulation, and being exposed at the same time to vicissitudes of temperature superadded to these. They are incapable under these circumstances of performing that part of their offices on which the formation of sound cuticle depends, and they produce instead, the delicate and ill-formed substance described.

The state of the circulation and system in all cases which come under our notice proves this view of the case to be correct. I have never seen a single case where want of energy was not apparent, and very few where the supply of this was not followed by speedy recovery.

Violent and distressing impressions on the mind,—original debility of constitution,—the depressing effects of long-continued illness in warm climates, &c. are found very commonly to have been co-existent with the first appearance of the disease.

The constitutional treatment which will be found most successful is that which is in strict accordance with the above principles. Measures which tend to invigorate the system will be always proper, if not forbidden by organic disease. Bark, steel, sea bathing, gentle exercise in the open air, ease of mind, nourishing food, and plenty of rest, constitute what is usually requisite on such occasions. Now and then the sulphur vapour-bath has been rendered necessary, the cutaneous vessels having failed to recover their tone, though the general health had been much improved.

When the scalp is much affected, and the scurf forms in considerable quantities, the free use of a solution of acetate of zinc, in equal parts of rose water and proof spirits, constitutes an agreeable and useful application. The scalp may be freely bathed with it twice a day with considerable relief.

M. Rayer, in his recent edition, appears to entertain very different notions of this disease from those which I have expressed, as also from those of Willan and Bateman. The definition of the latter is "irregular patches of thin bran-like scales, which repeatedly exfoliate and recur, but which never form crusts, nor are accompanied with excoriation." Possibly, M. Rayer may have confounded it with other affections, or perhaps truly describes it as it occurs in France, with just those features of aggravation commonly noticed in that country, to which I have alluded elsewhere. His description is rather diffuse, but it is quite in keeping with the rest of his production.

The treatment too, adopted by this author is at variance with that found most successful in England.

"General Pityriasis is one of the least frequent and most obstinate of the diseases of the skin; it almost always invades without precursory symptoms; patients complain, in those regions that are about to become affected, and most commonly in the extremities, of a violent feeling of itchiness, or rather of a painful and tantalizing prickling sensation, which seems to have its seat under the skin, between it and the flesh. If the part thus affected be examined, and it chance to be without hair, a number of very superficial *erythematous blotches* are perceived. The heat of the surrounding integument is almost uniformly increased, the subcutaneous cellular tissue is swollen, sometimes painful when pressed upon, and the soft parts generally seem to suffer from distension. It may be difficult to demonstrate this primary redness when the scalp is the part affected; but the phenomenon is always abundantly evident on other regions of the body when looked for early enough. Within a few days the blush diminishes in intensity, and soon disappears entirely; the epidermis then cracks, becomes less adherent, and a process of desquamation commences, which varies in its characters

according to the nature of the surface affected. On the outer surfaces of the legs and arms the cuticle is thrown off in foliaceous lamellæ, which continue to adhere for some time, in one case by their
centre, in another by either of their extremities, so as to appear
floating, as it were, on the surface of the skin. These lamellæ,
formed by the cuticle nowise thickened, vary from about three to
eight lines in diameter. Those parts of the skin from which they
have been recently detached, or from which they have been removed
by the action of the nails, the rubbing of the clothes, &c., are usually
of a rose colour ; farther, when patients have yielded to the impulse
to scratch, occasioned by the violent pruritus which accompanies
the disease, the parts of the skin which have recently shed their
cuticle pour out a serous, yellowish coloured fluid, similar to that
observed in the moist eczemas, and occasionally so abundant as
completely to soak the linen or other clothing with which the parts
affected happen to be covered.   When this adventitious circumstance
occurs to such an extent, it is apt to render the diagnosis of pityriasis obscure.

" Pityriasis is generally less severe on the insides of the limbs,
and the desquamation then always consists of smaller laminæ of the
cuticle ; it is often, indeed, pulverulent. Behind the ears, about
the clavicular fossæ, the axillæ, bends of the arms, wrists, navel,
prepuce, groins, and insteps, the inflamed skin has something of the
look of intertrigo, that is to say, it is rough, moist, and slightly
chapped in the direction of the natural folds of the skin which are
powdery on their edges. On the anterior part of the breast and
belly the desquamation always occurs in much smaller lamellæ than
on the posterior surfaces of the trunk. On the regions of the olecrana and patellæ, and especially on the palms of the hands and soles
of the feet, where the cuticle is naturally thicker, the exfoliation
takes place in broader and thicker laminæ than elsewhere ; on the
face and hairy scalp, on the contrary, it is almost uniformly powdery in its appearance.

" Pityriasis is accompanied with violent pruritus,* especially of
those parts where the eruption is recent, and of those where it is
accidentally exasperated ; the symptom then gains such a degree of
intensity that the sleep is disturbed, and is only changed into a sensation of the most distressing kind by the efforts of the patient to
allay it by scratching ; the satisfaction, however, the enjoyment
experienced by yielding to the impulse, is described by patients as
surpassing every other they have known. After the state of excitement or exaltation thus induced, has passed away, patients experience smarting sensations of considerable violence, and fall into a
kind of dozing state which is frequently followed by sleep.

" As to the principal functions, they are not in general very
remarkably disturbed ; yet it does occasionally happen that patients,

* This symptom is noticed by Willan and Bateman, as only occasionally taking
place in one form, the P. versicolor, and then only in deranged health of constitution.

labouring under extensive pityriasis, present functional disorders of the digestive organs ; I have seen all the symptoms of chronic inflammation of the mucous membrane of the stomach and bowels under such circumstances, and remember one case in which the severe suffering occasioned by acute pityriasis extending to the whole surface of the body, and the repeated serous evacuations that took place from the bowels even proved fatal. Amenorrhœa is occasionally observed to precede or to follow an attack of pityriasis. I have never observed anything like a true febrile paroxysm, except in those cases in which the eruption broke out over almost the whole surface of the body at once, or at those times when the disease seemed to increase in severity, or otherwise, when an inflammatory affection of the intestines was superadded to the preëxisting malady.

" The continuance of the symptoms of *general* pityriasis, and their mode of succession and increment, are subject to numerous shades of individual variety. The disease is frequently seen arising in situations where it never before appeared, at the same time that it is vanishing upon others where it had appeared fixed for a great length of time. The very appearance of the disease, indeed, is modified by these same circumstances : powdery or scaly, as in slight ichthyosis upon those points where the inflammation is declining but still lingers in a trifling degree, of a vivid red, and moist upon those where the inflammation has been accidentally increased ; whilst in those places where desquamation is no longer going on, the skin has a white and slightly yellowish tint.

" The cellular membrane is more or less tumified in those places where the inflammation possesses some acuteness, and in the lower extremities, even when there is little apparent redness under the squamæ. When the disease extends to the greater portion of the surface, patients can generally gather a considerable quantity of scales from their beds every day. Lastly, when parts provided with hair are attacked, the disease causes its partial fall.*

" I have already mentioned the principal varieties of appearance presented by pityriasis occurring on different regions of the body. The better to expose the characters of the local varieties of this disease, however, I here add a few particulars connected with each :—

" 1st. Pityriasis *capitis* is, of all these varieties, the most frequent, and that which has been longest known ; it has, however, been often confounded with the desquamations consequent on psoriasis, lichen, and eczema, and with those which take place without inflammation from the scalp in some individuals.

"Individuals attacked with pityriasis *capitis* suffer habitually from a considerable itching of the scalp, especially on the first attack of the disease, and at the period of subsequent exacerbations ;

* It is not possible to consider or inquire respecting the diseases of the skin, without being struck with this marked peculiarity of those affecting the scalp.

they are then led to scratch or rub the head with different degrees
of force, when a whitish powder is detached, consisting of minute
epidermic squamæ. The secretion of these goes on continually,
and a quantity may at all times be detached by rubbing with the
hand, or the action of a brush. By separating the hair in different
places, numbers of small, red, irregular, and very superficial patches
may be discovered under the squamæ, disseminated over the surface
of the scalp. The skin on these points is shining, dry, and rough
to the touch. The red patches, it is to be observed, are only very
distinct in those places where desquamation has but recently been
established; after having been long affected, instead of being red,
the scalp becomes on the contrary of a rather dull white in the places
affected.

" It rarely happens that the inflammation reaches a very high
degree of intensity in this variety of pityriasis. I have, however,
met with some patients who complained of a feeling of stiffness and
tension, of burning heat and insupportable pruritus in the hairy
scalp. When this is the case, besides the ordinary attendant epidermic desquamation, there is almost always an exudation of a thin but
glutinous fluid, similar to that secreted by surfaces affected with
eczema. This matter agglutinates the hair and scurf into masses,
and when the disease continues in this state for a month or two, the
head looks as if it were covered with a grey or whitish cap, composed of the hair and an abundance of scaly matter united, the
superficial layers of which, drier, and more friable than those that
are deeper, have the greatest resemblance in point of colour to the
mineral named asbestos (*teigne amiantacée*, Alibert). When an
attempt is made to isolate the particular hairs, they are found buried
as it were and lost amidst scaly masses ; and if, by means of pretty
strong pulling, a few locks be separated from the mass, they are so
strongly agglutinated, and so much mixed up with scales, that they
still form stiff bundles only to be divided into others of smaller
dimensions : the individual hairs can only be separated from each
other with extreme difficulty. When the hair is clipped off pretty
close to the scalp, after having been softened by means of poultices
and fomentations, the skin is found of a rather vivid red, on numerous points. The hair is only found detached from a spot here and
there : baldness forms no feature in this disease, in which *pediculi*
are also much more rarely encountered than in favus.

" Whether pityriasis *capitis* consists of a mere furfuraceous
desquamation, or of the asbestous cap that has been described, it
may in either case extend to the eyelids, and occasion the fall of
the eyelashes. In young children the disease is most commonly seen
on the upper part of the forehead and temples ; in the aged it frequently extends to the eyebrows, and when severe, almost uniformly
spreads to the face and different other regions of the body. Pityriasis
capitis is a disease that always lasts long ; it may exist for months,
and even for years. It is known to be approaching amendment when
the secretion of furfuræ becomes less copious, and when the serous ex-

udation, when it has taken place, no longer goes on. A new, smooth, and shining epidermis is formed when the recovery is complete.

" Pityriasis *palpebrarum* may also exist alone, and independently of any anterior manifestation of the disease on another region. The only peculiarity presented by this variety is the frequency with which it causes the fall of some of the eyelashes, and the propagation of the inflammation upon the conjunctiva; it differs, and may be distinguished from psoriasis developed on the same parts, by the erythematous look of its patches and flimsiness of its scales, which contrast strongly, with the papular elevations that announce the development, and the thick squamæ that afterwards characterise psoriasis.

"Pityriasis *labrum* is a variety that has hitherto been confounded with psoriasis, a disease, however, from which it differs in being evolved on the lips and surrounding skin, not as papular elevations, followed by thick squamæ, but under the semblance of minute red stains, to which succeed a general redness and a continual desquamation of the epithelium of the lips, and occasionally of the cuticle of the neighbouring skin. The desquamation goes on in the shape of little thin and transparent laminæ, very similar to portions of the healthy epidermis dried, or of the epidermis whose inner surface has imbibed a little serum previous to its desiccation. The lips, in this state, are affected with heat and tension; the epithelium gets yellow and thickened, it then cracks, and falls off in laminæ of considerable size. It frequently happens that these continue to adhere for some time by their centre, when their edges are loose and already dry, so that a new epidermis is formed under the one about to be detached before it falls; this new cuticle then grows yellow, cracks, peels off, and falls in its turn, to be succeeded by another which undergoes the same changes and shares the same fate. This is always a long-continuing and obstinate affection; every now and then it gets worse than usual, when the lips look swollen and of the brightest red. It is very different from that transient inflammation to which the lips are subject, attended with chapping and the detachment of the epithelium, which is induced by exposure to cold, or happens as a consequence of different acute diseases: this slight affection soon passes, whilst true pityriasis is always a lengthy and troublesome disease. The causes of pityriasis *labrum* are frequently obscure; I have observed it in two individuals, great talkers, who had a trick of always biting their lips.

" Pityriasis *palmaris* and *plantaris* are varieties which have hitherto been confounded with psoriasis affecting the palms and soles. The two diseases, however, may be distinguished by the following circumstances: Psoriasis begins in the shape of papular-looking elevations, the summits of which are soon afterwards covered with dry squamæ of a dull white colour; pityriasis, on the contrary, commences as small red spots or stains, irregular in their outline, which spread, and before long acquire a yellowish hue, probably in consequence of a slight exudation from the corion, which thickens the epidermis by penetrating its inner surface. The epidermis then dries, cracks, and is constantly peeling off in foliaceous lamellæ; this exfoliation may extend to the fingers and even take place under the nails, which are then occasionally detached. The skin almost always appears bathed in perspiration around the diseased points, which, on the contrary, are uniformly dry. I once attended a patient in whom this squamous affection appeared in the soles of the feet, a year after having attacked the palms of the hands. The heel and anterior part of the sole of the foot were painful when the erect posture was assumed, and during the act of walking.

" I have observed the *inside of the mouth* affected with chronic inflammation and habitual desquamation of the epithelium, especially about the base of the tongue, without any antecedent or concomitant affection of the pharynx, stomach, or lungs,—pityriasis *oris*. This state continued during five or six years with but brief intermissions, the principal functions being all the while performed with great regularity. At the time a desquamation of this kind was going on, one patient complained of heat, and often of painful sensations, difficult to define, in the interior of the mouth. In a woman, who was similarly situated, almost the whole of the mucous membrane of the mouth was habitually of a greyish white colour, and when the epithelium was thrown off from the tongue its surface presented several patches of a bright red colour, which continued until the investing membrane was either formed anew or again rendered thick and opake.

" The *prepuce* in the male, and *labia majora* in the female, are also occasionally the seat of superficial chronic inflammations which cause exfoliations of the epithelium, and an increase in the secretion of the follicular fluids ; several of these affections in their course, their principal phenomena, and their rebellious character, bear a singular affinity to pityriasis.

" *Causes.* General pityriasis is happily a disease of rare occurrence ; I have observed it more frequently among women than men ; patients are commonly altogether at a loss for any cause to which they can assign their disease. The causes of *local* pityriasis are also, for the most part, very obscure. It appears occasionally to be excited among men by the action of the razor on the chin. The repeated action of a rough comb or hard brush may possibly conduce to its development on the hairy scalp.

" *Diagnosis.* The natural exfoliation which frequently takes place from the skin of the infant, a few days after its birth, is of too short a duration, and appears under circumstances too particular to be confounded with pityriasis. The *scurf* of the scalp, often observed at the period of birth, is formed by a hard, yellow, and friable matter deposited on the anterior and upper part of the head, and bears a much stronger resemblance to an incrustation, than to an exfoliation of the epidermis. This scurf, indeed, continues without any appearance of desquamation for many months, unless it be got rid of by the aid of fomentations or other applications that soften and permit it to be removed : it no more consists, in fact, of the epidermis, altered and dried, than the scurf which is sometimes observed on the scalp of elderly individuals negligent of personal cleanliness.

" The skin of the scalp and of the extremities in some adults, and especially in some aged persons, is occasionally affected with an *habitual exfoliation* of the epidermis, which differs essentially from pityriasis by being attended with neither redness, heat, nor any other morbid sensation. Certain desquamations of the epidermis also, which follow acute diseases, differ essentially from pityriasis in their origin and transient nature. The desquamation, or rather the exfoliation of the cuticle that happens in pityriasis, differs in its character from that formation of squamæ which takes place in all the varieties of psoriasis, inasmuch as the cuticle, in the latter disease, is thickened, dry, rough, and of a dull white colour. But the feature that distinguishes these two diseases from each other, more than any other, is the circumstance of the red patches of psoriasis always rising above the level of the general integument, whilst those of pityriasis are not at all prominent. Farther, when the inflammation runs very high in pityriasis, the skin, especially when scratched, is very apt to pour out an abundant serous secretion, whilst in psoriasis it always continues dry. Lastly, in *acute* pityriasis the subcutaneous cellular tissue is often swollen and painful over a large extent of surface, a circumstance which never happens in psoriasis *discreta*, and which, in psoriasis *inveterata*, is only observed to occur in limited spaces. The heat and pruritus that accompany pityriasis are, moreover, much more troublesome than the same phenomena in psoriasis ; and pityriasis, when general, is much more frequently complicated with symptoms of constitutional disturbance and derangement of the digestive functions than psoriasis.

" When pityriasis is compared with lepra the same points of difference are detected, with two distinguishing features in addition ; the circular form of the patches of lepra, and their mode of recovery from the centre towards the circumference. The detachment of the cuticle in ichthyosis is not preceded by redness or morbid sensations of the skin. The desquamation that follows chronic lichen and eczema is preceded by the evolution of papulæ and vesicles. I shall by and by have occasion to contrast pityriasis with acrodynia and with pellagra, but I must here pause to expose the characters that distinguish it from chloasma (pityriasis *versicolor*, Willan), and from melasma (pityriasis *nigra*, Willan), diseases which I have felt called upon to transfer to another order, that, namely, of the adventitious *pigmentary discolourations*. In the first place, the most striking feature in the two diseases last mentioned, is undoubtedly the change of colour presented by the skin ; farther, if some degree of desquamation does take place at one period in the progress of these diseases, an habitual and abundant exfoliation of the cuticle forms no point in their history. Neither is there any of that serous exudation which I have mentioned when speaking of acute pityriasis ; lastly, the ease with which chloasma is cured, and the deplorable resistance of pityriasis, in almost every instance, to remedial measures of every kind, show an essential difference in

15*

the nature of these two diseases. As to melasma (pityriasis *nigra,* Willan), when desquamation has once taken place, it seldom happens that this phenomenon and the other symptoms of the disease return with an intensity, or prove of any duration.

"*Treatment.* When in *general* pityriasis the exfoliation from the epidermis is great, and when the skin looks vividly red and pours out an abundant serous secretion from various places, when the subjacent cellular membrane is swollen and the heat of the surface is increased, blood-letting, combined with the temperate mucilaginous bath, the antiphlogistic regimen and diluents, seldom fails to give considerable relief, at least for a season, a circumstance, however, which should not induce the neglect of this, the most important of these measures, when the patient appears able to stand it. The inflammatory affections of the bronchi, bowels, and genital organs, which occasionally make their appearance in the course of pityriasis, and especially at the periods of its paroxysms, are also relieved by the detraction of blood ; but they do not yield with the same readiness to this practice as the inflammation which is induced in these parts by external and appreciable causes. It would therefore be wrong to attempt to cut short these inflammations by the severest antiphlogistic means and repeated blood-letting. After a first or a second bleeding, which may have been very decidedly beneficial, a third generally appears to have no influence in modifying the disease ; it is even seen occasionally to return with all its original violence a few days after a freest use of the lancet. Another circumstance which ought to be noted is that the blood is often buffy in this disease, and continues so, unmodified by repeated venesections, during its whole course.

"In *acute general* pityriasis *opium* is often employed with advantage to procure sleep and lull the pruritus and heat of skin so much complained of ; this is also one of the best medicines we possess for combating the obstinate diarrhœa that occasionally comes on in the course of the disease. Purgatives and the preparations of *arsenic* cannot be prescribed with the same propriety in this as in the generality of the squamous inflammations, pityriasis being much more frequently associated with gastro-intestinal disturbance than any one of them, and this is a state which these medicines are apt to induce, and one which they would certainly aggravate did it already exist. Simple baths, emollient topical applications, and the vapour-bath, prove more generally useful than the sulphureous or alkaline bath, both of which are contra-indicated when the skin is of a vivid red and the subjacent cellular substance is painful and swollen. When the skin, however, scarcely looks coloured under the epidermic lamellæ, the vapour-bath and douche may be administered with advantage. But it is with *general* pityriasis as with almost the whole of the chronic diseases of the skin that are independent of appreciable causes,—a solid and enduring cure is only to be obtained by a general change of the constitution, brought about by dietetic means long and regularly pursued, effected naturally by the progress of years and the modifications undergone by the organization, or accidentally induced by some intervening disease, such as measles, scarlatina, &c.

" The *local* varieties of pityriasis frequently prove very rebellious also ; yet that of the hairy scalp of infants at the breast occasionally gets well spontaneously or by mere attention to cleanliness, after several months' continuance. The variety of pityriasis *capitis* which attacks adults and the aged, and is characterised by a simple powdery exfoliation of the epidermis and proritus, without serous discharge and matting of the hair, only requires attention to cleanliness, and the application from time to time of some soothing unguent. The severer variety, however, which occasionally occurs along with general pityriasis, but which may also exist alone, and is denominated *teigne amiantacée* by French writers, requires more active treatment. After having softened the mass of dried exudation, hair and squamæ, by means of poultices, vapour-baths, &c., the hair must be clipped off as near to the skin as possible with scissors curved on the flat. The vapour-douche must then be continued, and may be alternated with alkaline lotions ; purgative medicines should next be cautiously tried ; and whenever symptoms of excitement appear, when the scalp becomes of a bright rose colour in several places, and appears moistened with a serous exudation, leeches should be applied behind the ears. In a case of pityriasis of the lips, I have detailed the routine of treatment usually pursued. The white precipitate ointment and vapour-douche are the means which have generally appeared to me most efficient in curing pityriasis of the palms and soles. Pityriasis of the mouth, nipple, and genital organs, have as yet been too little studied to admit of any thing like a satisfactory view being given of the value of therapeutic means in regard to them."

SECTION III.

On diseases exerting a probably salutary influence on the system originally produced by, and usually symptomatic of, deranged digestive organs, and characterised by active inflammation.

CHAPTER I.

PORRIGO FAVOSA.

*Honeycomb-scall of Bateman and Others.**

THE diseases known by the names of Porrigo favosa, and P. larvalis, stand first among these, and the consideration of the pimples

---

* A totally different disease from P. lupinosa. Confounded by French authors, however, with the latter. The comparison with the structure of honeycomb to that of either does not convey an idea sufficiently correct, and is, in fact, useless.

of strophulus, lichen, &c. and uticaria, may with propriety follow. The vesicular herpetic affections, and the carbuncular boil of the cutis comprehend the remainder of this division.

The liberty which I have taken in the separation of the favous Porrigo from the other diseases bearing that name, may appear to require some explanation. From the views taken in the preceding pages of the pathology of the more obstinate forms of what has been denominated Porrigo, it will be observed, that I have particularly noticed the generally local characters of the complaint, and the dependence it frequently has for its support on the operation of locally irritating causes.

The causes and pathological history of Porrigo favosa and P. larvalis are obviously of a different character ; the latter have always a constitutional origin, and always requires the exhibition of remedies through the medium of the constitution.

The description of these affections given by a preceding author is as follows :—*

"On the scalp the pustules are large, soft, whitish, itching, and slightly inflamed at the base. At first they are distinct, and partially distributed, as on the side of the head, or about the occiput. When broken, they discharge a thick viscid matter, which gradually concretes into irregular brown or yellowish semitransparent scabs. The ulcerations gradually extend, with a constant and copious discharge, by which the scabs are kept moist, and the hairs are matted together. Under these circumstances, pediculi, which are bred in great numbers, produce an incessant irritation, and contribute to aggravate the disease, and to excite fresh pustules. The eruption finally covers the whole scalp, the pustules in some places remaining distinct, in others becoming confluent, so as to form irregular ulcerated blotches. From these, when the coverings or dressings of the head are removed, a sour rancid vapour is exhaled, which affects very disagreeably both the eyes and the organs of smell and taste in persons who examine or dress the patient.

" In many cases there are among the pustules small red smooth tumours, which desquamate at the top, and very gradually proceed to suppuration, in the same manner as scrofulous tubercles do on the arms, &c. Sometimes, large abscesses form near the vertex, or at the occiput, probably originating in lymphatic glands situated there. As soon as they are broken, and begin to discharge freely, the pustular eruption and small ulcers on the other parts of the scalp disappear.

" During the course of the Porrigo favosa affecting the scalp, the glands on the sides of the neck, and sometimes the parotids, harden and enlarge very gradually. They are at first like a series or chain of small hard knots, without discolouration of the skin, but some of them afterwards inflame and suppurate.

* It will be observed, that there are many points in this description that will not apply to *all* cases, or in fact to any occurring in decent society.

" The Porrigo favosa affecting the face sometimes commences about the lips or upon the chin ; but at other times extends thither from the scalp or from behind the ears. The pustules, in general, appear first at the corner of the mouth, without much tension, or inflammation of the skin. They are set near together, in an irregular cluster, and contain a straw-coloured fluid ; when broken, they become confluent, and discharge a clear, viscid matter, which afterwards concretes into a yellowish scab. Other similar ulcerations appear, soon afterwards, at the opposite corner of the mouth, on the lips, or about the chin. These blotches being attended with an incessant itching, children cannot be prevented from rubbing or picking them ; the consequence of which is, that their borders are kept sore, inflamed, and continually extending. The complaint has a most unpleasant aspect, when the ulcerations entirely surround the mouth, and are covered with large, elevated, irregular masses of scab, like honeycombs. There seems to be a considerable degree of acrimony in the matter discharged from beneath the scabs ; for the part of the breast, which comes frequently in contact with the diseased chin, soon turns red, and exhibits an eruption of pustules which terminate, as on the face, by a superficial ulceration. A similar appearance is produced on the arms of the nurse, who attends a child affected with the complaint."

In removing the scabs of this disease, whether existing on the head, or other part, we discover a reddened and inflamed surface pouring out, with excessive rapidity, a viscous transparent fluid, which speedily dries and forms fresh scabs of various shades of colour, from a transparent yellow, to a dark brown. An areola of inflammatory redness usually surrounds the part, as if the whole energies of the vessels of diseased spot, and adjacent cutis, were called forth in keeping up the fluid secretion.

This state of matters will be constantly found, whenever the scabs are removed ; in doing which, a small quantity of blood sometimes flows from the surface. The fluid secretion, however, at no time (except where constant irritation is kept up by picking the scabs, which children are accustomed to do) appears like pus ; for it is not opake, nor does its chemical analysis afford similar results. The surface is not ulcerated, but merely abraded ; the fluid, as I have before observed, being poured out from the open mouths of its vessels.

In the management of this disease there are few points of importance, beyond those which are comprehended in attention to the general health. The state of system under which it usually occurs, as a spontaneous disease, will be found more frequently to indicate the necessity for depletion and alteratives, rather than tonics, which have been recommended ; but many cases do undoubtedly occur, where the latter are imperiously called for.

A diligent inquiry into the history of the individual case will often enable us to discover, that the eruption first appeared in a state of constitution opposed to that existing when it first comes under our notice, thus the patient may be emaciated and weakly, when proper medical advice is applied for, and the cutaneous affection wearing the most unpromising character; a condition which is often brought about by the low living, and the excessive use of purgatives, very commonly resorted to on its first appearance,—one in which tonics and better living will be of primary importance.

I have lately seen a number of cases of this kind in a family of distinction, the children of which had been long suffering from the disease, which manifested no disposition to improve, till a decided tonic plan was had recourse to. If it be seen at its commencement, the mischiefs of such a plan of treatment may be avoided, and it will be found readily to yield to mild aperients and alteratives and a judicious system of diet.

Applications which allay irritation and diminish pain are useful auxiliaries ; and these together will be found adequate to the necessities of most cases, not occurring on the scalp.

On this latter part, however, obstacles intervene, requiring some alteration of treatment. Here the disease may be considered as constantly under the influence of irritation from the hair, while the glutinous secretion lodging, and being retained upon it, prevents any application to the affected surface, and becomes a source of mechanical irritation, sufficiently powerful to counteract the effect which internal remedies may be supposed to have in subduing it.

In protracted cases, such as those detailed by Alibert and others, where the mischief and torment of the disease are aggravated by myriads of vermin, and the accumulations of weeks and months of the secretions of the part—where neglect and filth have contributed from the beginning all their influence to support the disease ; the most material portion of our attention must be directed *solely* to the part.

To remove such collections of filth, at the risk of some pain to the patient : the scabs in the first place, by continued soaking in warm water, with a plentiful use of soap, and the hair afterwards, by the razor, are steps of absolute necessity, without which, our time and subsequent exertions may be fruitless.

In such cases, too, a considerable extent of ulceration is sometimes discovered, though not often sufficiently deep to affect the roots of the hair. But whether ulceration exists or not, the use of fomentations and poultices is necessary to subdue the inflammatory action of the vessels of the part ; and when this has been effected, a little attention to the general health is often all that is necessary to the cure.

The length of time during which local applications may be requisite, however, depends very much on the period which the disease may have existed ; for the action of the part acquires increased vigour with every week of its duration, and must frequently continue on the scalp a considerable length of time, after the causes originally producing it have ceased to exist : and even when much care is taken to remove the secretion as soon as formed, and allay irritation by soothing remedies, the effusion of viscous fluid on the part is now and then still kept up, apparently dependent on a morbidly relaxed state of its vessels.

An effectual application will be found in a solution of caustic,* or sulphate of copper, in such a case : these fluids should be applied with a camel's hair pencil, to the abraded surface two or three times a day, until the discharge ceases : they appear to act by constringing the relaxed vessels on the surface from which the discharge oozes. In some old standing cases, their strength may be sometimes considerably increased with advantage, or either of the preparations may be lightly rubbed over the surface in a solid form.

The remarkable distribution of small tumours among the favous pustules, when the disease appears on the scalp, noticed in the preceding quotation, is not confined to the P. favosa, but sometimes occurs in the others, where much irritation is present ; and here they do not appear to arise in the milder cases, but only where the diseased surface is very considerable, and the irritation proportionately great. The contents of these tumours after suppuration vary very much in character, being in some cases apparently healthy pus, and in others the curdy substance of scrofulous abscesses ; but never resembling the viscous secretion of the disease. The little abscesses which they terminate in usually heal without difficulty, and those which have not suppurated disappear as the irritation of the part subsides.

Like ringworm, the P. favosa spreads rapidly by infection through families of children, and it is not uncommon to see several of them inoculated from one child, around whose mouth one or two pustules may have appeared, and the contents have been applied to the lips and cheeks of its brothers and sisters, in kissing them. The breasts of the nurse are not unfrequently inoculated in the same manner. On these accounts it becomes a matter of prudence to separate the children from each other, when the disease once appears among them, or to prevent the infection of the healthy by more frequent ablutions than usual.

When a favous pustule has been produced by the application of matter in the foregoing manner, a speedy check to its course is often in our power. Its contents may be removed at once, and if the abrasion is of sufficient consequence to require, and the situation of it admits of such application, a small poultice may be applied. If this cannot be made use of, frequently washing the part with warm water will diminish the irritation, remove the secretion as fast as it is produced, and prevent the extension of the mischief.†

---

* R. Argent. nitr. ℈j ; Aq. distill. ℥j. M.—R. Cupri Sulph. ℥j ; Aq. ferv. ℥ss. M.

† In M. Alibert's original arrangement, immediately following " Les Teignes," the Plicæ are found. I venture to append the following note to the subject of Porrigo favosa, because I have frequently seen it among the poor in England almost completely corresponding with the description of that of the Plica Polonica. I have, in truth, no doubt of their identity.

In the absence of those advantages which personal observation only can give,

*Porrigo Larvalis.**

This disease differs very little from the preceding in any of its important characteristics, and its causes may be pretty uniformly ascer-

M. Alibert has done all that a compiler of the opinions of others can do, namely, collect all which has been written on the subject.

The Plica Polonica, so called because, as it appears, only more common to that country than to those adjoining, according to M. de la Fontaine, who published in 1792, is supposed to have been imported from Tartary in the early part of the fourteenth century. From this author, unfortunately, M. Alibert has collected a large portion of his subject. When I say unfortunately, and consider that M. F. was no less a personage than surgeon to the King of Poland, my readers will perhaps be surprised, but the fact is, that he has either given a most exaggerated picture of the disease, or the latter has undergone a total change of character. In the words of M. Fontaine, the disease is one of the most formidable, disgusting, and fatal. It is contagious and often congenital, its causes are constitutional, and its effects most destructive. Cured with the utmost difficulty, and, when cured, often followed by more formidable disease, requiring are production by inoculation of the original disease to save the life of the patient! To justify, therefore, my comments, it is only necessary to say, that Baron Larrey,† who personally investigated the disease among the poor of Warsaw, many years after the publication of M. de la Fontaine, satisfied himself that it was a *local factitious* disease, produced by diet and neglect, not contagious, and cured easily by attention to cleanliness, where the absurdities of prejudice and quackery were not allowed to prevail.

Baron Larrey's observations and opinions coincide with those of Dr. Chamseru, an army physician, whose opportunities of observation were equally extensive. The French army physicians, in their earlier campaigns, did not consider it as a disease peculiar to the country, but a degenerated form of syphilis. There is one circumstance confirmatory of Baron Larrey's opinion—in de la Fontaine's own description, abundance of lice, namely, is described as a constant characteristic. The latter, however, positively denies that cleanliness and combing of the hair are any defence against it, though he nevertheless states, that the Polish peasantry are of a different opinion. Inflammation and suppuration of the nails he considers an important symptom, a part of the disease; but the stronger probability would seem, that this was the simple effect of the irritating secretion lodged there in the operation of scratching. Still there is no doubt that the disease has been considered by the Poles themselves as a serious one; for in 1806, the University of Wilna offered a prize for the best essay on it.‡

Nearly a quarter of a century has elapsed since M. Alibert obtained and published the result of his inquiries on the subject of this disease, and I am not aware that any new fact of importance, either as regards its pathology or treatment, has been added to our stock since that period. The miserably unsettled state of that unfortunate country during this period accounts for it. When war comes in through the door, science flies out through the window. The reader must excuse this parody—the original has some gaiety in it; I would expunge it if it were less appropriate.

In the absence of practical acquaintance with the subject, I submit to my

---

* The Teigne muqueuse of M. Alibert; Impetigo larvalis, Crusta lactea, &c. of other authors; frequently named tooth-rash.

† Mémoires de Chirurgie Militaire et Campagnes de D. J. Larrey. Paris, 1813.

‡ The ancient University of Wilna, coeval with the first establishment of its kind in Europe, is abolished. Its splendid buildings are ceded to Greek priests for the trifling sum of 200,000 roubles in paper. There is, unfortunately, but little chance, therefore, that we shall know much more than we do at present of the disease in question.

tained to be the same. The difference in the size of the pustule before its contents are discharged, is, I think, satisfactorily explained

readers the ideas of the first dermologist of the age. It is a duty to the reader to give the substance of these entire, but it is equally a duty to divest them of a superabundance of dross, with which, from the best of motives, I doubt not, he has mixed them up. He occupies no less than a hundred and eight pages octavo with the subject. In justice to him, I omit to speak of the disease in the plural number, and, once for all, treat it as one individual peculiar affection, under the designation of Plica Polonica, omitting altogether the divisions into species, and their subdivisions into forms, as of no great pathological importance.

In this disease the hair of the scalp or pubis becomes glued and twisted together, and is embued and saturated with an adhesive fluid, in such a manner as to render it impossible to unravel it, forming meshes, tails, and tufts, or masses, according to circumstances. It is met with particularly in Poland; it is seen occasionally in Lithuania, Transylvania, and Hungary. From the source of the Vistula to the Krapact mountains, in Prussia, Russia, and Great Tartary, and has come so near to us as Belgium. It is sometimes, though rarely, seen in France.

The Plica is not confined to the human species. It attacks domestic animals of every description, dogs, cats, horses, cows, sheep, &c. in the same climates. The wolf and fox are particularly subject to it, but it has never attacked birds.

"I have seen three cases of this disease," says M. Alibert: "two of these were those of Polish beggars who had found their way to Paris, and the third a French woman from the North. The observation of these cases, of course, do not enable me to judge of the causes producing it, or any other important point in its history. Hence I have had recourse to the writings of foreign authors, whose means of doing so have been more extensive. There are two forms, the true and the false Plique; the one, as Polish physicians declare, is of constitutional origin, and a serious disease; the other is produced by excess of dirtiness, has no constitutional origin or connexion, and this is the Plique of which Larrey and Chamseru speak. The Polish physicians apply therefore as a distinction, the terms of *benign* and *malign* to the two different forms.

"There are as regards external appearance three forms of the Plique. In the first the hair is glued together, apparently divided into separate meshes. In the second it lengthens itself in the form of a tail, excessively enlarged; in the third it agglomerates in the form of tufts or balls. In the severer cases the hair is united from the roots downwards; in the milder, it is glued together near the points. If the agglutinated mass drops off, it often happens that healthy hair takes its place, and sometimes healthy hair grows among the diseased mass. One mass of agglutinated diseased hair may drop off and a new growth of hair take on the same diseased characters.

"Wandering pains and a general feeling of weakness and numbness in the extremities, spine, and head mark the commencement of this disease. In the evening there is fever, which increases during the night and terminates in a viscous and fœtid perspiration. The return of the paroxysm is accompanied by spasms of the muscles, tinnitus aurium, and intense headache, either over the orbits or in the back of the head. After some one of these attacks the hair is found united by the viscous fluid in a mass; in other words, firmly matted together. The nails some time after change their appearance, become deformed, dry, thicken and grow prodigiously; they become yellow, livid, or black, and break in pieces. The fluid discharge does not come merely from the pores of the skin of the scalp, but from the hairs themselves, as microscopic observations have proved. Even the extremities of the hair exhale a fluid or vapour which condenses and deposits itself in the intestines of the hair.

"Metastasis of this disease is not uncommon. Asthma, hydrothorax, hœmoptysis, phthisis, severe catarrh, palpitations, diarrhœa, and dysentery, are the common results!"   Mons. A.

by the greater degree of delicacy of the skin in infancy, during which period it occurs ; when, of course, the cuticle is sooner ruptured by any accumulation of fluid beneath it.

Mons. A. combats with great ingenuity and with the advantages of extensive reading, the opinions of Larrey and Chaınseru, apparently satisfactorily to himself; not so probably to his readers. He is not, however, unsupported in some of the views he has taken ; for M. Monton, Surgeon to the Imperial Guard, agrees with him that the Plique is a disease " sui generis," and not, as they imagine, the mere offspring of filthiness. The danger of cutting the hair, before alluded to, is, except under certain circumstances, according to M. Chaumeton, an army physician, great, and the measure only to be justifiably adopted when it begins to assume an aspect of health at its roots. The question of its communicability by contact is at present, as far as the evidence on each side goes, quite unsettled.

M. Alibert observes that we are utterly ignorant of the causes which influence the development of the Plique. We know only that parts covered with hair and the nails are the seats of it; but, he asks, do we know much more of syphilis ?

He considers the mode of treatment of Plique should be founded on considering it as a peculiar disease. A disease *sui generis*, and as one which like some of the species of Teigne acting by the discharge beneficially to the system, and often averting or preventing through life the attacks of any other important disease. "The poor," says he, "are fully aware of these advantages of the disease."

It sometimes ceases spontaneously, the matted hair dropping off, when the fluid secretion is discontinued. It is sometimes of hereditary origin, and newborn children occasionally show marks of it. A grandfather may have been the subject of it while the father of the child has been spared, as happens not unfrequently in the case of gout.

With reference to the external causes of the disease they are numerous. Humidity of atmosphere, and bad water, have been improperly reckoned among them ; for strangers who come to reside, and being therefore subject to the same, are hardly ever attacked. A superstitious notion is common among the poor of a supernatural origin. The influence of diet is great. Among the Polish Jews, who live chiefly on salted meats and raw spirits, it is decidedly most prevalent. Dirty habits are manifestly always in operation. Combing, washing, or brushing the hair are customs scarcely known among the poor. They wear warm fur or hair caps, which of course, by accumulating heat, add to the mischief. Unwholesome and crowded apartments only are occupied by the poor, and their rooms are often shared with animals of every description ; hence another constantly operating cause.

It is communicable by contact or by the transfer of the fluid secretion to a healthy part of the scalp. In coition it is conveyed from the pubis to the mons Veneris, and vice versa. It is sometimes produced by the different passions.

"A physiological consideration of the functions and nourishment of the hair is necessary to enable us to arrive at a knowledge of trichoma, which we could not attain without. The hair in different parts of the body facilitates the cutaneous perspiration, and conveys it more readily from the surface, serving as a sort of emunctory. The accumulation of scurf, as evinced in combing it, so much greater than on other parts of the body, shows also that something is going on in the action of the vessels connected with the hair not common to other parts.

"The post mortem examination of persons who have died while subject to Plique, discloses in the bulbs of the hair of the diseased part a considerable enlargement ; and when they (the bulbs) are pressed, a yellowish fluid exudes, similar to that which during life passes along the hair and glues it together."

M. Alibert then gives us the resuls of the Autopsie of different individuals dying with the Plique upon them, but it is a mere detail of organic diseases, beween which and the disease of the scalp, there is evidently little or no connexion. He then details a series of attempts by Vauquelin as a chemist, to obtain

Notwithstanding the formidable accounts of its occasional obstinacy and duration given by some authors, it is well known only to require treatment of the most simple description ; and in England, where cleanly habits distinguish almost the very lowest class of society, it should appear, that it cannot by possibility obtain the character of a formidable disease.

There seems to be a very near approach in similarity to impetigo in the disease under consideration, as regards its external appearances. In impetigo, the pustules appear rapidly, and in clusters ; go through their course, and are followed by scabby incrustations of a nearly similar kind ; the same sensations, or pretty nearly so, accompany it, though it occurs chiefly in other parts of the body, as the legs, arms, hands, &c., and is generally considerably more difficult to remove. Dr. Bateman's opinion may be adduced in support of these remarks ; but it does not appear that he considered the analogy so great between the favous species and this, from his proposition to denominate it impetigo larvalis : the points of difference, however, between the three, seem to form merely a scale of gradation from the most active excitement, copious and extensive secretion of the vessels of the cutis, to a state assuming more of the characters of chronic inflammation.

An eruptive disease so commonly seen in infants within the first year, in the lower classes of society, and forming the only one of

something illustrative of the disease by chemical experiments on the hair, &c., none of which have led to any important results.

Under the head "Methods employed for the cure of Plique," he observes again, that it often terminates spontaneously, but that very commonly the natives consider it a blessing instead of a curse, and will not submit to be cured at all ; and, adopting the creed of M. de la Fontaine, he is very reluctant to interfere with the disease as regards external applications of any kind.

Internal means for the cure grow apparently also on an equally barren field. To relieve the digestive organs by emetics, to institute a gentle and general perspiratory action, by decoctions of burdock, sassafras, guaiacum, preparations of antimony, &c. Cooling drinks of sorel, chicory, lettuce, dandelion, barley water, and honey, &c. are recommended as long as febrile irritation continues, and afterwards as tonics, quinine, gentian, and iron. All bitter tonics are useful, but a nourishing diet is necessary also, and is among the most effectual remedies.

Among external applications, fomentations and poultices, and means tending to subdue irritation, stand the first generally, but in some cases, stimulants, even to the extent of blistering, are necessary ; even the use of moxa has been had recourse to ' on dit,' with good effect.

The particular condition of the cases requiring this deviation from the first-mentioned, and apparently most reasonable measures, are not set forth.

Rayer has added nothing to the history of this disease, and I have been assured by the best authorities, that if it ever did really exist, which admits of doubt, it is not now to be found even in the lowest classes. The statement that acute pain is felt in simply cutting the hair is apparently fabulous. Its real character may be summed up probably in a very few words—an inflammatory disease of the scalp producing a viscid adhesive secretion, which copiously flows over and among the hair, matting and glueing the latter together in irregular bundles : the necessity of appending to this definition the existence of pediculi on the parts does not appear. *They* are always to be found in such a state of the scalp, or indeed in any approximating to it in even a slight degree.

importance to which they are subject, excepting the exanthemata,
appears scarcely to require elaborate description ; the very accurate
and scientific account of Dr. Willan, however, would sustain material injustice by any attempt at abbreviation.

"The Porrigo larvalis* generally appears first on the forehead,
in minute pustules with a whitish point, set close together, and producing a redness and inequality of the surface, attended with considerable itching. The pustules break in a few days, and discharge
a clear, viscid humour, which gradually concretes into thin yellowish scabs. From beneath these a discharge of fluid takes place,
from time to time, and forms additional layers of scabs, of a brown
or blackish colour, till the forehead is completely incrusted. The
scabs are in some places thick and rounded, though not very compact ; in others thin or laminated, and loose at the edges. They
do not separate at regular periods ; if any of them be detached, the
surface is presently covered by a new incrustation. The scab is
alternately dry and humid. Sometimes, from a fresh eruption of
the pustules, or from other circumstances, the discharge becomes on
a sudden, so profuse, that all the surface is laid bare, and remains
for several days in a state of ulceration, emitting a thin, viscid, and
acrimonious fluid from innumerable pores. Very young infants are
most liable to be thus affected, and they suffer extremely from pain,
itching, and irritation, when the complaint is extensive. On the
cessation of the discharge, brown or blackish scabs gradually form
again, and cover the ulcerated part. When the disease is about
to terminate, the scab becomes dry, and sometimes whitish, and at
length falls off, leaving a red, shining cuticle, indented with deep
lines, and very brittle,—hence it cracks and exfoliates, and is renewed perhaps three or four times before it acquires the usual
colour and texture.

"This complaint is not always confined to the forehead. In some
cases, it appears first on the hairy scalp, the pustular eruption being
preceded for some weeks by the dandriff, or pityriasis capitis.† In
other cases it may be first observed on the cheeks, or chin, on the
temples, or about the ears. Wherever the disease commences, it
usually extends, in the course of two or three months, to all the
parts above mentioned, and likewise to the neck or breast, so that
the whole face looks as if covered with a vizor,‡ the nose and eyelids
alone being exempt from the dark incrustation. The fluid, which
perpetually distils from among the scabs, diffuses a rank, unpleasant
smell, and is very acrimonious, for it excoriates the adjoining parts
where no eruption had previously appeared. The trunk of the

---

* Dr. Armstrong calls it the Tooth-rash, because it appears so frequently at
the time of teething in infants. "Sometimes," he says, "it spreads all over
them, and appears very much like the itch. Sometimes it is confined to the head
and face, putting on the form of large scabs or blotches, a good deal like the
small-pox just after they are turned."—On the Diseases of Children.

† On Cutaneous Diseases, p. 192.

‡ From this appearance, Dr. Willan denominated the disease, Porrigo larvalis.

body, and the extremities, are sometimes affected in this species of Porrigo. I have seen it on the back and loins, on the arms, thighs, and legs. An eruption of numerous small achores is succeeded by layers of brown or blackish laminated scabs, which nearly cover all the parts affected. After a few weeks, the scabs become dry and whitish ; and at length fall off, discovering a red, smooth, and shining cuticle : but the disease often returns in the same places, and exhibits the same appearance as at first. Although the eruption may commence in any of the situations above mentioned, yet it seldom remains long without affecting either the hairy scalp, the forehead, or some part of the face, where it finally settles. All the symptoms are milder in children somewhat advanced than in infants not a year old : there is less itching and irritation, and the discharge from the pustules is not so considerable ; the scab or incrustation is also drier and less extensive. The complaint is chiefly confined to the forehead, temples, or cheeks ; when it has disappeared from one of these places, a fresh eruption takes place in another, and sometimes the limbs are partially affected by it.''

The most distressing cases of this eruption which I have had an opportunity of observing are those in which it occurs in children of a full habit, and occupy the fore part of the neck, extending from the chin and angles of the lower jaw, on each side down to the clavicles ; the flexures and folds of the skin of the neck where the cutis is abraded, and pouring out the adhesive discharge in great abundance, often becoming adherent during quietude and sleep, and torn apart again on the slightest motion of the head. The pain, irritation, and suffering of the little patient under these circumstances, are easier conceived than described ; and the desperation with which it occasionally exerts itself to obtain relief from the itching, by tearing and scratching the part, truly painful to witness. Relief even of a temporary nature, is often under these circumstances not to be obtained. If unctuous substances be applied to prevent adhesion in the folds of the skin, the itching and irritation seem to be often increased, while cold or warm poultices, though of a little effect in allaying irritation, are often followed, in consequence of their disturbing and causing too early a separation of the scabs, by tedious and troublesome sloughy sores which are difficult to heal, and which sometimes leave scars behind them to a great extent.

With regard to the remark of Dr. W. in the foregoing description, as to the occurrence after, or connexion of this affection with pityriasis, it may be proper to observe, that the latter seems more frequently, when terminating in a form of disease more severe than itself, to be followed by confirmed scalled head, or a state resembling the advanced stages of ringworm. There is no question as to its frequent appearance in conjunction with pityriasis, because its constitutional causes are as liable to exist where the latter has established itself, as where the scalp is free from disease ; but it is only on the principle of local irritation that the former can act as a

16*

cause of any other affection, even on the scalp ; and it can by no means produce a disease so evidently of constitutional origin as P. larvalis. The disposition to *settle* on the scalp, according to the language of Dr. W., which is manifested in most instances of this disease, is merely the consequence of the irritation of the hair on the diseased surface, and the difficulty it affords to the removal of the scabs and diseased secretions.

The most experienced authorities have remarked, that P. larvalis makes its appearance with the strongest features of activity, and disposition to rapidity of extension, in strong and healthy children : but children who have been most liberally fed, and who appear to have digestive organs equal to the management of every thing administered to them, come under this denomination ; and there is every reason to believe, that so long as this continues to be the case ; so long as every portion of the food is converted into nourishment, and the action of the liver and bowels preserves its regularity ; so long will the appearance of this eruption be to be regarded as desirable, and as eminently serviceable in the prevention of more formidable disease.

On the other hand, it has been seen after long duration in emaciated and sickly children, with tumid abdomen and enlarged glands in different parts of the body ; and hence have arisen some doubts as to its original dependence on repletion. From the observations which I have been enabled to make, however, I am not disposed to think a low state of system capable of acting as a cause of the disease under any circumstances. Mere inattention to cleanliness neither, in impoverished subjects, has never appeared capable of producing a cutaneous disease attended with any thing like that rapid pustulation and excessive discharge which characterise this. When appearing under the above condition of constitution, therefore, it will generally be found to have made its appearance long before the distressing symptoms described have been noticed.

The itching, irritation, and restlessness belonging to it, together with the discharge, will always be followed by a proportionate reduction of strength, and, to a certain extent, consequently, it is capable of carrying into effect the grand and powerful remedial measure of nature, counter-irritation. The constitution of an over-fed infant is one of all others which renders the exercise of this principle most frequently necessary, and nature seems, accordingly, to provide for such necessity, by the institution of a disease like that under consideration. The common occurrence of glandular affections in conjunction with it should never be suffered to lead us to the notion of its analogy to diseases identified with original debility of constitution ; for it is well known that irritation of the cutis, even of the slightest kind, is capable of effecting mischief in every gland in its neighbourhood, and the occasional enlargement of the mesenteric glands only occurs as a consequence of the continuance of the irritation of the disease and the fever and debility induced by it.

Its connexion with the process of dentition has been already adverted to, and has been so commonly observed as to have obtained for it in its milder forms the popular appellation of tooth-rash. Alibert has stated, that it is never seen after dentition is complete; but cases do occasionally occur where it continues for some months after such period, though under such circumstances, disorder of the stomach, or fulness of the system, have been manifest, or as has been stated to take place sometimes with the P. favosa, so much debility has been induced as to render the constitution unequal to the task of reparation and the production of healthy cuticle.*

There is ample reason for the serious consideration of the influence of this affection on the constitution of the infant, with relation to the well-known disposition to determination to the head existing at this period of life ; it is more than probable, that in this point of view, its beneficial effects far exceed any thing which art can supply, and that death is averted by it in numberless instances.

With this view of the case, it will not be expected that notice should be taken of the vulgar ideas as to the mother's health, or the properties of the milk, to which nurses are accustomed to attribute mischiefs of great magnitude, if suspicion can possibly be attached, as causes of P. larvalis. It would be as great a deviation from propriety and common sense to spend many words on the discussion of the virtues of the remedy of Strack, alluded to by Dr. Willan.† We may therefore dismiss these points, and together with them all attempts to estimate the value of that important symptom, the harbinger of recovery from the disease, so much dwelt on by the former, namely, the urine of the patient assuming the odour of that of the cat.‡

It has been already stated, that under proper management, the disease is not found to assume an obstinate or tedious character ; and though it is described as recurring at short intervals, in some instances for eighteen months together, the scabs dropping off, and the affected parts continuing to be covered with fresh crops of pustules, yet such a case is by no means of common occurrence,

---

* In one of the instances to which I have alluded, of protracted P. favosa from the too frequent exhibition of powerful purgatives, the subject, a young female in whom the disease appeared about the mouth and face, had inoculated her sisters on the same parts, when this plan was instituted, and carried on for some time in conjunction with low living, without effecting the removal of the disease. On the contrary, the abraded surface, when the scab had separated, exhibited great languor of circulation, and the strength was much impaired. This state of things continued a considerable length of time, and was only removed eventually by tonics, sea-bathing, and a liberal allowance of animal food.

† Half a drachm of the dried, or half an oz. of the fresh leaves of the viola tricolor, to be boiled in half a pint of cow's milk, and taken night and morning. This is certainly prescribing on a larger scale for infants than we are accustomed to !

‡ There is an offensive smell in the secretions of all the forms of Porrigo, and some other cutaneous diseases, which should never for a moment be considered pathognomonic : remove the filth, the accumulated scabs, the vermin, the products of decomposition, &c. and no smell remains.

and when it happens, may most likely be attributed to the continuance of the original cause. Mothers and nurses are, moreover, with difficulty persuaded to employ any degree of assiduity in bathing and washing the part, from the pain which the infant suffers when it is touched, and hence the irritating secretions are allowed to accumulate and mat the hair together ; and thus a ceaseless cause of irritation and inflammation is produced, more than sufficient to excite fresh crops of pustules, and renew the mischief.

It has been considered an extraordinary circumstance, that even in the worst cases of P. larvalis, where the disease has extended over the whole scalp and face, no marks or seams of the skin should remain on the part after recovery ; an attentive observation of the pathology of the disease, however, fully explains this, as the discharge is only poured out from the mouths of the irritated vessels, on the surface, without the production of ulcerative absorption.   When the cuticle is first elevated and broken by the pustule underneath, a copious discharge takes place, not only on the particular point which the latter occupied, but the vessels surrounding it partake of the diseased action, and a more extensive surface of secretion is thus produced ; were this not the case, the quantity of discharge would be considerably more limited.

The constitutional treatment should generally consist of mild aperients exhibited in such doses as to keep the bowels gently acting, and effect a gradual reduction in the fulness of the habit.

Local applications should be such as are conformable to the obvious indications of nature.   Instead of powdery substances, or astringent ointments, poultices, if they can be applied, should be had recourse to, and unless the disease be situated on the fore part of the neck, preferred to all other applications : they relieve the irritation of the part by encouraging the discharge from the surface, and thus furthering the object of nature, the cure is much more speedily as well as safely brought about.

In the 15th vol. of the Med. and Phys. Journal, a case is recorded by Dr. Orme, where a strong hereditary disposition to the disease existed in a family, and became so generally troublesome, as to lead to the application of blisters to the back of the neck, at the period when in the infant it was expected to make its appearance.   The experiment appears to have succeeded ; and it may not be a bad practice, perhaps, in any case when the disease has established itself, and assumes an obstinate character.

CHAPTER II.

ON THE PAPULAR ERUPTIONS OF INFANTS AND ADULTS, DENO-
MINATED STROPHULUS, LICHEN, ETC.

1. Strophulus.

The above designation is applied to the eruption of pimples
occurring in infants from birth, up to and during the period of dentition. The trifling variations in character and extent which it
occasionally exhibits have led to a division into five species : an
attentive perusal, however, of the grounds of such distinction, and
some observation of the affection, enable me to see little reason for
adhering to it in the present discussion ; while, on the other hand,
attention to the important object in view of simplifying the subject
by abolishing unnecessary distinctions and multiplicity of terms,
demands their consideration under one head.

Dr. Willan does not seem to have been entirely aware of the
great influence of variation in the degree of original irritability of
skin in modifying the complexion of cutaneous diseases, or it is to
be presumed that, in the consideration of the subject before us, he
would have preferred following this plan in treating of them. It
does not require much attention to perceive that the different *species*
of Strophulus, unattended by constitutional affection, depend on this
circumstance, or, what would still less justify artificial distinction,
on the degree of local stimulus applied to the skin. The constitutional exciting cause is, of course, pretty nearly the same in all cases
where the affection, as in its severer forms, has such an origin ; but
there is every reason to think that the skins of infants are frequently
covered with pimples from the mere operation of local stimulus,
either in the form of rough clothing, or by exposure to undue
degrees of heat, without the slightest derangement of internal
organs.

It is by no means a common thing to see the skins of young
infants entirely free from clusters of red pimples resembling what
is represented under the disignation of S. intertinctus and S. confertus ; and that such affections should take on an active character
under bad management, or disordered stomach and bowels, is not
a matter of surprise. Inattention to cleanliness seems to be also a
grand cause of such aggravation in numerous instances, and, indeed,
there is reason to believe that where the latter is duly attended to,
and the state of the bowels, and quantity and quality of the food
carefully adapted to the age of the child, this affection rarely requires
the notice of the medical attendant.

With reference to the probable origin of the majority of these
cases previous to the period of dentition, and to their connexion
with local causes, it may be observed, that the habits of nurses are

such as lead them almost constantly to the exposure of parts of the body of the child to the heat of the fire in their frequent changes of linen, or at other periods, while the fear of exposure to cold leads them to suffer the skin to remain too constantly in contact with flannel ; and from what we often see of the inability even of adults to bear either one or the other of these without a great deal of itching and irritation, we are fully warranted in supposing that the delicate skin of the infant must often suffer very considerably.   In a climate like this, it ought to be always considered the safest plan to avoid exposure to cold too suddenly : but there is ample reason to suppose the opposite extreme ; *i. e.* inordinate heat of clothing and nursery-rooms are frequently adopted to the prejudice of the infant's general health, and the production of this disease in particular.

Any thing which leads to irregular determination of blood to the skin would be expected to produce this affection in the infant, and it seems, indeed, to consist of a condition of the part very much resembling that attending what has been heretofore termed miliary fever.   The appearance of the pimples of Strophulus in infants about the neck and shoulders only, being confined to these parts, is often the simple consequence of the irritation of flannel ; and it has, in repeated instances, disappeared as soon as the use of this kind of clothing has been forbidden.

In dividing the papular eruption of infants into so many species, and instituting such nice distinctions, it would appear that the difficulties to the inquirer in the study of these affections are materially increased, while the good which can, under any circumstances, arise from it, is exceedingly doubtful.   The information at present diffused as regards these subjects renders the first circumstance seriously objectionable.   The expense merely of Dr. Willan's work, as before observed, and now fully experienced, is calculated to prevent that extent of circulation requisite to enable every one to profit by his observations ; and it is too much, where such expense has been incurred, to find the subject encumbered with avoidable difficulties of the kind alluded to.

Instead, therefore, of adopting his plan of description, I should be inclined to substitute the following as being more simple, and equally well calculated to convey clear ideas of the characters of the complaint.

"Eruptions of pimples occurring during infancy, generally making their appearance first on parts most exposed, as the face, neck, and shoulders, hands, arms, &c. but occasionally on other parts (P. intertinctus).   When the pimples wear a florid red appearance, the term red gum is familiarly applied : if showing a minor degree of irritation and a paler hue, it receives a corresponding designation, and is termed white gum (S. albidus).   The affection, in its most common and simple form, consists of a few pimples irregularly distributed, now and then mixed with diffused patches of redness."

If circumstances favouring determination to the skin exist in an extraordinary degree, these patches are very numerous, and the pimples proportionately enlarged, the latter appearing now and then to be partly made up of effused serum, forming a minute vesicle, which soon disappears. When, to the common determination to the skin in the infant is added the general feverish excitement attendant on dentition, this affection assumes a more aggravated form (S. confertus), and occurs more extensively in different parts of the body. The pimples in this case appear in patches, accompanied by considerable redness of the skin, are harder, and exhibit marks of more active inflammation. Under circumstances of high derangement of the secretions, and where the child has been improperly fed, it has been sometimes attended with great itching, pain, and excoriation, and in this condition approaches on the lower parts of the body to the intertrigo of infants, or that abrasion and irritation of the cutis on the thighs and nates produced by the stimulus of the urine. Neglect in not frequently removing the napkins, or re-applying them after drying, saturated with the urinary salts, may be, however, capable of bringing about both the states described.

The occurrence of fever in conjunction with this eruption on a more extensive scale has led to another distinction, termed S. volaticus. Dr. W. denies that this combination is an unusual occurrence, and there appears ample reason for the opinion that the fever is accidental, and brought about by causes of a temporary nature ; and is therefore liable to occur wherever bad management prevails. The increased violence and extent of the cutaneous affection seems dependent on the latter, and is an effect analogous to that of the irritation of teething in the milder cases before noticed.

That the cutaneous affections of infants assuming the form of pimples should be so common is explained by the greater degree of vascularity of the cutis, and superior delicacy of the cuticle. In the formation of a pimple an unusual degree of activity in the vessels of the spot is necessary, and this condition is one which the vascularity in question is peculiarly calculated to supply. Turgescence to a certain extent seems likewise an absolute requisite. The cuticle covering the spot readily yields to the impulse of the circulation, and a minute effusion of lymph takes place under it. This I consider to be the correct idea of the formation of a pimple :—*it is produced by a minute escape of lymph from a distended vessel, and not by an enlargement of an original part of the cutis, or in other words, of a papilla, as hitherto supposed.*

The above peculiarities of the infant's skin may therefore be said to constitute the predisposing cause, though it explains by itself the production of the more simple forms ; while a course of management producing fulness of habit, or even what may be termed good health, rough and warm clothing, or any circumstances quickening the circulation and promoting determination to the skin, may be considered among the most prominent of those which excite it.

Enough may be collected from what has been said as to the proper plans of treatment. In a constitutional point of view little seems requisite beyond a modification of diet and clothing, and occasional aperients ; but the influence of (the warm or vapour bath in all cases where the eruption gives uneasiness, is both locally and constitutionally productive of the most decided advantage.   In aggravated cases, accompanied by febrile symptoms, Dr. Willan recommends an emetic to be given, followed by a mild aperient, after which the decoction of bark is found useful.

The accounts of the imminent peril of checking or repelling eruptions of the kind under consideration, given by different authors, require to be received with some degree of caution.   Any accidental determination of blood to internal organs must necessarily deprive the vessels of the skin of every thing like turgescence, and thus occasion the sudden disappearance of such trivial affections from the surface.   Thus their disappearance may be more frequently a consequence than a cause of internal disorder. Taking into consideration moreover the small extent of surface they occupy, and the small degree of determination to the cutaneous surface generally by which they are accompanied and supported, and also their frequent appearance from local irritation, it appears highly improbable that they are entitled to rank as efforts of nature to prevent internal mischief : consequently, that the apprehensions of suppressing them are not very well founded.

2. Lichen.

Eruptions assuming the form 'to which the above designation has been attached, in adults are usually ushered in by slight febrile symptoms, particularly when extensively diffused over different parts of the body.   They occur under similar circumstances with those of infants, the consideration of which occupies the foregoing pages, namely, where powerful causes of determination to the skin occur, and remain for some time in operation. Hence they make their appearance as the hot weather commences, and are followed by great aggravation, if the circulation is hurried, and perspiration brought on by exertion.

Dr. Willan defines this affection " an extensive eruption of papulæ affecting adults, connected with internal disorder, usually terminating in scurf: recurrent ; not contagious :" the more common forms, however, are not uniformly dependent on internal disorder, or followed by fever ; on the contrary, they seem to appear not unfrequently very suddenly, under exertion and perspiration, with every symptom of apparent good health, and disappear in the course of a short time, together with the itching and tingling belonging to them.

The analogy between this affection and strophulous is exceedingly intimate, as is proved by the similarity in the most important of their features ; and there is as little absolute necessity for its division into so many different species as in the latter. Hence the same mode of description may be resorted to with propriety.

Like strophulus, the milder cases of this affection may be considered indicative of pretty good general health, though it is stated now and then to occur under states of great constitutional debility, when the pimples wear an appearance in colour not much unlike petechiæ.

The description of lichen most conformable to necessary precision, without partaking of the complexity of Dr. Willan's arrangement, should be as follows :—

An eruption of pimples, occurring generally first in some part of the upper half of the body, as the face, neck, arms, &c. at seasons of the year when the circulation begins to be determined with unusual force to the surface of the body, as in the spring,* and the approach of summer. The extent of the eruption varies considerably, according to the degree of exposure to heat and violence of exercise or labour to which the patient is accustomed, being in some cases partially diffused, and merely attended with pretty severe tingling and itching ; while, in others, the eruption extends over almost every part of the body, and is accompanied by symptoms of general irritation and fever (L. simplex). The pimples are larger, paler, and showing appearances of less activity of inflammation in situations which are exposed than in others protected by the clothing, and when they occur on parts covered by the finer kind of hair, as on the chest, arms, legs, &c., they are often found to have one or more of such hairs growing from their centre (Lichen pilaris) ; the general disposition to irritability of skin appearing to be aggravated at the point where the hair penetrates it. When the pimples are so situated, and frequent friction is resorted to, to allay the itching and tingling, they appear to be speedily much increased in magnitude, and not unfrequently to pour out a bloody fluid round the roots of the hair. Now and then, from causes not clearly ascertained, the eruption, instead of being diffused, is limited to circumscribed patches, and hence the term (L. circumscriptus) has been employed to give this appearance its place as a species.

When any accidental cause of disorder of the digestive organs or exposure to cold has taken place, the eruption is of a more aggravated kind, and its appearance is accompanied by so much derangement of system as to give it the character of an exanthematous fever. " The papulæ are distributed in great numbers, without any certain order, chiefly on the arms, the upper part of the breast, neck, face, back, and sides of the abdomen ; they have a high red colour, and are surrounded by extensive inflammation, or redness of the skin, attended with itching, heat, and painful tingling. When the patient becomes warm in bed, the redness increases, and there is a strong sensation of burning and smarting for an hour or more, as if the parts had been severely scalded. The same effect is produced at any time by washing, especially if soap be used ; also by violent exer-

---

* The influence of the warmth of summer on the skin is too marked to be left out in a philosophical and just consideration of the affections of the skin. The strawberry or the fabled cherry mark on the neck of children, has been already alluded to as an illustration.

cise, or by drinking wine. In the morning, the papulæ subside, the inflammation in a great measure disappears, and no uneasiness is felt till after dinner, unless it be excited by the above-mentioned or by other similar causes.   Some small vesicles, filled with a straw-coloured fluid, are occasionally intermixed with the papulæ.   These arise dispersedly on the arms, about the roots of the hair, on the forehead or temples, and often on the fingers, but they soon dry and exfoliate.   By a long continuance, or by frequent returns, of the heat and redness, the skin is at length altered in its texture, becoming harsh, thickened, chappy, and exquisitely painful on being rubbed or handled.   The duration of this complaint and the modes of its termination are very uncertain.   It sometimes continues four or five weeks without any material change in its appearance : sometimes it has an earlier termination by slight exfoliations of the cuticle ; but in most instances the eruption appears and disappears repeatedly before the disease is removed.   It may also by improper applications be suddenly repelled from the surface of the body.   This incident is always succeeded by violent disorder of the constitution.   I have observed, in one or two cases, where it was occasioned by imprudent exposure to cold, that an acute disease ensued with great heat and thirst, and accelerated pulse, frequent vomiting, pain in the bowels, headache, and delirium.   After these symptoms had continued ten days, or somewhat longer, the patients recovered, though the eruption did not return.   The uneasy sensation of itching and tingling at the commencement of the disorder leads many persons to rub the affected parts too harshly, and thereby to produce fissures, or excoriations, with a considerable discharge of watery fluid.   At an advanced period of the eruption, similar effects take place from the violence of the cutaneous inflammation without external injury.   The ulcerated surface cannot, in either case, be readily healed by medicinal applications."*

The whole of the species of Dr. Willan may be said to consist merely of difference in the degree of violence of the disease, a circumstance which depends either on the state of the digestive organs at the time, or the share of irritability of skin of the patient ; the common summer rash, a local affection entirely, taking place where nothing particular exists in these respects, while the more aggravated form, as described in the quotation above, occurs under the circumstances of previously disordered constitution.

The appearance of these eruptions is stated to be sometimes followed by relief to old standing and obstinate complaints: pains in the stomach, in delicate females, are particularly specified among these, affording one among the many proofs which the study of cutaneous disease enables us to discover, of the value of counter-irritation in relieving internal organs.   When occurring in circumscribed patches after vaccination, Dr. Willan thinks it ought to be considered a proof that the system has been satisfactorily impregna-

* Willan.

ted with the virus—an observation, if correct, perhaps of some value ; but which experience, rather than reasoning, would be likely to enable us to confirm or reject.

In the treatment of these affections mild saline aperients are plainly indicated, but medicines which determine powerfully to the surface of the body aggravate the most troublesome symptoms ; the itching and tingling being particularly increased by them during their operation, while but little alteration in the disease seems to be ultimately affected. A low diet and regimen, and avoiding exposure to the heat of the sun and violent exertion, are also very necessary. After the disappearance of the eruption, the state of system is said sometimes to require the use of tonics, but I doubt whether any case can occur where these would be called for, if purgatives of the kind mentioned are judiciously employed. It is evident that a gradual reduction of the strength and fulness of system effected in this way would not be followed by debility, though violent cathartics may possibly produce it, and leave a state of the cutaneous affection resembling that alluded to in the discussion of Porrigo larvalis. Here, however, there is no abrasion of the surface, and the doubts as to the necessity of tonics receive additional strength from this circumstance ; in short, mild aperients, and the occasional use of the warm bath, comprehend all that is generally necessary.

From some observations which I have lately had an opportunity of making on the sulphur vapour bath, I am induced to think it a most powerful instrument in the hands of a judicious medical man in the treatment of lichen, though it should not be recommended till the bowels have been some time kept open, and the system, if the patient be of a full habit, has been a little reduced. The itching and tingling during its operation is rather severe, but it is followed by a much more tranquil state of the circulation in the cutaneous vessels, and the cure is altogether materially expedited by it.

The Lichen tropicus, or prickly heat of warm climates, is entitled, on some accounts, to a separate consideration from the foregoing. Dr. Willan has given an elaborate account of this affection from the pen of Dr. Winterbottom, and also quotations from Hillary, Bontius, Clark, Mosely, and others ; but the less tedious statement of Dr. Johnson,* who evidently writes from experience of the disease in his own person, appears fully adequate to all our purposes. It is moreover entitled to the preference on the ground of peculiar accuracy ; if the remembrance of my own feelings while a sufferer be at all correct. Persons a long time resident in India suffer considerably less than new comers, particularly if the latter are of a full habit, and disposed to indulgence in the luxuries of the table ; and it has appeared to me to occur with peculiar violence in persons who, after a course of mercury in the treatment of some one or other of the diseases of the climate, are going on very rapidly recovering their health and strength.

* See a Treatise on the Effects of Tropical Climates on European Constitutions.

"Among the primary effects of a hot climate (for it can hardly be called a disease) we may notice the prickly heat (Lichen tropicus), a very troublesome visitor, which few Europeans escape.

"This is one of the miseries of a tropical life, and a most unmanageable one it is. From mosquitoes, cock-roaches, ants, and the numerous other tribes of depredators on our *personal* property, we have some defence by night, and, in general, a respite by day ; but this unwelcome guest assails us at all, and particularly the most unseasonable hours. Many a time have I been forced to spring from table and abandon the repast, which I had scarcely touched, to writhe about in the open air, for a quarter of an hour : and often have I returned to the charge, with no better success, against my ignoble opponent ! The night affords no asylum. For some weeks after arriving in India, I seldom could obtain more than an hour's sleep at one time, before I was compelled to quit my couch, with no small precipitation, and if there were any water at hand, to sluice it over me, for the purpose of allaying the inexpressible irritation ! But this was productive of temporary relief only ; and what was worse, a more violent paroxysm frequently succeeded.

"The sensations arising from prickly heat are perfectly indescribable ; being compounded of pricking, itching, tingling, and many other feelings, for which I have no appropriate appellation.

"It is usually but not invariably accompanied by an eruption of vivid red pimples, not larger in general than a pin's head, which spread over the breast, arms, thighs, neck, and occasionally along the forehead, close to the hair. This eruption often disappears, in a great measure, when we are sitting quiet, and the skin is cool ; but no sooner do we use any exercise that brings out a perspiration, or swallow any warm or stimulating fluid, such as tea, soup, or wine, than the pimples become elevated, so as to be distinctly seen, and but too sensibly felt.

"Prickly heat being merely a symptom, not a cause of good health, its disappearance has been erroneously accused of producing much mischief; hence the early writers on tropical diseases, harping on the old string of 'humoral pathology,' speak very seriously of the danger of *repelling*, and the advantage of 'encouraging the eruption, by taking small warm liquors, as tea, coffee, wine whey, broth, and nourishing meats.'—*Hillary.*

"Even Dr. Mosely retails the puerile and exaggerated dangers of his predecessor. ' There is great danger' (says he) ' in repelling the prickly heat ; therefore cold bathing, and washing the body with cold water, at the time it is out, is always to be avoided.' Every naval surgeon, however, who has been a few months in a hot climate, must have seen hundreds, if not thousands, plunging into the water, for days and weeks in succession, covered with prickly heat, yet without bad consequences ensuing.

"Indeed, I never saw it even repelled by the cold bath ; and in my own case, as well as in many others, it rather seemed to aggravate the eruption and disagreeable sensations, especially during the glow which succeeded the immersion. It certainly disappears suddenly, sometimes on the *accession* of other diseases, but I never had reason to suppose that its disappearance *occasioned* them. I have tried lime juice, hair powder, and a variety of external applications, with little or no benefit. In short, the only means which I ever saw productive of any good effect in mitigating its violence, till the constitution got assimilated to the climate, were—light clothing—temperance in eating and drinking—avoiding all exercise in the heat of the day—open bowels—and last, not least, a determined resolution to resist with stoical apathy its first attacks. To sit quiet and unmoved under its pressure is undoubtedly no easy task, but if we can only muster up fortitude enough to bear with patience the first few minutes of the assault, without being roused into motion, the enemy, like the foiled tiger, will generally sneak off, and leave us victorious for the time."

Lichen tropicus, however, as it is termed, is by no means confined to the climates from which it derives its distinguishing cognomen, it being now and then seen here, though not in such a violent form. The pimples, in such cases, only exist a short time, making their appearance with the attendant itching and tingling when great exertion has brought on perspiration, and disappearing soon after the body becomes cool. Dancing seems to be the species of exercise most likely to bring it on, and it is generally more violent when the bowels are in a constipated state.

In the spring and summer seasons many young persons are severe sufferers from this troublesome disorder, and are compelled to forego their favourite amusement, or to pursue it in a degree not at all agreeable to their inclinations ; and I have known an instance of a public performer having been compelled to forego it, though the means of obtaining a livelihood, from the severity of the attacks of the itching, tingling, &c. belonging to the disease. Saline aperients, low living, and abstinence from beer, wine, and other stimulants, are under such circumstances absolutely necessary to obtain a reasonable degree of comfort.

M. Rayer observes that it still remains matter of uncertainty which of the elementary tissues of the skin contributes essentially to their (pimples) formation. "Several writers," says he, " have supposed that the nervous papillæ of the skin, enlarged in their dimensions, were the parts more particularly interested ; but this opinion appears by so much the less likely, as the fact is that papulæ occur but rarely on those districts of the skin which are most distinctly and abundantly supplied with papillæ, as on the palm of the hand and cushions of the fingers. *Papulæ again have been supposed by Mr. Plumbe to be produced by the effusion of a very minute quantity of lymph into the dermoid tissue, with which this fluid unites when it is not re-absorbed : it is true that by pricking the larger papulæ of strophulus deeply with a needle, and squeezing them firmly between the fingers, a very minute*

globule of transparent serum may occasionally be expressed ;*
but by resorting to the same procedure in cases of lichen and
prurigo I have never succeeded in forcing out any thing but a
drop of blood."

3. Prurigo.

The resemblance between Prurigo and some cases of lichen
renders it advisable to speak of it here, although it perhaps deserves
to be considered in the majority of cases, rather as constituted of
action of a chronic than active character.

The term employed to designate this affection is derived from its
chief distinguishing feature ; namely, the violent itching of the
papulæ : and it is in many cases accompanied with such a striking
similarity of phenomena to simple lichen, as to raise some doubts
whether any real difference exists.   It is, like the latter, most
troublesome in the spring, and at the commencement of summer,
and is similarly aggravated by circumstances which have a temporary
effect in inducing violent determination to the surface.   In the
milder cases, and where it exists only partially, the itching is not
constant, but varies according to circumstances, still, however,
preserving the disposition to become aggravated by exposure to
heat.   The pimples are described by Dr. Willan to be of a pale
hue, but this is by no means the case when the itching is not
present.   When the latter is troublesome, considerable redness and
heat of the skin surrounding the pimples is excited by the friction ;
and it is under these circumstances that the latter lose their florid
colour, appearing to be deprived of their usual supply of blood
by the irritation and increased demand of the skin surrounding
them.

Pimples, however, are by no means necessary to form the disease
called Prurigo, according to the general acceptation of the term ; mere
itching only existing in the greater number of cases coming under
the notice of the medical practitioner.   Like the majority of other
cutaneous diseases attended with itching, this sensation is much
increased by exposure to heat or irritation, whether produced by
exertion or by warm clothing ; and hence in Prurigo it is most

---

* I have not seen serum appearing on the surface when a pimple of strophulus
had been punctured and squeezed, because the practice in a young infant would
not be justifiable, and I must therefore have been misunderstood by the author
quoted.   The remark to which he alludes I applied only to the white elevations
of urticaria.   The papulæ which I described as very generally in obstinate cases
to be possessed of solid organization, were those of long standing prurigo, from
these of course a small drop of blood only could be expressed after punctuation,
and it is difficult to conceive that any other result should be expected.   A small
red pimple can only be supposed to contain blood.   A large semi-transparent
wheal or elevated pimple nearly of a pearly whiteness, surrounded by a vivid red
circle as in urticaria, invariably contains *nothing but* limpid fluid, and if punctured
carefully so as not to touch the vessels of the cutis, the fact is easily ascertained.

troublesome when the patient is warm in bed, and when he has been induced to resort to friction to relieve a trifling degree of it.

Forbearance is a grand point in all such cases, and should be suggested to the patient as a part of the management of his disease.

By far the greater number of cases of this affection which come under the notice of the medical practitioner are limited in extent, and confined to particular situations ; the most troublesome being those where the generative organs and their neighbourhood are affected ; and it seems that neglect of frequent ablutions of these parts is often instrumental in producing it.* When occurring in other situations, some deficiency on these points is usually discoverable, and it has appeared to me a circumstance of considerable weight in determining the question, as to the effect of lodgment of the secretions on the skin in producing it, that in persons usually atten-tive to these matters, where ablution can be easily applied, the affection is least troublesome, while between the shoulders and on the back, where some difficulty occurs in the use of the towel, the itching is very great, and the pimples are numerous.

Alibert, on the subject of Dartre squameuse, appears to have des-cribed the disease, as far as regards the tormenting itching and other symptoms belonging to it, with great accuracy and strength,

---

* "Lorry," says M. Rayer, "has given a picture as *faithful?* as it is animated, of the symptoms and sufferings produced by Prurigo and Lichen of the genital organ." The description so faithful and so animated ! is as follows. Surely utter disbelief and utter disgust must go hand-in-hand in the mind of the reader in perusing it. It cannot be a "faithful" record of a bodily disease. The patients must have been maniacs, and the recorder of the case at least not an authority to be quoted with confidence?

"Morbus ille adultos ut plurimum et primum pubertatis florem aggressos adoritur, eosque qui, castè vivéntes, urgenti tamen impetu ad venerem ferrentur ; mulieres etiam, sed maturiùs adoritur. Ejus ortus primò mitior est, et prurita totus continetur. At pruritui illi tùm in maribus, tùm in fæminis, jungitur ardor in venerem inexplebilis. Mores et præcepta repugnant, coercet virtus vivax, at manus indocilis ad has partes fertur, scalpendoque malum irritatur, et animus ipse in partem operis venit cum artuum tremore et palpitatione. Sedatur vulgò per plurimas horas malum, tuncque omnia tranquilla apparent, at recrudescit per paroxysmos, noctu potissimùm afficiens. Sævit autem eò vehementiùs, quò aut familiariter magis aut proximiùs cum fæminis mares, aut cum maribus fæminæ vixerint. Nec minores accepit vires à vino, piperatis, spirituosis, acribus alimen-tis, potu coffeæ, oleosorum spirituosorum, ità ut noverim viros qui nunquam similibus pruritibus, nisi una ex hisce causis accesserit, quas edocti experientiâ vitabant seduliùs. Progrediente malo partes ad aspectum maculosæ, maculis flavis vix supra cutem extantibus distinctæ sunt ; scrotum omninò rugosum est, ut et labia pudendorum in fæminis, et tempore paroxysmi prorsùs retractum. Erectio penis et libidinis ardens cupido 'mentem incendunt. Partes illæ non eruptione lichenibus simili afficiuntur, sed epidermis rugosa olet, et alluitur liquore unctuoso, non lintea maculante, non digitis adhærente, sed ad sensum lubrico. Increscente malo pruritus enormes fiunt, per paroxysmos et summè violentos, et frequenter redivives, ita ut nec pudor, nec reverentia regum à scalpendo divertiant, et sæpè per intervalla etiam paroxysmorum puncturæ acer-rimæ acubus inflammatis per cutem transactis morsu similes, in clamorum adigunt : hinc partes illæ rhagadibus atque fissuris manu factis undiquè hiant. Ardor semper inest, et ad quemvis levissimum incessum exhalat humor olentissimus, fervente intereà œstro venereo."

and observes, that it is well entitled to the names of Dartre vive, Lichen ferosus, &c. which it has obtained from other authors. The lips, ears, nose, nipples, anus, and pudenda, are particularly mentioned by this author as the most common seats of the affection. He moreover observes, that inattention to cleanliness always aggravates it, and sometimes gives occasion to its origin. The latter part of his description, however, does not appear to agree with the usual history of Prurigo, but the points of difference appear as usual to be the results of aggravation from the neglect of cleanliness.

A more than usual degree of irritability of skin seems to be peculiarly favourable to Prurigo, whether occurring extensively, or confined to the parts mentioned, and with the exception of an inveterate form, to be hereafter alluded to, it is generally most severe in young people enjoying pretty good health.

Cleanliness and the warm-bath are the most important remedial measures to be had recourse to in the treatment of Prurigo, as it occurs on the superior parts of the body in persons with unimpaired general health ; but from the neglect of the affection at its first commencement in the neighbourhood of the scrotum and anus, a chronic morbid action of the vessels of these parts is sometimes permitted to be established, in which such measures do no good, and the patient's life is not unfrequently rendered miserable by the almost incessant itching and irritation attending it.

Alibert and Dr. Willan prescribe sulphur in this affection to be exhibited internally ; it is probable, however, that this medicine has no advantage over other aperients ; in any case, indeed, it as a matter of question whether medicines possessing, as this does, a power of producing a determination to the skin, can be recommended on a proper principle, where the essential features of the disease to be cured are formed by the prevalence of this circumstance to a morbid extent, unless the energies of the circulation have been previously reduced by more active measures. This remark applies, of course, more particularly to the cases of young and otherwise healthy subjects, and does not admit of adoption as a guide in those cases assuming the chronic form, as in people more advanced in life.

The author of a pamphlet on cutaneous diseases lately published, has given a case of general Prurigo of a very obstinate character, which had resisted various medical treatment for a considerable length of time, and was ultimately cured by a plan of treatment which he had previously found successful in squamous and other affections.

" I dipped (says he) some lint into aromatic vinegar undiluted, and touched most of the prominent papulæ, bleeding as they were from the laceration of the comb, (which had been substituted by the patient for the nails in scratching the part,) till the sense of smarting was as much as the patient could bear, though at the height he declared it was preferable to the itching. I then sent him the following ointment :

R. Sulph. Sublim. Picis liquidæ. Axung. Porcin. a lbss. Terræ Cretos ʒiv. Hydrosulph. Ammon. ʒij. M. ft. Unguentum.

I desired him to apply this ointment liberally over the whole extent of the eruption ; to renew it every day, and wash it off every other day. I gave him four grains of pil. Plummer every night, and five drops of sol. arsenic three times a day. Wine, salt provision, shell-fish, and every stimulating article were forbidden.

"During the second night after these applications he slept above four hours, and the itching was considerably abated during the day. In three days the acid was applied a second time, and afterwards a solution of the argenti nitras every third or fourth day previously to the ointment.

" In less than three weeks the patient was very comfortable, the eruption having nearly disappeared, and the itching being entirely removed. The pills and solution were continued for three weeks longer, during which time a lotion of hydr. oxym. in sp. vini rectif. was applied two or three times a day instead of the ointment."— The success of this treatment may be considered as a proof that the organized state of the pimples of this disease hereafter to be more particularly alluded to is the chief cause of its obstinacy and long duration.

The pimples of Prurigo are stated to be observed in aggravated cases, now and then terminating in pustules resembling those of scabies ; and Alibert's description before referred to contains remarks on its assuming the character of extensive exfoliations of cuticle, leaving a moist, shining, reddened, and inflamed skin. Where an utter neglect of cleanliness in the habits of the patient at its commencement and inattention to proper medicinal remedies exist, this occurrence is not improbable : it is, however, from what information I have been able to collect, extremely rare, even in the very lowest and dirtiest classes of society. It may be remarked farther, that in cases where Prurigo has been much aggravated by inattention to cleanliness, the itch is a disease to be often expected, as a thing of course, arising out of the habits of the patient ; it is not, therefore, a matter of surprise that sulphur, as stated by Dr. Willan, should be the most effectual remedy.

The formidable and distressing cases of this disease, denominated P. formicans and P. senilis, do not appear to require a separate consideration, because there are no points of importance in which they differ, except that hereafter to be described as consisting of simi-organized pimples mixed with others : the degree of itching attending, and the tingling or sensation of creeping insects, from which the former term is derived, being often referred particularly to the pimples of this description. The consequence of scratching violently any of these pimples, namely, the formation of a small blackish scab, is similar to what happens from this step in common cases, nor does there appear to be any ground for the opinion, that constitutional causes are more frequently concerned in producing them than the milder forms.

Dr. Willan expresses his conviction, that Prurigo, at its commencement, is not allied to itch : Alibert has also drawn a line of distinction between them ; and from the common effect of determination to the skin being to produce itching, though in a minor degree, it should seem that no reasonable ground of suspicion of such connexion could have existed on this account ; it is moreover now pretty well understood that itch is a disorder of the skin entirely dependent on the habits and operations of an insect to which Providence has allotted this part of the body for occupation and subsistence.   It is not improbable, however, that many cases answering pretty correctly to the description given of the worst cases of Prurigo may have been those of genuine itch.

There is great variety in the plans of treatment recommended to be instituted for the relief of Prurigo.   The state of constitution under which it occurs should always be borne in mind : if it be such as to admit of the use of purgatives to a liberal extent, the local affection will be much relieved by them.   The milder cases, and for the most part those occurring in young people, are of this description ; but alterative and sudorific remedies are scarcely at first admissible in any instance.   Bleeding has been followed by temporary good effects in some cases.   The Harrowgate bath has rarely failed to produce a decided improvement.*

The suspicion which has been entertained by Dr. Willan, that Prurigo is produced by a peculiar insect, does not appear to have been confirmed by subsequent observation.†   If it had been correct, indeed, it is evident that a considerably less equivocal effect would have followed the application of sulphur in the treatment of the disease.

The parts particularly noticed as the seats of violent itching sensations usually unaccompanied by papular eruptions, are the verge of the anus, perineum, scrotum, prepuce, and glans penis in men, and pudendum in women : and in the latter it sometimes particularly affects the extremity of the urinary passage.   The itching of the nose is another instance of this kind ; its causes, namely, disordered states of the stomach and alimentary canal, are not unfrequently ascertained to have given origin to the affection in the before-mentioned situations.

Attentive examination of the parts when the irritation is present, (for it is in no instance, even of the most aggravated kind, incessant,) always discovers much redness, and scurfiness, as well as fluid secretion.   In the neighbourhood of the anus particularly, there is very generally more moisture secreted than is consistent with health; and the folds of the skin at the extremity of the rectum, in aggravated cases, pour out a glutinous fluid apparently possessing properties of a highly irritating nature.   The perineum and posterior part

---

* Artificially made as follows :—R.   Sodæ Muriat. ℔ij.   Magnes. Sulph. ℥iij. Potass. Sulphuret. ℔j.   Aquæ cong. xxxiv.   The salts must be first put into two-thirds of the water cold, and when dissolved, the sulphuret of potash added; then the remainder of the water boiling : to be used at 98.—*Wilkinson.*

† The insect has never been really detected, and I believe its existence entirely imaginary.

of the scrotum partake of the disordered action, till at length, by the constant friction employed, the whole of these parts become abraded, and a copious discharge of fluid takes place, which for a time somewhat diminishes the itching, and substitutes for it a considerable degree of smarting and tenderness. More trifling degrees of pruritus than these detailed are unquestionably produced sometimes by ascarides, but these are easily enough removed : the state described appears to be dependent on a certain derangement of the fluid secretion of the part, of the causes of which we at present possess but little information.

If no obvious cause exist in the constitution requiring alterative aperients or tonics, and local applications appear to hold out the only chance of advantage, they should be selected from among those known to possess the properties of correcting diseased secretions without acting merely as an astringent or sedative. It should be remembered that the result of more recent experience at present recorded justifies the opinion entertained by Dr. Lettsom, that the sudden suppression of affections of this sort are deleterious to the constitution, and even capable of producing fatal effects. Applications of the latter kind, moreover, do not appear to do any permanent good in allaying the most prominent symptom.

Lime water, solutions, or ointments of opium, tar ointments, &c. have been used with various success ; but ointments of any kind are inferior to lotions, inasmuch as their application favours the accumulation of the diseased and irritating secretions on the part. Lotions of Prussic acid, particularly where abrasion has been produced by scratching, are very beneficial ; but their effects are in general only temporary.

The most uniformly beneficial system of management consists of a simple and unirritating diet, and saline purgatives, with the local application of lotions of calomel and lime water. When the pruritus is concentrated about the verge of the anus, a pledget of lint dipped in the lotion should be introduced into the rectum, and retained there, while the adjacent parts are frequently wetted with the same. The folds of integument at the extremity of the rectum are more frequently the origin of the diseased secretion than is generally supposed, and if these can be kept by ablution free for a few days from irritating secretions, and the above application be satisfactorily made, the perineum, scrotum, &c. soon return to a healthy state.

Now and then, however, in old standing cases, where the morbid action has been a long time established, much difficulty is experienced ; in which event, if we expect to effect a permanent recovery, it must be by exciting a degree of active inflammation of the part which shall be adequate to the bringing about a total change in the action of its vessels. The best application with such a view is a pretty strong solution of the hydr. oxymur., and it may be persisted in till the skin is excited to a blush of deep red, with heat and smarting. Excoriation and a little vesication may be produced in this way, and the greatest inconvenience attending it is the confinement of the patient for a few days to his room.

The Prurigo preputii of Dr. Willan as being merely the common consequence of suffering the secretion of the glandulæ odoriferæ to accumulate on the part and become irritating, may be passed over here, as well as his remarks on the means of destruction of the insects vulgarly termed "crabs."*

Dr. W. has introduced some observations of Dr. Sims and others on the subject of symptomatic pruritus, or the itching of the labia, vagina, &c. sometimes attending hemorrhoids, scirrhous uteri, fungous excrescences, and other diseases in the neighbourhood, which would direct our attention to these points for inquiry whenever such parts are the seat of the irritation. Such symptoms, however, are by no means so constantly attendant on these diseases as to be of much value in their diagnosis.

In a former page, referring particularly to lichen and strophulus, I stated my opinion, that a minute effusion of coagulable lymph was stated to be necessary to the formation of a pimple. When such effusion has taken place, the turgescence of the vessels concerned being diminished, the inflammatory action giving rise to it is also reduced, and the pimple is rapidly absorbed: the general disposition to determination to the skin still continuing, however, other vessels in the neighbourhood undergo the same process, and in this way successive crops of pimples continue to be produced. In infants this state of things is often allowed to be passed over without notice, till it assumes a character of material interference with the general health ; when the injudicious wrapping with flannels, heated apartments, cramming with food, and consequent disorder of bowels, or the increased irritation of teething, are discovered ; and the disappearance of the affection is generally the result of measures directed to these points : but from the very gradual manner in which Prurigo comes on, from its being for a considerable time alleviated by friction, and consequently treated lightly or neglected, and from the continuance of the local or constitutional cause, the pimples here undergo a change, which amply explains the obstinacy which sometimes characterizes it. A degree of chronic inflammatory action in the vessels becomes established, and the pimples, to a considerable extent, instead of being rapidly absorbed, undergo a kind of semi-organization, and become permanently the seat of tormenting itching, much increased by exposure to heat. Pimples of this kind contribute, as before observed, largely to make up the eruption of the aggravated cases described under the designation of P. formicans and P. senilis, and are only to be speedily removed by measures which destroy their organization. Pencilling with caustic appears to be the least painful method of effecting this object ; a little patience in its application will enable us to make it very effectual, and it does not appear to be productive of any kind of inconvenience : but in the case of Mr. Wilkinson, quoted in a former page, another plan is detailed of carrying the principle into effect.

* The most expeditious and safe method of destroying these is by the free use of the common mercurial ointment. Its use for a single night is sufficient.

It is highly probable that the terrific description given by M. Alibert* does not apply to any even of the most severe cases occurring in this country ; and I do not recollect that any case has been recorded elsewhere of its existence from birth, of its extending through families, or being hereditary ; unless any particular state of skin may be the sole cause indeed, one would suppose this improbable. With few exceptions, Mons. A. has found it occurring most violently in people either of sedentary occupations, or whose habits were marked by great inattention to cleanliness ; and the opinions of Dr. Willan and Dr. Lettsom, as to the danger in suddenly checking the disease, are confirmed by his experience; its disappearance having been sometimes followed by delirium, and by marks of congestion in internal organs, and also by general anasarca in cases under his notice in the hospital of St. Louis. In attending to one or two of the milder cases of this disease, I have been led to suppose, and it is an idea which the observation of the authors quoted tend much to confirm, that the employment of the mind in an active manner is useful in warding off many of its attacks : in one instance, indeed, the patient himself was so convinced of its efficacy; as to suffer it to guide his conduct. He found that any conversation or reflection on the subject in which his mind was much engaged would be sure to bring on the attack, while unusually interesting business, or diversion, procured him a respite as long as they lasted.

CHAPTER III.

ON URTICARIA, OR NETTLE RASH.†

THE name given to this disease is a rare instance of felicitous appropriateness in the nomenclature of skin diseases. The designations usually employed, convey no information as to the external character or history of any. In the instance before us, the comparison of the sting of the nettle illustrates the first appearance of its evanescent character, together with the sensations by which it is accompanied in its progress, continues the resemblance till it passes away and is forgotten. It resembles no other disease of the skin, and those who have once seen the sting of the nettle, can by no possibility confound it with any other. Still, however, it stands forward apparently to rebut or reduce the effect of the hacknied axiom, "the knowledge of the disease is half the cure." To see and know it, is not to cure it. In the milder and trifling forms produced, by sympathy with a disordered stomach, the practitioner seldom sees much of it ; and

---

when he does, the remedy is easily found ; but in what may be called the chronic or remittent forms, I know of nothing so unconquerable.

The classification of Dr. Willan places the varieties of nettle rash among the exanthemata, while the Eczema Mercuriale, a disease almost as frequently accompanied by febrile symptoms, and considerably more regular in its progress and decline, depending too, very frequently, like Urticaria, on causes affecting the skin, through the medium of the stomach, is placed with small-pox, &c. in the order vesiculæ. In this arrangement, one instance is presented out of many which will occur to the reader, of the difficulty of classifying cutaneous diseases, so as to comprehend at once their constitutional causes and symptoms, and local characters. Even if our knowledge of the precise states of constitution under which they occur was complete, it is much to be questioned whether an arrangement could be formed upon it of superior utility to that of Dr. Willan, notwithstanding the objections which have been raised against it.

With a view of adhering to a general arrangement according to constitutional causes, I have placed Urticaria here. It is obviously allied in this point of view to the preceding affection, consisting chiefly of that species of determination to the skin, produced by sympathy with the stomach, or by local irritation, and may be also equally beneficial to the constitution.

The appearances of Urticaria are generally pretty well known. It is an affection so familiar, as rarely to give occasion for the attendance of the medical practitioner, except accompanied by symptoms of constitutional derangement, and in by far the larger proportion of cases this does not happen.

It consists in its more common form of patches of inflammation, distributed here and there on different and sometimes distant parts of the body : these patches are in some cases small in extent and number ; in others they are large, and each patch occupies a considerable portion of the skin. In the centre of the inflamed and reddened skin, which itself is not elevated, is situated a white irregularly formed spot, considerably raised above the surrounding cutis, which varies greatly in figure, being sometimes long and narrow, at others broad or round. It is perfectly destitute of colour, and severe sensations of itching, smarting, and tingling, are commonly referred to it. If the finger be drawn lightly over it, it presents much irregularity of surface to the touch, and its margin is also very irregular. The smaller inflamed patches not unfrequently exhibit this white protuberance in the form of small, distinct, circular tubercles about the size of a spangle ; but, generally, if more than two or three of these exist on a particular spot, they run into each other, and produce a lengthened stroke or wheal. Many, if not all, of the larger spots or wheals, are formed by the clustering together of the circular tubercles mentioned ; and, in proportion to their size, they are surrounded by a more or less vivid inflammatory redness, and a corresponding degree of increased heat, itching and smarting, attend them. In whatever part of the skin they occur, they are seldom found to continue long, and are frequently so evanescent as to appear in one part of the body, and be visible only for a few minutes, leaving no vestige of their existence behind them ; while, in a short time, a distant part will be found occupied by them to a considerable extent. The efflorescence forming the greater portion of the patches is evidently the result of simple excitement of the cutaneous vessels, but the white, elevated spots in the centre, which resemble the effects produced by the sting of the nettle, have never been accurately examined or described ; and I am not aware that a fluid exudation from the vessels of the surface of the cutis has ever been considered necessary to their formation. This fact, however, is sufficiently manifested on puncturing them with a finely pointed instrument ; the fluid readily escapes and allows the elevated cuticle to collapse and fall into contact with the cutis. The difference between these spots, therefore, and the vesicles of herpes, appears simply to depend on the circumstance of the fluid exudation of the former not being sufficiently extensive to separate the cuticle entirely from its attachment to the cutis.

Notwithstanding the very general origin of Urticaria in temporary or accidental disorder of the system, it often occurs as an idiopathic affection, or, in other words, as a consequence merely of extraordinary irritability of skin ; and hence some persons can produce it with slight friction on any part of their body, even where no suspicion of the slightest derangement of the function of any organ can be ascertained.

Though now and then occurring in debilitated constitutions, and in persons affected with visceral disease, a state of pretty good general health seems to be most favourable to its appearance. A full habit of body is one of the circumstances which act as an exciting cause, where the preternatural disposition to irritation in the skin alluded to is known to exist.

In the severer and more extensive cases suddenly produced by substances received into the stomach, the tumefaction is often greater about the neck and face than other parts, not unfrequently closing up the eyes, and obliterating every vestige of the natural features of the patient, a state frequently referred to the eating particular kinds of fish ; but, in more trifling cases, the substance, or liquid to which it is attributed, is often such as others, and perhaps the patient himself, may have been in the habit of partaking of on former occasions with impunity. So that it may depend on mere idiosyncrasy, or a state of stomach, or of the secretions, existing only temporarily. Much difficulty will exist in such cases in ascertaining to which of the articles forming the patient's diet the mischief is to be attributed ; and there seems to be no more speedy method of arriving at this discovery than that of omitting first one and then another for a day or two, and watching the state of the complaint during the period such changes are being made.

The high degree of tumefaction above-mentioned is not generally produced by any of the ordinary articles of food or drink, but usually follows the reception into the stomach of such as the patient is not constantly accustomed to. Almonds, mushrooms, cucumbers, honey, fruits of different kinds, opium, &c. have been particularly noticed among these ; but the poisonous properties of fish, in a particular state, not well understood, is its most common cause. Muscles and lobsters seem to have been most frequently possessed of this property, and several fatal instances have occurred of their deleterious power. There is ample reason to believe that the commencement of the putrefactive stage in the fish generates the poisonous properties in question : and from my own observation in one or two cases, as well as from the report of a medical friend, who has many years resided in the neighbourhood of Thames-street, I am indeed inclined to consider it in the light of an established fact. The lower classes of people, both men and women, attending the fish markets, are constantly affording cases of this kind, and. indeed, it is so common as to excite but little alarm. It is generally ascertained to have been produced in such cases by eating the refuse of the market, or the cheaper kind of shell-fish, particularly the muscle.

The poisonous properties in question are by no means confined to the fish of this climate ; it is more common, as well as more violent, within the tropics, both in the neighbourhood of land, and many degrees out at sea, and hence an important obstacle arises to the belief of its dependence on any particular species of sustenance which the fish may have obtained.

Dr. Burrows[*] has given a list of the different kinds of fish which are found to have been occasionally poisonous : the yellow-billed sprat, however, has been noticed as the only species which has produced immediate death within the tropics. From the result of the consultation of different authors, who have been quoted by Dr. B.. with regard to the supposed seat of the poison in particular parts of the fish, it appears that no foundation exists for the belief that any part of it can be eaten with safety, when another is tainted. Two cases of Urticaria, alluded to in the third volume of the London Medical Repository, were supposed at first to be produced by merely eating the skin of the dried herring, but the patients were subsequently ascertained to have eaten the fish itself ; and two inferences of some importance have very properly been drawn from these cases . by Dr. B.; namely, that as two persons, *not related*, ate of the fish, the effect was less likely to be the result of idiosyncrasy, and, that the process of pickling and drying does not destroy the poison. The gall-bladder and liver, stomach and intestinal canal, have been supposed, with as little apparent reason, to be its exclusive seat ; and the weight both of evidence and reasoning seems to be decidedly in favour of its distribution over every part of the fish, and of its not being dependent on disordered secretion of any particular organ.

---

[*] An account of two cases of death from eating muscles, with some general observations on fish poison, by G. M. Burrows, &c. &c.

The idea that impregnation of the circulation and solids of the animal with copper is consistent with life is in itself exceedingly absurd, and yet among the various strange conjectures as to the cause of the poisonous properties of the fish, this is spoken of by scientific men in systematic works on the science of medicine, as the most rational explanation of the fact. It is true that copper, in the form of solution, received into the stomach produces many of the symptoms attendant on fish poison ; but, in such cases, the grand characteristic of the latter is absent ; there is no urticaria : and if the stomach takes the alarm, and instantly rejects the obnoxious fluid, (and it so constantly does so, as to justify the use of the solution, under particular circumstances, as an emetic,) no symptoms of disorder remain even for a single hour after it has been administered.

The evidence brought forward by Dr. B. in the publication above alluded to is in itself, I think, very conclusive against the theories of the origin of this affection in the secretions of any particular organ, or of the residence of the poison in any particular part of the fish. There is, moreover, ample reason to doubt that even a disordered state of the health of the aminal during life (except as increasing the disposition to putrefaction after life is extinct) can have influence in inducing the mischief. The following facts may be reckoned among those which justify such doubts.

In February, 1813, the East India ship Lady Castlereagh arrived at St. Helena, and was detained a considerable period waiting for convoy. During this time the crew were employed occasionally in catching mackerel, and they usually obtained a liberal supply for their daily consumption. For the space of several weeks, under the continuance of this practice, no instance of disorder occurred ; but when the time for sailing approached, a large number of the men were induced to make attempts at pickling and drying, and thus obtaining a store for their homeward-bound passage. They succeeded in this plan very much to their satisfaction, but were induced to regale themselves with a dinner off the produce of their industry before the ship sailed, when no less than sixty were seriously affected. Vomiting, with extreme violence, general febrile symptoms, and urticaria about the face and neck, with immense tumefaction, occurring in every case. The officers and persons who only partook of the fish when fresh caught were never affected. Hanging the fish up in the moonlight, in the drying process, was the only explanation occurring to the minds of the seamen of the cause of this accident.

The following symptoms are stated to have occurred in fatal cases from eating muscles :—Sickness and vomiting of a great quantity of dark-coloured matter, and subsequently of a dark-green fluid ; the urticaria covered the skin, and was attended by intolerable itching ; great difficulty of breathing came on the second day ; tormina, intense thirst, and swelling of the abdomen and face followed, and the extremities gradually became cold and benumbed ; the countenance became of an ashy paleness, and the pupils were extremely dilated ; respiration difficult, insatiable thirst, quick, low, and tremu-

18*

lous pulse, and subsultus.   In one instance death was preceded by severe convulsions.

The fish were obtained in these cases under circumstances unequivocally justifying the conjecture that the putrefactive stage had fully commenced.*   The severer symptoms did not make their appearance till the day after the fish had been eaten; but in ordinary cases vomiting soon comes on, and it is therefore not improbable, that death is averted in many instances, by the readiness with which the stomach rejects the poison.

" In ordinary cases, an hour or two after it has been received into the stomach, a sense of weight and oppression is referred to this organ, with nausea and vertigo, universal uneasiness, numbness of some part or parts of the body, constriction of the throat, a sense of heat about the head and eyes, quickly followed by urticaria and great tumefaction ; immoderate thirst, and an eruption and itching on the skin in other parts, and vomiting, and now and then diarrhœa."

The most direct means of relief consist of assisting nature in her efforts ; first, by unloading the stomach with emetics, and afterwards, by the exhibition of brisk cathartics.   If the symptoms are violent, the sulphate of copper, on account of its quickness of operation, seems best adapted to accomplish the first of these objects, and powdered jalap, for the same reason, should be selected for the second.

The lower class of people who work in and about Billingsgate are, as before observed, very subject to this accident ; and a general idea prevails among them, that large quantities of vinegar are calculated to check the violence of the symptoms : they are accordingly accustomed to drink it to a great extent as soon as they are made aware of the commencement of the disorder.   I am not aware whether sugar has been tried as an antidote in any instance in this country, though from Dr. Burrows's account, it has been employed with success by one gentleman in numerous instances.   Ether, in doses of twenty, thirty, or forty drops, given every half hour, is also mentioned as a remedy of great efficacy.   In the slighter cases, produced by eating fish, if the disease is left to itself, the irritation and tumefaction begin to subside at the end of thirty-six or forty-eight hours, and it never reaches the duration of common febrile nettle-rash occurring from other causes.

The febrile nettle-rash is thus described by Dr. Willan.

" The symptoms preceding the eruption are pain and sickness at the stomach, head-ache, great languor or faintness, with a disposition to sleep, a sense of anxiety and increased quickness of the pulse, and a white fur on the tongue; in two days, or sometimes later, after these symptoms, the wheals appear, with an efflorescence in patches of a vivid red, or sometimes nearly of a crimson colour.

---

* For further particulars of these cases, as well as for a very able inquiry into the subject of fish poison generally, I must refer the reader to Dr. Burrows's pamphlet.

They are preceded by fits of coldness and shivering, and are attended with a most troublesome itching or tingling, which is greatly aggravated during the night, and which prevents rest for many hours. In order to avoid this inconvenience, I have known many persons sleep on a sofa without putting off their clothes, as their distress begins immediately on uncovering the body. The patches often coalesce, so as to produce a continuous redness : they appear on most parts of the surface, but they are diffused particularly on the shoulders, loins, nates, thighs, and about the knees. They extend likewise to the face ; and there is sometimes a red circle round the palm of the hand, accompanied with a sensation of violent heat. They appear and disappear irregularly, first on one part, then on another, and they may be excited on any part of the skin by strong friction or scratching. During the day the efflorescence fades, and the wheals in general subside, but both of them return with a slight feeble paroxysm in the evening. The red patches of efflorescence are often elevated above the level of the adjoining cuticle, and form dense tumours, with a hard distinct border : the interstices are of a dull white colour. When the patches are numerous, the face, or the limb chiefly covered with them, appears tense and considerably enlarged. At the latter end of the disorder, the eyelids are red and tumefied, and there is often a swelling and inflammation on the sides of the feet. On the appearance of the eruption, the pain and sickness at stomach are in general relieved ; but when it disappears, those symptoms return. The whole duration of the febrile nettle-rash is seven or eight days. As the eruption declines, the tongue becomes clear, the pulse returns to its usual state, and all internal disorder ceases : the efflorescence exhibits a light purple or pink colour, and then gradually disappears, being succeeded by slight exfoliations of the cuticle."

In the febrile nettle-rash, though the cutaneous affection is very severe, the constitutional condition should obtain the greatest share of the notice of the medical attendant. A more than usually severe attack of the febrile symptoms, in a constitution impaired by hard labour and intemperance, proved in one instance, recorded by Dr. Willan, very suddenly fatal, the abrupt disappearance of the eruption being followed by increase of the fever and delirium.

The sudden disappearance of the eruption of urticaria, where even slight symptoms of general irritation exist, may generally be considered an exceedingly unfavourable occurrence, particularly if the former has been very extensive, as there is no reason to doubt that in such cases it has a most important office to perform in the economy. Of many other cutaneous eruptions it may be safely considered, that the degree of danger of a sudden check to their progress depends somewhat on the existence or absence of disposition to mischief in any internal organ : such organ being that which evinces its latent disposition in this respect under such circumstances ; but with regard to extensive eruptions of urticaria, from what I am able to collect, a considerable aggravation of the general febrile symptoms, with delirium, is the most common bad consequence, where their sudden suppression has been observed. It has been usual for medical writers on the subject of suppressed eruptions, to direct measures for the restoration of the latter to the surface ; and sudorifics and the warm bath comprehend what is usually thought advisable with this object in view. If, however, we are to consider eruptions of the kind under consideration as efforts of nature to avert evils of greater magnitude ; in other words, as means by which a dangerous determination of blood to internal organs is to be prevented, such measures are obviously inadequate, and too tedious in their operation for the purpose, because when the determination to the surface ceases, that to the organs alluded to may very generally be supposed to have begun : it should seem, therefore, that vesicatories of such kind as are quickest in their operation are entitled to a preference over other measures, as being capable of producing a state of determination to the skin approaching in similarity to that constituting the original disease of the part. As will be hereafter noticed in treating of Herpes, blisters may be rendered very useful in anticipating, as it were, the intentions of nature : and there does not appear any sound reason, why they cannot be employed with propriety where she has so obviously made an effort to accomplish an object, as is exemplified in the institution of Urticaria. Where, therefore, this affection has been suddenly suppressed, one of the first objects ought to be to provide against harm by the application of blisters, and this principle may be advantageously acted on in all other points, which the circumstances of the case are composed of.

The state of the febrile symptoms, where the eruption is unrepelled, will point out such measures as the case requires ; but it should be always borne in mind, that morbidly increased determination to the skin constitutes some of the most troublesome parts of the disease ; and medicines, therefore, which increase this, such as sudorifics, ought not to be employed. The bowels should be copiously evacuated, and kept open by sulphate of magnesia exhibited at proper intervals, perhaps every four hours, alternately with the nitrate of potash, in pretty large doses. By this plan of treatment, the cutaneous irritation, as well as the febrile symptoms, will be kept under, and any cause of the affection connected with fulness of system gradually got rid of : the principles on which it is founded apply to almost every case of nettle-rash, whether accompanied by fever or not ; but it is of essential importance that a corresponding diet and regimen should be observed at the same time.

Distension of the stomach, even if effected by the mildest and least irritating substances, or liquids, has appeared, in many cases, to be a cause of nettle-rash ; and it is not improbable, that in delicate females, it has given occasion to the most obstinate and protracted cases of this affection, where the medical attendant has been defeated in every attempt to find a remedy. I have lately had an opportunity of seeing a case of this kind in a young female answering to this description, who, as I had been informed, had consulted and followed the prescriptions of various medical men of reputation, for the last three years, without material change in the troublesome character of her disease. Every variation in diet had been had recourse to, and various plans of medical treatment had been tried : she had been at one time much reduced by aperients and low living, at another directed to live on animal food and take wine. A vegetable diet had been found most consistent with comfort, but had not been attended by a complete eradication of the disease. Inquiry into her habits, as regarded her appetite, and the frequency of her meals, led to the idea above mentioned, and she was directed to eat oftener, and limit the quantity at each meal to four or five ounces ; to use the tepid bath twice a week, and have recourse to a saline aperient occasionally. Under this plan of management she has been secured from any further annoyance, though now and then admonished by a spot or two, after any infringement of its rules. One of the cases of an obstinate character detailed by Dr. Willan, seems of a similar kind to the foregoing, though it is possible that preternatural irritability of skin may have stood in the situation of a predisposing cause.

Cases now and then occur, where the white prominence remains, accompanied by itching, after the inflammatory redness of the cutis disappears ; and these have given rise to the designation by Dr. Willan of Urticaria perstans. Another species (U. conferta) seems to have no peculiarity of importance,—"the eruption is full and extensively diffused ; the wheals in many places coalesce, or are indented by close contact ; they have very irregular forms ; when they are singly considered, however, their size and elevation are, perhaps, less than in other species of Urticaria." Moderate doses of the aq. kali puri in the former, and the application of ung. calcis hydr. alb. in the latter, are stated to have been productive of benefit.

The U. evanida, as it has been termed by the last mentioned author, derives its distinguishing epithet from its temporary appearance and sudden removal ; it is the most trifling form which nettle-rash assumes. The observations as to the treatment generally, contained in the few preceding pages, comprehend every thing necessary on these variations.

The U. subcutanea and tuberosa appear to demand particular consideration on some points in their characters ; but affections answering the descriptions given under these names, by Willan and Bateman, are extremely rare. "The Urticaria subcutanea is a sort of lurking nettle-rash that is marked by violent and almost constant tingling in the skin, which, from sudden changes of temperature, mental emotions, &c. is often increased to severe stinging pains, as if needles or sharp instruments were penetrating the surface. These sensations are at first limited to one spot in the leg or arm, but afterwards extend to other parts.' It is only at distant intervals that an actual eruption of wheals takes place, which continues two or three days without producing any change in the other distressing symptoms.   In persons so affected, the stomach is frequently attacked with pain, and the muscles of the leg are subject to cramps. It is relieved by repeated bathing in warm sea-water, and gentle friction.

" The Urticaria tuberosa is marked by a rapid increase of some of the wheals to a large size, forming hard tuberosities, which seem to extend deeply, and occasion inability of motion and deep-seated pains.   They appear chiefly on the limbs and loins, and are very hot and painful for some hours ; they usually occur at night, and wholly subside before morning, leaving the patient weak, languid, and sore, as if he had been bruised or much fatigued.   It seems to be excited by excesses in over-heating by exercise, and the too free use of spirits, and is often tedious and obstinate.

" A regular light diet and a course of warm bathing are to be recommended, with occasional gentle laxatives, where the organs of digestion appear to be deranged.''

I have observed a few cases of a disease answering to this description in most points, but never found the measures above mentioned of essential service.   It is truly an obstinate form of disease, and one which, under my own observation, has often resisted every plan of treatment which sound and rational principles would justify us in adopting.   The misery attendant on such cases would appear to justify experiments.   I have seen obstinate cases of this kind cured by salivation.

I am not aware that any thing has been added to the above observations on these singular forms of Urticaria, since the publication of Dr. Bateman.   With respect to the first, it seems clear that the deep-seated pains described are the consequences of impeded determination to the surface ; and with this view of the case, it appeared to me probable that they would be much diminished by the occasional use of small blisters in different parts of the body.

I have adopted this plan in several instances with success, and am induced to consider it as of considerable value as a local remedy : there is no question, however, that the measures chiefly to be relied on are those which act through the medium of the constitution.

As a general rule in chronic cases, it is well for the patient to direct his assiduous attention to investigate the cause by a series of experiments on his diet ; noticing that kind of food which is not followed by signs of indigestion or discomfort, and however unpalatable to him, persisting in it ; avoiding, on the contrary, all which may produce the slightest sign of disturbance in the digestive organs. Our patients will sometimes feel the impossibility of following this advice, and where they do there is little hope of a cure.   The unnatural system of living in the present state of society, probably, in a majority of cases, constitutes the predisposing, exciting, and all other kinds of causes !

The disease, as far as it has come under my observation, is seldom seen in persons of spare habits and abstemious practices as to diet and drink, and whether in the higher or lower classes, this observation applies with equal effect. Repletion seems to be almost constantly its companion, and if this cannot be provided against by the joint attention of patients and physician, the disease is seldom found to disappear.

I cannot, perhaps, leave this subject with propriety, without naming those articles which have been so obnoxious to the stomachs of individuals as to be capable of producing attacks of this disease, where it has occurred under my own observation, and adopting *comparative* terms, conveying to the reader the best information I am able.

1. Opium. . . . . }
2. Calomel . . . . } not frequently.
3. Raw vegetables. Mushrooms and }
   cucumbers seem to be the most }
   frequent causes of the disease as } often.
   regards the vegetable world . }
4. Fish of different kinds. No description of solid animal food has ever been known to produce it.

That stale or putrid fish *of any kind may* produce it is obvious enough, but it is doubtful whether the state of the fish is the sole cause. The Billingsgate porters, who should be practical judges of the matter, declare that it is only by chance that they are attacked after a meal of fish, and that the attack has often occurred when they have partaken of one which was alive and placed almost before their eyes in the frying-pan? They generally run to the nearest shop to obtain vinegar, which they drink largely the moment the disease attacks them, and are cured.

Shell-fish are, in nine cases out of ten, found to be the articles which have been swallowed by the patient antecedent to the attack.

" Febrile urticaria," says Rayer, " is often induced by the ingestion of various articles,—such as shrimps, lobsters, crabs, the roe of certain fishes, and, above all, muscles. Salted fish, that is dried or smoked, and various other substances,—such as the white of egg, mushrooms, honey, oatmeal gruel, bitter almonds, and the kernels of stone-fruit generally, raspberries, strawberries, raw cucumbers ; several medicines—such as valerian, balsam of copaiba, &c. may also occasion this eruption in persons predisposed to it. It would, moreover, appear to be demonstrated that it is neither to any diseased condition of the muscles that cause urticaria, nor to any change they may have undergone, as mentioned by Burrows, nor to any poisonous substances upon which they have been believed to feed at times, nor to the presence of the *cancer pinnotheres*, a small species of crab they often contain, nor to the black sediment or scum —the *crasse marine*, of which Lamouroux speaks*—nor yet to

* Orfila. Toxicologie gener. tom. ii. p. 45.

the sea-stars on which the creature appears to live, according to the interesting researches of Beunie, from the month of May to that of August, that their property of exciting this disease is to be attributed, but rather to *peculiar susceptibility,—to an individual predisposition on the part of those who are affected."*

"An hour or two, and sometimes much sooner, after the ingestion of one or other of the articles mentioned, a weight is felt at the epigastrium ; nausea, general sinking, and giddiness, are then complained of ; the skin next becomes hot, and the eruption appears on the shoulders, the loins, the inner sides of the forearms, the thighs, and about the knees, generally characterized by red or whitish elevated spots, surrounded by an areola of bright crimson.   The spots are commonly of an irregular shape, but sometimes circular, raised above the general level of the skin, and of very various sizes.   When they happen to be extremely numerous or actually confluent in any quarter, the skin presents a general red tint, and the face and limbs are stiff and swelled in this case—(Urt. *conferta*, Willan).   The eruption is attended with itching, and a sense of prickling of the most intolerable kind, especially during the night, or when the parts affected are exposed to the air.   In some cases this variety of nettle-rash is complicated with erythematous blotches.   Besides the symptoms already mentioned, it is often preceded by vomiting and by purging ; and spasms, choking sensations, and convulsions, have occasionally been seen added to the list ; nay, there are several cases on record in which this kind of poisoning even ended in death.*— At the end of twenty-four or thirty-six hours, in the generality of cases, the eruption declines in intensity, and soon only leaves but very faint traces of its presence on the skin, which are completely effaced a few days afterwards."

CHAPTER IV.

HERPES.

I now approach a part of my subject which will help me still further to illustrate the soundness of my position as to the principle of classification I have adopted.

M. Rayer, following the example of others, has adopted the order vesiculæ or vesicular inflammations.   He considers under that head Herpes Eczema, Hydrargyria, Scabies Miliaris Sudatoria and Sudamina, Vesicular Syphilis, and Artificial Vesicles ! !

Now the soundness of a doctrine may be generally easily tested

---

* Foderé, Méd. légale, t. iv. p. 85—Vancouver's Voyage of Discovery, vol. ii. p. 286.

if we can bring facts to bear upon it, and the value of the evidence is equally easily estimated. In like manner the value of a classification may be tried. It is easy to find fault with one, but not easy to substitute another : but it is easy to be candid, and that M. R. has not been, as will be seen from the following quotation. Speaking of "vesiculæ," he says, " *The vesicular character of scabies has been disputed by Bateman, who ranks it among the pustular affections!* The mistake he committed has been exposed by M. Biett. On the other hand, Bateman has classed vaccinia, aphthæ, rupia, and varicella, among the vesicles. But the vaccine pock is indisputably a *pustule ;* aphthæ cannot be counted among the diseases of the skin ; and rupia is a bullous affection. With regard to varicella, I grant that of the three or four varieties which the disease presents, severally designated under the names *chicken pox, swine-pox,* and *modified small pox,* one at least, the *chicken pox,* is perfectly vesicular in its form ; but it is also certain that the other varieties mentioned, and particularly the *modified small pox,* are invariably pustular diseases. Varicella, therefore, by this double character, may be held as forming the link of transition from vesicular to pustular eruptions. Feeling myself free to attach it to one or other of these groups, I have preferred classing it among the *pustules* with a view of approximating it to variola, of which it is a mere modification !"

Who can doubt that the whole of the diseases here mentioned do regularly go through the stages of pimple, vesicle, pustule, scab, and exfoliation ? But M. Rayer must have wofully mistaken Bateman, when he asserts that the latter disputed the vesicular character of scabies ; and, forsooth, the mistake has been exposed by M. Biett ! What are the words of the *definition of the disease merely* as given by Bateman ? " a contagious eruption of minute pimples, papular, vesicular, pustular, or intermixed according to circumstances, and terminating IN SCABS !" In truth, as has been before urged, the sooner this system of nosological arrangement according to the appearance of scabs and scales is abolished the sooner will the diseases be understood.

The import of the term Herpes is now generally understood to be an eruption of clusters of vesicles of various sizes, from the minutest distinguishable to the naked eye, to that of two or three barley-corns, situated upon a red and inflamed areola of skin. The term, as applied by the older writers, comprehended an extensive variety of other cutaneous affections very materially different ; and Tilesius, whose observations have been inserted by Dr. Bateman, in the 11th vol. of the Med. and Phys. Journal, speaks of the eruption as partly consisting of papulæ. The best authorities in England, where cutaneous diseases have been most successfully studied, however, limit it to a " vesicular disease, passing through a regular course of increase, maturation, and decline, and terminating in from ten to twelve or fourteen days. The vesicles arise in distinct but irregular clusters, which commonly appear in quick succession, and they are set near together on an inflamed base, which extends a little way beyond the margin of each cluster. The eruption is preceded, when it is extensive, by considerable constitutional disorder, and accompanied by a sensation of heat and tingling, and sometimes by severe deep-seated pain in the parts affected. The lymph of the vesicles, which is at first clear and colourless, becomes gradually milky and opake, and ultimately concreted into scabs : but in some cases a copious discharge of it takes place, and tedious ulcerations ensue. The disorder is not contagious in any of its forms."

This vesicular disease occurs in different parts of the body, and is generally preceded by marks of some constitutional derangement. Slight febrile symptoms most commonly prevail for a few days before the appearance of the eruption, but they are often not of so much importance as to attract particular attention. The patients themselves, in some cases, do not feel sufficiently unwell to complain, nor is there much apparent disorder of any natural function. It has been often seen preceded, in delicate constitutions, by great thirst and a quickened pulse, which have prevailed for a day or two, notwithstanding the tongue may be clean, and the bowels regular ; but in some instances pains referred particularly to the epigastrium, and other symptoms of disorder of the digestive organs, are complained of for some time before it makes its appearance. Languor and loss of appetite, rigors, head-ache, and sickness, if accompanied by heat and hurried circulation, are symptoms particularly connected with its severest forms. When the eruption appears, these symptoms are rarely in any respect materially mitigated, but continue generally so long as the inflammation extends, and fresh vesicles continue to be produced. The deep-seated pains specified above usually take the course of the eruption, continuing to be referred to the regions of the liver and stomach, when the latter extends in this direction round the waist, as in the common shingles, or to the scapulæ, spine, and os humeri, if situated on the shoulders, and extending down the arms.

The attention is usually attracted to the cutaneous affection by a sensation of heat and tingling in the part. On examination, a blush of bright redness is discovered, in the centre of which a few small vesicles, varying in size, appear to have been recently formed ; near to it, a smaller patch of inflamed skin is discovered with a smaller number of vesicles, and, perhaps, a short distance from this another, without a vesicle upon it, or with a minute resemblance of a pimple, which, in an hour or two, becomes a perfectly-formed vesicle,* and is surrounded by others somewhat less advanced. If slight pressure by the finger or any accidental means be applied at this period, no great degree of tenderness is evinced, but a sensation of pricking is experienced, which is evidently the result of distension of the vesicle ; but as the disease advances, the accidental contact of any sub-

---

* It is probable that these incipient vesicles have given rise to the opinion of the existence of papulæ mixed with vesicles as forming this disease.

stance is apt to produce a great deal of smarting pain. Within twenty-four hours, the vesicles first appearing attain the size of small pearls, and contain a clear transparent fluid, when two or three, situated in the thickest part of the cluster, run into each other, forming a larger vesicle of an irregular shape, the areola of inflamed skin being also increased.

The inflamed spots and clusters of vesicles described extend in an irregular line from the spot in which they first appear to distant parts ; thus, if the first spot is discovered on the shoulder or back of the neck, the successive eruptions and redness appear along the dorsum of the scapula and down the back of the arm every successive day for the first four or five, disclosing fresh spots, till the disease reaches below the elbow, by which time the transparent vesicles, described as being first formed, become opake, shrivel up, and terminate in a scab, the centre of which is in a few hours brown, and gradually adds to its extent the surrounding collapsed vesicles, which undergo the same change. The inflamed skin on which they are situated undergoes a corresponding alteration in appearance, and from a florid red becomes of a bluish colour, and loses all appearance of heat and irritation. In fourteen or fifteen days from the first appearance of the disease, the scabs fall off, leaving a tender discoloured state of the skin, which gradually disappears.

From this description, it will be obvious that the duration of the complaint depends entirely on the extent at which the eruption may arrive, and hence the answer to be given to the impatient inquiries of the sufferer should be particularly guarded till a day or two has elapsed without the appearance of a fresh crop of vesicles, for it is only under such circumstances that an opinion can be formed as to the cessation of the excessive smarting pain and tenderness belonging to it. If the eruption is completely effected in two days, and the constitution be strong and generally healthy, the space of time elapsing between the commencement of the vesicle and falling off of the scab may not exceed fourteen days ; but this is an instance of rare occurrence, the ordinary time elapsing between these periods being sometimes protracted to from twenty to twenty-five or twenty-seven days. When the constitution is much disordered, or improper applications have been used, it sometimes happens that little white sloughs occupy the sites of the vesicles, and that permanent indentations, or marks, are left in the skin after the healing process is complete.

The foregoing remarks are applied more particularly to that form of the disease familiarly termed Shingles, or Herpes zoster, the only peculiarity of importance belonging to which is its disposition to extend in a line from the spot in which it makes its first appearance, and from which its common designation is supposed to have been derived. The parts of the body on which the eruption takes this form are the back of the neck and shoulders, and the waist ; but when the eruption appears indiscriminately in other situations, it has obtained a distinguishing appellation, "Herpes *phlyctænodes.*"

In Herpes *phlyctænodes*, "the eruption has no certain seat; sometimes it commences on the cheeks or forehead, and sometimes on one of the extremities, and occasionally it begins on the neck and breast, and gradually extends over the trunk to the lower extremities; new clusters successively appearing for nearly the space of a week." When occurring so extensively, the vesicles do not attain the size common to more limited forms of the eruption; they dry up, and the scabs fall off much quicker, though from fresh eruptions continuing to appear for a longer period, the duration of the disease is ultimately nearly the same. This variation in the course and character of the eruption is probably entirely accidental, and may be considered as affording another instance of the impolicy of increasing the number of distinctive appellations for mere shades of difference in cutaneous disease, where the correct pathology and best methods of treatment are precisely the same.

Willan and Bateman have omitted to notice the occurrence of this disease in infants during the period of dentition: I have seen a great number of such cases. As far as regards the pathological character of the eruption, the description of H. phlyctænodes applies to it tolerably correctly; the tingling, itching, and heat appear, however, to be absent, for the infant does not appear to feel any inconvenience from it. The extent of the patches does not usually exceed that of three or four vesicles, situated close to each other; but the former are distributed more or less thickly over the trunk and extremities. As some of the vesicles collapse, form scabs, and fall off, others succeed them, and the disease generally does not disappear while the gums are inflamed and the teeth advancing.

Alibert, in allusion to herpetic vesicles, speaks of a belief which some entertain of its being infectious, and takes some credit for the assurance that such is not the fact; but there does not appear any sound reason for such a suspicion in the history of either of the forms of the affection before mentioned.

The disease under consideration, whether answering the description of H. phlyctænodes or H. zoster, is materially, as well at its commencement, as during every stage of its progress, unlike any other cutaneous affection; nor does it either in its mildest or severest forms approximate to, or terminate in, any other which has been separated from it by the classification of Willan: its being confounded, therefore, with Eczema, Impetigo, or Erysipelas, as remarked by Dr. Bateman, could only have been the result of great inattention to its generic characters. The red inflamed areola on which the vesicles appear, and the uniformly transparent character of the fluid which they contain, together with their regular progress towards exsiccation and ultimate falling off of the branny scales which they form, constitute sufficient grounds of distinction from all others.

Among the causes of Herpes, Tilesius mentions the suppression of hemorrhoidal or menstrual discharge; sudden change of the habits of the patient, more particularly from an active to a sedentary life, &c. He has also observed it most frequently among dirty people, and natives of warm climates, and in those whose diet consists largely of oil and fish. It occurs, also, he states, very frequently in marshy neighbourhoods, in autumn. " Young people, from the age of twelve to twenty-five, are most frequently the subjects of the disease, although the aged are not altogether exempt from its attacks, and suffer severely from the pains which accompany it. Sometimes it has appeared critical, when supervening to bowel complaints or to the chronic pains of the chest remaining after acute pulmonary affections. Like erysipelas, it has been ascribed by some authors to acute paroxysms of anger."*

From what has been observed on the subject of the symptoms which usually precede the eruption, little doubt can be entertained as to the dependence of Herpes on constitutional causes, though the nature or character of such causes are involved in the greatest obscurity. It is at one time supervening on general irritation of system, marked by quick pulse, thirst, and heat ; and unaccompanied by determination to any particular organ ; while at another it is preceded by unequivocal indications of disorder of the stomach, liver, &c. ; and it is proper to observe, that such symptoms, whatever they may have been, are not usually observed to continue after the eruption has gone through its course. Irregularities in the habits of the individual, we are justified in pronouncing to have little effect in inducing the disease, as it is not often seen in those whose conduct is even marked by extravagance in this respect ; but a sudden change in the diet, where great uniformity, as regards exercise and exertion is observed, is in a great number of cases ascertained to have occurred. Thus persons who have been long resident in warm climates, and who during a lengthened sea voyage have enjoyed uninterrupted health, are subject to severe attacks soon after their arrival.

In the absence of any data regarding the manner in which disorder of vital organs may produce cutaneous eruptions, we are fully warranted in supposing from the symptoms usually preceding Herpes, and commonly disappearing after this eruption has gone through its course, that mischiefs of some importance to such organs are averted by it; in short, that it exerts all the influence, and is entitled to the estimation of, a natural counter-irritant. In this character it has in two or three instances, in delicate females, within my own observation, checked the progress of symptoms which had given rise to great anxiety respecting the state of the thoracic viscera.

The treatment of Herpes comprehends but little. It would be as useless to attempt (and perhaps nearly as dangerous to succeed in such attempt) to repress it, as the eruptions of small pox, &c. The tingling, smarting, aching and burning heat belonging to it at its height, may however be considerably reduced, and the patient's

* Bateman.

19*

feelings rendered more comfortable by the free use of sedative applications. I believe it is a common practice among the generality of medical men to direct such applications ; and as far as inquiry goes, I have not been able to ascertain that any mischief has arisen from their use. Ancient opinions on such points are now submitted to the ordeal of unrestrained scientific reasoning and attentive practical observation ; they are canvassed with that confidence which solid acquirements and observation have given to the majority of those who, at the present day, practise the medical profession ; they are not permitted to restrain us from extending relief to our patients, merely from their antiquity ; but on the contrary, are often only alluded to for the purpose of pointing out their absurdity.

Like the prickly heat, the eruption of Herpes can never be checked by any medicine or medicinal application ; and those, therefore, which relieve the sufferings of the patient, ought not to be neglected. Solutions of ceruss. acet., or the liq. plumb. acet. dil. with the addition of alcohol, may be applied with advantage by means of wetted linen ; they lessen the pains in question, *but never check the eruption in its course.* The vesicles ought not to be cut or rudely broken, such a proceeding generally rendering the separation of the scab considerably more tedious than it would otherwise be ; but if care be taken to puncture each individual vesicle early, so as to allow of the free escape of the fluid, the pain is much diminished, and the irritation sooner subsides.

The constitutional treatment consists only of mild saline aperients and low living. Sulphur, I believe, really possesses the properties attributed to it of equalizing the determination to the skin, it may therefore be preferred to any other kind of medicine.

The decoction of dulcamara taken internally has been stated to be very efficacious in the cure of Herpes. The oil of the walnut kernel, the juice of the rhus radicans of Linnæus, which is highly acrid and corrosive, as also the juice of the husk of the cashew nut, of similar properties, are recommended by different writers as applications ; they can never, however, be applied with propriety, except in those trifling spots unattended with much vesicular development, constituting some of the cases termed Herpes circinatus ; where substances possessing slightly caustic or powerfully astringent properties speedily remove the affection. It is on this principle that ink, solutions of sulphate of copper, &c. are found of considerable efficacy in removing the herpetic ringworm.

French writers entertain great dread of metastasis from Herpes : *according to Alibert*, it has been known to extend to the mucous membranes of the nose, throat, and larynx. He mentions several instances of formidable determination to particular organs, some of which evidently were connected with cutaneous affections of a very different kind from those under discussion. According to him, these mischiefs are so common, and the symptoms supervening so distinctly marked, as to have enabled him to know at once from their characters, that repulsion had taken place, though no external marks exist ! It may be as well to observe with respect to this latter assertion, that the entire disappearance of the vesicles of genuine Herpes in a few hours, or in any way besides that of exsiccation and scabbing, is an event exceedingly improbable.

As regards the pathology of Herpes, it may be said to consist of effusion of serum from the minute extremities of vessels on the cutaneous surface. It does not seem to be in any case dependent on impediments existing locally to the proper functions of the skin. The cuticle in a strong and healthy state is elevated by the fluid, and forms pretty strong parietes for each individual vesicle.

Looking at the disease as a counter-irritant, and satisfied, as we must be, of the important office it has to perform in the animal economy, it should seem that any interference with its progress would be injudicious ; and if, as observed by Dr. Bateman, it was of so trifling a character as to allow, in all cases, the patients to proceed about their occupations, or to feel little or no personal inconvenience from it ; perhaps no attempt to relieve could with propriety be made ; but in a very large portion of the cases denominated shingles, the pain, burning heat, and tenderness, and great extent to which it often spreads, are, to say the least of them, inconveniences of a serious character, and such as it would be desirable to diminish, as far as may be consistent with safety, and due attention to the obvious indications of nature.

It is fair to suppose that vesication is calculated to effect the purposes of nature in that state of system in which shingles occur better than any other artificial means. When, therefore, the disease is observed at its first commencement, and where only a few vesicles are formed, the course which it appears disposed to take being sufficiently manifest, a sound principle of reasoning would seem to justify the anticipation of the object of nature by the application of a strip of blistering plaster upon this part ; in this way substituting the trifling and temporary inconvenience of a small blister for a tedious and painful complaint.

Acting on this idea, I have in two or three instances applied small blisters to the uninflamed skin on the side of the eruption on which the latter seems disposed to extend, not only with the effect of checking such extension, but of producing a shrivelling of the vesicles already formed, and cutting short its progress altogether ; avoiding at once its tediousness, and all the pain attending it. The smarting and tenderness in the vesicles near the blister are soon diminished, nor has it in the cases alluded to appeared that the blistered surface healed less readily than under other circumstances. Care should, however, be taken not to apply the blister to the vesicles themselves, as it not only fails to raise the cuticle, but irritates the cutis, and is followed by a superficially sloughy surface of the latter on the site of each vesicle, forming so many little irritable indentations, which heal very tediously.

The grand distinguishing feature of Herpes, namely, the limited
size of the vesicles, seems to depend on the cuticle being bound
down to the cutis by the adhesive inflammation, during that state
of the vessels of the skin, marked by great heat and redness, which
precedes actual effusion of serum. Thus, when effusion takes place,
it does not elevate the cuticle generally and extensively, as in erysipelas or pompholyx, but only on the precise spot occupied by the
mouths of the vessels, from which the fluid escapes. Hence the
pearl-like elevated form of the vesicle, and the obviously thinned
and distended state of the cuticle forming its parietes.

The sense of pricking and pain on touching the apices of the
vesicles is the consequence of the extreme distension existing : the
tender surface of the cutis is affected, and the adhesion round the
margin of the vesicle disturbed by the slightest pressure.

That the above explanation of the peculiar form of the herpetic
vesicle is correct, is, I think, proved by the effect of blisters already
alluded to. These applications do not fail to elevate the cuticle, and
produce effusion on the surrounding uninflamed skin ; but they
never have this effect between the vesicles, or close to the margin
of the cluster.

When the eruption is very slight and limited in extent, it is apt
to assume a circular form, and hence has obtained the popular appellation of ringworm. Sometimes in such cases the vesicles are extremely minute, and then they generally dry up, and the cuticle
falls off in a few days in the shape of small exfoliations, leaving a
reddened scurfy areola. Spots of this kind occur in different parts
of the body, which are seldom attended to, and speedily disappear
by the use of any of the applications mentioned in the preceding
page.

Dr. Bateman* has confounded this form of Herpes with a different
affection of the skin, which assumes this circular figure, and which
is distinctly communicable by contact. I have described the latter
in a former page, and have established its identity with common
ringworm of the scalp : the pustular form which it assumes among
the hair I have endeavoured to explain in the chapter on Porrigo
scutulata.

The whole of the varieties of Herpes are found to be more severe
in warm than in cold climates : the remark applies with equal correctness to most cutaneous affections, and it is evident that the more
free determination to the skin in hot weather affords an adequate
explanation of this fact.

Herpes labialis and preputii, so designated by Willan and Bateman, from their situation, form two other species according to their
arrangement. The first of these, a well known symptom attendant
on catarrhal fevers, and appearing often towards their termination,
requires but little attention. Though now and then troublesome
for a day or two, and extending round the margin of the lips, it goes

* Synopsis, page 234.

through its course in a week or ten days, the vesicles becoming first turbid and yellow, then drying up on the part, and ultimately falling off in the form of scabs. Sometimes it supervenes on mere disorder of the digestive organs unattended by fever, and frequently accompanies bilious fevers, dysentery, and other acute diseases. The posterior part of the fauces is sometimes affected when it appears suddenly after checked perspiration and cold, these parts having a few vesicles distributed upon them, surrounded by an erysipelatous inflammation.

The affection of the prepuce is described by Bateman to be so closely resembling chancre as to be "liable to a practical mistake of serious consequence to the patient." The mistake to which he alludes, however, is by no means an occurrence to be apprehended, where much professional knowledge exists on the part of the surgeon or physician. At the present day no man who knows what he ought to know of the science could possibly commit such a blunder.

" The attention of the patient is attracted to the part by an extreme itching, with some sense of heat ; and on examining the prepuce, he finds one and sometimes two red patches, about the size of a silver penny, upon which are clustered five or six minute transparent vesicles, which, from their extreme tenuity, appear of the same red hue as the base on which they stand. In the course of twenty-four or thirty hours the vesicles enlarge, and become of a milky hue, having lost their transparency ; and on the third day they are coherent, and assume an almost pustular appearance. If the eruption is seated within that part of the prepuce which is extended over the glans, so that the vesicles are kept constantly covered and moist, like those that occur in the throat, they commonly break about the fourth or fifth day, and form a small ulceration on each patch. This discharges a little turbid serum, and has a white base, with a slight elevation at the edges, and by an inaccurate or inexperienced observer, it may be readily mistaken for chancre ; more especially if any escharotic has been applied to it, which produces much irritation, as well as a deep seated hardness beneath the sore, such as is felt in true chancre. If no irritant be applied, the slight ulceration continues till the ninth or tenth day nearly unchanged, and then begins to heal, which process is completed by the twelfth, and the scabs fall off on the thirteenth or fourteenth day.

" When the patches occur, however, on the exterior portion of the prepuce, or where that part does not cover the glass, the duration of the eruption is shortened, and ulceration does not actually take place. The contents of the vesicles begin to dry about the fifth day, and soon form a small hard acuminated scab, under which, if it be not rubbed off, the part is entirely healed by the ninth or tenth day, after which the little indented scab is loosened, and falls out."

Mr. Evans, to whom the profession is much indebted for his "Remarks on Ulcerations of the Genital Organs," observes, that this disease, except when occurring on the inner surface of the prepuce, rarely comes under the notice of the surgeon. When on the outer surface, in consequence of improper interference on the part of the patient, or from the effect of friction of the clothes, it is often first seen in the form of an ulcer, with a yellow or white and plain surface ; the scab described in the foregoing extract having been partially or entirely removed.

I believe this is the state in which the affection usually comes under the notice of the surgeon, whether situated on the external or internal surface of the prepuce ; and if the vesicles have been recently broken, the white specks on the cutis, marking the situation of their bases, are not found to be surrounded by thickening. Nor are they to be considered, correctly speaking, as ulcerations, the only essential point in which they differ from simple abrasion consisting of the whitened appearance of their surface. This whitened appearance of the surface of the sore may be readily pronounced to indicate a state of vessels of the part approaching in similarity to those on the surface of sloughy ill conditioned sores situated elsewhere ; and while the inflammatory areola continues to surround it, it often rapidly extends and obtains a similarly thickened and elevated edge. In this state of things, it is not surprising that the influence of a prevailing fashion to consider every kind of affection of the genital organs venereal should have led to the mistake alluded to by Dr. Bateman. Since the publication of the observations of Mr. Abernethy, Mr. Evans, and others, however, it appears that not only this disease, but a great variety of others occurring on these parts, which have been also hitherto treated as venereal, are more readily cured by common sedative applications and alterative medicines than by mercurials.

The disease termed by Mr. Evans venerola vulgaris, is said by that gentleman to be most frequently confounded with herpes on the prepuce ; but he has given us, under the head diagnosis,* ample means of distinguishing them. The herpetic disease is marked at its commencement by distinct vesicles ; and if not seen at this period, or till the vesicles have been broken, and the scab rubbed off, it consists of the superficial white specks described. If it has been any length of time in existence, and subject to irritating and improper applications, or as is often the case, passed over with neglect, while its original cause remains in full operation, it assumes the character of an irritable superficial sore.

The venerola vulgaris is in every point of view a much more important disease ; it commences with the formation of a pustule, the contents of which undergo exsiccation on the spot, and form a scab of much greater solidity and dimensions than that which follows the vesicles of herpes. This scab, instead of speedily separating and leaving a superficial sore, adheres to the surface, and if its base be raised up and minutely examined, it seems to be attached by means of a stringy slough, approaching in similarity to that described as attaching the scab of rupia in a preceding page. A pretty copious secretion of matter is also found under it, which concretes on the

* Remarks on Ulcerations of the Genital Organs, p. 28.

scab already formed, and gradually enlarges it ; and when the latter separates, a concave ulcer with a raised edge is disclosed, which subsequently heals by distinctly new granulations.

The cause of herpes on the prepuce is generally understood to be disorder of the digestive organs, which may be either habitual, or brought about by temporary causes. Alteratives and aperients exhibited for a day or two, with the local applications of the liq. plumb. acet. dil. are generally found to comprehend every thing necessary to the cure. If, however, as now and then happens, the process of cicatrization is not completed by the time this plan has subdued the usual marks of irritation in and about the spot, the black lotion* may be substituted for the former with advantage.

The lymph of the vesicle taken from the prepuce, and inserted under the cuticle of the arm, in the part usually chosen for inoculation, has been in one instance seen by Mr. Evans followed by a vesicle of much larger dimensions than the original ; " but in several later experiments the lymph has altogether failed in producing any effect ;" it does not appear, therefore, that the causes of the affection here are different from those producing it more generally on other parts of the body.

Mr. Evans's observations as to the causes of H. preputialis do not confirm the opinions of Mr. Pearson or Mr. Copeland ; the former of whom supposed it to be connected with the previous use of mercury, and the latter† as sometimes the consequence of an irritable state of the urethra, or actual stricture of this part.

Herpetic vesicles sometimes occur on the edges of the eyelids, accompanied by a considerable degree of smarting and itching, followed in a day or two by inflammation of the conjunctiva ; the vesicles being usually very small, and distributed among the hairs of the eyelids. Cases of this kind are often considered and treated as simple inflammation : on minute examination, however, the character of the complaint becomes sufficiently manifest.

The course and duration of the disease in this situation are similar to that of the smaller kinds of vesicles, situated on delicate membranes elsewhere, and it does not generally require more attention. As a matter of precaution, it is advisable to employ mild aperients and sedative lotions : I have seldom found or heard of anything further being requisite.

CHAPTER V.

OF THE FURUNCULUS, OR BOIL.

Dr. Bateman has omitted the consideration of this disease of the cutis, on the ground of its occupation of a place in the works

* R. Hydr. subm. ℥ij. Aq. calcis ℥vj. ft. lotio.
† Bateman's Synopsis, page 240.

of surgical writers. Notwithstanding, however, the attention it has
obtained from Richter, Richerand, Dr. Pearson, and others, there
are some points to which the reader's attention may be directed,
in a work like the present, with some prospect of advantage.

There are two species of boils. One involves the cutis and cellular tissue beneath to a great extent, forming a tumour, sometimes
attaining the size of a pigeon's egg ; the other confined to, and
only involving in the suppurative process, the substance of the cutis.

The former of these makes its first appearance in the form of a
painful, red, circumscribed tumour, excessively tender to the touch,
and generally approaching to a conical form as regards that portion
of it which is raised above the surrounding uninflamed skin.   On
the apex of the cone and centre of the tumour a little white speck
or slough is generally observed, which is sometimes picked off by
the patient, with the hope of extricating the matter underneath,
but which, under such circumstances, only discloses an excavation
of a corresponding size.   There is reason to suppose that matter is
formed in boils of this kind in two or three days after the commencement of the disease ; but from its being deeply imbedded in the
cutis, it is prevented from making its way to the surface by the
thickening of the superincumbent structure produced by the adhesive inflammation.   The matter being confined in this manner, the
ulcerative absorption extends downwards rather than upwards, and
involves the parts beneath the cutis, forming a much more extensive
collection of matter than naturally belongs to the disease, and con-
cealing the circumstance of its cutaneous origin.

The second form is less painful, but equally tender to the touch,
very small in extent, and does not involve the cellular tissue ; it is
strictly confined to the substance of the cutis, and is vulgarly termed
the blind boil.

Mr. Fosbrooke* has given a very valuable, and, except in one
point, very correct account of this form of boil, with useful observations on the best method of treatment.   He says "the features of
this little disease are a dark red lenticular swelling ; the inflammation
is intense and concentrated ; and if these little tubercles undergo
pressure in the early stage of suppurative inflammation a transparent serum is effused, probably coagulable lymph.   It is slow in
maturation, and pus first appears in a minute yellow elevation in the
centre, the surrounding space circumscribed by the inflammatory
process still continuing in a slow state of that action, particularly
hard and solid.   The quantity of matter discharged is not very
great, and, in most cases, where the inflammation is not very obvious
on the surface, it appears to act so vigorously beneath, that the
cellular membrane partakes of the disease, and comes away in a
small eschar."   After the excavation formed by the slough is filled
up, and the inflammation has subsided, it requires some time for the
absorbents to remove the thickening of the part which has been

* Edin. Med. and Surg. Journ., No. 66.

produced ; some languor of circulation usually remains also, marked
by a blue appearance of the thickened part.

Mr. Fosbrooke thinks this disease allied to true carbuncle : perhaps the same remark may be made with great propriety of any
form of boil, the character of the disease being really the same, and
the difference consisting only in extent. He does not advise "an
early exoneration of the central fluid by the lancet; the knot of vessels which are throwing out an effusion in the early stage are merely
irritated into increased action thereby ; and parts in such a state,
according to the common laws of surgery, are not exactly likely to
to assume, if wounded, a healthy action."

Large doses of the diluted sulphuric acid seem to have been very
efficacious in the hands of this gentleman. He commences with
twenty minims, gradually increasing it to one, or even two drams
twice a day, very largely diluted, and he describes the effect to be a
deadening of the pain, and a gradual absorption of the swelling,
without suppuration.

This form of boil does not terminate usually in the formation of
any considerable quantity of healthy pus, and if squeezed at any
period, and the little white slough in the centre be extracted, the
inflammation sometimes subsides. The slough in question does not,
as Mr. F. supposes, involve any part below the substance of the
cutis.

The part of the body in which it appears most numerously is the
trunk, particularly on the abdomen, while the larger boil is most
commonly seen on the arms, thighs, and nates.

From the tedious progress of this kind of the disease under consideration, and from the continued appearance of fresh inflamed
tubercles as fast as the first which have appeared subside, the local
surgical treatment is of but little importance ; the most obvious
indication being to correct the state of system which produces it.
Months will in some cases elapse before the disposition to the affection subsides, though moderate and even low living be adopted
and strictly persevered in. The communication of Mr. F. above
stated appears, therefore, to be of the greatest value. If the practice
he mentions be followed by generally similar results, much pain to
the patient, and trouble to the medical attendant, will be saved.
The free incision even into quarters of the tubercles : poultices,
fomentations, leeches, and applications tending to disperse them,
however diligently applied, have been usually found far less productive of advantage than he has described the sulphuric acid to be.
It would be a source of very considerable satisfaction, to be able to
confirm his statement, for the disease is, in truth, of a very obstinate
and intractable nature. It may possibly happen, that the effect of
this medicine is more marked where a pure and healthy atmosphere
is breathed by the patient while under its influence ; but it has
been employed very liberally in a great variety of cases in this
metropolis without any apparent advantage.

The exhibition of aperients and alteratives to the extent of a few doses, from day to day, have sometimes appeared to suspend, or entirely put a stop to, the disease ; but in a great majority of cases they certainly failed to do so.   I have been compelled by the results of observation and experience to place little reliance on any medicine or plan of treatment, the effects of which are limited to a short period, and hence, probably, am induced to offer a better opinion of the compound decoction of sarsaparilla and Plummer's pill than many will think they merit.   They are, however, certainly efficacious when taken for two or three weeks, and often effect that which no other medicine or medical treatment appear to be equal to. Change of air, and active exertion, have, in the greater number of cases which I have observed, been evidently required with relation to the general health.

Some constitutional affection is generally produced by the pain attending extensive formations of boils, and indeed the occurrence of a single boil of the first species is sometimes productive of much febrile irritation ; such symptoms, however, do not in any case usher in the disease, and are therefore to be considered, whenever they occur, simply as its consequences ; they usually rapidly disappear as soon as a free opening is made, and the matter discharged.

The same causes produce the two species or forms described : indeed it will be seen that there are no pathological differences of importance between them ; the circumstances on which the extension of the mischief downwards, and consequently the increased size of the former, being, perhaps, merely accidental.   Young people of full plethoric habits, and those enjoying good health, and what is termed good living, are most subject to them ; and like other cutaneous diseases, they appear in their worst forms during the spring and summer quarters, when the determination to the skin is much increased.

The same principles of treatment also are generally applicable to both, except that, as regards the smaller species, it is sometimes advisable not to wait for the tedious process of unhealthy suppuration, attempts at dispersion of the tubercle being now and then found effectual.

The degree of pain fully justifies, and sometimes amply repays, the trouble of applying a leech or two to its centre ; but if the disease be extensive, this plan is hardly practicable to a sufficient extent.   Laying open the substance of the tubercle, with a lancet, however freely done, does not appear to be speedily followed by ease, as in the other species, nor does the thickening and hardness much more rapidly subside, than if it had been left to pursue its course.

With respect to the common boil, where the inflammation is more extensive, and produces an enlarged tumour, but little good is to be obtained by any management but that which encourages suppuration. If it be cut into before the latter process has fully taken place, the cure seems to be rather retarded than expedited, and other boils are much more likely to occur in the neighbourhood.   When suppuration has occurred, and the matter has found its way to the surface, and been freely discharged, a few days only are necessary to fill up the excavation, and restore soundness to the part.

If sulphuric acid be a medicine of such efficacy in the small carbuncular furuncle of the cutis, it will probably be found of some service in the constitutional treatment of this ; but heretofore the treatment generally had recourse to has consisted of purgatives and low living. The precise state of system, however, will always be found the best guide in the constitutional management of any local disease.*

SECTION IV.

On diseases of a mixed character essentially dependent on active inflammation, with which the constitution is not necessarily connected.

CHAPTER I.

IMPETIGO.

By the term Impetigo is meant a state of active inflammation of the cutis, on which minute vesicles are speedily formed, the contents of which are at first transparent, but which become shortly after opaque. When the cuticle is broken so as to allow of the escape of the fluid, the latter dries on the part, leaving scabs or scales of a yellowish brown hue, varying in thickness and adhesiveness according to the quantity of the fluid discharged.

In speaking of diseases of a mixed character, it is intended merely to apply this phrase to their local appearances. Impetigo, in different cases and their stages, exhibits vesicles, pustules, and regularly-formed scales, somewhat resembling those of psoriasis. Vesicles and pustules also alike characterise Scabies and Eczema.

With respect to the latter disease, it must be confessed, that its claim to introduction here, is somewhat equivocal, if the more formidable cases of it, from the use of mercury, be borne in mind : the reader will remember, however, that my object is an arrangement with a view to perspicuity in the present treatise, rather than the substitution of a new classification for that of preceding authors.

Under this head, therefore, I propose to consider the varieties of Impetigo, of Scabies, and of Eczema ; the whole of which, in by

---

* Rayer, with all the industry he has manifested, adds nothing to the foregoing facts, as regards the history of the disease. He has given, in plate IX. of his work, illustrations of this disease, and also of the carbuncle. The reader will hardly derive information from them—they are utterly unlike the one or the other of these diseases, as they are seen in England.

far the larger portion of cases coming under our notice, may be referred to local causes, while, as regards their local characters, they very much resemble each other.

The descriptions of Impetigo, which have been heretofore given, do not appear to have been founded on sufficiently minute observation of its origin and progress, and apply only to cases where the diseased secretions are allowed to lodge on the spot, to become dried and hard, and, consequently, a cause of considerable aggravation of the irritation, heat, &c. which properly belong to it.

The influence of frequent ablutions, with warm water, and the removal of the secretions as fast as they are produced, effect a strikingly important change in its characters ; and the progress of any case, under such circumstances and management, would have been more than sufficient for Dr. Willan to have founded a distinct species on, and even warranted the impression at first sight, that a different disease existed.

There is, generally, infinitely less of that fiery redness and heat —successive crops of pustules much less frequently occur, and the duration of the disease is altogether shortened when the soothing application of tepid water, to the extent of clearing away the morbid secretions, is diligently attended to.

There are five species of Impetigo spoken of by Willan and Bateman ; four of which differ from each other, merely in their degree of activity and extent ; circumstances which are probably entirely accidental, and dependent either on the share of irritability of skin of the individual, or the state of the digestive organs or constitution.   These are the sparsa, figurata, erysipelatodes, and scabida.

In the first, denominated I. sparsa, " the pustules are at a distance from each other : and the eruption extends, without any certain order, along the back of the hands, the arms, neck, shoulders, thighs, or legs.   After a few days the pustules break, and discharge a thin humour, which gradually concretes into yellowish, laminated scabs. The cuticle, as far as the eruption extends, becomes reddish, rough, or scaly ; and a slight discharge from rhagades or chops in various places, as well as from beneath the thin scabs, continues through the complaint ; the duration of which, in the upper extremities, is seldom more than two or three weeks.   When the lower extremities are affected with this eruption, it continues a long time.   Small yellow pustules first appear on the instep, and then on the ancle and leg, with a violent itching.   They are most numerous on the foot and ancle, and when they are broken, a considerable quantity of humour issues from small pores, around which the cuticle is rough, reddish, shining, and a little elevated.   The parts affected are for some weeks covered with thin scabs, but not sufficiently so to prevent the watery discharge.   When the surface appears to be healed, and the scabs are about to separate, a fresh eruption of pustules often takes place, and the discharge recommences with great heat and irritation.   After several returns of the eruption, ulcers are sometimes formed on the fore part, or sides of the ancle. The ulcerations discharge a clear ichor ; they exhibit a considerable, but unequal cavity, and irregular edges surrounded by the pustules. In sedentary persons, who have passed the middle period of life, the edges of the ulcers are blackish, or of a purple hue, and the limbs become œdematous. The small pustules diffused over the surface are of nearly the same colour, and sometimes the intervening skin appears livid, or speckled with livid and red.

"The Impetigo sparsa is most troublesome when the yellow psydracia are intermixed with small irregular vesicles, as frequently happens on the upper extremities. The complaint commences about the knuckles, and spreads along the thumb and fingers to the nails ; likewise along the back of the hand, and round the wrists, to the forearm. Both hands are usually thus affected about the same time, and the eruption extends in some cases to the bend of the elbows, the upper arm, the neck, and the cheek. It is always succeeded by a little watery discharge, and by the formation of laminated scabs ; when these fall off, the cuticle beneath remains for a long time scaly and chopped, and in this state of it fresh pustules arise, with heat, soreness, and violent tingling. Thus by repeated suppuration, and scabbing, the texture of the skin becomes, in many places, rough, harsh, and inflexible.

" This disease generally appears in autumn, and continues through the greater part of the succeeding winter. It disappears in many cases during the summer, but returns at the latter end of the year. The eruption is preceded by some disorder of the constitution, as headache, indigestion, and pain in the stomach, violent pains in the limbs and back, and sometimes cramps of the lower extremities. Children, and even infants, are occasionally affected with this disease ; it occurs, however, much more frequently in adults, than in children, or in persons of an advanced age. A predisposition to it is communicated hereditarily ; and in those who are predisposed the complaint appears after intemperance, violent exercise, or exposure to sudden interchanges of heat and cold."

When the disease appears in the form of circumscribed patches of pustules of an irregular figure, and situated at a distance from each other, it answers to the description of the I. figurata. The hands are most frequently the seat of this more limited state of the disease, and it seems almost constantly to occur from the influence of locally irritating causes. The character and progress of the complaint are similar in other respects to that described in the foregoing quotation, though it is sometimes, from the necessary exposure of the part, rendered more tedious. The blotches do not always spread beyond the back of the hand, up the arm, or into the palm, though the pustules often occupy the interstices of the fingers at the roots of the first phalanx, and are rendered more painful and irritable by the motion of the parts, particularly if scabs are allowed to form in this situation.

The connexion of this form of Impetigo with constitutional causes is not often to be traced : and I have not been able from my own observations, or the reports of others, to confirm the remarks of Dr. Willan, of its being preceded by pains of the stomach, headache, &c. Indeed, as I before observed, I believe the greater number of cases are produced by local irritation.*

The description of the two foregoing species, or states of the disease, would comprehend, as nearly as any general description can do, the majority of cases which occur : the erysipelatous form occurring on the face being merely accidental, while the scabida is evidently that of an aggravated case from disordered general health and neglect. With respect to the species termed rodens, Dr. Bateman states, that he never had an opportunity of seeing it, and that it is probably of a cancerous nature.

Impetigo rarely comes under the eye of the medical observer at its first commencement ; and I am inclined to think, that, at this period, the term pustule is improperly applied to its chief feature, the fluid which the vesicles contain being transparent, though the change to opacity takes place in a few hours. The vesicles are sometimes broken at this period, when the fluid which they contain concretes on the edges of the little excavation which they occupied ; and if the part be at this time minutely examined, the excavation will be found lined with this incrustation, while other vesicles are forming by the side. In this way the disease extends in a circular or other form, till a sufficient abrasion of surface is produced to furnish the materials of a scab.

In the treatment of Impetigo, the frequent removal of the diseased secretion has never been considered of sufficient importance : the benefit of this step, (I must be pardoned for a little repetition,) if carried into effect by frequent ablution of the part with warm water, is incalculable. By this plan, in conjunction with the exhibition of simple alteratives, entirely rejecting any thing in the shape of ointments, or other greasy applications, the disease will be often readily subdued. The part may be kept in a state of moisture at other times by covering it with oilskin, or by the application of soft linen wetted in the liq. plumb. acet. dil.

If these means prove inefficacious, the Harrogate waters, or the internal exhibition of sulphur in any other form, are recommended. Plummer's pill, the decoction of dulcamara, &c. present other means of obtaining success.

On the subject of the Harrogate water, it may be proper to observe, that the custom of transmitting it in bottles to distant parts when the convenience of patients does not admit of their visiting the wells, renders it liable to some deterioration, if great care be not taken. "It loses its transparency when exposed for about two hours to the air, at first acquiring rather a green hue, and after long standing, by transmitted light, a slight reddish colour. It gradually

---

* The most intractable cases I have ever met with, have been caused by the imprudent use of strong alkali to the skin, for the purpose of removing particular stains.

loses its sulphuretted taste, and then has the flavour of a strong solution of common salt. We found by experiment, that the sulphuretted hydrogen gas undergoes decomposition by exposure. The oxygen of the atmosphere unites with the hydrogen, and the sulphur is precipitated in a state of minute division, the precipitate being of a light ash colour. Hence the turbid appearance of the water. It is, however, extremely worthy of observation, that this water bottled at the spring, and immediately corked and sealed, retains its gas and all its virtues for a long time."*

When the benefit is expected to be derived in Impetigo from the use of local applications, the latter, as before observed, ought not to be had recourse to in the form of ointments, unless the parts have been entirely divested of morbid secretions : and even then they are seldom beneficial. Independent of the difficulty of applying remedies in this form, from the slowness with which they penetrate the scabs covering the diseased parts, it is impossible to avail ourselves efficiently at the same time of the sedative effects of cold, which, particularly in the local varieties of the disease, is of the greatest importance. I have had reason, moreover, often to suppose that greasy applications, even of the simplest kind, are productive of increased irritation—of aggravation rather than benefit. Cooling sedative lotions constitute the chief and most powerful agents in the treatment of Impetigo.

In the more trifling cases the mixture of a little alcohol with water, applied by means of a bit of linen rag, will be sufficient, if the habit of the patient be not full, but if the opposite state of constitution exist, it will be apparent that active depletive measures must be chiefly relied on.

Of all the external applications which have been used in the form of lotion, I have not found any equal in effect to the formula below.† The Hydrocyanic acid, combined with alcohol, has a power in subduing the irritation of this disease which is really surprising. Pieces of linen wetted with it should be kept constantly applied. It is much more efficacious in subduing the marks of irritation which exist on and surround the diseased spot, than any other application which I know of, if care be taken to clear away the secretions, so as to admit it directly to the diseased surface ; but if the latter point be not attended to, its superiority is not so distinctly marked. It sometimes rapidly removes every vestige of the disease, but if the patient begins to relax in the application, it returns with redoubled violence.

The grand and predominant features of Impetigo are extreme irritation and active inflammatory action, accompanied or followed by a proportionate degree of relaxation of the vessels of the part involved. The objects in view, therefore, should be, first, to diminish such irritation, and secondly, to supply the loss of tone which

---

* Scudamore on Mineral Waters, p. 92.

† ℞ Acid Hydrocyanic, ʒiij. Aq. distil. ℥viiss. Alcohol ℥ss. m. ft. lotio.

the vessels have sustained.   It is evident, that the common sedative applications are not possessed of these properties conjointly, and hence arises the *temporary* effect only by which they are followed. The desiderata evidently consist of means which will at once relieve the turgescence and irritation, and correct the relaxed state of the vessels of the part.   It is upon these principles that the sulphur vapour bath acts occasionally as a curative measure with the greatest effect.

A very common affection of an impetiginous character, confined to the cutis of the ears, and generally most violent on their posterior surfaces, prevails among females in advanced periods of life.   In the milder cases, if it be not very minutely examined, an apparently abraded state of the part, with much redness, and a slight fluid secretion, are noticed, and much itching and heat are usually complained of.   If the surface be particularly examined, however, the cuticle will be found to have been partly removed by the formation of a number of minute vesicles, which have been broken down, and their contents discharged.   In other cases, the small vesicles and pustules are distinctly seen very copiously distributed, but being covered by the delicate cuticle of the part, are ruptured by the slightest roughness.

This affection is exceedingly obstinate, and does not readily yield to any application or general method of treatment.

Constitutional remedies, whether tending to give energy, or to produce an opposite effect, are equally uncertain as to the benefits they may be expected to produce ; the disease sometimes occurring in full habits, and at others, in those of an opposite character.   Common sedative washes do little good, and the only applications which have, under my notice, brought about any permanently beneficial effect, are the black lotion and the prussic acid lotion before spoken of ; and sometimes where the one or the other fail in effecting a cure, they may be made use of alternately with good hopes of success.   Strong cathartics, such as calomel and jalap exhibited twice a week, but not to the extent of materially reducing the strength, constitute the best part of the internal management ; and as far as my observation goes, they are equally advantageous in thin and spare habits,* and the opposite.

* The substance of M. Rayer's very industrious compilations on the subject will be found in the following note.   I fear there is a vast portion of the details, if applied to the history of the disease as it occurs in England, which will be considered imaginary.   It is evident that the author has used, in a most unlimited manner, the works of our countrymen, and converted mole-hills into mountains, as usual, in his descriptions.

   "The disease is observed on the neck, trunk, and extremities, but even more frequently on the face,† and almost always on the middle of the cheeks; from thence it is apt to spread to the whole of the malar region, and to the commissure of the lips, in such a way as to form a circle upon the face.

   "When impetigo is evolved on the face, and it is watched from the period of

_____

† In England it is never seen on the face.   It is certainly a very different disease from that which has been proposed for the cognomen of Impetigo larvalis.

Of Scabies.

The most common form in which Scabies makes its first appearrance is that of minute vesicles, containing a transparent colourless

its invasion, one or more small red and very superficial blotches are first perceived ; these gradually become more conspicuous, and are soon affected with a considerable degree of pruritus. By-and-bye these patches rise and become covered with small yellowish pustules, confluent or agglomerated, and but little elevated above the general level of the integuments. These clusters, of various dimensions, generally of a circular shape, and surrounded by a rosy circle, may continue isolated, or become connected by the development of fresh pustules in their circumference, or by the skin acquiring an erysipelatous tint in the spaces between them. The eruption is occasionally accompanied with severe pruritus and a degree of heat and tingling that amounts to smarting.

"After the lapse of three or four days, and occasionally sooner, the pustules burst, pour out a yellowish fluid, which dries quickly and turns into thick crusts of a bright or greenish yellow colour, semi-transparent, slightly furrowed, very friable, and in appearance very similar to portions of candied honey, or to the concrete gummy exudations poured out by different trees. A considerable discharge continues to go on under the crusts originally formed, the thickness of which is thus gradually very much increased, whilst their dimensions extend greatly beyond the limits of the pustules that first produced them. The skin in the circumference of the incrustations is red, and frequently presents several small and unbroken pustules, the contents of which are scarcely consistent. The integuments under these crusts are of a bright red, and occasionally appear denuded of epidermis.

" When impetigo is evolved in a young and well-constituted individual, or when the disease is slight, its duration is not necessarily protracted beyond an interval of more than three or four weeks : the heat of the skin gradually abates, the morbid secretion lessens by slow degrees, and before long ceases entirely ; the incrustations grow constantly drier, and being detached at length, leave one or more red spots or marks which commonly continue visible for more than a month, and are followed or not, according to circumstances, with a sensible desquamation of the cuticle. Small miliary specks, of a dull white colour, are at times observed upon the remaining red spots, these are owing to a number of sebaceous follicles distended with hardened secretions, or of which the walls have become altered and thickened.

" Impetigo is occasionally met with confined to the eyelids, upon which prominent and conical-shaped incrustations are then produced. This variety of the disease is commonly complicated with a particular species of ophthalmia, or with an inflammatory affection of the follicles of the ciliæ.

"I have seen this form of impetigo prolonged downwards on either side of the under lip in a very regular manner ; and I have remarked it forming a streak upon the upper lip so as to appear like a pair of thick mustaches.

" Impetigo of the face may become chronic under two forms : 1st, the development of the psydracious pustules is at one time successive ; fresh groups are thrown out in the neighbourhood of the yellowish crusts produced by the desiccation of the primary pustules or of the secondary crops developed in the circumference of the first pustular or incrusted clusters, the dimensions of which they increase. In the latter case, the desiccation and cure commence in the centres of the several groups.

" 2nd, Instead of spreading superficially, the inflammation in impetigo may penetrate the whole thickness of the skin, and even affect the subcutaneous cellular tissue. After the detachment of the incrustations a fresh discharge may give rise to the formation of new scabs of the same kind ; and they may thus fall and be reproduced several times in succession, becoming, however, thinner and thinner on each renewal. The surface of the integuments under them is of a

fluid, intermixed with small papulæ, which, in a few hours, if suffered to remain undisturbed, assume the character of vesicles also. The attention is usually first directed to the part by the inordinate pre-

vivid red colour; it subsequently becomes furfuraceous, and the inflammation then seems to acquire something of a squamous character.

" When chronic impetigo *figurata* thus approaches its decline, if it be treated by an ill-timed recurrence to stimulants, or if the constitution happens to suffer from any cause, the disease may be fastened, as it were, upon the skin for many months, and even for several years. The parts of the skin that have in this way been repeatedly affected with inflammation, become chapped, and occasionally even present superficial excoriations.

"The pustular groups of impetigo *figurata* of the face, although usually situated on the malar regions, may be observed occurring in other regions : they are occasionally thrown out on the upper lip, immediately below the septum of the nostrils, and upon the alæ nasi. In the latter case, the matter of the pustules may dry in such a way as to produce a conical-depending scab, compared by Alibert to the stalactitic formations observed in certain caverns (*Dartre crustacée stalactiforme*).

" In impetigo *figurata* of the limbs, the groups of psydracious pustules and the scabs that succeed them, usually circular on the forearms and hands, are of a larger size and less regularly round shape on the lower extremities. The pustules in these situations are evolved in the same manner as on the face, and are soon replaced by thick crusts of a greenish or brownish yellow colour. When the disease has declined into the chronic state, it frequently happens that no untouched pustule is any where to be seen; but the partial eruptions that take place from time to time, and the particular shape of the crusts and of the red marks they leave behind them, are sufficient at any time to characterize this variety.

" When recovery takes place, the heat and itchiness of the skin diminish; the discharge becomes less in quantity ; the crusts, last produced, are of superior thickness, their edges dry completely, and are occasionally marked out by a white epidermic border. Lastly, after the crusts have been detached and are no longer reproduced, the skin of a deep red in the first instance, becomes furfuraceous, and by slow degrees regains its natural colour and appearance.

" Instead of being arranged in circumscribed groups, the pustules and crusts of impetigo may appear irregularly scattered over the limbs, neck, shoulders, face, and external ears. This constitutes the second variety of disease—the impetigo *sparsa*.

" Impetigo *sparsa* of the lower limbs is always a disease of long duration. One of the extremities may be attacked singly, or both may be implicated together, or in succession.

"The disease is characterized by small yellowish pustules which appear on the instep, ancles, and especially, on the outer part of the leg. The evolution of these pustules is accompanied with insupportable pruritus. They soon burst and pour out a sero-purulent fluid, which is gradually changed into yellow-coloured laminated scabs of less breadth and thickness than those of impetigo *figurata*. In the spaces between them, the skin is reddish, and the cuticle looks rough and shining ; a considerable discharge takes place from the pustules for some time ; but, by-and-bye, this becomes smaller in quantity, under and in the vicinity of the crusts, which consequently acquire a more consistent and drier appearance. At the very time when they seem about to be thrown off, however, it frequently happens, that a fresh eruption of pustules takes place, accompanied with heat and violent pruritus ; and secondary pustular eruptions of this description, may go on occurring from time to time, at various intervals, until the whole of one, or of both legs, from the knee to the ancle and dorsal aspect of the foot are implicated. A sero-purulent fluid then flows abundantly from the surface of the skin, and, by drying, encases the limbs in crusts. The crusts often acquire a great degree of thickness in the aged, and among individuals of shattered constitutions. They

valence of the sensation from which the disease takes its name,
which, leading to frequent friction and scratching, is soon followed
by inflammation of the skin, and rupture of some of the vesicles,

are then of a deep yellowish-brown colour, and might very properly be compared
to the bark of certain trees (impetigo *scabida*, Willan). The legs are moved with
pain and difficulty; the incrustations split, the legs frequently become œdematous,
and, before long, are furrowed by fissures, running in various directions and of
different depths. A yellowish, sero-purulent discharge exudes from these cracks,
and forms additional incrustations that appear to gird and enclose the leg. If
this hardened exudation be removed in part or completely, by the continued use
of emollient fomentations and cataplasms, the denuded corion which is then exposed, speedily furnishes a fresh supply of discharge, which before long concretes into a new incrustation.

" Arrived at this stage, impetigo *sparsa* of the lower extremities is a very
obstinate disease, especially when it attacks the aged, the weakly in constitution,
or the infirm and broken down in health. The inflammation occasionally extends
to the toes and secreting matrices of the nails. The nails then become altered,
and finally loosened from the skin (onychia *impetiginodes*). An œdematous infiltration of the legs, and ulcers, which commonly appear about the ancle, are
common consequences of this affection. The surface of these ulcers is uneven,
and discharges a sero-purulent fluid; their edges are irregular, purple, or livid,
and frequently crowned with small pustules full of sanguinolent serum; or, otherwise, they are covered with yellowish crusts of varying thicknesses.

" When the progress of this inflammation is successfully arrested—the incrustations become dry, and, once detached, are not again reproduced. The skin
in some places maintains a blueish, or purple-red tint, and in others, where it
has been attacked with ulceration, it presents reddish or violet-coloured indelible
cicatrices.

" Impetigo *sparsa* of the superior extremities is most frequently found attack-
ing the forearms; it does not differ from the disease situated on the legs, save in
being less severe, and more rarely complicated with œdema and ulcers when it
has advanced into the chronic state.

" In acute impetigo *sparsa* of the face, the greenish yellow incrustations dispersed over the cheeks and clinging to the beard in the adult, are not long of
being loosened from the skin. If the inflammation extends to the nose, as it
frequently does in childhood, the nostrils become plugged up with thick and dry
incrustations; the nose swells, and the disease then commonly passes into the
chronic state.

" Various phenomena frequently appear connected with the local symptoms of
impetigo: the lymphatic glands in the vicinity of the pustules may become swollen
and painful; the pruritus and morbid heat are occasionally so troublesome, that
they prevent the approach of sleep, and impede the due performance of different
functions of the economy. The disease, in fine, is frequently seen associated
with a gastro-intestinal affection; it is much more rarely complicated with any
other form of internal lesion.

" *Causes.* Impetigo is not transmitted by way of infection, and its causes are
very obscure. Children, at the period of teething, especially those of a lymphatic
temperament and scrofulous constitution, are frequently attacked with impetigo of
the face and hairy scalp (teigne *granulée*, Alib.), or with the eczematous form of
the disease, which is then generally entitled *crusta lactea ;* this complaint is more
especially and frequently observed among the poor, ill-lodged, badly fed, and
filthily disposed classes of society.

" Young persons of a sanguineous and lymphatic temperament, with a fine
and delicate skin, are occasionally attacked with impetigo of the face, when
they have been exposed to the bright rays of the sun in the spring and heats of
summer.

" In young females whose catamenia are irregular, and among women arrived at
the critical age, impetigo is apt to show itself either on the face or on the limbs;

232   PLUMBE ON DISEASES OF THE SKIN.

and, consequently, the fluid or the insect they contain possessing
the power of infection, by the extension of the mischief to the surrounding parts.

in these circumstances, it very frequently attacks the upper lid immediately under
the septum of the nostrils.

" Impetigo seems occasionally to be induced by the presence of some other inflammatory affection of the skin, particularly by repeated attacks of lichen agrius.
The disease seems also, occasionally, to coincide with some derangement of the
digestive functions ; and this is a complication that is met with very frequently
in children during the periods of teething. To conclude, the small pustules of
impetigo have now and then been observed to follow excesses of every kind,
violent muscular exercise, acute and prolonged mental affections, grief, &c.

" *Diagnosis.* Impetigo may present itself in the shape or stage of pustules and
incrustations ; or the disease may be reduced to its traces—red marks covered
with squamæ, or stains of a yellowish-red hue ; it may farther be discovered in
different places under each of these different degrees or appearances. The minute
pustules of impetigo are readily distinguished from the large pustules of ecthyma,
and from the artificial pustules produced by the tartrate of antimony and the
inoculation of purulent matter. I have already given the elements of the diagnosis
between impetigo, and acne and rosacea. Impetigo of the hairy scalp is not
liable to be mistaken for disseminated favus (porrigo *lupinosa,* Willan), nor with
that which appears under the form of circular patches (porrigo *scutulata,* Willan).
The pustules of impetigo discharge, whilst those of favus, deeply situated
within the substance of the skin, are rapidly changed into dry, yellowish-coloured,
cup-shaped scabs. The crusts of impetigo are brown, or of a dull grey, like
small pieces of dirty plaster, and never present those broad, thick, and continuous incrustations observed in favus confertus (porrigo *scutulata,* Willan).
Lastly, impetigo of the hairy scalp is not contagious, and does not implicate the
piliferous bulbs, like favus.

" It is more difficult to distinguish impetigo of the scalp, from eczema *impetiginodes* of the same region. The principal diagnostic features of each of these
inhere in the dissimilar aspects of their incrustations.

" The pustules of sycosis, often isolated, and always prominent, are larger in
size, and not so yellow in colour as those of impetigo when it attacks the chin ;
the impetiginous eruption, also, is always very much crowded, and secretes
abundantly. The scabs of sycosis are drier, and of a deeper colour than the
incrustations of impetigo ; they are likewise only reproduced after a fresh eruption of pustules. The crusts in impetigo are of a greenish-yellow hue, thick,
semi-transparent, and reproduced without any renewal of the eruption. Tubercles and indurations are encountered in sycosis,—alterations that are never seen
in impetigo.

" When the vesicles of scabies become purulent, or when they are accidentally
complicated with pustules, they are always much larger, and more elevated than
the small psydracia of impetigo.

" The red and scaly spots consecutive to the formation, or to the fall of the
crusts of impetigo, may be distinguished from inflammatory affections originally
squamous in their nature, such as lepra, psoriasis, and pityriasis, by the circumstances of the squamæ in these latter maladies being accompanied by no discharge,
and by their having been preceded neither by pustules nor incrustations. Those
pigmentary yellowish-coloured spots, so frequently observed after the cure of
syphilitic eruptions and confluent psoriasis, are very seldom seen after the invasion
of impetigo. The pustules of impetigo can never be mistaken for those of syphilitic origin, covered with black and firmly adherent scabs which conceal
ulcers to which indelible cicatrices succeed (vide *Syphilis*). Lastly, impetigo
with its pustules, disseminated or collected into clusters, and its thick, rough,
and yellowish-coloured incrustations cannot be mistaken for eczema in any of
its forms, with its vesicles, its lamellar crusts, or the thicker squamæ of its latter
stages.

" The prognosis in impetigo is generally more favourable than in eczema, lepra,

In the lower classes of society the disease is rarely seen before it has been some time established, and has extended itself over most parts of the body. In such cases, it is most virulent and trouble-

lichen, &c. The disease, with the acute type, wherever situated, commonly gets well in the course of two or three weeks. The duration of the chronic form of impetigo is influenced by the number of eruptions, the state of the constitution generally, and the existence of other particular conditions, such as amenorrhœa, pregnancy, period of life, presence of scrofula, &c. I have seen the disease appear in a woman each time she was pregnant, resist the most energetic remedial measures, and get well spontaneously a short time after the confinement.

" Chronic impetigo of the hairy scalp, of the upper lip, and other regions covered with hair, is often a very intractable disease, especially when the patient is advanced in life, when he is of a scrofulous constitution, or when his health has suffered in any way.

" When impetigo appears with a character of acuteness on the face or hairy scalp of a child during the process of dentition, it is generally advisable to restrict remedial measures to simple attention to cleanliness: the eruption occurring under these circumstances is frequently accompanied with a signal improvement in the state of the constitution, which might be prevented from taking place by any thing like active remedial measures. I have seen the ill-timed medication and cure of these impetigos, then entitled crusted tetters (dartres crustacés), followed by diseases of different degrees of severity ; in other cases, the appearance of the pustular eruption has seemed to me to act advantageously on various old-standing diseases of an obstinate character.

" On the other hand, the treatment of certain impetigos is to be conducted on the principle of effecting some modification in the constitution. I have had opportunities of satisfying myself on this point, by treating successfully with the preparations of iron, sulphur, and iodine, several impetiginous eruptions befalling individuals labouring under scrofula. In other cases where the impetigo has been preceded by amenorrhœa or dysmenorrhœa, I have derived great advantages from the preparations of iron and other emmenagogues.

" When the state of the constitution affords no particular indication of cure, impetigo on its appearance, and as often as the eruption is attended with a great degree of redness of the skin (impetigo erysipelatodes), or is distinguished by the abundance of its pustules, is to be treated by means of general blood-letting in adults and individuals in the flower of life, and by leeches in children and the weakly in constitution. Should amenorrhœa or dysmenorrhœa complicate the disease, the vena saphena should be opened, or leeches should be applied around the external organs of generation, especially when emmenagogues have been tried in vain. These bleedings will occasionally require to be repeated. They usually prove detrimental to individuals of a scrofulous or weakly habit of body. Whatever the temperament of the patient, the blood abstracted is almost invariably found to be buffy.

" The local or general simple tepid bath at a low temperature (24° or 25° Reaum.); and frequent ablution with cold water, milk, decoction of bran, almond emulsion, decoction of mallows, digitalis, poppy-heads, &c. are used with advantage in this first stage of impetigo. At a later period, aluminous, saturnine, or alkaline washes, and the application of the ointments of the oxide of zinc, and of the acetate of lead, contribute to accelerate the cure, which often takes place without the necessity being felt of having recourse to any other measures.

" In acute impetigo of the region of the beard, or hairy scalp, the hair must be removed with sharp scissors, and the diseased surfaces laid bare. This variety, like all the others, requires to be dealt with by means of emollient applications, and occasionally of blood-letting. Depilation, the utility of which is incontestable in favus, is always injurious in acute impetigo of the hairy scalp*, or of the

* Impetigo of the hairy scalp does not occur in England.

some in situations well covered by clothing, subject to friction. The trunk, margins of the arm-pits, and flexures of the joints elsewhere, usually exhibit marks of the greatest irritation ; while in more recent cases, in cleanly persons, it is usually first detected about the fingers, wrists, and backs of the hands.

It is generally understood that this disease is in the majority of cases produced by the direct application of the contents of the ve- sicles by contact with an affected person, or by handling and making use of their clothes, sleeping in the same bed, &c. It is also gener- ated by the neglect of personal cleanliness, and inattention to fre- quent changes of linen, and hence has been said to prevail much among the natives of cold and mountainous countries.

When it has been of long standing, and the skin of the patient is very irritable, it often exhibits at one view the appearance of pim- ples, vesicles, and pustules of different dimensions interspersed, a condition, which, together with the excessive itching and irritation, will enable us readily to distinguish it from any other disease. The degree of irritability of skin, indeed, seems very much to influ- ence the character the eruption assumes : where this property is

skin; neither is this measure ever necessary in the chronic impetigo of these regions. Saline purgatives, such as the acidulous tartrate of potash, the tar- trate of potash and soda, the sulphate of magnesia, the sulphate of soda, &c. in doses of from two drachms to half an ounce daily, are also frequently employed with advantage.

" When impetigo has lost its acute character, or when it has become decidedly chronic, the incrustations are best got rid of by means of the steam douche, which seems very often to have the effect of warding off a fresh eruption. It is even found advantageous to recur to the vapour douche shortly after the forma- tion of the crusts of impetigo in any stage of the disease whenever the skin appears but little inflamed. These douches directed upon the parts affected, before the formation of scabs, that is to say, during the pustular state of the eruption, or when any considerable degree of inflammation still lingers around the incrustations, are almost always injurious. I have frequently substituted, with good effect, the simple bath and ordinary emollient cataplasm, at a very moderate temperature, for the vapour-bath and douche generally recommended in these cases.

" It is seldom that local blood-letting and emollient and sedative applications are very actively employed in cases of chronic impetigo ; although the practice, assisted by gentle laxatives, when the state of the digestive organs offers no bar to their exhibition, is one which in analogous circumstances has unquestionably been very frequently crowned with success.

" When the skin is but slightly inflamed, and not very irritable, the artificial or natural sulphureous baths of Baréges, Loüesche, Canterets, Bath, Harrowgate, &c. are often resorted to with advantage at a temperature of from 28° to 30° of Reaumer's scale, and this is not only in the cases of the elderly and weakly, but in those of individuals in the spring and vigour of life, possessed to all ap- pearance of an excellent constitution. The time of continuing in these baths ought to be gradually prolonged, and ultimately carried the length of several hours each time.

" The sea-water and the alkaline bath are generally less beneficial ; yet it does sometimes happen that taken every day, or on alternate days, along with the fresh-water bath, they act more beneficially than the sulphureous bath. Alkaline washes are usually prescribed at the same time; and these may be combined and alternated with the use of acidulous lotions."

particularly marked, rapidly running from the papular form to vesication, the vesicle in the course of a short time terminating in a
change to opacity of its contents, and all the characters of a pustule
of considerable size.   Hence the vulgar distinction of watery and
pocky itch, which, though it does not seem to be of much practical
importance, has furnished the ground-work of Dr. Willan's division
into five different species.

The disease seems to be totally unconnected at its origin with
any constitutional cause ; but when it occurs where the system is
debilitated, or the habit udhealthy, it sometimes leaves a troublesome impetiginous affection on the skin of the parts, long after its
specific character has been eradicated, a state which appears to answer
to the description of S. cachectica.

The existence of an insect in some of the vesicles and pustules
of itch has been discovered in repeated instances by the assistance
of good glasses, and it is therefore supposed, with apparent good
reason, that the disease is the mere result of their operations in the
skin.   On the other hand, however, it has been maintained, that
such insects are the consequences rather than the cause of the disease, and that they are not found in the majority of cases where
inquiry has been instituted.   The first of these opinions prevailed
very generally until the publication of Dr. Bateman's work, and
he expressed his suspicions that the insect is not to be found in all
cases of the disease.   He brought forward the authority of Dr.
Heberden and of Baker and Canton, the two latter of whom had
the advantage of great experience in the use of the microscope, in
support of his opinion.   Since Dr. Bateman's work was published,
the subject has been several times canvassed by French pathologists ;
among whom, Lugol, Mourouval,* and Dr. Suriray, of Havre,†
contended for the non-existence of the insect ; while the experiments of Dr. Gales, the well known inventor of sulphur fumigations,
were supported by a jury of medical men, and ought to have been
considered conclusive in favour of the original opinion.‡   The three
former have by no means adduced that weight of evidence which
Dr. Gales's statement contains, and it must be confessed as more than
probable, that the superior opportunities of inquiry which the situation of the latter gave him would enable him to arrive nearer the
truth than his opponents.

The hospital of St. Louis, in which Dr. G. was a long time resident, constantly contains a large number of cases of itch : he states,
that he has examined some hundreds of the insects in question ; his

* Nouvelles Recherches et Observations sur la Gale, faites à l'Hospital St.
Louis, à la Clinique de M. Lugol, pendant les annees 1818, 1820, 1821, et recueillies
par I. F. I. Mourouval. 1 vol. 8vo.
  † Letter to Dr. Mark, Journal de Medecine. Aug. 1813.
    ‡ The Acarus Scabiei is not a solitary instance of the domicile of animalcula
in the substance of the skin, and of consequent production of much irritation and
disease.   The Furia infernalis of Siberia, the Chigre and the Dranunculus are
analogous cases.   Fatal effects have arisen from each of the three latter in a
variety of instances.

description of which exactly corresponds with that given by Linnæus. He succeeded in producing the disease by confining the insect on his own skin repeatedly ; and his experiments have been witnessed by some of the most eminent medical characters in Paris, who have been satisfied with the manner in which he has conducted, and with the conclusions which he has drawn from them.

The differences of opinion on this point admit of ready explanation. A short time since, in the course of my inquiries, I had an opportunity of seeing great numbers of the insect readily extracted with finely pointed needles ; not, however, from the centre of the vesicle or pustule, but from their sides.   The situation of the insect in such circumstances is marked by a minute speck exterior to the margin of the vesicle : and with care it may be always extracted alive. In the situation in which it is detected, it is evidently making its escape, from the fluid which its operations have produced, and in which it would probably be unable to live.   The greateast care does not enable us to find it in those vesicles which are of enlarged dimensions, or which have become opake.   When it has penetrated the cuticle, and obtained sufficient nourishment for its present wants, it proceeds to extricate itself from the spot in search of a new field for its operations, leaving the irritation it has excited to form an enlarged vesicle which undergoes the regular change to an opake state, and ultimately to the formation of a scab.   It is fair to presume, that those who contend for the non-existence of the insect have been unfortunate enough to select the vesicles which it has evacuated, for their researches, for, indeed, every one would probably be induced to do so, merely on account of their size and number.

A great variety of applications have been made use of in the cure of itch, notwithstanding our knowledge of a decided specific.   The unpleasantness of the smell of sulphur would render it very desirable to find a substitute for it, and many trials have been made, but hitherto without effect ; the undermentioned* have, however, been found sometimes adequate to the removal of trifling cases of the disease : but there is no doubt that sulphur is the safest and most expeditious remedy.

The most common form in which it is employed in this country is that of ointment : but the importation of the French fumigating-baths has afforded us some opportunities of avoiding the unpleasantness of this plan, and though somewhat more tedious in bringing about a cure, they are sometimes entitled to preference on account of the superior degree of cleanliness the patient is enabled to observe during the treatment.

The history of the sulphur vapour-bath appears to have commenced with the experiments and researches of Dr. Gales, before referred to.   The comparatively rude and unpleasant expedient of saturating blankets with the vapour, by means of a warming-pan, was first

* Solutions of potash, muriate of ammonia, oxymuriate of mercury, arsenic ; decoctions of hellebore, digitalis, and tobacco ; sulphuric and oxygenated muriatic acid properly diluted, &c. &c.

adopted, on the burning coals of which the sulphur was strewed when the vehicle was introduced into the bed, until the combustion was complete. The patient was then directed to enter it naked, and was covered up to the throat. Seven repetitions of this process in as many nights was found to be adequate to the cure ; and it seems probable, that in very delicate habits, or in the cases of pregnant women, it is entitled to preference over the vapour-bath at present in use.

The vapour of the sulphur is unquestionably the agent by which the cure is brought about, and there seems as little doubt that the destruction of the insect constitutes its *modus operandi*. Dr. Horn, of Berlin, and Dr. de Carro, of Vienna, appear to have been next to Dr. Gales in the use of the bath. Subsequently to these, Mr. Wallace, of Dublin, has published his observations ; and step by step, by the joint improvements or suggestions of the observers, the instrument, from having been inconvenient and uncomfortable to the patient, is now become not only an important and decided remedy of great value in some cutaneous diseases, but an absolute luxury as regards the patient's feelings.

The number of fumigations necessary to the cure of Scabies differs much, according to the virulence of the disease, the degree of irritability of the skin, &c. ; but if the clothes of the patient be suspended in the bath during each application, which is always prudent in the cases of the lower class of people, the cure will be much expedited.

Besides the advantages of the bath already alluded to, it is never followed by that irritation of the skin, which long continued use of the ointment is often found to produce, and which has, not unfrequently, led to the continuance of the remedy long after the disease has been really subdued.*

CHAPTER II.

OF ECZEMA.

THE disease termed Eczema, as it is most frequently seen according with the general definition of Dr. Bateman, is the simple effect

---

* The foregoing article is a mere reprint from the last edition of this work, and has not the recommendation of a line of novelty. Baron Alibert has recently offered in Paris a reward for the discovery of the Acarus Scabiei, and the successful candidate obtained some hundreds of francs. A physician of London has also *discovered it !* Bonomo, and *he only*, is entitled to the credit of first detecting and giving an accurate account of it, which latter was published in the Philosophical Transactions nearly a century ago! The curious inquirer may find in the different workhouses of London, numbers of *old women* who will supply him with plenty of specimens at a few minutes' notice at a very small expense.

of the application of heat to the skin.*   I am not aware that irritation applied in any other form is capable of producing an equally
diffused eruption of vesicles, " with little or no inflammation round
their bases ;" though it is stated, on the authority alluded to, to be
produced by a great variety of other irritants in persons whose
skin is constitutionally very irritable.

Among the varieties as they have been termed of this disease,
the most important is the E. rubrum, and it differs from the above
definitions in having the vesicles much closer set together, and their
interstices of a bright inflammatory red colour, and also in being
much more generally and extensively diffused.

In cases produced by the heat of the sun even, it should be remarked, that considerable tumefaction and heat of the part accompanies or precedes the formation of vesicles ; the definition alluded
to would otherwise seem likely to lead to mistakes ; it is, indeed,
equally applicable to the appearances of some cases of itch, the
tumefaction alluded to being at first sight the only mark of distinction.

This affection has obtained in its more formidable form, as produced by mercury, the attention of many experienced observers,
from whose published remarks it appears, that though attended with
much local irritation and a copious discharge from almost every part
of the surface of the body, it is generally unattended with danger.
It subsides according to the violence and extent of the eruption at
its commencement, apparently uninfluenced by medicinal treatment,
in the space of from three weeks to two months.

The heat of the sun is, in delicate and irritable skins, capable of
producing its more insignificant forms on the hands, neck, face, and
other exposed parts in a very short period, and, consequently, such
forms prevail most among field labourers in the time of harvest.
With the same state of skin, it may of course be produced by the
application of heat in any other way.

It is generally, when occurring to a small extent, and from the
operation of local causes, a distinctly vesicular disease ; and in a
day or two, if the part be defended from heat and irritation, it usually
disappears.   Now and then, however, when due attention has not
been paid to it, a healthy state of parts does not so speedily follow :
and after it has been a little time established, the new vesicles which
form are of larger dimensions, and their contents become opake
before they break, thereby giving the disease an affinity to Impetigo.
When obviously produced by the heat of the sun, it has been named
accordingly ;† and when, from the circumstances mentioned, it
approaches to Impetigo, it has received a distinguishing designation ;‡
the understanding, however, that when occurring to a limited extent,
and from the operation of local causes, its variations of character are

* An eruption of small vesicles on various parts of the skin, usually set close,
or crowded together with little or no inflammation round their bases.   Bateman's
Synopsis, p. 252.

† E. solare.   ‡ E. impetiginodes.   Bateman.

so minute and unessential, will be sufficient to justify us in declining the use of these terms altogether. From some experiments which I have instituted, I have no doubt that the minute vesicular form which the disease assumes, when occurring from exposure to the sun, is materially dependent on the degree of heat applied ; that a blistered state of skin would occur if the heat was increased, while a minor degree of the latter would only be followed by erythematous redness ; that it is, therefore, to be considered rather as an accidental injury, and treated on surgical principles, in a manner similar to a slight burn or scald, than to be spoken of as a disease. It is, moreover, a state of parts, which the abstraction of heat, by cooling sedative washes, even though the irritation may be great, will enable the efforts of nature to correct in a few hours.

The vesicular character of the disease, when it occurs on the backs of the hands and fingers from exposure to the sun, has been considered as approaching scabies so nearly, as to be likely to lead to mistakes ; and points in which dissimilitude exists have been stated by Dr. Bateman in the work so often alluded to. The patient is, however, generally too well aware of the cause, to require to be set right on this point ; but should this not be the case, the swelling and heat attending the eczematous eruption, with the absence of the characteristic itching of scabies, will enable us to form a correct opinion. In scabies, moreover, of any standing, the vesicles are larger and more irregular in their size, and intermixed with distinct pustules, and with the scabs which form on the bases of the latter which have been ruptured by scratching.

Like most other cutaneous diseases, the eczematous eruption, sometimes by long and repeated application of the exciting cause, becomes more permanent and established, and much less ready to yield to the applications mentioned. In this state the vessels of the part have become debilitated and relaxed by constant excitement, and the addition of a few grains of alum, or of the acetate or sulphate of zinc to the spirituous lotion, will be found necessary. This form of application is generally more useful in those cases which approach to the impetiginous character than any other. The use of ointments, or any other kind of greasy application, cannot be too much condemned.

When Eczema is produced by mercury, the patient is first made sensible of the approaching disorder by a feeling of burning heat, itching, and tingling extending over the greater part of the body, but more particularly severe in the flexures of the joints, on the inner surface of the thighs, in the groins, and about the pubes and axillæ. A considerable degree of roughness of the skin of these parts speedily follows, with a deep inflammatory redness, not at first unlike that of scarlatina.

On the second day, the roughness is increased, and is easily observed to be produced by an immense number of minute vesicles, pretty regular in their size, and distributed closely and equally upon the parts particularly mentioned. On the third day, the more exposed parts are also covered in a like manner, by vesicles, containing a transparent fluid, while those previously formed on the thighs, groin, &c. begin to turn opake and milky. On the fourth, many of these latter break, and the disordered surface is covered by a copious exudation of viscous fluid, having an unpleasant odour, with which the linen is speedily imbued and stiffened ; the latter, in this state, adding to the unpleasantness of the situation of the patient, by farther irritating the parts with which it comes in contact. On the fifth day, the cuticle desquamates in large patches over the greater part of the body ; the inner surface of the thighs, as well as the groins, scrotum, and margins of the axillæ, are quite raw, and covered with the same fluid. There is much pain attending every attempt to change the position, and the smarting is excessively severe on the groins and thighs, if the patient attempts to extend himself ; the most easy position being that in which the knees are kept considerably elevated.

The only marks of constitutional derangement commonly distinguishable are a weak and quickened pulse, and a slightly furred tongue. The patient usually complains of weakness, but his appetite is not impaired. The bowels are regular, and the urinary secretion not much affected.

This state of things continues for many days, a succession of new vesicles continuing to appear wherever patches of unimpaired cuticle remain, till the greater part of the cutis has been denuded. In those situations which have been particularly noticed as favourable to the disease, the newly-formed cuticle on the recently abraded and inflamed surface is soon elevated and destroyed by more minute and delicate vesicles, which are ruptured in a few hours after their formation, and spots where the disease appeared to have subsided are thus again found perfectly denuded, and pouring out the same discharge as others. The protracted character which the disease assumes arises from these continued interruptions of the formation of new cuticle, and it not unfrequently happens, that this structure, in a new and delicate state, is destroyed and re-produced repeatedly, in the course of twenty-four hours.

As the disease begins to subside, the quantity of fluid secretion gradually diminishes ; the latter, however, appears still to mix with, and half dissolve, the ill-formed and delicate cuticle ; and at this period, and under these circumstances, scales of considerable thickness, and fissures of corresponding depth, are produced, from the latter of which the discharge is kept up, while along their course much pain and irritation continues to be felt. Repeated exfoliations of this mixture of cuticle and diseased secretion continues, till, as the inflammatory action subsides, the cuticle becomes more perfectly formed, and obtains its original strength and flexibility.

It has already been stated, that the duration of this affection is uncertain. It may even, though occasioned by mercury, be very limited in extent, and cease in a few days : and I have not seen a case even occurring in the most plethoric and healthy states of system (and it is in such that it shows itself most formidably) continue longer than five weeks, though the formation of solid unbroken cuticle may not be effected for a much longer period.

The foregoing description applies most particularly to the more formidable species of the disease produced by mercury. To a much more limited extent, and in a milder form, it is not unfrequently produced by opium, antimony, bals. copaibæ, &c.; in which instances, the contents of the vesicles rarely become opake, but are absorbed in a day or two, without rupturing the cuticle, being only followed by slight exfoliations of scurf.

Dr. M'Mullin* has divided the disease into three different stages. The first of these, however, consists of a train of constitutional symptoms by no means occurring in the majority of cases, and the doctor himself observes, that "whilst the eruption is making its appearance in one place, another part may have arrived at its most advanced form, so that all the different stages may be present at one time in the same individual;" an observation which has been confirmed by all who have had opportunities of seeing the disease. The accompanying remarks, that it is attended with typhus through its entire course, and that it is the peculiar effect of mercury, have not been by any means confirmed by subsequent inquiry, as may be seen by referring to Dr. Rutter's† and Dr. Chisholm's‡ papers on this subject, as well as to Dr. Moriarty's tract.§

From the facts which have been recorded, it appears that it is only when the strength of the patient has been originally not very great, and when debility has been brought on by excessive discharge, that typhoid symptoms occur. Under these circumstances only is danger to be apprehended ; and in such a state of things diarrhœa has occasionally come on, which has resisted every description of remedy, and ultimately destroyed life.

Sometimes, as observed by Dr. Rutter, a state of inflammation of the nares, trachea, and bronchiæ, form important features of the disease, which will of course add much to the danger of the case when combined with the debilitated condition described ; this, however, is only of accidental occurrence, and does not, in a great majority of cases, rise to such a height as to attract much attention.

A large proportion of the cases which have been recorded have occurred after the appearance of catarrhal symptoms, while the system was under the influence of mercury ; and hence has arisen the opinion, that it is produced by taking cold under such circumstances : but as it has been justly remarked by Dr. Spens, this accident is much too common to justify us in supposing that nothing else is necessary to the production of the disease, while the latter occurs in numerous instances where symptoms of catarrhal affection, or any other obvious disorder, have not made their appearance. Such an opinion was, however, we are informed, held by Dr. Gre-

* Edinburgh Medical and Surgical Journal, vol. ii.
† Medical and Physical Journal, vol. xxi., and Ed. Medical and Surgical, vol. v.
‡ Ibid. vol. viii.   § A Description of Mercurial Lepra, 1801.

gory, Dr. M'Mullin, and others, on the grounds above stated ; but the weight of evidence adduced is directly in favour of the conjecture, that previous idiosyncrasy must have existed.

In the treatment of this disease it will now and then happen, that though the bowels may be sufficiently open, the state of tongue and secretions may make it desirable to have recourse to a mercurial alterative ; or this medicine may require to be speedily renewed to meet the exigencies of the case, where the patient is afflicted with syphilis. It would therefore be desirable to ascertain the propriety of this step, and probable risk attending it, of bringing back, and re-establishing the eczematous eruption. It appears to have been done in one or two instances without any mischief of this kind occurring ; but in other cases, much aggravation of the inflammation, and fresh crops of vesicles have been immediately produced. So susceptible, indeed, is the constitution sometimes found, that the smallest portion of mercury exhibited internally is injurious ; and in one instance recorded by Dr. Crawford,* even the application of ung. hydr. nitr. to a tender part was followed by a return of the disease with its original violence. On the whole, it would seem advisable to refrain from the use of mercury in any form, except under circumstances of the greatest emergency.

There is no evidence in the cases to which I have referred, as to the precise state of system under the influence of mercury, in which the attack has commenced, and it is probable, that in the greater number of instances, the eruption appears where no previous disorder of the system has been effected. That it does often occur simultaneously with catarrhal fever, is unquestionable, but it is a matter of doubt whether the latter may stand in the situation of cause or effect ; for it may be just as reasonable to suppose that the catarrhal symptoms arise from the extension of the cutaneous inflammation constituting the disease along the membranes of the nares, trachea, and bronchiæ, as that these symptoms, with their accompanying fever, produce the eruption. In a case lately under my notice, the subject of which was a stout, muscular, healthy man, no trace of disorder existed at the period of the attack, and he went through a disease of two months' duration without the slightest appearance of any of the symptoms alluded to.

Dr. Bateman's plate 57 appears to bear a great similarity to some cases of the affection which I have noticed under the head of Impetigo, and I have reason to think it is the same disease in a more aggravated form than usual. Neglect of cleanliness, or preternatural irritability of skin, are at all times fully adequate to the production of greater variations than between the plate in question and the description referred to.

In the treatment of Eczema Mercuriale, sufficient attention does not appear to have been directed hitherto to the use of the warm bath. In the case described by Dr. Marcet,† it seems to have been efficacious, not only in allaying the irritability of the skin at the

* Ed. Med. and Sur. Journal, vol. xvi.
† Med. Chir. Transactions, vol ii.

time, but in preventing so frequent a return of the disease as the patient had been previously subject to. At any period of the disease, this measure will be productive of much advantage, in lessening the irritation, and relieving the surface of its turgescence, in removing the fluid adhesive secretion, and making the patient infinitely more comfortable. If repeated twice a day, for three or four successive days, it has appeared to lessen the duration of the complaint very materially ; the secretion on the surface being much reduced in extent, and the formation of healthy cuticle being thereby much earlier accomplished.

The occasional use of mild saline aperients, and confinement to an unirritating, but nutritive diet, with softly sponging the most tender parts occasionally with warm water, constitute all that is necessary or useful in the treatment of the disease, beyond the use of the bath. If the irritation is so great as to disturb the patient's rest, the use of opiates may be necessary, and do not appear objectionable. When, at the termination of the disease, the state of system requires tonics, it is not of so much importance what kind may be employed : but if the strength is so much exhausted before the discharge has ceased as to require such medicines, those which are chosen should be of the simplest and most unirritating kind.

The discolouration of the skin which frequently exists in cases of Eczema from mercury for some time after recovery, has occasionally led to mistakes of an unpleasant nature in medical treatment. I have known it in two instances considered and treated as of a syphilitic character, and leading to a renewal of the use of mercury, by which the eruption has been reproduced. It is, however, quite unlike any other cutaneous affection, and the history of the case, if attentively inquired into, will generally be enough to guard us against such an occurrence.*

*The substance of Rayer's compilations and evidence on this subject is as follows. It will be observed that there is the usual quantum of exaggeration and prolixity in the part quoted, but it is still worthy of a place in our attention.

" In this generally very mild variety, *E. simplex*, the skin, covered with vesicles, almost always preserves its natural colour between the clusters. There is neither heat nor tumefaction ; the vesicles, extremely minute, contain a globule of limpid serum, and usually correspond with the minute projections whence the hairs issue, and which may be very distinctly seen by examining the insides of the arms and thighs with attention. When the serum of the vesicles is re-absorbed, the cuticle that concurred in their formation shrivels, and is detached in the shape of a very minute plate or scale. More frequently still, the vesicles, after having existed for several days, burst, or are ruptured by scratching, when the drop of serum escapes and gives place to a yellowish coloured speck which, being before long thrown off, leaves a little pink spot now dry, now moist and surrounded by a whitish circle, upon the skin. When the spot is moist, a very minute pore is perceived, whence a small quantity of serous fluid distils, which drying up forms a scab the size of a pin's head. Occasionally also, layers of the cuticle, altered in their nature and thickened from the adhesion of the dried fluid of the vesicles, are detached from the skin. And it is at this stage that, frequently, and without known cause, a new eruption of vesicles takes place, which follows in every particular the course of the first, when the eczema becomes chronic.

" A variety of eczema *simplex* has been described by one of my pupils, Dr.

Levain, which is not noticed by Willan, nor by any of the other pathological writers who since his time have given particular attention to the diseases of the skin. This variety is distinguished by clustered patches of vesicles, the dimensions of which vary considerably. The vessels of these clusters are numerous, very small, and in all respects similar to those of the other varieties of eczema, consequently they are much more minute than those that characterize herpes *phlyctenodes.* The clusters are scattered over the skin, which only appears red in the places affected. On the red patches covered with vesicles, the cuticle may sometimes be raised and removed in a single piece. Its inner surface looks moist, and covered with small whitish or deep yellow points produced by the fluid of the vesicles. The corion beneath is red, but not ulcerated, the clusters of vesicles here bear a resemblance to those of the herpes *præputialis ;* and the variety of eczema we are describing, seems the link of connexion between these two genera of vesicular eruptions.

" Eczema *simplex* often extends to the whole surface of the body, especially in children, young persons, and subjects of an irritable constitution. Disease in general soon gets well, and relapses are not frequent. The diseases with which it may most readily be confounded are *certain vesicular eruptions artificially produced* by the action of the sun's rays, and lichen *simplex.* To avoid mistakes in regard to the latter, it is enough to remember that the vesicles of the eczema contain serum, whilst the elevations of lichen are solid, and yield a drop of blood when they are punctured.

" The inflammation of the skin is occasionally more intense than in the variety just described, and the disease is then entitled eczema *rubrum.* The part which is about to be affected with this eruption, swells, becomes hot, red and shining as in erythema and erysipelas. It is soon covered with small confluent vesicles, transparent at first, but speedily becoming milky, which burst anon, and pour out a little red-coloured serum. At a later period the cuticle, saturated with this fluid inspissated, becomes softened in some points and detached in others, when it dries into yellowish laminæ of little thickness, which are soon replaced by slight incrustations proceeding from the drying of the fluid poured out by the diseased surfaces. Lastly, the skin here and there presents small pink points, around which the cuticle forms a true border with a jagged edge, indicative of the dimensions of the vesicles.

" When eczema *rubrum* is very intense, the heat, redness, and tension continue, or even increase during several days ; the vesicles are enveloped, and burst with great rapidity ; the fluid they pour out irritates still more the parts already very painful, and by its contact gives rise to excoriations of varying extent. The skin, stripped of its cuticle, and inflamed, appears beset with a multitude of pores, each of which might be covered with the head of a small pin, from whence a red-coloured fluid exudes, sometimes in such profusion as to soak the clothes of the patient.* At other times the small vesicles unite, become blended together, and form irregular bullæ, analogous to those observed in certain cases of erysipelas. The epidermis, detached over a considerable space, bursts at length, a torrent of serum escapes, and the sub-epidermic layer now exposed and greatly swelled, besides the pores that have been already mentioned, presents false membranes of a whitish colour and soft consistency, which adhere slightly to the structures beneath. The serous exudation soon becomes less in quantity and ceases entirely ; the cuticle, moist at first and slightly adherent, becomes of a yellowish or greenish colour, by being soaked in the fluids exuded ; it then dries, falls, and is replaced by other laminated incrustations of a firmer and more permanent description. The skin loses by insensible degrees its tension and increased heat : the redness also declines, and the parts slowly recover their natural condition, the return to this being announced by the formation of a new and healthy cuticle. It frequently happens, however, that fresh eruptions break out, and the eczema *rubrum* becomes chronic.

" Eczema and impetigo have between them many strong points of resemblance, as well in reference to the parts of the body most commonly affected, as perhaps

---

* " The disease in this form is described by French authors under the title of *dartre humide.*"

in regard to the constituent element of the skin, the follicles, in which they are both evolved; it is not, therefore, uncommon to meet in the same individual with impetigo affecting one quarter, and eczema developed in another. It often happens also that we find a mixture of the vesicles of eczema and of the pustules of impetigo covering surfaces of the integuments of greater or less extent, and still more frequently do we find the vesicles of eczema becoming purulent and giving occasion to an anomalous variety of the disease which has been described by Willan, under the title of eczema *impetiginodes*. When this variety makes its attack in an acute form, the tension, heat, and redness are considerable; it is not now mere tingling and itching that are complained of. but shooting and violent smarting pain. The vesicles now pass rapidly into the purulent state; the cuticle, raised in large flaps, is impregnated with the fluid effused, and acquires the appearance of greenish coloured laminated scabs, which being before long detached, a surface is exposed of as bright a red as carmine. When the eruption is considerable, the ichorous fluid secreted is so profuse that dressings of every kind, and even the bed-clothes and bedding, become drenched with it; the smell of this matter too is as offensive as possible; it is faint and sickly, and something like that which a large burned surface in a state of suppuration diffuses. Around these *impetiginous* eczemas we commonly observe a tumid red circle, the surface of which is studded with small vesicles, transparent, milky, or dry, according to their ages, and in all respects analogous to those that characterize eczema *rubrum*. The vesicles and incrustations are occasionally renewed successively, and the disease becomes chronic.

" Eczema *impetiginodes* may last several weeks, be transferred from one place to another, or lastly attack almost the whole surface of the integuments; most commonly, however, it implicates but a single region. When it shows no tendency to pass into the chronic state, all the symptoms decline, the inflammation lessens, the laminated incrustations fall off, the cuticle is reproduced, and the skin, of a violet colour, is unaffected afterwards, save by a slight exfoliation.

" The three acute forms of eczema that have now been described, present shades of extreme variety. Most usually the morbid symptoms do not extend beyond the parts affected, or the structures in the immediate neighbourhood of these. Nevertheless when the eruption is very extensive, it is accompanied by disordered actions of a general nature : the pulse becomes frequent, there is thirst, anorexia, and the sleep is disturbed. The pain is increased by the heat of the bed ; motion at times is impossible or attended with extreme suffering. The most common complications are inflammations of the lymphatic glands in the vicinity of the affected parts, and in some cases, especially in children, inflammatory disorders of the stomach or intestines.

" The three varieties of inflammation of the skin which constitute *acute* eczema may occur with the *chronic* character; this is even, it must be noted, the tendency of the varieties entitled *rubrum* and *impetiginodes*. After the bursting of the vesicles the inflammation often increases in severity, extends to the deeper structures of the skin, and even to the subcutaneous cellular tissue. Irritated by repeated eruptions of vesicles, and by the contact of an acrid ichor, the skin becomes excoriated, and presents chaps and fissures which every movement tends to make deeper and more extensive, especially if the disease occurs between the fingers, on the nipples, verge of the anus, or in the popliteal regions. In the greater number of cases the affected districts of skin present at first the appearance of a blister in a state of suppuration, and pour out a purulent serous fluid, of a disagreeable smell, which quickly penetrates any dressing that may be applied. These humid eczemas occasion intense pruritus, accompanied by severe smarting pains : the skin, highly inflamed, is stained with blood, looks of a violet colour, and seems beset with an infinity of minute pores, from which a sort of serous dew distils. Tormented by pruritus of the most violent description, patients thus affected talk of nothing but *heat of blood, inward fires*, &c. They cannot forget themselves in sleep ; their sufferings, lulled for an instant, often return suddenly and without appreciable cause; nothing can then prevent or moderate the energy with which they begin to scratch themselves; a bloody serum flows from the torn surface of the skin ; but nothing assuaged, the pruritus continues as unbearable as before, particularly when the perineum, orifice of the vagina, or verge of the anus, are the parts affected; when left to itself, this cruel state often continues for months and even for years.

"When the inflammation declines in severity, *chronic* eczema assumes another character. After the lapse of a longer or shorter interval, the vesicular, or vesiculo-pustular eruptions become rarer, and even end by not appearing at all; the scabs, at first moist and thick, and reproduced as soon as detached, grow thinner and thinner, drier, and more adherent to the skin, which at length appears covered by small yellowish coloured scabs,—the *dartre squameuse ou furfuracie* of some authors—among which, several bloody incrustations, the consequence of the excoriations caused by the nails of the patient, may be detected. The serous exudation is replaced by a simple epidermic exfoliation, to a greater or smaller amount. The more severe these eczemas have been, and the longer they have continued in this state, the longer is their complete disappearance expected, even after amendment has begun, and longer still are certain sequelæ, by which the previous existence of the disease may be certainly recognized, of being completely obliterated. Should a fresh eruption of vesicles chance to appear on surfaces which either have been or are still affected with eczema. the new eruption bursts more quickly than that which is evolved on regions that have never before been attacked; these fresh crops scarcely continue entire above five or six hours in the former of these cases. a circumstance which is undoubtedly owing to the tenuity of the newly formed epidermis. Lastly, it happens occasionally that slight vesicular eruptions are thrown out under the epidermis, thickened and altered by prolonged disease.

"1st. *Eczema of the hairy scalp* (*teigne muqueuse*, Alibert; *Porrigo larvalis*, Willan). Extremely frequent among children at the breast, of three, five, and eight months old, and at the period of the second teething, it not uncommonly attacks young persons of both sexes. especially such as have fair hair and a fine and delicate complexion, with a scrofulous taint and lymphatic constitution. This variety, which has been separated erroneously from the group of eczemas, and variously placed among the *tineas* and *porrigos*, appears at one time on a portion, and at another invades the whole surface of the scalp, extending occasionally even to the ears, nape of the neck. forehead. and face. In very young infants the vesicles of this eczema spread over the scalp and temples, and soon become covered with thin scabs that increase in thickness as the exudation continues. The swollen hairy scalp. indeed, pours out a profusion of a viscid fluid which glues the hair into masses or layers, and in drying, forms yellow or brown lamellar incrustations. In this acute state the head is hot, and the scalp appears injected and tense; children are then tormented with a pruritus of the affected parts, the violence of which cannot be expressed by words, and which seems to gain in intensity when their heads are uncovered, and exposed to the air; they rub them violently upon their shoulders. and if their hands be at liberty they scratch themselves with the greatest imaginable eagerness, though the blood constantly follows the nail.

"When the hair has been cropped with care, and the scabs got rid of by means of emollient poultices, the hairy scalp appears to be covered with a sort of cheesy matter. Occasionally the inflammation extends to the subcutaneous cellular membrane, which forms small prominent tumours attended with very severe pain, usually ending in suppuration. The lymphatic glands of the nucha and parotid regions swell and grow painful. In some cases the vesicles of the eczema are mixed with the pustules of impetigo, and the incrustations formed are then much thicker and more adherent than wont. An immense quantity of pediculi usually appear on the scalp at the same time.

"Eczema of the hairy scalp often extends to the forehead, temples, face, nape of the neck, and shoulders.

"If the children attacked with this disease be carefully attended to, if the incrustations be gently removed by means of lotions and soothing cataplasms, the inflammation of the scalp declines, and the exudation from its surface usually ceases within one or two months at farthest. If these measures be neglected, the caps and other articles applied to the head, become impregnated with the fluid secreted by the inflamed surface, and increase the pruritus ; the inflammation be comes chronic, and extends more deeply ; the bulbs of the hair inflame, and often cease from their functions over a considerable extent of surface ; the scalp at the same time assumes a furfuraceous appearance on some of the inflamed points (teigne *furfuracée*, Alibert).

" When the ichorous exudation ceases suddenly, either naturally, or in conse-quence of ill-timed medication, and the incrustations grow hard and friable, children become dejected, taciturn, restless, and evidently unwell. On the other hand, when the exudation is very abundant, the principal functions are frequently performed with the most perfect regularity, and the health of the little patients seem occasionally even to improve during the whole period that the disease con-tinues. I shall add farther, that those children who labour under eczema of the face and hairy scalp whilst they are teething, rarely suffer from convulsions or ob-stinate diarrhœas. This remark is in accordance with the result of M. Billard's observations, who tell us that at the Foundling Hospital of Paris he noticed a con-siderable number of infants at the breast attacked with eczema of the scalp (*teigne muqueuse*), who after the slow and natural cure of the disease, were re-markable for the freshness of their colour, and their excellent state of health. Among adults, *chronic* eczema of the scalp seizes particularly on individuals of a lymphatic and scrofulous habit of body. Women at the critical period of their lives are more frequently attacked than men. The greater number of these cases of eczema, *humid* and *secreting* at first, become at a later period *squamous* and *furfuraceous ;* the swelling, redness, and heat of surface are then almost wanting, and the scalp, freed from the squammæ that covered it, appears somewhat red and shining. The squamæ are occasionally of a silvery and pearly lustre, and very much resemble the pellicles that envelope the sprouting feathers of young birds. Occasionally tufts of five or six hairs are bound together, as it were, at a short distance from their roots and free ends by these squamæ. In this state the disease is not accompanied by any great degree of itchiness, and the head has no particular odour.

" Chronic eczema of the scalp now and then spreads to the ears and eyebrows ; it also occasionally attacks the margins of the eyelids, causing the fall of the eye-lashes, and a chronic ophthalmia of a very intractable description.

" Eczema of the hairy scalp is a disease that is very rarely met with among the aged, in consequence probably of the alterations that have taken place in the organization of the skin. I have oftener than once seen the disease coincide with a more than usually copious secretion of cerumen.

" 2d. *Eczema of the face.* In young children eczema of the face often accom-panies that of the hairy scalp and of the ears ; many authors have described the affection under the title of *crusta lactea.* It usually appears on the forehead, the cheeks, and the chin ; the small vesicles that characterize it are arranged in irregular clusters, and scarcely rise above the level of the skin, which soon assumes an erythematous blush ; within four or five days these vesicles burst, and pour out a viscid and yellowish fluid, which concretes and turns into thin yellowish-green coloured scabs ; fresh vesicles are before long developed around the circumference or in the immediate neighbourhood of these clusters ; the fluid they contain is shed on the surface of the skin, at the same time that a consider-able exudation takes place below the first formed squamæ or scabs, which adds farther to their thickness and extent. If this disease be left to itself many eruptions occur one after another, until the whole countenance is covered with yellowish laminated incrustations. The serous or sero-purulent fluid secreted, is often very copiously shed under the laminæ and scabs ; the skin is of a vivid red, and appears beset with a multitude of minute pores which are covered with slight false membranes of a milky white colour ; it becomes chapped and excoriated on the cheeks, at the angles of the mouth, and in the furrow between the lips, and the chin, and these tender places are all made worse by the act of sucking, crying, and the contact of the tears ; the disease in this state has all the dis-tinguishing features of eczema *impetiginodes.* At a still more advanced period eczema of the face presents all the characters of a chronic inflammation ; the vesicles are few, the discharge becomes smaller in quantity, ceases at length, and the skin is endued with dry and greyish-coloured scabs, which are thrown off without being reproduced; the diseased surfaces, covered with a cuticle of extreme tenuity, continue long to show an erythematous blush, and are then affected with a furfuraceous desquamation, which also ceases at last. The excoriation and fissures which attend eczema of the face never leave cicatrices behind them. Those that do occasionally remain are effects of the wounds which children inflict on themselves with their nails; to prevent them from doing mischief, therefore, it is proper to confine their hands during the night, for without this precaution I have seen many who made their faces bleed with scratching.

"Eczema of the face sometimes spreads to the margins of the eyelids, to the mucous membrane of the mouth and nasal fossæ, and to the conjunctivæ; the epithelium, where it exists, is rapidly destroyed, and is replaced by small patches of whitish false membrane. When the eczema of the eyelids extends to the conjunctivæ we have all the symptoms of acute ophthalmia: the eyes become red and injected, watery, and sensible to light; the free edges of the eyelids are swollen and œdematous. When eczema occurs in the nasal fossæ, it causes a very troublesome sense of itching, and a very copious flow of a serous fluid. This disease seldom attacks the mouth; I have seen it confined to the under lip, round which it formed a kind of ring; it happens occasionally that the mucous membrane of the mouth, which appears to be generally red and swollen, presents here and there small superficial ulcers like aphthæ, and children then excrete a great quantity of saliva.

"In adults, eczema *rubrum* and *impetiginodes* of the face is often attended by a general swelling of the features, and an œdema of the eyelids similar to that which is observed in phlegmonous erysipelas. Eczema of the face differs from this form of inflammation by being essentially a disease of long continuance, and by the skin, instead of exhibiting a simple exanthematous inflammation complicated with occasional phlyctenæ, presenting a vesicular, or a vesiculo-pustular eruption, generally accompanied with a severe itchy heat of the surfaces affected. When eczema of the face has passed to the chronic state, the serous exudation is almost insensible, the face becomes covered with bran-like scales, which fall off, and are renewed repeatedly; the eyebrows and eyelids now and then lose their hairs. I have seen this variety, which is very intractable, more especially in young girls of a lymphatic temperament, whose menstrual discharge is irregular, or in whom this evacuation is not yet established. It is very seldom seen among the aged.

"3d. Eczema *Aurium*. This is one of the varieties most commonly met with in the two sexes at all ages. It often occurs to women at the period of the menstrual cessation. Lorry has specified its characters accurately (De auribus suppurantibus). Infants are attacked with the disease at a very early age; and it frequently coincides with the eczema of the scalp, or face, the development of which it occasionally precedes or follows. It must not be confounded with intertrigo, a kind of erythema of the posterior parts of the ears, attended with chapping, and some slight exudation. I have also seen many cases of this affection among young women from fifteen to twenty years of age, who either had not yet menstruated, or who had menstruated very irregularly. When this eczema appears with the acute character, the ears become red and swelled, to such a degree, that their size is often doubled; a reddish fluid flows rapidly from the vesicles, chaps and fissures are formed, and the inflammation is propagated to the meatus auditorius, around which small purulent abscesses are occasionally formed that prove excessively painful. The sense of hearing is either perverted or lost for a time; the lymphatic glands in the vicinity swell. This eczema most usually becomes chronic; the skin is covered with lamellæ of a deep yellow colour, very similar to cracked strata of yellow bees'-wax; a reddish fluid, the flow of which is increased by pressure, exudes from the fissures. Often, when the diseased parts appear to be returning to their natural state, all at once and without known cause, a new eruption appears.

"Eczema of the ears is usually a very obstinate disease when it attacks females at the critical age; it gets well, on the contrary, readily and naturally in children, when it has broken out during the process of teething. Pieces of sponge or tents of lint have by some been recommended to be placed in the meatus auditorius to prevent the contraction of this passage; but the precaution has more inconveniences than advantages. When the eczema is acute in its character, more good is done by blood-letting, the application of leeches, and the use of pediluvia and aperients; and when the disease is of a chronic nature, the swelling of the subcutaneous cellular membrane seldom goes so far as to make any precaution of the kind mentioned necessary. It rarely happens that eczema of the ears does not extend to the parotid regions and hairy scalp; the two ears are also most commonly affected at the same time; but the disease seldom attains the same degree of severity in both.

"4th. Eczema *Mamillarum*. Eczema of the nipple is a much more uncommon disease than the varieties I have hitherto described; I have never met with it in young children. M. Levain collected several cases of its occurrence in young women who were nursing for the first time. It is of consequence not to confound this affection with erythema and chapping of the nipples, which are much more frequent complaints with young nurses. It is occasionally observed, especially in the chronic form, in young girls and grown women who have never given suck. The inflammation is at times transferred from one nipple to the other; very violent itching is felt, and a yellowish or reddish serum flows abundantly from the affected parts, and rapidly penetrates any dressing that may be applied. The mucous membrane of the nipple, inflamed over its entire surface, although in an unequal manner, presents small excoriations like linear scratches; some points are of a bright red, moist, studded with sero-sanguinolent drops, others are covered with yellowish scabs, which are thick in the middle, and decrease towards their circumference. This complaint is usually attended with severe pruritus, which increases on the approach and during the continuance of menstruation. The nipples remain scaly for a long time after the inflammation has subsided: they are moist and exuding one day, dry and scaly the next. At length, after many recoveries and as many relapses, the pruritus ceases, the serous exudation appears no more, and the parts are covered with a new epithelium, smooth and uniform like that which covers the healthy surface. I have never observed eczema of the nipple in the male. It is important, as I have already said, to distinguish eczema of the nipple from simple chapping or cracking of the part; and, above all, not to mistake it for a syphilitic affection; it is usually very intractable, may continue for years, and requires active treatment on its first invasion.

"5th. Eczema *umbilicalis*. The skin of the umbilical region bears great affinity to that which surrounds the natural openings. Eczema of the umbilicus consequently very much resembles that of the nipple and vulva. It has been mistaken for a syphilitic blenorrhœa. In newly-born infants, the pulling which the cord has undergone, the ligature of this part and the use of unguents, give rise now and then to the development of minute vesicles and to slight excoriations, distinguished from those of true eczema by their short continuance.

"6th. Eczema *of the insides of the thighs, prepuce, scrotum, verge of the anus, and lower end of the rectum in the male*. These varieties of eczema are all very rare in early life; they are more frequently met with between the thirtieth and fortieth year than at any other age. The eczema may begin in any one of these regions, and then creep on to the others in succession, or else attack them all at the same time. The sleep is broken; tormented by incessant pruritus, the patients become restless and irascible; the vesicles either break or are violently torn at the moment of their formation, so that it is often impossible to find a single one untouched; the skin is bedewed with an ichorous fluid; the patients tear themselves with their nails; crevices are formed from which a sero-sanguinolent exudation takes place; the penis, scrotum and perineum are extensively excoriated; the lint and dressings applied to these parts are speedily soaked with discharge; walking, and the friction of the clothes, the heat of the bed, and occasionally the presence of pediculi, add to the amount of irritation, which is already excessive; erections, and the act of emptying the bladder and rectum, are often attended with pain. In the great majority of cases, this form of eczema becomes chronic. It proves at all times a protracted and obstinate disease, and one in which patients are willing and eager to submit to the most energetic treatment, in order to get rid of their misery : there are cases, however, in which this ought to be employed with great discretion.   One of my patients, who had laboured under a chronic eczema of the margin of the anus for twenty years, became accidentally affected with a very severe inflammation of the gastric and pulmonary mucous membrane, which yielded to a rigid adherence to low diet, the use of asses' milk, mucilaginous diluents, and the insertion of an issue. During the acute and most severe period of this disease, the eczema of the fundament disappeared completely, but broke out again after the cure of the gastropulmonary inflammation.   This interchange of internal and external inflammations is well worthy of engaging the attention of the pathologist and therapeutist.

" Eczema of the scrotum, perineum, &c., and especially the fissures it occasions in the skin, have occasionally been confounded with lichen *agrius* and syphilitic sores.

" 7th. Eczema *of the inner parts of the thighs, of the vulva, of the verge of the anus, and of the mucous membrane of the vagina and rectum in the female.*   Children are seldom attacked with eczema of these regions.   In the adult female the disease may commence in any one of them and be propagated successively to the others ; or it may seize on all of them at once.   Like the disease in the male it begins with heat and intolerable itching ; the vesicles burst as soon as formed ; the pain becomes unbearable, excoriations take place, the disease spreads to the labia majora, to the mucous membrane of the vagina, to the verge of the anus, and to the rectum.   The scalding heat and pruritus are extremely violent; the passage of the urine is painful ; a discharge of an offensive smell takes place from the external organs of generation.   The vagina and inner surface of the labia present slight superficial excoriations ; patients occasionally addict themselves to masturbation with a sort of fury as in prurigo *pudendi ;* the sexual act is either impossible or exceedingly painful.

" This variety of eczema has been sometimes taken for a syphilitic affection ; and it is often difficult, when it is accompanied with a leucorrhœal discharge, to determine whether this flux be the cause or the effect of the vesicular eruption.   Discharges from the vagina, however, give rise much more frequently to intertrigos than to true eczemas.

" 8th. Eczema *of the upper and lower extremities.*   Eczema of the fore-arm, of the arm, and of the thigh, present nothing peculiar ; those of the legs, in elderly persons, have been described under the title of *tettery sores (dartres ulcereuses).* They usually begin in a chronic form, and are occasionally accompanied with varices of the veins and with ulcers.   Eczema of the legs often presents the characters of eczema *rubrum.*   The skin, of a livid hue, tense, not very hot, and sprinkled with numerous pores that pour out an ichorous reddish fluid, presents excoriations of a bright red, the surface of which is dotted over with points of a deeper shade of the same hue ; other parts present yellowish laminated incrustations, fissures, or extensive excoriations.   The vesicles are very rarely to be seen entire.   The eruption is occasionally propagated to the dorsal aspect of the feet, to the toes and the integuments between them, and then the same phenomena are observed as when the disease attacks the hands and fingers.   Farther, it is requisite to distinguish a primary form of eczema followed by sores, from those vesicular eruptions that are produced by the contact of discharges from ulcers of an older date.   The cure of these eczemas is accomplished with difficulty; and even when the excoriated places are healed up, when the serous exudation no longer takes place, and the fall of the scabs has been completed, an epidermic exfoliation, and scaly state of the skin continue for a long period ; the skin too retains a reddish, livid and shining appearance, and the slightest irritation brings back the disease in greater severity than it possessed even on the first invasion.

" 9th. Eczema *of the bend of the arm, of the axillæ and hams.*   These varieties, in their evolution and in their progress, have a great resemblance to those that appear about the margin of the anus and vicinity of the genital organs in either sex ; they are, however, much less painful.   Those of the axillæ are the most uncommon, and have frequently the appearance of the eczema *impetiginodes.*   In these regions the heat is usually considerable ; they are in an habitual state of moisture; the follicles are numerous, and the motion of the parts incessant: hence the violence of the pruritus, the copious discharge of serous fluids, and the occurrence of excoriations and chaps so difficult to cure. It is important to distinguish these varieties from confluent lichens.

"10th. Eczema *of the hands*. Eczema *simplex* occasionally appears between the fingers, on the backs of the hands, and on the anterior parts of the wrists. I have seen the vesicles of this disease as large and as pointed as those of scabies, so that it was matter of difficulty to distinguish between them. It is true that one of these affections is contagious and the other is not; but the experiment that must be made to establish this diversity of character cannot of course be recommended to be instituted on purpose. Eczema *rubrum* very frequently occurs on the backs of the hands and fingers: occasionally between the fingers and around the nails; the vesicles being extremely crowded in these cases may give rise to the formation of bullæ of various dimensions. When confined to the circumference of the nails, as I have several times seen the disease, it stimulates onychia very closely; and when it gets chronic in its nature, the skin on the backs of the hands becomes covered with large thick scabs of a yellow or brown colour, and the spaces between the fingers present deep fissures, the bottoms of which are highly inflamed, whilst their edges are covered with laminated scabs. From these fissures a sero-purulent fluid is continually exuding, especially when the parts are bent or used in any way. When the hand is passed over the diseased surfaces they appear as rugged as the bark of an aged oak. Long after the disease is to all appearance cured, the skin remains hard, dry, and scaly, and is very liable to chap and split.

"In *general* eczema, after several successive crops of eruption have been evolved, the inflammation occasionally spreads to the palms of the hands. The fluids effused are then detained under the thickened cuticle, and not shed upon the surface, a circumstance that modifies the appearance of the eruption notably.

"*Complications.* The numerous cases which I have watched myself through all their stages, prove that eczema may co-exist with the greater number of the diseases that attack the external integuments. Pustules of impetigo are very commonly seen in the middle or near vicinity of a district attacked with eczema: this is even the most natural as it is the most frequent complication. I have seen eczema coincident with lepra. The vesicles were disseminated in the intervals between the scaly patches, and got well, whilst the leprosy continued unchanged; as if these different affections had severally depended on some peculiar and distinct condition of the system. I have seen a young woman labouring under very extensive psoriasis *guttata*, and a *moist* eczema of the face and ears at the same time. I have observed eczema co-existing with scaly syphilitic affections. The bullæ of rupia, the pustules of ecthyma, and boils are occasionally observed in individuals labouring under chronic eczema of different districts of the body. Certain forms of eczema of the sexual organs and hairy scalp are attended with the production of pediculi which incessantly excite the patients to scratch themselves. Eczema is occasionally observed to precede scabies; at other times it seems to be induced by the unguents employed in the treatment of this disease.

"Among children, inflammatory affections of the mucous membranes frequently coincide or alternate with the development of this eruption. When it is general, and the secretion is abundant and continues long, cough and diarrhœa are apt to supervene, especially among the aged, and individuals weakened or worn out by previous disease. I have seen eczema of the scalp, ears, and face, spread to the mucous membranes of the eyes, meatus auditorius, and nostrils, causing intense ophthalmia, severe otitis, and chronic coryza, followed by profuse and fetid discharges. Pregnancy may cause the development of eczema, and complicate or stand in the way of its proper treatment; occasionally also the disease has appeared in nurses as a consequence of weaning. Among children it now and then appears on the mucous membrane of the mouth. Eczema of the legs in old people is rather frequently accompanied with the appearance of petechiæ around the inflamed parts of the skin, also with œdema, varicose tumours, and ulcers which protract or impede its cure. I have seldom seen eczema occur among the consumptive; on the other hand it frequently coincides with gastric, enteritic, and bronchial inflammations, especially among children, in whom it is often replaced by convulsions. The practical inferences to be deduced from these observations are, that the treatment of eczema is to be modified according to the ages and the idiosyncrasis of individuals, and above all, according to the nature and intensity of its different complications.

"*Alterations of structure.* The follicles of the skin are the parts essentially affected in eczema. And we observe, in fact, that if the disease may and does attack almost every district of the integuments, it shows itself in preference on the inside of the thigh, at the bend of the arm, in the hams, the axillæ, the groin, on the scrotum, labia, margin of the anus, and generally in those situations in which the follicles are the most largely developed, and the most numerously bestowed. The disease is very frequent in the scalp among children, and there, in early life, the follicles are very numerous and very large; in this situation, on the contrary, it is rarely met with among the aged. Further, the disease seldom appears in districts where the existence of follicles is matter of doubt, as in the palms of the hands and soles of the feet, the neighbourhood of the patellæ and olecrana. And to conclude, in eczema *simplex* the affection of the follicles is evident to the eye. In eczema *simplex* the papillary body does not appear to be injected; it is so however in the variety entitled *rubrum*, in which the deep layers of the corion and the subcutaneous cellular tissue are also occasionally inflamed. When excoriated, the skin may become covered with pseudo-membranous deposits, analogous to those which are observed on blistered surfaces. I have even seen the skin of the leg, in old subjects labouring under eczema rubrum, presenting small holes which might have been filled with the head of a pin, and little sinuous ulcers, very irregular in their course, which did not affect the whole thickness of the corion; the skin had an appearance very similar to that which a piece of mahogany board worm-eaten on its surface might be supposed to present. In chronic impetiginous eczema I have noticed small nipple-like projections on the surface of the skin, which were owing to a sort of elongation of the natural papillæ. The cuticle itself undergoes various alterations in this disease. In the chronic forms it is resolved into a kind of farinaceous dust, or is thrown off in little laminæ, the centres of which, of a yellowish grey colour, are more adherent than their circumferences. In certain eczemas of the hands the cuticle becomes dry and falls off in large yellow scales which are thick in proportion as they have been infiltrated with diseased secretions. Lastly, the cuticle is entirely destroyed in certain cases, and the corion is either exposed or is covered by incrustations of various thickness. The fluid secreted by the diseased surfaces, serous, limpid and almost inodorous in one case, is turbid, yellowish or greenish, and more or less consistent in another; it has a faint and sickly smell in eczema *impetiginodes*, especially when pustules of impetigo occur mingled with the vesicles of the eczema. In drying, this secretion gives rise to the formation of scabs which have some resemblance to those of impetigo. When the inflammation runs high, it may implicate the bulbs of the hair and the matrix of the nails, and cause these appendages of the skin to be detached. In brief, the primary seat of eczema is in the follicles of the skin; but other elements of this tissue are affected in eczema *rubrum* and *impetiginodes*. The papillæ, the entire thickness of the dermis, the subcutaneous cellular membrane, and the lymphatic glands are occasionally effected one after the other : hence those small abscesses that form in children who are attacked with eczema of the head, and those painful inflammatory swellings of the glands which occasionally end in suppuration. Other diseases of the skin (acne, rosacea, mentagra, and impetigo), also affect, and indeed more evidently affect the follicles than eczema. There are consequently essential differences between the various diseases of the skin, which must be sought for elsewhere than in the immediate affection of the several anatomical elements of which this tissue is composed.

"*Causes.* I have met with several cases of eczema that appeared exclusively during the period of uterine gestation, that yielded with this state, that returned in a second and a third pregnancy, and got well immediately after delivery, although various plans of treatment had been vainly tried before this event.

Among children the process of teething and the quality of the nurse's milk; among women the states of amenorrhœa and dysmenorrhœa frequently exert a notable influence upon the development of eczema.

" Our inability in a multitude of cases to discover any evident or even probable cause of the disease, often leads us forcibly to the conclusion that eczema is most generally evolved and kept up by some hidden alteration of the fluids and solids. In this disease, as in almost all inflammatory affections, independent of external causes, the blood is buffy.

" If a patient who has recovered from one attack of eczema is seized a second time, it is almost certain that the disease will be of the same genus and species as before.

" In infancy and youth, eczema appears more particularly on the head ; in riper years on the breast and belly, but especially on the genital organs ; and in advanced life on the lower extremities, and about the margin of the anus. According to Billard, eczema is also frequently observed on the trunk and extremities of the newly-born infant ; he tells us he has seen the disease in infants scarcely a day old. M. Levain having delivered a woman labouring under impetiginous eczema of the whole body, observed that the child two days afterwards presented several clusters of vesicles on the left forearm, on the neck, and in the hams, and by-and-bye eczema *rubrum* appeared on the forehead and hairy scalp. It is at the periods of the first and second dentition that children are more especially attacked with eczema. Women are more subject to the disease than men, particularly at the critical period of life. Chronic eczemas of the hairy scalp, ears, eyebrows, and eyelids are common in scrofulous subjects. Eczema is not a contagious disease, but under certain circumstances, especially when the discharge is copious, the contact of the secretion may determine vesicular eruptions upon healthy parts. M. Levain has seen acute eczema of the labia in a woman whose husband had long laboured under a similar eruption of the scrotum ; the woman, in this case, had a rapid recovery. I have collected several cases of the same kind.

" Relapses are observed to happen in almost all diseases, but in none are they more frequent than in eczema. I have collected many instances of unlooked-for returns of the complaint in consequence of variations of atmospheric temperature, errors of diet, affections of the mind, &c. I have seen the disease get well and recur twelve or fourteen times within the space of a few months, in despite of the most scrupulous and undeviating attention to regimen. These relapses happen particularly among individuals of an irritable and nervous constitution.

" *Diagnosis.* Eczema is, of all the diseases of the skin, that which presents the greatest variety in its appearance, for as it is acute or chronic, simple and unmixed, or complicated with pustules, it may be characterized by vesicles with or without redness, by moist or running excoriations, by laminated incrustations, or by small furfuraceous scales.

" Willan and Bateman have both connected to the group of eczemas several artificial or accidental vesicular eruptions, which both in their nature and means of cure are totally distinct. Such is the *sun-fret* (eczema *solare* of these authors), which is observed among labourers in the time of harvest, and the inhabitants of towns who flock into the country during the fine season of the year. The skin of the face, neck, hands, and other parts usually left uncovered, is attacked with erythema, to which succeeds a plentiful eruption of vesicles, analogous indeed in appearance to those of eczema, but speedily ending in a slight furfuraceous desquamation of the cuticle.

" Some preparations of mercury excite a vesicular eruption which has been connected with eczema, and which resembles this disease considerably in its external characters ; but in its nature and progress it bears a much nearer affinity to inflammations artificially produced. Various other substances, such as Burgundy pitch, adhesive strap, plasters of cicuta and opium, the juices of certain plants of the family of euphorbiaceæ, the oil of the croton tiglium, sulphurious lotions, diluted acids, &c. have also the property of causing minute vesicles to be developed on the skin, which in their form, arrangement, and size, approach more or less nearly to the eczemas. All these vesicular inflammations, however, differ essentially from this group in their nature, and get well with an ease and a rapidity that make a strong contrast with the inveteracy and tendency to return of eczema.

" The vesicles of herpes are globular, and arranged in clusters surrounded by an inflammatory areola, of much greater extent than that of eczema. The minute serous vesicles that are occasionally observed along with rheumatic affections, puerperal peritonitis, the furuncular disease of the bowels (dothinenteritis), &c. are devoid of inflammatory characters, and are very different from the vesicles of eczema.

" When eczema *simplex* appears between the fingers, on the wrists, bends of the arms, hands, and fore parts of the belly, it is at times difficult to distinguish it from scabies : it was a mistake to say that the vesicles of eczema were always flat and agglomerated ; I have seen them as large as those of scabies, dispersed and separate like them, and like them, too, a little pointed ; but scabies is essentially contagious, eczema is not; this is almost always acute, that as constantly chronic in its nature; the pruritus of eczema is accompanied with a kind of smarting, that of scabies is rather a pleasant than a painful sensation. The small red solid and itchy papulæ of lichen do not contain any serum like the vesicles of eczema *simplex*. In confluent and inflamed lichen (lichen *agrius*, Willan,) when the papulæ, crowded together in large patches, have been torn by the nails, the skin, in a raw and bleeding state, pours out a sero-sanguinolent fluid, which in drying assumes an appearance intermediate between squamæ and scabs, that might readily cause the disease to be mistaken for eczema *rubrum ;* this advanced and very severe form of lichen has consequently been approximated by M. Alibert to the excoriations of eczema, and mixed up with his description of the *Dartre squameuse humide.* The small psydracious pustules of impetigo from their very first appearance contain a thick greenish-yellow fluid. Eczema *impetiginodes* in its commencement either presents transparent vesicles which become rapidly purulent, or, more rarely, a mixture of the proper vesicles of eczema and of the pustules of impetigo. The scabs of eczema *impetiginodes* are not so thick, but drier, and more compact than those that follow the rupture of the pustules of impetigo, the scabs of which are of a greenish-yellow colour, rough and uneven aspect, and not very unlike the masses of gum that concrete upon the cherry-tree. Eczema of the labia and vagina causes a discharge that might be mistaken for a gonorrhœa; it is rare, however, that a few untouched vesicles may not be discovered in the neighbourhood of the affected parts to serve as guides in the diagnosis.

" The absence of pruritus in syphilitic affections distinguishes these sufficiently from the eczemas of the sexual organs, in which the itching is almost insufferable. It is occasionally difficult to discriminate between chronic eczemas grown scaly, and old lichens and prurigos, situated on the genital organs. Lorry appears to me indeed to have confounded these three obstinate 'forms of cutaneous affection in his description of *prurigo pudendi :* nevertheless, before falling into the chronic state, eczema of the genital organs is attended with a copious exudation which is never observed in the other diseases just mentioned. Eczema of the scalp in the state of desquamation is not always easily distinguished from psoriasis and pityriasis *capitis :* it is, however, seldom that no remains of yellowish incrustations and scabs, characteristic of this disease, are to be seen on any part of the scalp, or ears, or person at large, a circumstance never observed in pityriasis, a disease essentially furfuraceous, and one that is never accompanied with a discharge.

" In children eczema of the scalp and face is often a salutary eruption. When it appears during the process of teething, it will hardly yield to treatment until the teeth have appeared. In young women whose menstrual function is irregular, eczema of the ears and scalp is an intractable malady, and seldom gives way before some favourable change is effected in the state of the general health. The disease is always subdued with difficulty in women arrived at the critical period of life ; and when it appears during pregnancy it can rarely be subdued until after delivery. When eczema is hereditary the cure of the disease is very frequently followed by a relapse. Eczema of the hands among cooks, hat-makers, dyers, &c., is always difficult of cure. In elderly persons eczema of the legs, attended with œdema, and a varicose state of the veins, is often altogether incurable. Eczema resists curative means so much the more obstinately as it occupies a more extensive surface, as it is of older date, of a severer kind, and as it appears on the lower extremities or on the hairy scalp. When children and the aged are the subjects of eczema, it often proves a disease *which it is dangerous to cure.*

" *Treatment.* In calling to mind the influence which teething, amenorrhœa, dysmenorrhœa and pregnancy exercise on the production of certain eczemas, we are naturally led to the indications of cure. Some varieties of the disease get well in time under the influence of a regulated diet: others require means of greater potency; and there are a few which are either intractable, or which it would be dangerous to interfere with.

" A considerable number of cures attributed to the use of medicines of little power, ought to be almost entirely ascribed to the influence of *regimen, rest,* and *time,* which are frequently very powerful, especially among the poorer classes leading laborious lives, who are mostly admitted as patients into public hospitals.

" I have seen children at the breast attacked with eczema of the hairy scalp who recovered by changing the nurse. I have seen many persons of mature years labouring under chronic eczema of the scrotum, verge of the anus, and other parts, whose malady was constantly aggravated by the slightest irregularity in point of diet. It would be idle to ascribe too great an influence to the farrago of *cooling* and *cleansing* decoctions and drinks so constantly recommended in cutaneous affections, and especially in eczema; but, on the other hand, it is perhaps not going too far to affirm, that of late these means, or rather the diatetic plans of which they formed a part, have been too much neglected. In no case, however, would it be reasonable to restrict individuals, otherwise in perfect health, to a lowering system of diet; their constitution might suffer from such a course.

" I have met with several cases of eczema in which a vast variety of therapeutic agents had been fruitlessly employed, whilst the patients went on with their usual avocations, and took active exercise, but which were successfully attacked by the same remedies from the moment that these individuals consented *to lay themselves up.* Time also modifies eczema at length, and occasionally accomplishes its cure; so that individuals labouring under the disease in a chronic form, have now and then got well without recurring to any medical treatment whatever.

" The simple or emollient cold or tepid bath is frequently of the greatest service in the different acute species of eczema, even when the affected parts are not immersed in the water. In the decline of these diseases, when stiffness and dryness of the skin are alone complained of, and in the chronic eczematous affections of the backs of the hands, fingers, &c., the vapour bath, and better still, the steam douche to these parts, is found to be useful. When eczema has passed into the squamous state, the warm sea and alkaline bath are efficient in freeing the skin from the layers of epidermis accumulated on its surface; but they almost always increase the redness, and the squamæ are rapidly reproduced. Local baths, repeated several times a day, and fomentations of decoction of linseed, marshmallow flowers, poppy-heads, and milk, are often of advantage in eczema of the genital organs; in these cases, tepid hip-bath twice every day, always gives the greatest relief. If the disease have extended to the mucous membrane of the vulva, injections of althea root decoction, with or without the addition of a little acetate of lead, are generally found to be soothing. Sulphureous baths have also been tried in the advanced stages of eczema, especially when the aged and enfeebled were the subjects of the disease. These occasionally cause new eruptions, and have been found effectual in restoring eczema, the disappearance of which, either spontaneously or obtained by art, has been followed by unpleasant or serious symptoms. The waters of the baths of Louësche have been frequently recommended with effect for this purpose. They have also occasionally seemed to make old standing eczemas run their course more rapidly, and thus to hasten their cure. I have seldom found artificial sulphureous baths produce such good effects, except in the chronic eczemas of elderly people, and some middle-aged persons, when they now and then seemed to lessen the redness and discharge from the skin, after having exasperated these morbid states for a time; scrofulous subjects alone received invariable advantage from the use of these baths. I have occasionally seen beneficial effects from the use of a sulphureo-alkaline ointment. Sulphur exhibited internally has never seemed to me to exert any appreciable influence on chronic eczema, except when it acted as a purgative.

"In running eczemas of small extent, emollient fomentations have been found of service. When the disease is followed by painful and extensive excoriations, and the skin is red and swollen, or covered by yellowish-looking scabs of considerable thickness, soothing-washes must be replaced by poultices of floury potatoes, of ground rice, and of crumb of bread, softened still farther with decoctions of althea and poppy-heads. These cataplasms are greatly preferable to such as are prepared of linseed meal, which are observed occasionally to induce artificial vesicular and even pustular eruptions. When the parts of the skin affected are covered with hair, these various topical applications must be used folded up in a fine muslin rag.

"When poultices are employed in the treatment of children labouring under eczema of the hairy scalp and of the face, care must be taken to keep the head well covered, especially during the first few days, a precaution without which otitis and ophthalmia of greater or less severity are extremely apt to supervene. *Depilation* is an absurd and cruel practice during the acute period of eczema of the hairy scalp ; neither is it ever to be recommended even when the inflammation has passed into the chronic state.

"Moderate *compression* by means of a roller properly applied, is often employed with advantage in the cases of elderly persons affected with eczema *rubrum* of the lower extremities, when the disease is complicated with œdema or a varicose state of the veins, or when patients are obliged habitually to keep the erect posture without much motion.

"Slight *escharotics*, solutions of the *nitrate of silver, diluted muriatic acid, &c.* have been recommended with a view to change the actions of the skin, when eczema has passed into the scaly state, and has continued for several months or years. With like intentions ointments of the *red precipitate*, of the lesser *celandine, clematis, spurge, &c.*, and even *blisters*, have been applied to the whole of the affected surface of the skin. If the cure of circumscribed chronic eczemas have occasionally been obtained by such means, these diseases have also been frequently and seriously aggravated by them. In general, when eczema has passed into the squamous or furfuraceous state, *soothing* ointments are the best local applications ; the good effects of the ointments of oxide of zinc, and of the protochloride of mercury are certainly due in great part to the hog's-lard with which these substances are then largely incorporated.

"M. Alibert has given the details of a case in which the disappearance of an eczematous eruption (*dartre squameuse*) was followed by insanity. I am in the habit of employing issues in obstinate eczema of the hairy scalp and genital organs ; and when we have succeeded in bringing about or are anxious to accomplish the cure of the eczemas of elderly persons which have existed long, or of individuals who have previously suffered from a chronic affection of the viscera, it is advisable to institute and keep open an issue or a blister in one of the arms. If the theory of counter-irritation still requires clearing up, it is enough that by the practice we can relieve the mind of the patient, especially from doubts and fears, and therefore it ought not to be neglected.

"I have said that eczema of the scalp, face, and ears, was occasionally a salutary disease in childhood. It is therefore proper to inquire in the first instance, whether it might not be dangerous to attempt its cure ; the best and safest termination being that which is accomplished naturally. Facts enough prove the danger of discussing these eruptions. On the other hand, inflammatory affections of the eyes, ears, and viscera have been seen to disappear on the eruption of certain eczemas ; in such cases it would be improper to attempt the cure of the cutaneous affection in any other than the slowest and most gradual manner. These remarks also apply to other ages, when the disease appears under similar circumstances.

"The *vegetable acids*, diluted with water ; sherbets of the *sulphuric* and *muriatic acid*, with or without the addition of gum ; or milk mixed with barley-water or gruel, for those individuals with whom acid drinks do not agree, are usually recommended in the treatment of acute eczema. Such means are of less avail when the disease is chronic in its nature. When eczema is acute, and the pruritus is exceedingly troublesome, and the inflammation runs high, as in the eczema *rubrum* or *impetiginodes*, it may be necessary to abstract blood once or oftener. I have had occasion to prove the utility of bleeding in a great number of cases, even of chronic eczema. When one bleeding has been followed by a notable improvement in the symptoms, it is commonly an inducement to repeat the operation after a few days have elapsed. I frequently make such experimental bleedings in the treatment of diseases of the skin. Cases of eczema, however, frequently occur, which resist this powerful means, or which even continue to advance in spite of its employment. It is, therefore, difficult to lay down precise rules for the management of every case, or to specify those in which blood-letting will be found serviceable or detrimental. It is almost always hurtful to individuals of an irritable constitution and spare habit, and in whom the cutaneous affection has supervened or been increased after some violent affection of the nervous system. Hereditary eczema is usually a very obstinate complaint, and we must beware of persisting in endeavours to effect its cure by means of blood-letting. In adults and individuals of mature years, the *general* is constantly to be preferred to the *local* abstraction of blood. The last is the only form of blood-letting that can be practised in regard to young children. In eczema of the face and scalp, of the pudenda and margin of the anus, a number of leeches are often applied with good effect in the neighbourhood of the affected parts. The aged bear bleeding badly; yet the measure occasionally becomes necessary when the parts implicated are severely excoriated and discharge abundantly, and when the disease is accompanied with violent pain and sleepless nights.

"In chronic forms of eczema, especially affecting the scalp and face, certain waters are valuable; the sulphates of soda and magnesia, or the tartrate of potash administered so as to procure one or two liquid evacuations daily without inducing colic, or even exhibited in cathartic doses twice a week during two or three months, are found to be useful when they only cause a temporary state of irritation in the digestive passages, without exerting any lasting ill effect on the state of the general health. The use of these remedies must be immediately suspended when pain, a continual feeling of uneasiness, and febrile symptoms, give us reason to fear the excitement of inflammation in the stomach or alimentary canal.

"Purgatives are constantly had recourse to in the treatment of the eczemas of childhood. They are injurious to women who are pregnant or nursing. Individuals of a nervous temperament, and habitually confined in their bowels, always derive benefit from this class of medicines. Calomel alone, or in combination with jalap, is a good form of purgative when prescribed in frequent small doses, however, its use is almost constantly followed by painful inflammation of the mouth. Eczema, in some of its forms, is so painful a disease, and causes such distressing insomnia, that recourse must often be had to medicines of a narcotic character."

SECTION V.

On diseases dependent on debilitated and deranged states of system, and consequent diminished tone of the vessels of the cutis.

CHAPTER I.

PURPURA.

UNDER this title I propose to consider the varieties of Purpura, and the cutaneous affection attendant on scurvy, or, in other words the variations of constitutional condition under which actual rupture of the minute extremities of the vessels of the surface occurs —that state of relaxation in which the serous portions of the blood only escape, forming by its extravasation under the cuticle what has been termed bullæ or blebs—the diseases designated Ecthyma and Rupia, and Erythema nodosum.*

* I have attentively watched the communications published from time to time in the different periodical works, but have not been able to see sufficient grounds for changing the opinions expressed in the last edition of this work as regards this disease, its true pathology, and the most successful methods of treatment. On the contrary, as time has passed on, I perceive nothing but confirmatory evidence in the recorded cases of it in the various foreign and English Journals. I assume, therefore, that the view I formerly took of it is correct. It is still, however, my duty to point out to the reader where other sources of information may be found.

Dr. J. Johnson's† Medico-Chirurgical Journal abounds with the details of cases of the disease, some of which clearly show the danger of taking away blood ; others proving the value of alterative purgatives, or in other words, the system of treatment I have from the beginning endeavoured to advocate.

There is no difficulty in renovating the powers of the system when the secretions have become healthy ; there is no danger of inducing increased debility by the use of purgatives of the class described : the idea that the pain in epigastrio, or in the hypochondrium, indicates inflammation, and requires the abstraction of blood, is a mere delusion. That congestion exists as a cause of such pains it is impossible to doubt ; but the removal of that congestion does not appear to me to be likely to be accomplished by reducing the powers of the circulation. There are not the usual evidences of inflammatory action afforded by the appearance of the blood in three out of four cases when this measure has been had recourse to in Purpura. Moreover, many instances are placed on record where it has been evidently the direct means of destruction of life.

General repletion may be, and perhaps sometimes is, co-existent with the disease. Gorging and habits of indolence, inasmuch as they tend either to the oppression of vital organs or to their over excitement and exhaustion, will sometimes, like *ignis fatuus*, lead the practitioner astray. Extreme caution is requisite in all such cases, and I have recently seen an instance in which the free use of the lancet had very nearly destroyed a life in every respect of the highest value.

Whether this disease be observed in the case of the half-starved mariner in

† The Index of the different volumes is the best guide for the reader.

The exclusion of the influence of the state of constitution in giving a character to cutaneous diseases, apparent in the attempts at classification which have been made heretofore, has been much noticed,

the coldest climates, who has tasted no food but that which lacks its proper quantity of nourishment because of the length of time it has been pickled. Whether it is seen in the delicate female in the better walks of life at home, and such cases are not uncommon, or in the case of the indolent epicure, active purgatives, combined with a liberal portion of calomel, constitute the most successful plan of treatment.

The alvine evacuations produced by medicines of this description are almost invariably worse than in any other kind of disease. The most depraved kind of biliary secretion often forms the largest portion, and blood, which it would seem had undergone the putrefactive process, constitutes the remainder. Occasionally blood is evacuated in a dark state in considerable quantities, but the circumstance furnishes no ground for apprehension of the effects of purgatives—no objection to their being continued and pressed. The effect of persisting in them is almost uniformly good; they seem to purge the system of its debility, while the bowels only are supposed to be under their influence.

I have appended the substance of M. Rayer's compilation of this part of my subject as in former instances. There does not appear to be any new light thrown on the pathology of the disease by him, but his references and cases will probably be considered entitled to the reader's attention. Some of the remarks are of sound practical value. I have not removed the tabular view of the results of the different kinds of treatment, for in truth those results have continued on an average nearly the same. If I had remodelled it, and taken care to avoid prolixity, I could only have substituted the names of the Journals and recorders of the cases since it was first published.

" Several of the sanguine congestions which are observed on the surface of the body depend on an impeded or retarded venous circulation. Such are those which may be produced at pleasure by the application of a tight ligature round any of the limbs; of the same description are those also which occur in the feet and in the extremities, in diseases of the heart, in the asphyxia of new-born infants, &c. The sanguine congestions which appear over the cheek bones, in pneumonia, and the livid blotches which are observed on the posterior parts of the trunk, at the moment of dissolution, or after death, are in like manner attributable to impeded venous circulation.

" It is well known that these livid blotches (*maculæ morientium*) are chiefly found occurring in those parts on which the body has lain during dissolution, or after death, and that they are sometimes observed to extend over the whole posterior surface of the body and limbs. Their bluish tint is in general not so deep as that of ecchymoses.

" When the skin of the regions thus marked is incised, it is found gorged with dark blood. It is sometimes even possible to cause these livid marks to disappear, by giving the body at the time of dissolution, or immediately after death, a position different from that in which they had been formed.

" Congestions again, which are occasionally preceded by morbid paleness, appear to be owing to an abnormal influence of the system; such is blushing or the redness of the face which is caused by emotions of various kinds, and that which is observed in the second stage of intermittent fever.

" Whatever the cause of sanguine congestion of the skin, whether the afflux of blood be transient and accidental, intermittent or continuous, it is distinguished from every form of exanthematous inflammation by the latter being constantly accompanied with morbid heat, or followed by furfuraceous desquamation of the cuticle.

" Sanguine congestions of the skin are not in themselves of a serious nature; but they are often symptomatic of highly dangerous affections of the heart, lungs, &c. Congestions do not require any treatment other than that proper for the diseases which produce them.

" In persons labouring under contraction of the auriculo-ventricular orifices of

and objected to with good apparent reason, and hence but little surprise will be felt at my including under one head diseases which form parts of no less than four different orders of preceding authors.

the heart, or of the pulmonary orifice of the right ventricle, under congenital or accidental communications of the right cavities with the left, or with the principal arterial trunks, &c. a peculiar sanguineous injection of the skin, and mucous membranes, is observed; these present a livid bluish tint, and the state is known by the name of *cyanosis.* It is distinguished from other congestions by its causes, its generality, its continuance, and by a more or less marked disturbance in the functions of respiration and circulation.

"The diseases in which blood is deposited on the surface of the skin, within its substance, or in the cellular membrane underneath it, have received different denominations, according as the extravasations are altogether local or general, or phenomena super-added to some other affection more or less serious in its nature. Farther, those effusions of blood which are occasionally observed from the surface of the skin, when accidentally deprived of epidermis, and even when still covered with the cuticle in new-born infants, have been designated by the name of *dermatorrhagiæ*.

"*Petechiæ* are minute red or violet-coloured spots formed by small quantities of blood deposited within the substance of the skin. The name of *ecchymoses* has been given to spots of a larger size; these are generally of a ruddy violet-colour, often livid and sometimes quite black; they are commonly of a deeper hue in the middle, than in their circumference, and they vary in extent from a diameter of a few lines to one of several inches. These two varieties of spots are to be observed in all cutaneous hæmorrhagiæ, whatever their causes and characters.

"The sanguineous infiltrations which take place in scurvy and in petechial typhus,† as well as those which occur incidentally in certain diseases of the skin, in severe erysipelas, in malignant scarlatina, and in the eczema rubrum of the inferior extremities, can only be studied in connection with these diseases. The ecchymosis traumatica, or ecchymosis which occurs in consequence of a blow, &c. is described in all treatises of surgery.

"PURPURA.—I comprehend under the general name of *purpura*, several diseases whose common and generic character is to manifest themselves internally by hemorrhage, and on the external surface of the body by petechiæ or ecchymoses, independently of outward violence.

"This group comprises two *species*, which are very distinct from each other in their progress, and in the symptoms associated with the common hemorrhagic phenomena which characterise them; viz. *Purpura sine febre* and *purpura febrilis*.

"*Purpura sine febre* itself comprehends three varieties (purpura *simplex*,

"* M. Gintrac (*Observations et recherches sur la cyanose*, in 8vo. Paris, 1825) has published a very good monography of cyanosis; and has collected and compared a great number of facts for the purpose of determining the organic conditions, which may give rise to this phenomenon. M. Louis (*De la communication des cavitées gauches du cœur*, Mémoires d'anatomie pathologique, in 8vo. Paris, 1826), and Messrs. Bertin and Bouillaud, (*Traité des maladies du cœur et des gros vaisseux*, in 8vo. Paris, 1824), have made some observations on this disease which deserve to be consulted.

"† Petechiæ frequently supervene in typhus between the second and tenth day. According to the report of Messrs. Raikem and Bianchi, out of a hundred and ninety-four subjects attacked with typhus at Volterra, in 1817, a hundred and fifty-six exhibited petechiæ. In typhus, the petechiæ generally show themselves on the lateral parts of the neck, on the shoulders, thighs, and especially on the interior of the fore-arms from the elbows to the wrists. True petechiæ are rarely observed at Paris in typhus fevers—(dothinenterites), they supervene more frequently in variola and scarlatina. They are sometimes observed with ecchymoses in those animals, into whose veins putrid animal matters have been injected.

The objection to the classification alluded to, that the particular state of constitution or derangement of important organs, on which the disease may depend, is omitted, appears to be well founded,

purpura *urticans*, purpura *hæmorrhagica*), to which purpura *senilis*, and purpura *cachectica*, must be added as sub-varieties. *Purpura febrilis* may present the external and hemorrhagic characteristics according to which the varieties of purpura apyretica have been established. The disease which has been described under the title of *febris hæmorrhagica* must be associated with febrile purpura.

"Purpura *simplex*. This disease almost always commences independently of any known causes, and without marked derangement of the principal functions. Some patients, nevertheless, complain of lassitude and dejection some days before the appearance of the spots. These are generally true *petechiæ*, sometimes intermixed with *ecchymoses*; in some very rare cases ecchymoses only are to be distinguished on the exterior of the body.

"When the eruption is petechial* only, the disease may present a considerable variety of appearances according to the number of the spots, as they are disseminated over nearly the whole surface of the body, or limited to a certain number of regions, and as the petechiæ have all been evolved at once, so as almost every where to present the same hue, or as they have appeared successively, during the course of several days, when they present a mixture of recent and old spots, and vary in colour from a reddish brown to a clear yellow.

"In ordinary cases the petechiæ show themselves principally on the legs; a great number are sometimes seen upon the face, which, at a certain distance, appears pricked with blackish and yellowish points, as if it had been bitten by a host of flees. In this case the conjunctivæ are apt to exhibit a number of ecchymoses.

"The distinguishing character of petechiæ, which are, as has been said, formed by a very small quantity of blood shed into the substance of the skin, is that they undergo no change of colour under the pressure of the finger. In this, consequently, they differ from other small red or rose-coloured spots observed in typhus and typhoid fevers, which are immediately effaced by pressure and return when it is removed. The greater number of these petechiæ are not quite so large as a flea-bite; for the latter, besides the small ecchymosis produced by the suction of the insect, is surrounded by a narrow pink areola, capable of being obliterated by pressure, which is not observed in the spots of purpura. Some spots of larger dimensions, and even proper ecchymoses, are often interspersed among the ordinary petechiæ. These petechiæ and ecchymoses are developed without local heat or pain, or any implication of the principal functions; children affected with them continue their play, and persons of more advanced age do not cease in general to occupy themselves with their ordinary affairs; the pulse remains undisturbed, and the digestion, respiration, excretions and nervous functions continue to be performed as in a state of health.

"Shortly after its formation, each petechiæ undergoes a change of colour; from a reddish brown it passes through various intermediate shades to a yellow; and, unless where there is a cachectic state of the constitution, the blood which

"* Sauvages describes simple petechial purpura, under the title of phœnigmus petechialis: Est exortus macularum purpurearum in universà cute—sine pruritu, tumore, alioque symptomate. Differt à pulicum morsu quòd in maculis phœnigmi non sit puncturæ vestigium, ut in pulicum, apum morsibus, quos delet cataplasma ex farinâ, aceto et oxymelite. Hunc affectum vidi aliquoties in pueris ob calorem æstivum enasci sina ullà notabili functionum læsione, et potu refrigerante et dein levi cathartico adhibito, inter paucos dies evanescere. Illust. vero professor Haguenot similem observavit in tribus mulieribus, quas quidem ille morbus defœdabat sed aliundè nullatenus molestabat, quique intra aliquot dies spontè recessit. (Sauvages. Nos. meth. t. ii. p. 594-95.)—Acta phys. med. nat. cur. 1757, p. 386.—Cusson. diss. de purpurà, 1762. Monspel. (Purpura apyreta).—Parry. Edinb. Med. and Surg. Journ. vol. vii.—Bree. Medic. and Physic. Journ. vol. xxi.

because it is quite evident that no useful result can accrue in classing diseases according to their local characters, if the state of constitution on which they depend, and through which only any

has formed it is generally re-absorbed in the course of a fortnight. This re-absorption almost always proceeds from the circumference towards the centre of the spots ; I have, however, seen it take place in a contrary direction, so that the spots, in the last stages of their duration, assumed the appearance of little yellow rings or arcs. These spots are not prominent, except in those very rare cases in which the blood is not only shed into the skin, in the form of a dark point, but where a minute drop of the fluid is effused beneath the epidermis, which it raises ; this little eminence, of the size of a pin's head, dries up into a small dark crust.

" When the eruption of petechiæ is considerable, they occasionally appear collected in different places into irregular clusters. This disposition is very rare. Finally, in a few very rare cases a kind of marbling, of a light violet colour, like the eruption in *rubeola nigra*, is observed disseminated among the petechiæ.

" The petechial eruptions which characterize this form of purpura *simplex*, of infinitely more frequent occurrence than the other eruptions of the same nature, are not accompanied by epistaxis, hematemesis, hemoptysis, nor by any other internal hæmorrhage ; they are, however, pretty frequently intermingled with real ecchymoses of the skin and sub-cutaneous cellular membrane.

" There are even cases in which purpura *simplex* is characterized by considerable numbers of ecchymoses disseminated over the body and limbs, on which very few or no petechiæ are to be seen. When the blood is thus more widely effused into the *substance of the skin*, it is almost always over the dorsal surface of the feet and hands, and on the internal aspects of the limbs that the broader spots are encountered. These never exhibit anything like regularity of shape, and sometimes bear no slight resemblance to the marks left by the strokes of a whip, or violent bruises. In the intervals between the ecchymoses the skin preserves its natural colour, temperature, and sensibility. The colour of these spots, which is a dark or livid blue, remains stationary for several days, and fades more slowly than that of the sub-cutaneous ecchymoses. They are sometimes succeeded by an exfoliation of the epidermis, particularly when a certain quantity of sanguinolent serum has been effused under it, so as partially to detach it; this happens to such an extent in some cases that the skin appears beset with a number of sanguinolent bullæ or phlyctenæ.

" When the blood is effused into the *sub-cutaneous* and *intermuscular membrane*, the skin commonly presents a number of broad and slightly prominent spots, dark in the middle and of a greenish yellow towards the circumference. These spots are little, if at all, painful ; but when the blood has been shed in large quantities into the sub-cutaneous cellular substance of the lower extremities, as sometimes happens, more especially in purpura hæmorrhagica, the tumefaction and tension of the parts may be so considerable as to produce a rather intense degree of pain. In this case, the entire skin of the inner aspect of the leg is of a greenish-yellow hue, which only disappears with extreme slowness. A sort of œdematous swelling has been observed to occur on the back of the hand in certain cases of purpura *simplex*, characterized by similar ecchymoses.

" In purpura *simplex*, the petechiæ and ecchymoses sometimes return suddenly in a single night, just as the disappearance of a first eruption seemed to indicate the complete re-establishment of the health. These hæmorrhagic phenomena may be repeated at periods little remote from one another: some of the former spots are then yellow or nearly effaced; others, very recently evolved, are of a reddish brown; while others of intermediate standing are of a paler red. It often happens, too, shortly after an apparent cure has been accomplished, that a new eruption shows itself; so that the duration of purpura *simplex* can never be predicted ; it varies from a few weeks to several months.

" There is a last variety of purpura *simplex* in which the extravasation of the blood is preceded by the formation of reddish oval or circular-shaped spots,

remedy can be conveyed, is overlooked. In the consideration of the diseases which form the subject of the present chapter, this observation applies with peculiar force, for the constitution is most

prominent and accompanied by smarting or tingling sensations similar to, but much less decided, than those of urticaria (purpura *urticans*). These little spots, usually of the size of a lentil, sink at the end of two or three days to the level of the surrounding skin; their colour, which was pink at first, becomes at the same time deeper and livid. New spots appear while the first are going off. They appear most frequently upon the legs, and sometimes in other regions of the body, intermixed with true petechiæ; the lower extremities in these cases are often œdematous; or they become affected with a sensation of stiffness or heaviness. The eruption is generally of a month's duration, but may continue beyond that term.

" Purpura *urticans* can only be distinguished from *hæmorrhagic* urticaria, *i. e.* urticaria in which a little blood is effused into the wheals, when numbers of true petechiæ are found interspersed among the larger raised spots. Cases of urticaria have been met with, which were followed by an attack of purpura, sometimes of a pretty severe character.

" The spots of purpura *urticans* are broader than those of *petechial* purpura; but are not so extensive as the ecchymoses of purpura *hæmorrhagica*, which are besides more irregular in their outline. The spots of purpura *urticans* sometimes exist on the skin without being accompanied with true petechiæ and without primary ecchymoses; but there are cases in which these three appearances are found united in the same individual.

" When purpura *simplex* shows itself under the form of ecchymoses, it is generally a more serious disease than when it appears under the *petechial* form. Purpura *urticans* is the least important of all the forms of purpura.

" Purpura *hæmorrhagica* (*morbus maculosus*, Werlhofii). This species of purpura is of a much more serious nature than the species first described. It is sometimes preceded for several weeks by a sensation of lassitude and debility, by pains in the limbs, &c.; but it often makes its attack suddenly* upon persons apparently in the enjoyment of good health.

" Purpura *hæmorrhagica* is characterized by the same external appearances as purpura *simplex*, sometimes by petechiæ, often by ecchymoses, still more frequently by both of these hæmorrhagic forms at the same time, but very rarely by an actual *dermatorrhagia*.† In the majority of cases the ecchymoses appear before the petechiæ. The body has been seen to become covered with livid spots similar to those that follow bruises; in children actual thrombus occasionally forms under the hairy scalp, and the blood has even been seen exuding from behind the ears and the vertex, and the eruption of such extent as to cover nearly the whole of the skin. The disposition to hæmorrhage is so great in some patients that the mere act of feeling their pulse, the pressure of a bleeding fillet, or that occasioned by the weight of the body in sitting or lying down, are

---

" * Johannes Dolæus speaks of a child : ' Cujus omne corpus, absque dolore, febre aut lassitudine prægressâ, subito unà cum facie, labiis, et linguâ, ubi mane adsurgeret, numerosissimis maculis lividis et nigerrimis obsitum fuit, etc.' (Ephem. nat. cur. Dec. ii. Ann. IV. Obs. 118). Zwinger (Act. Nat. Cur., vol. ii. Obs. 79). P. G. Werlhof (Commerc. liter. Norimberg. Ann. 1735. Hebd. 7 and 2) have reported analogous facts. Sauvages mentions this disease under the title of Stomacace universalis. (Nosol. Meth. t. ii. p. 296.)

" † I have not observed this transudation of blood through the skin. It is asserted that in the majority of cases, the skin, from the surface of which the blood flows, is not sensibly injured. Nevertheless, in a case reported by Doctor Whytt, the end of a finger, whence the blood flowed, was painful, and exhibited a red spot. Ploucquet (Art. Hæmatidrosis, Sudor cruentus) mentions several authors who have observed these blood-sweats or hæmorrhages through the skin. Fournier cites two examples of the kind. Art. Cas rares. (Dict. des Scien. Méd.)

certainly the medium through which any efficient remedial measure can be employed ; local applications being rarely required or admissible.

sufficient to produce actual ecchymoses. In these cases the slightest punctures, the most trifling injuries that merely graze the skin, are always followed by a far greater amount of hæmorrhage than usual ; blood also occasionally flows copiously from the surface of blisters, issues, ulcers, &c. in such subjects. The number of petechiæ and ecchymoses in purpura hæmorrhagica is likewise almost always more considerable than in purpura *simplex;* and they recur with more promptitude and intensity, and continue longer than in the latter.

" But the essential or distinguishing feature of purpura *hæmorrhagica* consists in the hæmorrhage from the viscera or internal membranes, which invariably precedes, accompanies, or follows the ecchymoses or effusions of blood into the skin or sub-cutaneous cellular membrane. The disease is always complicated with epistaxis, intestinal hæmorrhage, hemoptysis, hematemesis, hematuria, and, in females, with menorrhagia.

" Of all internal hæmorrhages, epistaxis is that which occurs most frequently, particularly in children ; menorrhagia in females, and pulmonary and intestinal hæmorrhage are common in adults. Several of these varieties of hæmorrhage often occur simultaneously, or they alternate at different times in the same patient.

" The general characters of the disease, which are always striking, are modified by diversity in the local symptoms, and, indeed, these vary themselves, according as the hæmorrhage occurs from organs of greater or less relative importance in the economy, as the blood lost is small in quantity or very large, as the flux is frequently repeated, or occurs at distant intervals, or as the hæmorrhage takes place from one point in succession or from several simultaneously.

" 1st. Purpura *hæmorrhagica* with *epistaxis* is the most common of all the varieties. Bateman has seen it accompanied by hematemesis, and followed by death from sheer loss of blood.

" 2d. Purpura *hæmorrhagica* with hæmorrhage of the throat, or only of the *amygdalæ,* is rather a rare variety ; sometimes the whole fauces appear of a deep red and the blood issues from every part ; after the hæmorrhage has ceased, the parts appear dark. In one case, in which there existed but an inconsiderable number of petechiæ on the skin, I have seen purpura complicated with angina *membranacea;* this complication has even been observed under an epidemic form.

" 3d. Purpura *hæmorrhagica* with hæmorrhage from the *mouth* and *gums,* is often accompanied by *epistaxis* and *hematemesis.* The gums are livid, spongy, and the blood exudes from their free edge ; the tongue is livid and blackish, bleeding and fungous in appearance, and twice its natural size ; the inner surface of the cheeks presents some blackish and soft patches, and the palate is covered with blackish spots. Children have been seen to die in a single night after excessive hæmorrhage from the mouth or nose ; but they more commonly sink from the effects of bleeding, which, though less abundant at any one time, has recurred frequently during several weeks. The epithelium is sometimes raised on the tongue, palate, inside of the cheeks, lips, &c. in consequence of ecchymoses. It then forms irregular phlyctenæ or bullæ filled with blackish blood ; by-and-bye the epithelium bursts, and the blood flows from the surface of the excoriation ; the mucous membrane often ulcerates more deeply, and hæmorrhage to a greater amount is apt to take place.

" 4th. Purpura with *hæmorrhage from the stomach* is sometimes accompanied with pain in the left hypochondrium, and an increase in the size of the spleen ; symptoms which have been more particularly observed in individuals who have been attacked with purpura after having suffered from intermittent fever.

" 5th. Patients more frequently pass blood by *stool;* the blood is rarely pure, and of a bright colour ; it is often a blackish-looking matter, of a sooty colour ; this variety of purpura is less serious than the preceding one.

" 6th. When the blood flows by the urinary passages, the urine is tinged with

The disease termed Purpura, in its more formidable shape, accompanied by a train of symptoms immediately threatening the existence of the patient, merits, and appears to have obtained,

the blood, or the fluid is passed unmixed, partially coagulated, and sometimes in considerable quantities, without even the urinous odour. Blackall has found the urine coagulable by heat, or nitire acid, in four cases of purpura without hematuria; all the patients had the legs slightly infiltrated with serum. In one case of *purpura hæmorrhagica*, with febrile symptoms, Doctor Coombes remarked, that the urine, which was very coagulable in the height of the disease, became less so after blood-letting, and abundant loss of blood; these evacuations of blood were of evident advantage, and when the cure was effected, the urine ceased to be coagulable.

"7th. In purpura with hæmorrhage from the *uterus*, the *vagina* or the *pudenda*, patients often experience pains in the loins. These hæmorrhages of the uterus are almost always serious; I have seen a case in which the bleeding was mistaken for a miscarriage, and the patient died. It very rarely happens that such hæmorrhages are salutary, or critical; nevertheless Bateman met with a woman affected with purpura *simplex*, who recovered after an attack of uterine hæmorrhage.

"8th. When hæmorrhage takes place from the lungs* patients have fits of coughing, spit blood or bloody matter, and feel great pain in the precordia and chest. Duncan and Bateman have seen the description of hæmorrhage under review prove rapidly fatal.

" 9th. Finally, there are cases in which hæmorrhage occurs successively, and in a few days from the nasal fossæ, from the mouth, from the lungs, from the stomach, the intestines,† &c.; in some cases the bleeding recurs at particular hours, in others there is a slow and almost continual exudation of blood.

" If hæmorrhage recurs frequently or proves very abundant, from whatever channel it may flow, the lower limbs become œdematous, the face pale, and the body generally assumes a livid and yellowish hue; the petechiæ and the ecchymoses, greater in number, have a deep brown tint, the blood becomes more and more serous, the extremities cold, the patient experiences weakness, the pulse becomes small and frequent; nervous symptoms, sometimes convulsive movements, trembling of the whole body, and fainting fits occur; and if the hæmorrhage is repeated, the patient dies ex-sanguine, unless febrile symptoms supervene and give rise to some other form of fatal termination. In fact, after a certain number of attacks of hæmorrhage, an acute fever, with symptoms of a typhoid character, sometimes comes on, and patients frequently sink on or about the twelfth day.

" Purpura *febrilis*. This description of purpura may be sporadic or epidermic;‡ its ordinary duration is from two to three weeks. Purpura *febrilis* attacks persons of all ages, and of every variety of constitution.   " A

" * Kift. Edinb. Med. and Surg. Journal, vol. xxvii. p. 71—Planchon. Journ. de Méd. Paris, Ann. 1770 (two cases).

" † Latour. (Ouv. cité. tom. ii. p. 180. Case from Horst).—Ibid. t. ii. p. 20, Obs. 621.—Ibid. t. ii. p. 498. Case of Leroy, erroneously quoted as a case of scorbutus.

" ‡ Lordat observed, in the prison at Montpellier, a hæmorrhagic petechial fever, which appeared in the spring of the year 1800, lasted five months, re-appeared towards the middle of September, 1804, and continued till the month of January, 1808. This description of fever attacked nearly one half of all the women who were confined there, and no more than two men. It began by a shivering fit, followed with intense fever and depression, pain in the head, acute suffering in the epigastrium, redness of the face, white tongue, scanty and scalding urine, constipated bowels, &c. This state continued for three days; on the fourth, the fever abated, and an eruption of petechiæ took place on the neck, breast, upper part of the arms, trunk, &c. After the eruption had appeared, the fever and the hæmorrhage, which was generally from the nose and the uterus, subsided at the same time; towards the ninth day the spots usually disappeared. After several days of apparent convalescence, a relapse often oc-

at various periods, the attention of some of the most distinguished characters in the history of medical literature ; and the names of Parry, Bree, Buxton, Duncan, Harty, and others, are scattered through

"A feeling of great lassitude and depression, shivering fits, of longer or shorter duration, followed by heat, pains in the back and limbs, head-ache, and oppression, sometimes a sensation of great heat over the whole of the body, nausea, retching, quickness of the pulse, and other febrile symptoms, are the precursors of the petechiæ or ecchymoses. They appear from the third to the sixth day, sometimes without hæmorrhage from the mucous membranes, or the viscera (purpura *febrilis simplex*).

"Frequently in purpura *febrilis*, after the initiatory fever, the eruption of petechiæ is preceded by *exanthematous spots*,* analogous to those of urticaria *febrilis*. The skin, already red, is then covered with petechiæ of a purple colour, the dimensions of which vary from that of the head of a very small pin to that of the tip of the little finger. These last are slightly prominent.

"Purpura *febrilis* may appear under the form of *ecchymoses*. M. Ollivier d'Angers† observed a curious case of this kind in an infant three years of age, which he gave me an opportunity of seeing, in whom the cutaneous ecchymoses made their appearance on the limbs at the same time that these became œdematous. The skin was hot and painful, the pulse from 120 to 130 in a minute ; the child experienced pain in the abdomen ; the pressure of a garter, or of the finger was sufficient to produce an ecchymosis. This child was cured after a month's illness.

"In purpura *febrilis*, after the primary fever, hæmorrhages occasionally occur from one or other of the different passages of the body, at the same time that petechiæ and ecchymosis are formed in the skin, and sub-cutaneous cellular tissue (purpura *febrilis hæmorrhagica*‡). In some patients the urine is highly

curred, and these relapses were sometimes repeated ; no one died. Some women, after several relapses, became decidedly scorbutic. (Latour. Op. Cit. p. 170.) Vandermonde Journ. de Méd. t. vi. p. 339, *maladie noir d'une espèce particulière*, has recorded a remarkable case of *purpura febrilis*. We must assimilate to these cases the accounts we have of a disease observed among deserters who were brought by forced marches to a place of confinement, and whose bodies became covered with petechiæ and ecchymoses, and in whom nothing succeeded in stopping the nasal and intestinal hæmorrhages. (Latour. Op. Cit. ii. p. 469.) See a case of Sporlius, cited by Fabricius Hildanus (Obs. Chir. Cent. vi.) Th. Schwenck, *Sang. Hist.* p. 130.

"* During the summer of 1797, Latour saw at the Hotel Dieu of Orleans, a great number of reapers, the whole of whose bodies, after great lassitude in the limbs, irregular shivering fits, and vertigo, became redder and hotter than usual, whilst the head was heavy and painful, the pulse very full, and the beating of the carotid arteries extremely violent ; the tongue was at the same time red and dry, the thirst urgent, the urine scanty, the bowels relaxed, the breathing occasionally interrupted by sighing, &c. On the second and third days of the attack, exanthematous patches appeared over the skin, which looked as if it had deen whipped with nettles. Towards the fifth or sixth day innumerable brown and black petechiæ, of a lenticular shape, appeared in the middle of the exanthematous blotches ; the fever commonly subsided about the fourteenth or fifteenth day, and always before the one and twentieth, and the petechiæ vanished, becoming yellowish, like ecchymoses, as they declined ; the disease was not fatal in its tendencies. Latour (Hist. philosoph. et med. des hæmorrhagies, 2 vol. 8vo. Paris, 1828, tom. ii. p. 172, and sequent) relates another case of this species of hæmorrhagic nettle rash, which, after beginning with delirium, and other serious symptoms, terminated happily on the fifteenth day. See an analogous case published by Johnston, (Edinb. Med. and Surg. Journ. vol. xviii. p. 402.) who very justly remarks that this case of purpura febrilis hæmorrhagica differs materially from the ordinary purpura hæmorrhagica.

"† Ollivier d'Angers. Développement spontané d'ecchymoses cutanée avec œdème aigu sous-cutané et gastro-entérite. (Archives génér. de Méd. t. xv. p. 206.)

"‡ Reil (Observationes quædam de hæmorrhœa petechiali.—Memorabil. Clinic. fasc. V.) relates three remarkable cases of purpura febrilis hæmorrhagica.

the volumes of the periodical press, in connexion with communications regarding it, not inferior in correct observation to their more distinguished works.   It is a disease which, in its more important

tinged with blood ; the pulse, small and contracted, in the first instance, sometimes acquires more strength and softness after a first hæmorrhage ; but upon further loss of blood, all the symptoms which I have mentioned in treating of purpura *hæmorrhagica non-febrilis*, may occur.

" Purpura *febrilis* may present from the commencement very serious symptoms, and terminate in a few days in death, when blood is extravasated in large quantity into the tissue of the lungs or substance of the brain.*

" We must assimilate a species of *hæmorrhagic fever*† to purpura *febrilis*, in as much as it only differs from purpura *febrilis hæmorrhagica*, by the absence of ecchymoses and petechiæ.   In this fever, after symptoms of a more or less violent character, hæmorrhages from the nose, mouth, intestines, urinary passages, &c. take place.   These *general hæmorrhages* are sometimes not accompanied with fever in the beginning, but it generally appears after an interval of a few days.

" These general hæmorrhages have been seen to alternate with determinations of blood to particular organs, as to the tonsils, &c."

Notwithstanding my earnest desire to avoid encumbering my work with the cases or " *Observations*" of French authors generally, the following seem to be too instructive to be omitted with strict propriety.   The reader must judge whether they tell for or against the views I have taken of the disease, or whether they affect the question of its etiology at all.

" Case 1st. *Of the disease preceding small-pox.*   Sorel, three years and nine months old, admitted into the *Hospice des Enfans Malades* on the 19th of August, 1825, had enjoyed perfect health till within two months of the above date, when measles made its appearance, without, however, presenting anything unfavourable in its character.   During the last fortnight, the patient has had a slight cough, and the skin has been burning; but she has made no complaint, although evidently dull and listless ; for six days she has experienced great difficulty of breathing ; in the evening of the 9th and following day, she complained of pain in the epigastrium *(four leeches applied*); on the 11th, the difficulty of breathing increased *(blood taken from the arm ; gum water*), no improvement; the 12th, an eruption of purple spots over the whole body, which still continue to-day, the 14th : the variolous eruption was not remarked till the 13th.   The following is a description of the patient's state at the time of her admission : The variolous pustules, in inconsiderable numbers, are small, blanched, shrivelled, and centrally depressed.   On the surface of the skin, and particularly on the forehead, upper eyelids, back, and posterior aspect of the limbs, a number of spots are observed, irregularly scattered, rounded and well defined, but not prominent; some, of a bright purple colour, are about the size of a lentil; others, of a larger size, are of a deep violet hue.   The lips and nostrils are covered with a dark sanguinolent incrustation.   It is impossible to examine the inside of the mouth, so as to ascertain the state of the gums.   The orifice made in the vein of the arm three

" * Zacutus Lusitanus speaks of a patient whose whole body was livid, and who during two days was in a general bloody sweat, and became covered before death with a perfectly black eruption.   In another place, he relates that several persons, who had been cupped in this disease, bled so profusely that it was impossible to stop the hæmorrhage by any means, so that they all died. [Praxis medica miranda, Obs. 41, 42.]

" † M. Littré showed me a young man who for several days had laboured under hemoptysis, and hæmorrhage from the nose, intestines and urinary passages. [Gazette Medic. de Paris, 1833, p. 263.]   Petrus Poterius, (Obs. et Curat. Insig. Cent. iii. 60.) has published a similar case.   Morgagni states, that in the year 1200, a great number of men died in Etruria and Romagnia, in twenty-four hours, in consequence of a flow of blood from the nostrils.   [De sedib. et caus. morbor. Epist.]

form, frequently paralyses the hand of the physician, and suspends in doubt and apprehension the decision of the most practised and fearless mind. Its distinguishing feature is, on the one hand,

days ago, is still open, and discharged a little very serous blood this morning. The pulse is very languid, the extremities cold. The child retains its consciousness notwithstanding the extreme depression (*sinapisms*). Death at three o'clock in the afternoon.

"*Sectio cadaveris*, at eleven o'clock on the morning of the 15th. *External appearance.* Cadaverous stiffness of the inferior members, none of the superior. The colour of the skin and of the spots is precisely the same as during life. On incising the skin it is easy to perceive that all the spots do not penetrate to the same depth; some are very superficial, and situated directly under the epidermis; others occupy the areolæ of the corion; finally, there are some (and these are the largest and deepest), whose seat is entirely in the sub-cutaneous cellular membrane. All these spots are formed by extravasations of blood, coagulated in the larger and darker kind, and liquid in those of smaller size. Examined with the assistance of a powerful glass, the vascular ramifications in the neighbourhood of these ecchymoses and petechiæ are not observed to be more developed than in the ordinary state. If, after laying bare one of these effusions, a stream of water is directed upon it, the blood is soon washed away. A piece of skin, put into water to macerate, did not present any appearance of spots on the following day. There are no ecchymoses in the deeper strata of the cellular membrane generally; the right arm, however, where the puncture made in a recent venesection still continues open, is the seat of an universal sanguineous infiltration, to which its tumefaction and livid appearance are owing. The veins and arteries of this extremity, traced into their most minute ramifications, do not present any visible alteration; the median cephalic vein, on which the operation of bleeding was performed, does not even exhibit any redness on the edges of the orifice; its walls are thin and transparent, and its external membrane is smooth and greyish as in the healthy state. There is no ecchymosis in the substance of the scalp. Limpid serum, in small quantity, appears on the surface of the cerebral hemispheres; the sub-arachnoid vessels are empty; the cortical substance is pale, the medullary substance firm; the superior longitudinal sinus empty; the sinuses of the bases of the cranium are filled with a liquid vermillion-coloured blood; traces still remain of a slight sanguinolent effusion, converted into dark incrustations, upon the lips and nostrils. The tongue and gums are pale and blanched; the palate is violet-coloured; the epiglottis swelled, as well as the edges of the glottis; the latter are covered with a very thin false membrane; the tracheal, and bronchial mucous membranes, are apparently healthy. The external surface of the lungs presents a great number of bright red circumscribed points, and several ecchymoses of a deeper colour; one of these, on the summit of the left lobe, is four lines in diameter; there are three smaller ones on the lower part of the same lobe, which also presents another much more extensive at its base. These spots are the more striking on account of the lungs retaining their natural greyish colour. On examining the spots, they are found to correspond to a kind of dense circumscribed nucleus, which, on being cut across, presents a reddish brown, homogeneous, granular tissue, in which the blood appears as if combined with the proper substance of the lungs. The parenchyma of both lungs contains several formations of the same kind, circumscribed in like manner, and situated in the midst of a perfectly crepitating tissue, from which, when incised, a quantity of blood, mixed with serum, escapes. Several of the glandular bodies of the bifurcation of the bronchi are red and enlarged. The pulmonary artery, at its origin, and the right ventricle and right auricle of the heart, present three lenticular ecchymoses of a bright red, approaching to the natural colour of these parts, and another violet-coloured one, nearly of the size of a sixpence, with irregular edges, situated on the posterior aspect of the apex of the heart. These effusions are situated entirely in the sub-serous membrane, and do not extend to the muscular tissue of the heart. A sanguineous suffusion

according to received notions, identified with the lowest degree of debility : on the other, it is ushered in with symptoms which cannot subsist many hours uncontrolled by active treatment, without the

two inches in breadth, exists under the pericardium on the right side. The substance and cavities of the heart, as in the healthy state ; the internal membrane of the aorta and pulmonary vessels perfectly healthy. The stomach appears contracted ; its mucous membrane very much wrinkled, particularly in the direction of the great curve, and beset with numbers of small red points, of a vermillion tint, like the pricks of a pin. Within three inches of the pylorus, there is a lenticular softened spot, of a dark yellow colour, limited to the thickness of the mucous membrane, under which there is a small ecchymosis of the same size ; the duodenum presents for the space of an inch from the pylorus, a number of small petechial spots, closely set together like those of the skin ; they are far fewer in number towards the lower portion of this division of the intestines. The mucous membrane of the small intestines, is of the natural dirty grey colour; the glands of Peyer are slightly developed ; the ileum contains green, flaky, muco-bilious matter. From the ileo-cœcal valve, the great intestines present a general violet red colour, which increases in intensity towards the sigmoid flexure of the colon ; they are covered by an infinite number of red points; some, which are white in the middle, appear to be follicles ; others are very small ecchymoses, which, towards the commencement of the colon, follow the circular folds of that intestine ; in this situation they are covered with a greyish secretion, and are much less numerous than in the transverse and descending colon. Numerous trichocephala in the cœcum and its vermiform appendix, which is rather red ; the fecal matter contained in the large intestines firmly moulded ; mesenteric glands of small size, intensely red, brownish in their centres. Some purple marblings are observed on the liver, but no spots; spleen small and healthy ; kidneys pale ; bladder contracted and healthy.

Case 2.—*Pulmonary tubercles, chronic pneumonia and peritonitis ; purpura characterized by epistaxis, sub-cutaneous, sub-mucous, sub-pleural effusions of blood,* &c. Ferd. Hélène, seven years of age, of a very feeble constitution, was admitted into the " *Hôpital des enfans,*" on the 14th of March 1825. He had been a long time ill, was subject to diarrhœa and cholic, and the belly was distended and painful. For the last four or five days the cough had increased, the fever was higher, and the thirst more intense. At the time of entering the hospital, this patient was labouring under chronic pneumonia and peritonitis. In spite of the active measures immediately employed, the patient continued to suffer from frequent cough, constant diarrhœa, and intense fever.

. After being about twenty or twenty-five days in the hospital, the respiration became very laborious and short ; the respiratory murmur was scarcely to be heard on the left side, and posteriorly to the right it could not be perceived at all. A blister applied a few days previously to the right side of the chest, became gangrenous, and was surrounded by erysipelatous inflammation of the most vivid red. At the same time a number of small violet-coloured circumscribed spots were remarked on the upper and lower extremities, some circular, others oblong, and about a line, or something less, in diameter. The patient had had several attacks of epistaxis ; his pulse was very weak, the extremities were cold, (*mucilaginous decoction of bark, lavement of the same.*) The patient died in the course of the day. *Sectio cadaveris.* Externally a number of dark purple spots were observed on both fore arms, some the size of flea-bites, others of grains of millet. There were also several on the legs and thighs, but these were much less dark in colour, and smaller in size. On cutting into, and dissecting off the skin of the fore-arms, hands, thighs and legs, we perceived that the sub-cutaneous cellular membrane was much injected, and very red, and presented a number of small sanguineous effusions, corresponding to the spots upon the skin, though none of these small ecchymoses were seated in the tissue of that membrane. The lymphatic glands of the armpit were injected and swelled. The sub-cutaneous veins of the extremities were pale and void of blood ; externally they were

most imminent peril to internal organs. In one point of view, the probability exists of extinguishing life by extracting a few ounces of blood ; in the other such a measure appears a matter of absolute necessity.

white ; the muscles were healthy. *Respiratory organs.* Larynx, trachea and bronchi in a healthy state ; inter-bronchial ganglions tubercular, softened in the centre and very voluminous ; agglomeration of tubercular ganglions in front of the trachea. *Right lung.* Upper lobe healthy, crepitating, spotted internally by an infinite number of small ecchymoses ; middle lobe hepatized, and filled with pus, which flows out when the part is compressed ; several hepatized points in the interior lobe : two cavities in its centre, of the size of a hazel nut, containing yellow pus. *Left lung.* Superior and inferior lobes crepitating, but filled with a much greater number of ecchymoses than the right lung. On several points the surface of both lungs presents spots analogous to those observed upon the skin ; sub-pleural ecchymoses. Some miliary tubercles were found in the lungs. *Digestive Organs.* The mucous membrane of the mouth is pale. Several small sub-mucous (similar to the sub-cutaneous) ecchymoses are remarked at the root of the tongue. Pharynx and œsophagus healthy ; stomach healthy. The mucous membrane of the large and small intestines presents few spots not much injected. The liver and spleen are healthy. The peritoneum is thickened at all points, and its folds adhere together. A number of small whitish tubercles, some of which are soft, occur at intervals between the layers of the epiploa. The great epiploon adheres throughout its whole extent to the abdominal parietes ; all the intestines are agglutinated, and it is very difficult to separate them. The mesenteric glands are swelled and purple coloured. *Urinary Organs,* healthy. *Nervous system.* The membranes of the brain are in a healthy state, the cerebral substance is of the natural consistence ; there is very little serum in the ventricles, the cerebellum is perfectly healthy.

Case 3.—*Inflammation of the mouth ; amygdalœ, with pseudo-membranous deposits. Purpura hæmorrhagica.* A pale and delicate girl had been in bad health for a long time ; she was in a very feeble condition, when red, livid spots made their appearance on her body ; the amygdalæ at the same time became covered with dark spots, and a species of false membrane, of a black colour, and rather pulpy consistence. This young person had hæmorrhage also from the mouth and nose. The dark hue of the amygdalæ gave reason to suspect her of being affected with gangrenous angina ; the left tonsil was scarified, and the right touched with hydro-cloric acid and water in equal proportions. The abdomen was not painful, the respiration was unaffected. On the evening of the 23d July, 1829, the skin was hot, burning, dry ; the pulse quick and weak ; the face swelled, particularly in the lower part and the region under the jaw. This swelling was owing rather to a sort of œdematous puffing, than to inflammation ; the skin was straw coloured ; lymphatic glands could be felt under the swollen angles of the jaw, more to the right than the left. The lips were swelled, thick, and glossy, particularly the lower one, the mucous membrane of which, from its origin to the gums, was black and swelled. In the same situation a salient spot of the size of half-a-crown was observed, the centre of which was of a darker colour than the circumference. This prominent spot was formed by a kind of false membrane which could be removed with forceps without causing pain, excepting at the middle of the patch. This false membrane, of a greyish or dirty white colour, extended over the gums, diminishing in thickness, and less and less adherent to the parts beneath ; it was easily removed piece-meal. On each side of the mouth, a dark, irregular skinny substance was observed attached to the gums by one extremity. These false membranes did not emit the offensive gangrenous odour. Beneath them, the tissue of the gums was red and glossy like the mucous membrane of the lip. The colour was not very bright. On the false membrane being removed, blood was seen to flow from the surface of the lip, and to coagulate in a short time afterwards. The teeth were white, and not loosened ; and the false membrane insinuated itself between

To ascertain the causes of this apparent inconsistency of symptoms, and to reconcile it with a systematic line of treatment, seems at first sight a matter of much difficulty ; and though the original

several of them ;* the upper lip was swelled in like manner, and also presented some dark points ; the tongue was covered with a dark thick mucus. The patient was unable to open her jaws ; a purulent, ropy liquid which had nothing of the gangrenous odour, flowed from her mouth. Over the whole surface of her body, were scattered red, livid spots of the size of a lentil, which could not be effaced by the pressure of the finger ; there were also many others of smaller size, which appeared of more recent formation. *Draught with pulvis cinchonæ : drink acidulated with muriatic acid ; astringent gargle.* Died at one o'clock in the morning of the 24th. *Autopsia* 18 *hours after death.* The skin presented here and there spots of a deep colour, formed by the effusion of blood into the substance of the skin only. There were none in the sub-cutaneous cellular membrane. The tongue, right amygdala, and gums were covered by a thick black false membrane, which was removed without difficulty, the tissue of the amygdala was yellow, and soft, internally it was *black* ; there were a kind of flaky filaments on its surface. The left amygdala was swelled in like manner; its tissue was yellow ; the neighbouring cellular membrane was *black* ; the amygdalæ did not emit the gangrenous odour. The epiglottis and glosso-epiglottic ligaments, presented *black spots* under the mucous membrane ; there was an effusion of dark blood into its substance ; the glands of the neck on both sides were swelled, without being the seat of any effusion ; the nasal cavities did not contain blood.

" The larynx, the trachea, and bronchi were healthy ; the lungs presented several black spots on their surface, varying in size, but never exceeding the dimensions of a sixpence ; they were evidently owing to blood deposited underneath the pleura, and infiltrated superficially into the tissue of the lungs. The lungs were crepitating except at their base, where they were loaded with fluid. The whole surface of the heart, more particularly in front, and more on the left ventricle than elsewhere, as well as under the serous membrane, presented purple spots of the size of a pin's head, in very close contact. The pleura under the third and fourth ribs, near the sternum, presented a considerable patch tinged with effused blood.

" The stomach, rather livid, presented a small dotted patch ; its mucous membrane presented the natural thickness and consistence. The intestines, liver, and spleen were healthy. The upper part of the right kidney was a little spotted with black under its proper membrane ; the left was pale and less consistent. The uterus and bladder were healthy.

" On the dura mater was found a large red irregular patch, formed by blood deposited under the arachnoid membrane ; on the side of the left lobe of the brain, between the arachnoid and the pia mater, a slight sanguine infiltration ; the corpus striatum and corpus callosum presented a dotted appearance. A similar infiltration was remarked in the whole of the right lobe; the same disposition in the left lobe of the cerebellum. The muscles were red and healthy ; the cellular tissue exhibited no traces of extravasation.

" Case 4.—*Purpura hæmorrhagica febrilis ; petechiæ and ecchymoses, preceded by a kind of erysipelatous attack.* &c.—Madame Robert, 70 years of age, a washerwoman, of a sanguine temperament, with a coarse ruddy skin. This woman inhabits a healthy lodging, lives well, leads a very regular life, and has not suffered from illness for a great many years. She began to be affected with headache in the beginning of May, and afterwards by general indisposition, and itching of the lower extremities, which swelled slightly. Some days afterwards the face looked puffed, and she became a patient in the hôpital de la Charité on the 31st of May, 1834. Both cheeks were swelled, tense and red as in erysipelas, and strewed with specks of a darker red, the size of a lentil, and not to be

---

* This false membrane is nothing more or less than a film of coagulum formed by the blood which oozes from the gums during sleep.

opinion of the character of the complaint ought not to be impeached, merely because symptoms of apparent inflammation occur at times in conjunction with the cutaneous affection, yet it appears to be clear,

effaced by the presure of the finger. These dark ecchymotic specks were also discovered on the nose and eyebrows : the eyelids were infiltrated with black blood, as well as the lobe of the right ear, which was mottled like marble. There was also a large black ecchymosis under the chin, but without any elevation of the epidermis. The body only presented two or three black spots on its posterior surface ; both the superior and inferior extremities were swelled, hard, and covered with large black patches, the epidermis of which was raised by a sero-sanguineous liquid, and surrounded by a pink areola. These patches existed principally on the dorsal aspect of the fore-arms, and on the anterior surface of the thighs. On the latter there was also a hæmorrhagic petechial dotting, particularly on the right thigh. The palms of the hands and soles of the feet were livid or black, swelled, and slightly œdematous ; an effusion of blood had taken place under the corion, and even into its outer surface, but without any detachment of the epidermis ; the patient complained much of pricking sensations in these parts.

" All the spots had existed for six days. The lips were swelled. The gums were without any appearance of redness, and did not bleed. The patient had had no epistaxis, hemoptysis, or hematemesis ; neither had she ever remarked blood in her stools or urine. The tongue was moist and natural in colour. There was no sensible derangement of the functions of digestion. The respiration was free ; there was no cough. The sound of the chest on percussion, however, was somewhat dull ; and the respiratory murmur in one particular spot of the posterior surface of the left lung was only heard very indistinctly. The patient was feverish (the pulse 88 per minute). The temperature of the skin was higher than natural. The beat of the heart was regular ; near its apex a single sound could alone be distinguished ; the bellows sound (bruit de soufflet) at the base of the organ ; the same bellows sound was heard synchronous with the first sound of the heart ; the second sound was natural. Under the sternum and near its right margin both sounds of the heart were heard as in a state of health. The cerebral functions were intact. (*Sherbet of sulphuric acid for drink ; half an ounce of Epsom salts.*) The bowels acted once during the day. On the following morning, the 1st of June, a great number of new hæmorrhagic patches made their appearance on the limbs ; they were purple, irregularly circular, and formed by sanguineous serum effused under the raised epidermis ; some pain in the throat was complained of ; the pulse was very full and hard ; the skin hot ; (*half an ounce of Epsom salts, drink as before, decoction of rutanhia acidulated with nitric acid ;*) tormina, and several stools in the course of the day. On the 2d the tumefaction of the cheeks and limbs had diminished a little : that of the hands only still continued as before ; but fresh hæmorrhagic spots had been thrown out upon the limbs, principally in the line of extension. *The pulse continued full,* the skin hot ; the pain in the throat was not abated. *Drink as before, venesectio.* Little blood was obtained, the swelling of the arm having rendered the operation difficult.

" On the third the fever continued, the left lung gave a dull sound on percussion ; posteriorly, a slight bellows sound was heard ; the expectoration was mucous without any mixture of blood ; (the same drink ; fifteen leeches to the epigastrium). The leeches drew little blood, yet gave relief ; the spots of the fore-arm disappeared from the centre towards the circumference, as in lichen circumscriptus hæmorrhagicus ; the fever had abated, (the same drink ; leeches to the anus). On the 5th the amendment continued, the swelling of the extremities subsided, the sanguinolent fluid effused beneath the epidermis, had either escaped or been absorbed ; the epidermis had become dry on some of the patches ; and the violet red of the skin had faded into a yellow : the redness and tumefaction of the face had diminished greatly, the pain in the throat had ceased ; a little cough still remained.   " On

that want of tone in the vessels of the surface is perfectly compatible with unimpaired strength of the system at large, and capable of being produced to such an extent as under the ordinary impetus of the circulation to admit of the rupture of such vessels ; forming, in some cases, the Purpura simplex, or petechiæ sine febre : and in others, the more serious and formidable P. hemorrhagica.

The best descriptions of the purpura simplex concur with each other in attributing it uniformly to a state of general debility ; but it is nevertheless sometimes seen as well as the P. hemorrhagica, where such a state of system does not exist.

The appearance of the spots or petechiæ is almost too well known to require description ; they are " small, distinct, purple specks and patches" usually distributed over every part of the body. Although classed by Willan and Bateman in the order Exanthemata, this form of the disease is rarely attended with fever, though not unfrequently with manifest disorder of the digestive organs.

The small purple specks constituting the most simple forms of the disease may be sometimes mistaken for flea-bites, which, in appearance, they very much resemble. On attentive examination, however, they are found to be unattended with inflammatory redness, and are consequently not raised above the surrounding surface. They are, moreover, distributed more extensively over the skin ; and if febrile symptoms, or much disorder of the digestive organs exist, and more particularly if vomiting has recently taken place, are dispersed very thickly over the face, neck, and superior parts of the body. Frequently two or three appear to join and run into each other, forming larger spots, which still, however, preserve their distinctive character, and are circumscribed.

The more formidable cases, namely, those where the cutaneous affection shows itself in the form of large purple coloured blotches on the surface, and with which a few of the specks above described are intermixed, claims the most important attention, and may be considered a seriously aggravated form of the disorder. In such cases the effusion does not appear to take place from two, three, or more minute vessels, but from a comparatively extensive space of the surface, the extravasated blood diffusing itself under the cuticle,

" On the sixth the internal surface of the epidermis raised by the sanguineous serum, was observed to be covered by a reddish coating, while the surface of the dermis retained its natural colour. The dark or yellow discoloration of those patches, the epidermis of which had not been raised, continued to disappear from the centre towards the circumference, which assumed a jagged appearance. The same drink was continued through the following days; the spots from purple became yellow, and they afterwards disappeared completely. The swelling of the extremities subsided, but the cough and mucous wheezing or rattle in the left lung continued. There was still some expectoration. These trifling symptoms ceased some days afterwards, and the health was improving, when the hands became swollen afresh, but without any new hemorrhagic spots. A few doses of rhubarb put all to rights again, and the patient was in good health when she was discharged from the hospital. Auscultation of the heart gave the same results as when the patient was admitted. At no stage of the disease did the urine, treated with nitric acid, let fall any albuminous precipitate.

and giving the appearance of contusion, the margins of the blotches being considerably lighter than the centre, and here and there assuming a yellow or greenish hue.

There is every reason to admit the truth of the position, that loss of tone in the vessels of the surface is requisite to admit of that rupture of their extremities and consequent extravasation necessary to the production of petechiæ ; and also that the cause of such loss of tone must have a constitutonal origin.   It is not true, however, that an obvious wasting of the solids and other symptoms of debility always precede it ; or that deficiency either in the quantity or quality of the food taken, or that excess of laborious bodily or mental exertion, is necessary to its production.   It unquestionably does occur now and then under such circumstances, but it is also very frequently seen in persons who a day or two previously have been in apparent good health.   It is usually accompanied with sensations of languor and lassitude, even in these ; there is a furred yellow tongue and uneasiness in the head, nausea and constipation, and diminished appetite ; a train of symptoms, in short, indicating the highest degree of derangement of the digestive organs.   If this state of disorder exist, it has appeared to me, not unfrequently, that the most robust and full habits of body are more liable to petechiæ, though appearing partially, than those of an opposite character, and I have been led to form this opinion from having witnessed several cases, of which the following may be taken as a specimen.

A young female usually enjoying very good health, of a robust habit of body, and florid healthy complexion, had complained for two or three days of nausea, constipation, uneasiness about the head, loss of appetite, &c. ; with a view of removing which, she had, the day before I saw her, taken some aperient medicine.   The lattter not appearing to have given the expected relief, she, on the following morning, had recourse to an emetic of ipecacuanha : after a few efforts to vomit had been produced. the persons about her were alarmed by the sudden appearance of petechiæ of very considerable size, distributed exceedingly thick over the whole of the face, neck, and shoulders, while the conjunctiva was almost entirely suffused with blood.   Her complexion had been previously exceedingly clear, and the sudden and extraordinary change which the appearance of the disease effected excited the highest degree of alarm.   Her pulse was at this time hard and quick, and she complained of deep-seated pain in the chest with considerable oppression in breathing : having, however, seen other cases of a similar kind, where bleeding was dispensed with, and speedy relief effected by the use of cathartics, I depended entirely on the latter, making calomel an important part of what was exhibited.   The offensiveness and pitchy colour of the evacuations, as well as the immense accumulation of undigested aliment which they contained, amply satisfied me that this was the best and most direct plan of treatment.

It has been maintained by some writers, and among others, I believe, by Dr. Mills, Dr. Parry, and Dr. Combe, that venous congestion is necessary to the production of petechiæ. However true this may be in the petechiæ of low fevers, it may be doubted as regards those occurring under symptoms of general debility, unaccompanied by fever, as well as those arising from disorder of the functions of the abdominal viscera, where fulness of system is not decidedly marked. It must be admitted, however, where the latter exists, as in the case above detailed, during the act of vomiting, that such congestion is, sometimes at least, temporarily established, and precedes the appearance of the disease.

Purpura simplex, or petechiæ sine febre, may be said to be *always* preceded, whether in debilitated or other constitutions, by some disorder of the digestive organs ; and in all the cases which I have had an opportunity of observing, where symptoms of such disorder were not plainly discernible, the state of the evacuations only, fully justified this opinion. I am inclined, therefore, in all cases where debility is present, to place this and the disorder mentioned, in the relation of predisposing and exciting causes of the disease. With respect to the treatment of simple petechiæ, it is evident no harm can arise in commencing it by some alterative aperient, and this course is always entitled to the preference over one which rests solely on the exhibition of tonics. Purgatives will frequently remove every vestige of the disease without having recourse to tonic medicines at all ; but the same cannot be said of the latter, for it often produces a considerable increase of the number of spots and great febrile excitement : but of course it is not meant to be stated that the latter can be dispensed with with propriety, if symptoms of debility remain after the secretions have been restored to order. As medicines, bark and the acids seem to be entitled to the preference, where a tonic plan of treatment requires to be instituted.

The very correct description of Willan and Bateman of the P. hemorrhagica discloses no points of dissimilarity between the latter and P. simplex but such as constitute mere difference in degree both of the cutaneous affection and the constitutional symptoms. In P. hemorrhagica, a few of the smaller petechiæ are intermixed with large spots of extravasated blood, varying greatly in extent, according to the situation of the part, being largest, and often appearing first where the return of the blood to the heart is effected with the least facility, or where a greater degree of warmth is produced by the covering of the part, or other circumstances. The lower extremities affording instances of this, while the spots on the face and neck are proportionably less. From the same cause partly, so far as heat is concerned, and partly from the greater degree of delicacy of the cuticle in these situations, the inside of the cheeks and lips, the surface of the tongue, and membrane of the labiæ pudendi and vagina, frequently poor out dark-coloured blood in considerable quantities ; while the motions are often discoloured by this fluid to an alarming extent. The slightest degree of pressure applied to the surface of the body appears to break down the relaxed and delicate veins of the spot, and produces a mark of contusion considerably larger than what may be supposed to have been covered by the compressing substance which has been applied : even the pressure of the finger, as in feeling the pulse, has been found adequate to this effect. In examining the gums attentively, the edge in contact with the tooth sometimes appears to have lost here and there its florid complexion ; to have become changed to a livid venous hue, and to have blood of the same colour oozing from it between the teeth. In some instances, the formation of vesicles of considerable size has been effected, containing extravasated blood, the quantity of the latter having been equal to the distension and elevation of the cuticle ; an occurrence, for obvious reasons, more common on the membrane of the mouth and lips, than on other parts.

"The same state of the habit which gives rise to these effusions under the cuticle produces likewise copious discharges of blood, especially from the internal parts, which are defended by more delicate coverings. These hæmorrhages are often very profuse, and not easily restrained, and therefore sometimes prove suddenly fatal. But in other cases they are less copious ; sometimes returning every day at stated periods, and sometimes less frequently and at irregular intervals ; and sometimes there is a slow and almost incessant oozing of blood. The bleeding occurs from the gums, nostrils, throat, inside of the cheeks, tongue, and lips, and sometimes from the lining membrane of the eyelids, the urethra, and the external ear ; and also from the internal cavities of the lungs, stomach, bowels, uterus, kidneys, and bladder. There is the utmost variety, however, in different instances as to the period of the disease, in which the hæmorrhages commence and cease, and as to the proportion which they bear to the cutaneous efflorescence.

"This singular disease is often preceded for some weeks by great lassitude, faintness, and pains in the limbs, which render the patients incapable of any exertion ; but, not unfrequently, it appears suddenly in the midst of apparent good health. It is always accompanied with extreme debility and depression of spirits : the pulse is commonly feeble, and sometimes quickened ; and heat, flushing, perspiration, and other symptoms of slight febrile irritation, recurring like the paroxysms of hectic, occasionally attend. In some patients, deep-seated pains have been felt about the præcordia, and in the chest, loins, or abdomen : and in others a considerable cough has accompanied the complaint, or a tumour and tension of the epigastrium and hypochondria, with tenderness on pressure, and a constipated or irregular state of bowels. But in many cases no febrile appearances have been noticed ; and the functions of the intestines are often natural. In a few instances frequent syncope has occurred. When the disease has continued for some time, the patient becomes sallow, or of a dirty complexion, and much emaciated ; and some degree of œdema appears in the lower extremities, which afterwards extends to other parts."

The pains described in the foregoing passage from Dr. Bateman are noticed in almost every case recorded, both before and since the publication of the latter ; and an attentive perusal of the cases will also generally discover an intimate connexion between them, and high degree of disorder of the digestive organs : the evacuations, where they have been at all attended to, exhibiting proofs that the secretions have been for some time in so vitiated a state as to be quite incompatible with generally healthy sensations to the patient, or with the due and efficient carrying on of the processes of digestion, and consequent support to the system. In the cases recorded by Dr. Harty, of Dublin, Dr. Buxton, Mr. Rogerson, and others, to which I shall refer more explicitly hereafter, these facts were most particularly noticed, and were allowed to point out a line of treatment which was followed by the most gratifying results. It may be true that, in many cases, as stated in the preceding quotation, " no febrile appearances have been noticed, and that the functions of the intestines are often natural." The remark, however, does not apply satisfactorily to any which have come under my notice, nor are the cases on record, which I have been able to refer to, calculated to support such a statement.

Influenced by consideration of the obscurity in which the pathology and treatment of the disease is involved, as well as by the great importance of the question as to its dependence on the same causes as scurvy, I have been induced to arrange the following table of ten reports, recorded at different periods within the last fifteen years. I have been anxious at one view to present a correct notice of the most prominent constitutional symptoms co-existent with the cutaneous affection in the different cases, together with the plans of treatment adopted, and their various results.

The obvious similarity of the phenomena of Purpura in many essential points to that of scurvy leads us to the opinion of its being merely a modification of the latter, and the result of the treatment of a case detailed in the medical cases and observations of Dr. Duncan,* with that of another to which this author alludes in a following page, from the pen of Dr. Graaf, of Goetingen, go far to the establishment of such an opinion ; and hence, probably, has arisen the occasional adoption of the term land scurvy. This view of the matter seems to have been very generally taken and acted upon up to the period of Dr. Willan's publication.

The ideas of the latter author, together with his rather unqualified recommendation of bark, wine, acids, and good living, seem to warrant the conclusion that the cases which had chiefly come under his notice were similar to those above alluded to. Dr. Parry, however, in the paper noticed in the annexed table, has adopted a different, or rather directly opposite opinion ; he considers the cases he has detailed as confirming his opinion that " in various diseases,

* Medical Cases selected from the Records of the Public Dispensary at Edinburgh, with Remarks and Observations, by Andrew Duncan, M.D. F.R. and A.S. Ed. 1784.

among which may be reckoned inflammations, profluvia, hemor-rhagics, dropsies, exanthemata, and other cutaneous eruptions, and even the generality of nervous affections, there is one circumstance in common, which is *an over distension of certain blood-vessels,* arising probably from their relative want of tone, or the due con-traction of their muscular fibres, and that the cases he has detailed were entitled to be denominated active hemorrhagies." It is to be observed, however, that no notice has been taken in their details of any hemorrhage from internal organs, exudation of blood from delicate membranes, or of deep-seated pains. The state of the stomach, bowels, and secretions, are also omitted, the tongue only being noticed as a little furred, so that it must be admitted that this was a case of the mildest and most insignificant form, and not con-stituting evidence of much weight in the question.

With reference to the cases detailed by Dr. Bree, it would appear that the conjoint influence of bleeding and purgatives rescued the patients from apoplexy and paralysis, as well as the disease under consideration. In every instance, however, the simple detail of the facts affords abundant evidence, that most essential benefit was derived from the use of purgatives, though the character of the evacuations is not noticed by the author, and notwithstanding the opinion he has given, that "Purpura may arise from compression of the brain, giving occasion to the want of a contractile power in the fibres of the extreme vessels, we cannot, with due regard to that rigid examination of evidence necessary to lead us to correct ideas on a subject of so much importance, fail to observe, that in the first case, petechiæ had shown themselves three weeks before a debility of the whole system had increased, so as to bear the cha-racter of paralytic weakness :" or that in the second, intended to be explanatory of the theory of the cause of Purpura above quoted, the patient had been several times affected with slight symptoms of paralysis which had *given way to purges of calomel; that she had been oppressed with bile, and relieved by copious discharges from the bowels in the early part of her complaint;* that although this state may have been followed by confusion of the head and affection of the organs of speech *accompanied by diarrhœa,* reason-able doubt may be entertained that that diarrhœa, even though it may have been relieved by bleeding, depended on the same state of vessels on the surface of the bowels, or internal organs, as that which admitted of actual extravasation of blood on the surface of the body : not that we mean to question the possibility of relaxations of the vessels of the mucous surface of the bowels or elsewhere, as produced by oppression of the brain, but simply to observe that some-thing more than mere relaxation is necessary to admit of that actual extravasation of blood on which Purpura is well known to depend; and that it appears probable that the apoplectic state of Dr. B.'s patients depended on the same cause as the petechial spots; namely, a high degree of disorder of the digestive organs. The absence of any notice, however, of the state of tongue, or character of the secretions, leaves us much to wish for in the shape of evidence.

Dr. Harty's communications are of a different character from the preceding, and when added to the evidence produced in the cases which follow them regarding the uniform efficacy of purgative medicines, and the character of the evacuations, amount almost to proofs that this formidable disease, whatever may have been the predisposing causes, is immediately brought about by obstruction in the hepatic circulation, and consequent impediment to the functions of the stomach and alimentary canal. In no instance does it appear out of the whole recorded, where any attention to the state of the evacuations was paid, that the latter did not exhibit the most unequivocal proofs of congestion in the liver, and highly disordered biliary secretion, such as have been noticed by Dr. Harty. As the abbreviated manner in which the details of Dr. H.'s cases have been worded, enables me to give them at length, their sterling worth and great importance in the question under discussion render it an imperative duty on me so to do.

"CASE 1. The subject, an unmarried female servant, about thirty years of age, rather corpulent ; having, till of late, enjoyed apparent good health. Indigestion and a gnawing pain at the pit of her stomach constituted her chief complaints when I first saw her. In about three weeks petechiæ appeared, soon followed by slight and occasional hæmorrhage. The case, both in its commencement and progress, so closely resembled the very accurate description of Purpura hæmorrhagica given by Dr. Willan in his Reports, p. 90, that I need not employ a long detail of symptoms. Entertaining a high and well-merited respect for Dr. Willan's authority, I strictly followed *his mode, i. e.* tonics and stimulants, of treatment. The kindness of friends liberally supplied her with nourishing diet and tonics ; and, with the advantages of country air, I have reason to believe that all my injunctions were fairly complied with. In vain :—the hæmorrhage became daily more and more profuse and difficult of suppression. After suffering immense losses of blood, from every organ successively, she gradually sunk under the struggle. The fifth month terminated her sufferings. I was refused permission to examine the body after death : a refusal I have not yet ceased to regret.*

"CASE 2. The subject, a delicate woman, about the same age, worn down by frequent parturition, poor diet, bad air, deficient clothing, want of cleanliness, and confinement to a cold, damp ground-floor. The disease set in with a severe attack of cholera :

* It is much to be regretted that practitioners generally do not oftener follow the example of Dr. Harty, by recording cases in which they *fail*, instead of constantly blazoning forth their extraordinary success in others. A system of treatment which has been unsuccessful in any particular disease is surely valuable as a landmark to warn others against repeating it.

in two days after, petechiæ appeared, quickly followed by hæmorrhage from the mouth, nose, and stomach.   Convinced of the inefficacy of mere tonics in bad cases, and forcibly impressed by the occurrence of cholera previous to the appearance of petechiæ, and by a recollection of the *remarkable pain in the epigastric region,* which so generally precedes them, I determined, in this case, to direct my whole attention to the state of the abdominal viscera, and accordingly prescribed a brisk purgative of calomel.   From the good effect of the first, I directed its repetition for a few successive nights.   To my surprise, the hæmorrhage soon ceased, the spots rapidly disappeared, and in less than ten days the patient recovered, under every possible disadvantage of constitution, of air, and of diet.   Encouraged by the unexpected result of this unpromising case, I now no longer hesitated in employing purgatives, and trusting to them only in both species of the complaint.

" CASE 3.   A boy about three years old, on whose face and body large purple spots appeared, when three weeks convalescent from scarlatina.   There was occasional, though not profuse, hæmorrhage from the nose, gums, and fauces.   Calomel and jalap were liberally administered for four successive nights.   The fæces were black like pitch, and highly offensive.   After each purgative there was evident improvement.   The fæces assumed a more healthy appearance ; and, by the sixth day, a single spot was not to be traced on the whole body.   These are the only cases of Purpura hæmorrhagica that have fallen under my care.

" Of the Purpura simplex, I have seen about ten cases, all of which readily yielded to the same plan.   The most obstinate case was that of a girl of eleven years, on whom the petechiæ appeared without any previous illness, if we except a slight degree of languor and heaviness.   The spots were more numerous, and more generally diffused than I have witnessed in any other instance ; in two days they had reached their acmé, and were then accompanied by headache, quick pulse, and foul tongue, (the only case in which I had seen fever present).   It was necessary to purge this patient to a greater extent, and to employ much stronger doses than usual, to effect that purpose.   The purgatives were continued for eight successive days, at the expiration of which period every symptom of disease had disappeared ; and without the aid of other medicines, the patient was restored to better health than she had hitherto enjoyed.   All the cases of Purpura simplex were under fifteen years of age, and all among the children of the poor.   In some, the eruption had continued three or four weeks before purgatives were tried : and, in all, the stools were dark-coloured, though not so black as in Purpura hæmorrhagica.   From my own experience I can add nothing further on this subject ; but I can state, that, having made an early communication to the other physicians at the Dispensary, the practice was adopted by some of them with a success equally marked and rapid.   One of them (now no more) employed calomel only.

A physician, attached to another Dispensary, informed me of his having successfully employed purgatives in one case, attended with enlarged abdomen, and complaints of pain in it—circumstances which led to the use of that remedy."

In the first of Mr. Rogerson's cases, nothing was observable in the constitution of the patient which could lead to a supposition that mere debility of system was the sole cause of the disease ; if that had been the case, indeed, the result would have proved the fallacy of the opinion, as recovery took place without measures of a tonic kind having been had recourse to : the pain in the head, oppression at the stomach, nausea, and colour of the tongue, together with the constipated state of bowels, were sufficient indications of disorder in the functions alluded to, to lead to the exhibition of purgatives, from the action of which the relief was so immediate, " that even her attendants were convinced of their utility." Cold ablutions were resorted to in both the cases which he details with apparently good effect ; but the rapid disappearance of the whole of the threatening symptoms from the time the bowels were freely opened, leaves little doubt that the life of the patient was saved entirely by the adoption of measures leading to this object. In the second case no ground exists for believing that the abstraction of blood at all contributed to the recovery of the patient ; nor is any notice of a beneficial change following it alluded to by the writer. As in the former, improvement rapidly went on after the powders (calomel and jalap) "had operated freely, and produced many stools of a dark colour, and intolerably offensive ;" nor was any idea of the necessity of tonics to the cure suffered to suggest the use of even a little porter and animal food, till the petechiæ and all other symptoms had entirely disappeared. In both cases, the most dangerous set of symptoms had established themselves ; the sudden recovery from which, after the operation of purgatives, does not appear to admit of explanation, except by referring them directly to congestion and disorder of the liver, stomach, &c. Indeed the character of the secretions merely is sufficient to establish that fact.

Dr. Buxton's case, as will be seen by reference to the table, approaches very nearly in the chief distinguishing features of the disease to those of Dr. Harty and Mr. Rogerson. The same beneficial effect followed the use of brisk purgatives, and the evacuations were of a dark olive green colour : a colour, according to Dr. B.'s impression, not dependent on any mixture of blood with the secretions. Two important facts worthy of observation in this case are, that the tongue was tolerably clean, and the bowels regular. The absence of symptoms of such importance in the indication of disorder of the digestive organs might have led to a different practice from that adopted, had not the appearance of the complexion led to the suspicion of the truth.

The use of calomel and jalap is as unequivocally followed by rapid improvement in the case recorded by the Editors of the Medical Repository, as in the foregoing ; and the same doubt seems to hang over the point as regards the benefit or necessity of bleeding, as in the last case of Mr. Rogerson : the evacuations here also were of a dark green colour. and copious.

The next in succession are those of Dr. Nicholl, whose experience of the utility of the oil of turpentine in a variety of other affections has led him to submit it to trial in that before us. When the patient first-mentioned came under the doctor's notice, he had been well purged by the directions of another medical man previously in attendance. This discipline was ordered to be repeated by means of calomel and jalap, and on the following day he began to take the turpentine, night and morning, and rapidly recovered.

The second case is precisely similar to the first in all essential points : no notice is taken of the character of the evacuations, except during the recovery of the second patient, when they are stated to have become natural. Nor is the *modus operandi* of the turpentine either theoretically or practically alluded to, and we are left to conjecture whether it is to its stimulant, diuretic, or cathartic properties that we are to consider ourselves indebted for the cure. In the first case, a great part of the object seems to have been accomplished, if our ideas, which the preceding cases have suggested, are correct, before the turpentine was employed : it is to be supposed, however, that the latter kept up the purgative action of the bowels in both cases.

In the very violent and speedily fatal case given by Dr. G. Johnson, *no passage through the bowels had taken place for a week previous to the first visit of this gentleman to the patient;* and the symptoms probably were rendered the more violent on this account. Injections, large doses of the pil. colocynth comp., and of calomel and jalap, produced no effect on the bowels of the patient, and fourteen ounces of blood taken from the arm was followed only by temporary relief. *The patient died without having had the bowels excited.* The blood drawn exhibited only slight traces of coagulable lymph ; nor did the *post mortem* examination lead to any discovery of the products of inflammation in any internal organ. The petechiæ were distributed over the heart, stomach, and other viscera in great abundance.

Dr. Duncan, jun. has added to the case last mentioned in the Ed. Med. and Surgical Journal, one which he himself treated some time previous, by referring to the abstract of which in the table, a similar striking disappointment in the expectation of permanent benefit from bleeding will be noticed. This operation appears to have been had recourse to on Dr. D.'s patient, in consequence of its having been apparently beneficial in the cases of Dr. Parry, and to have been repeated on the supposition that the " return of the hæmorrhage was from the pulse, *à posteriori*, being greater than the tender state of the vessels could bear. It appears more than probable that this last abstraction of blood was very prejudicial, it having been followed by great prostration of strength at a period .

when a great demand for the latter was made on the constitution of the patient, for the purposes of bringing on a healthy state of parts after a gangrenous slough in the throat. The effect of purgatives, whenever employed, was the production of black and fetid dejections, the character of which never changed for the better after the first appearance of the spots and hæmorrhage.

The case more recently detailed by Mr. Pretty is remarkable on many accounts. Before the petechiæ appeared, the tongue was furred, there was pain in the epigastrium, and sickness, and the bowels had not been opened for two days, when several evacuations were produced by purgatives, which however did not relieve the pain, though they appeared to have abated the sickness. The breathing became more frequent, the pulse harder, and the fever increased ; when bleeding was had recourse to, and nitrate of potash with antimonial powder prescribed. The next day, symptoms of formidable congestion about the chest and head had taken place, *but the bowels were free.* The petechiæ had increased, but the symptoms of congestion appeared to call for a repetition of bleeding, which was performed to the extent of 10 or 12 ounces, and followed by syncope and temporary relief : again the fever and bad symptoms returned, and the mineral acids were had recourse to under the direction of Dr. Johnson : death, however, took place, preceded by the symptoms of the last stage of pulmonic inflammation. A part of the blood drawn at the second bleeding only exhibited a buffy surface. The *post mortem* examination disclosed no vestige of recent active inflammation of any internal organ, but merely the appearances of congestion. It is much to be regretted, and this remark is made without the slightest intention to condemn any part of the proceedings in this case, that the state of the secretions had not been more fully noticed, and the use of purgatives carried to a greater extent ; for the hepatic congestion and stomach disorder indicated by pain in epigastrio, sickness, constipation, and furred tongue, which ushered in the disease may be, in this as in other cases, justifiably considered in the situation of an exciting cause to all the subsequent commotion and mischief which occurred.

Dr. Duncan concludes the recital of his patient's case, by observing, that the possible modes in which he can conceive this disease to arise, are

1. Increased tenuity of blood, allowing it to escape from the superficial extremities of the minute arteries.

2. Dilatation of the mouths of these arteries allowing natural blood to escape.

3. Tenderness of the coats of the minute vessels giving way from the ordinary impetus of the blood.

4. Increased impetus of the blood rupturing healthy vessels.

5. Obstruction in the vessels causing rupture, with natural impetus, and without increased tenderness.

6. Two or more of these causes may act simultaneously or successively.

Most of these points have undergone much learned disquisition
and theoretical argument with relation to scurvy and other diseases.
The third idea of Dr. Duncan above quoted is unquestionably correct as regards the formation of the cutaneous spots of purpura.
That this tenderness is the result of deficient nourishment in the
superficial vessels, is perhaps equally clear ; and it may fairly be
suspected that such deficiency is consequent on the congestion in
the hepatic and gastric circulation.

On a review of the foregoing cases, it will be observed, 1st, That a
striking uniformity of symptoms indicating hepatic congestion and
general disorder of the digestive organs ; of those organs, on the proper
performance of the functions of which, the formation of blood capable
of conveying nourishment and the materials of growth to distant parts
of the body chiefly depends—occurs in all of serious importance. 2d.
That these symptoms, consisting of constipation, dyspepsia, oppression of the chest, pain and tenderness in epigastrio, headache, &c.
have been pretty clearly ascertained, both by the state of pulse
during life, and by examination of these organs after death, as well
as by the absence of important marks of inflammation in the blood
drawn by venesection, and also by the effects of blood-letting, not
to depend on a state of actual inflammation.   3d. That the constant
effect of purgatives in dislodging vitiated secretions, consisting
chiefly, apparently, of accumulations of black biliary matter, has
been pretty uniformly followed by the most strikingly beneficial
change in the symptoms, while recovery took place in no instance
without free purging.

The effects of remedies exhibited on principles founded on particular theories are usually tolerably correct tests of the truth of such
theory ; and though instances often enough occur where the constitution bears up against both disease, and medicines given under
incorrect and dangerous notions of disease, or what is as bad, given
without any knowledge or care about the matter ; yet such instances may be considered only as exceptions to a general rule.   It is
only necessary after this remark to direct the reader's attention to
the comparative good or evil of blood-letting—of the opposite system of tonics,—and the use of purgatives, as exhibited in the table,
taking due notice of the cases which have been extracted from Dr.
Harty's paper.

The direct cause of P. hæmorrhagica does not appear in any case
of which I have been able to discover the records, or which have
come under my observation, to depend on debility of system merely.
Privation of food had not been experienced to any considerable
extent in any of those immediately before us, and it may be doubtful whether a certain degree of energy of circulation may not be
requisite to cause the escape of so much blood as is necessary to
produce the enlarged blotches on the surface which form the distinguishing feature of this disease, and from which it derives its
classic name, and to explain the profuseness of the hæmorrhage
from surfaces covered with more delicate membranes.

The chief questions for decision seem to be, whether any thing like energy of circulation is consistent with that degree of debility and relaxation of the vessels of the skin where the tenderness of their coats disqualify them to resist the common force of the circulation ; and if it be, upon what circumstances can the latter depend. Is the highest degree of hepatic and general visceral congestion and obstruction in the abdomen with which we are acquainted capable of so impeding the functions of digestion and chylification, as to become a cause of such reduction in the nutrient properties of the blood, as to render this fluid unequal to the efficient nourishment of every part of the system ? If it is, in what parts of that system would the debility consequent thereon be first manifested ?

With respect to the first of these questions, if a positive answer cannot be readily given in the affirmative, it is at least to be considered not improbable. For the second, it will occur to us, that parts already built up by the previous healthy action of vessels, and not dependent on the latter every hour for their vitality ; parts in a state of quietude and rest are not those in which such debility would be expected first to appear. The vessels themselves, in ceaseless action, and constantly under the influence of a distending power, would, reasoning on common principles, of necessity, be the first to suffer. The *vasa vasorum* supply the coats of the vessels themselves no better than the latter supply other parts, and these, therefore, being called on to make greater efforts in resisting the impulse of the circulation, first disclose the general deficiency by the rupture of their extremities.

The occasional temporary relief to the hurried and laborious breathing and pain in the chest, experienced from bleeding, rather confirms than opposes this view of the case, necessarily followed as the operation must be by diminution of the congestion in the lungs, which these symptoms denote, and which may be fairly concluded to be the consequence of impeded circulation in the liver.

Another case equally instructive on these points is recorded by Dr. Bateman, * in which the decided superiority of evacuating remedies,† and the danger of bleeding, are both strikingly exemplified.

* Edinburgh Medical and Surgical Journal, vol. ix.

† Among many cases which I could produce there is the following, in illustration of this remark. This child being of indolent habits, and addicted to eating and drinking in an inordinate degree, still continues, from time to time, to experience an attack more or less severe : it is always, however, readily subdued by a few doses of calomel and extract of colocynth.

A youth about ten years of age was attacked with a considerable degree of fever, accompanied by the eruption of Purpura, the large purple patches of the disease being mixed with petechiæ, and distributed over every part of the body, particularly the lower extremities. Great difficulty of breathing, a rapid and very hard pulse, severe headache and pain in epigastrio, accompanied by much tenderness on pressure. He had been previously occasionally subject to great irregularity of the action of the bowels, accompanied by a febrile disorder, restless nights, loss of appetite, and great disposition to drowsiness, which usually yielded to calomel and active cathartics without much trouble, the bowels at

25*

There are two other species of Purpura described by Dr. Willan, the P. urticans and P. contagiosa.   The P. contagiosa is merely the petechiæ of low fevers.

The Purpura urticans, as its name implies, is somewhat allied to nettle rash.   It seems to be chiefly, if not entirely, confined to the poorer classes of children, between the ages of four and fourteen ; and cases of it are not uncommon among the patients of the Metropolitan Infirmary for Children.

If this form of the disease be seen at its commencement, it will be found much more to resemble urticaria than common Purpura. It usually appears, indeed, as urticaria accompanied with fever, the purple hue of the spots not taking place for many hours after the attack.   It is generally of a distinctly intermittent nature, or more properly speaking, the febrile symptoms by which the rash is ushered in continue only for a few hours, and return every day, or every other day, with tolerable regularity, most frequently in the evening.

such periods evincing proofs of having been some time much overloaded, and the evacuations being exceedingly offensive.

The bowels at the period of this attack were supposed to have been regular, but it was evident, on examination of the abdomen, that they were in a loaded state.   The case was one which would, on a superficial view of it, appear to have called loudly for immediate venesection, and the pain alluded to, with the state of the pulse, &c., would, under other circumstances than their accompaniment by the eruption, have been thought to denote a serious degree of inflammatory action.   The stomach was extremely irritable, and rejected the smallest portions of liquid.

The day after the appearance of the purple spots and petechiæ on the skin the gums began to bleed ; hæmorrhage also took place from the nose to a considerable extent, and blood passed from the bowels.

The state of the stomach precluded a possibility of employing any fluid medicine, and six grains of compound extract of colocynth, with two of calomel, were given in the form of pills every three hours.   This dose was four times repeated before the bowels began to act ; at the end of this time a copious evacuation of horribly offensive secretions, having but little appearance of feculent matter, took place.   This was soon followed by others of the same nature, to the speedy relief of the patient from the pain and other severe symptoms.

The two following days were spent in the exhibition of the same purgative at more distant intervals, and the occasional use of a little chicken broth by way of sustenance, the stomach after the first evacuation having become considerably less irritable.

At the end of ten days the secretions had become tolerably healthy, the purple spots had nearly disappeared, the sites of some of the larger being only occupied by yellowish discolourations, similar to what take place after contusions.

He was at this period removed into the country, but in a few days relapsed, no doubt in consequence of some oversight either in what he had taken as food, or from the bowels not having been kept sufficiently open.   He here had the misfortune to have a different view taken of his case, and the antiscorbutic plan put in full operation upon him.   Calomel purgatives, which had been formerly found effectual, were suggested by his parents ; but his medical adviser unfortunately considered them to have been the chief causes of his present disease. All his symptoms daily became aggravated under the employment of tonics, and his situation was one of serious peril.

The free use of the same purgative was ultimately had recourse to, and he gradually improved : he became anasarcous, however, to a monstrous extent, and remained for some time afterwards extremely delicate.

A slight degree of chilliness is first complained of; this is soon followed by heat and fever, and then the wheals and spots of nettle rash rapidly come out, accompanied by the usual feelings of heat, tingling, itching, &c. These sensations end with the febrile excitement, the white and red wheals subside, and the spots which they occupied become in the course of a few hours of a dark purple hue, a few petechiæ being generally interspersed, and both disappear very slowly. When the febrile symptoms return, a fresh formation of the white wheals and spots accompanies it, and when it subsides, these also become blue; and thus by repeated attacks the whole body is sometimes nearly covered. In the absence of fever there is no itching or tingling; and the case, as regards the local characters of the disease, is one of Purpura. It is very rarely, however, accompanied by hemorrhage from the lining membranes, as in common cases of this disease.

The treatment most successful, and which is indeed always adequate to the cure, consists first of the free use of calomel purgatives, and secondly, of the sulphate of quinine in the manner in which it is employed in common ague.

The latter medicine may generally be depended on for the prevention of a second attack, if the secretions have been previously rendered healthy by a proper use of purgatives.

The question as to the identity of Purpura with the common scurvy appears, on account of the support it has derived from the authority of Dr. Willan, to require some notice in this part of our subject. On a first view of the cases which I have selected to enable the reader to form his opinion of Purpura, it would appear that, except as regards the cutaneous affection, but little resemblance existed.

Whatever may have been the particular theories of individuals as to its causes on board ship; whether putridity, in any thing like the literal sense of the word, may really be capable of existence in the living body, as supposed by Pringle; whether simple debility from privation of nourishment, according to Drs. Lind, Blane, and Milman, or privation of oxygen, according to the theories of Trotter, Goodwin, Beddoes, and others, may appear to have had the power of inducing scurvy; yet it must be remembered that this disease has always been most distressing in cold weather, even when the circumstances, by which any of such states may be brought about were much less in action than when the ship in which it has made its appearance has been much longer out, if arrived or cruising in warmer latitudes: a fact, which is of serious importance in the investigation of the pathology of scurvy, and one which every surgeon of an East Indiaman, making a tedious voyage to India, may have opportunities of observing, sometimes on a large scale.

The very trivial forms, in which the disposition to scurvy makes its appearance in well-regulated ships at the present day scarcely enables us to identify it with the formidable disease described by the authors alluded to; but on such voyages as those named, sufficient opportunities sometimes occur to show that the minor degree of nourishment contained in salt provisions ought only to be considered in the light of a predisposing cause.

On such voyages, where the provisions had not been carefully selected, a period of seven weeks or two months at sea, about which time the ship arrives off the Cape of Good Hope, brings generally, a few cases of scurvy on the sick list. Stiffness of the hams and legs, languor, debility, purple spots, and a vesication or two, usually about the knees, with uneasiness at the pit of the stomach, bleeding of the gums, a doughy thickening of the skin and integuments of the lower extremities, and constipation, are, however, the only symptoms noticed ; and as the ship approaches the termination of her voyage, these usually disappear.

It should be observed, that lime-juice is always furnished on board the ships in question ; but this is not sufficient to protect the crews from the slight attacks described.

The facts that the cutaneous affection is accompanied by the same symptoms of hepatic congestion, *i. e.* uneasiness in the epigastrium, constipation, and languor, which ushered in the phenomena of the old scurvy, that it importantly resembles, in its essential characters, the spots of the latter, and that like it, it rapidly disappears, if an increased flow of perspiration and healthy biliary secretion be induced; are quite sufficient to establish an intimate analogy between them ; and while these favourable changes continue to take place on board ship on the approach to warmer latitudes, where no variation in the diet or circumstances of the patient can in other respects have taken place, it is fair to presume that, as in Purpura, the direct exciting cause must be hepatic congestion and deranged intestinal function. In short there can be no doubt that the terms "land and sea scurvy," constitute distinctions without a particle of difference.

It has long been matter of notoriety, that the employment of lime-juice is not to be depended on as a preventive of scurvy, notwithstanding its former high character : and but little opportunity of observation is now necessary to show, that its good effects depend less on its antiseptic properties, than its power of exciting that gentle and healthy action of the bowels, which is the effect of most vegetable productions which are used as food.

It need not be observed that a diet consisting chiefly of meat which has been long salted, and consequently suffered a material diminution of its succulent and nutritious properties, is more speedily followed by scurvy than that which has recently prepared ; but it is also a fact, that the disposition to constipation is most marked and obstinate in such cases, long before other symptoms of disease make their appearance.*

* In the season of 1815 and 1816 these remarks were exemplified in the cases of two Indiamen, with troops on board for Calcutta. By some mistake, part of the beef and pork had consisted of old naval stores unfit for such a voyage. The cases of scurvy, while the men were fed on these, became alarmingly numerous

The light which has been lately thrown over the pathology of dysentery by Dr. Johnson,* Mr. Bampfield,† and others, enables us to see a considerable similarity between this disease and scurvy, hepatic obstruction and deranged biliary secretions alike distinguishing both; and it may not be improbable, that in some cases the inflammatory symptoms of dysentery supersede those of scurvy, only because a greater degree of susceptibility to irritating causes exists in the constitution of the patients. Thus an additional link is added to the chain of effects which leads us from the common dyspeptic sensations of every day's experience, down to formidable and destructive disease.

It has been already stated, that the cutaneous affections of scurvy, as the disease occurs at the present day, are a thickened and hardened state of the skin of the lower extremities, with dark-coloured blotches, and frequently the formation of a bulla containing bloody serum. The situation of the latter is usually about the knee, and I have been induced to think it generally produced by kneeling, or by some slight contusion against the rigging in going aloft. The cuticle of the vesicle is generally broken when it first comes under the eye of the surgeon, and part only of its contents remains. Its striking similarity to the Pompholyx, together with the fact that the latter is often intermixed with petechiæ, and originating in debilitated habits, establishes the opinion as to their dependence on states of the constitution somewhat similar.

It is not difficult to conceive that a degree of relaxation in the vessels of the cutis, somewhat less than that which renders the mere impulse of the circulation equal to their rupture, as in petechiæ and vibices, will admit of the escape of the serous portions of the blood only under the cuticle; and this appears to be the manner in which both the vesicles of scurvy and the common pompholyx originate. In the former, however, blood to some extent always escapes, and is mixed with the serum, giving it the appearance of bloody water.

The thickened appearance of the skin is, I believe, rather the conjoint effect of an interruption of the action of the absorbents of the part, with a slight escape of serum from minute vessels of the cellular membrane. Neither of these affections admit of any alleviation, except through the medium of the constitution.

and severe; and the commander of one, after expending the whole of his live stock in remedying the mischief, thought it unsafe, from want of hands, to attempt the passage round the Cape of Good Hope. In the other the mischief was less extensive; but constipation, hepatic congestion, and black evacuations, were observed in most instances, followed by dysentery of a severe character. Two fatal cases of the latter occurred within a short time of each other under my own observation. I had the advantage moreover of a post mortem and leisurely investigation with every convenience.

* On the Influence of Tropical Climates on European Constitutions.
† On Tropical and Scorbutic Dysentery.

Name of Reporter.	Description and state of System, &c. under which the Disease appeared.	Hæmorrhage, where from.	Appearance of the Tongue.	Pulse.	Internal Pains, their Seat.
1. Dr. PARRY.	A female, aged 60. —Full habit, slight febrile symptoms.	None.	Little furred.	Full.	None noticed.
2. Dr. BREE.	A female.—A disposition to corpulence, with slight symptoms of determination to the head.	None.	Not stated.	70 full, but not hard.	None noticed.
3. Dr. HARTY. 1.	A corpulent female, aged 30—Pretty good previous health.	Slight hæmorrhage, gradually getting more profuse through the case: where from not stated.	Not stated.	Not stated.	Severe pain gnawing at pit of mach; indigestion
2.	An opposite state of system, and bad general health.	Hæmorrhage, from the mouth, nose, and stomach.	Not stated.	Not stated.	Pain in epigastr
3.	A boy 3 years old.	Nose, gums, and fauces.	Not stated.	Not stated.	Pain in epigastr
4. Mr. ROGERSON, 1.	A young female.— Debilitated habit, and supposed disposition to phthisis.	Stomach and nose.	Flesh-coloured.	100 full, and intermiting.	Oppression at mach, nausea, thi constipation, he ache.
2.	A man who had been some time ill, and in a state of debility.	Bladder, the urine coloured with blood.	Covered by dark-brown fur, offensive breath.	. . . .	In the head, al men, and chest, c stipation.
5. Dr. BUXTON.	No apparent debility or fever.	Spitting of blood from the velum palati and throat.	Tolerably clean.	Quick, soft, and small.	Pain not notic bowels regular, dingy skin.
6. EDITORS of the MED. REPOSITORY.	A boy 12 years old. With anasarca from temporary debility.	From the nose.	Not stated.	Quick.	In the epigastric with tenderness a oppressed breathi
7. Dr. NICHOLL.	A man aged 37.— Very good general health, but lately weakened by an attack of fever.	From the mouth and bowels.	Not stated.	Not stated.	No pain noticed.
8. Dr. G. JOHNSON, of Berwick.	A married woman. —Stout and robust.	Blood oozing from the nose, and mixed with the sputa; bloody urine.	Covered by a dark brown and white fur.	Very quick and weak.	Head and back, w oppression about t chest, constipation
9. Dr. DUNCAN, Jun.	A young man, by trade a tailor, of debilitated habit.	Mouth and nose.	Not stated.	100 and small, bowels not confined.	Abdomen, w symptoms of dyspsia; pain in epig trium.
10. Mr. PRETTY.	A little girl, 7 years old. — Pretty good health, but subject to cough.	Labia pudendi.	Much furred.	Quick, but not hard at first, but subsequently rapid and inflammatory.	Oppression about t chest, pain in the s mach, nausea, hea ache, constipation.

Whether bled, and the Characters of Blood drawn.	Medicines exhibited.	Effects of Purgatives when had recourse to.	Termination of the Disease.	Journal in which the Case is recorded.
. S. Blood. Cupped c buffy.	None noticed.	. . . .	Recovery.	Ed. Med. and Sur. vol. v.
. S. Blood dense, but n inflamed, apparently beficial — repeated afwards from the disposi-f to paralysis, and sub-iently cupping.	Jalap and calomel often repeated.	Purging kept up during the disease, but the state of evacuations not noticed.	Recovery.	Medical and Physical Journal, vol. xxi.
leeding not had re-rse to.	Tonics and good living, and the advice of Dr. Willan strictly followed.	. . . .	Death.	Ed. Med. and Sur. vol. ix.
leeding not had re-rse to.	A brisk purgative of calomel and jalap every night.	Purging kept up, but state of secretions not stated.	After each purgative evident improvement, and by the tenth day not a single spot remained.	Ed. Med. and Sur. vol. ix.
leeding not had re-rse to.	Calomel and jalap every night for four nights.	Free evacuation of fæces, black, like pitch, gradually improving, and the purging continued.	Rapid recovery, complete in six days.	Ed. Med. and Sur. vol. ix.
leeding not had re-rse to.	Calomel and antimony to the effect of purging kept up.	Black offensive motions, directly followed by improvement and relief of pain.	Rapid recovery from the time the bowels were freely opened.	Med. and Phys vol. xlii.
V. S.	Calomel and jalap.	Copious evacuations of a dark colour; and very offensive; as the operation was kept up, the stools improved, till they became healthy.	Recovery.	Med. and Phys. vol. xlii.
leeding not had re-rse to.	Brisk purgatives continued through the case, with acids.	Evacuations copious, dark olive-green colour, no mixture of blood.	Recovery.	Med. Repos., vol. xix
led once to 8 ounces, h relief to the breath.	Calomel and jalap, the action of which was constantly kept up.	Evacuations copious, and dark-green.	Recovery.	Med. Repos., vol. vi.
leeding not had re-rse to.	Had previously taken purgatives. Calomel and jalap, and next day ol terebinth. regularly continued twice a day.	Evacuations denoting deranged secretions.	Recovery.	Med. Repos., vol. xvi

CHAPTER II.

OF APHTHA, OR THRUSH.

In justice to Dr. Bateman, it is proper to observe, that on the subject of the above affection but little has been recorded by ancient, or discovered by modern writers, which has not been duly noticed in his Synopsis. It is a disease, as occurring among infants, usually of trivial importance, and rarely capable, in its most aggravated forms, except where extreme debility has been induced, of producing any seriously bad effects.

Infants of all classes seem to be more or less subject to it; but it occurs much more frequently where deviations from the natural and proper manner of feeding have been observed, it being more common, as well as more violent, in children who have not been properly suckled, than in those who are supported entirely by the mother's milk. Even in the latter case, however, it sometimes occurs, and is then usually supposed to depend on some derangement of the health of the mother.*

It is usually spoken of as a distinctly vesicular disease, as formed of "whitish or pearl-coloured vesicles appearing on the tongue, lips, and interior surface of the mouth and throat;" though its analogy, at its commencement, to the pimples of strophulus has been suspected, and alluded to by one or two preceding writers. I have never been able to see this disease in the mouth of the infant in the form of distinct vesicles, and am inclined to think that the appearances which have given rise to this idea of its character depend entirely on the peculiarities of the part, and that each white speck constituting part of the disease, is produced as the pimples of strophulus are, by a minute effusion of lymph under the delicate cuticle of the part. The peculiar delicacy of the latter in the infant, and the state of constant moisture and friction in which the tongue and other parts covered by the membranes of the mouth are kept in the infant, will readily explain why the substance of the pimple should be so soon disturbed and rubbed off, leaving that minute circle, or irregular and partly-detached white crust of cuticle which distinguishes the affection.

If the opinion of the vesicular character of thrush were correct, the disease would appear to be somewhat analogous to herpes; but as far as we are able to judge, there is by no means equal derange-

---

* Rayer has hastily confounded the Thrush of infants with the herpes preputialis of adults : there is no approach to identity in the character of these. They are utterly unlike as regards their etiology or pathology ; it is strange, beyond measure, that so bulky a work as Rayer's should have omitted the large and free consideration of the thrush of infants. This is, without doubt, a disease of the cuticular surface, and, what is more, of those parts of such surface as when diseased interfere with the healthy functions of life even to a fatal extent.

ment of the system acting as a cause, nor any thing like the same degree of pain and irritation attending it in its course ; nor does it ever happen that herpetic vesicles on other parts of the body are co-existent with the disease under consideration, situated as before described.

But, on the other hand, the pimples of strophulus are often, nay, generally, visible to some extent on the skin of other parts where the Thrush makes its appearance ; which observation renders the identity, or, at least, close analogy of the two affections much more probable. As an argument, perhaps, it is of little importance to allege, that the causes of both, namely, improper feeding and derangement of the stomach, are the same, for, in truth, this assertion may be made of a large portion of the diseases forming the subjects of our notice.

The unimportant character of Thrush, where it is not accompanied by diseased mesenteric glands, emaciation, or appearances of petechiæ and other marks of great debility in the infant, will excuse the following abbreviated description and notice of what has been recorded by different authors respecting its pathology and best methods of treatment.

The notice of the nurse is generally first attracted by apparent inability on the part of the infant to draw the milk in its usual satisfactory and contented manner, the effort being accompanied with more or less of pain and crying. The heat and irritation of the disease are soon communicated to the nipple of the mother, producing excoriation and excessive tenderness. On examining the infant's mouth, the lining of the cheeks, angles of the mouth, sides and dorsum of the tongue, have small white specks, more or less thickly distributed over them, and sometimes they are so closely set together, as to furnish a white incrustation down the centre of the tongue. The detached specks resemble and are occasionally mistaken for minute portions of curd, and these commonly separate earlier than in parts where the disease is more thickly distributed, leaving a florid and rather unhealthy-looking state of the part. If the original causes of the disease remain in operation, fresh crops of the minute specks are apt to occur, leading to much irritation and exhaustion, and sometimes considerable wasting of flesh.

Every thing which disorders the stomach and bowels, or reduces the strength of the system, is capable of producing Thrush. So that the habits and health of the nurse may be occasionally reasonably suspected, where no feeding by the hand is had recourse to, or where the latter has received an ordinary share of attention. An unhealthy atmosphere, want of cleanliness, and food in which a great quantity of sugar has been employed, and in the preparation and exhibition of which little attention is paid to consistence or quantity, are often noticed as its causes.

The duration of this affection depends on the degree of disorder of the general health, stomach, and bowels of the infant. If this be such as to admit of easy correction, the local disease lasts but two or three days, and does not become extensive; but if no alteration is made in these respects, it may continue much longer, and become a powerful cause of additional debility, by the irritation and pain, and interference with the exhibition of nourishment, attending it.   Children brought up by hand are most subject to its severer forms, and when these are sickly and delicate, inattention to their food is often productive of serious forms of the disease ; but there is no doubt that restriction as to quantity or deterioration of the quality of the milk is equally capable of bringing it on.

Delicate females sometimes have their infants suffer considerably by persisting in suckling them, though assisted by what is supposed an adequate quantity of spoon victuals ; but as soon as all attempts at suckling are forbidden, and the latter is 'entirely depended on for the nourishment of the child, the disease disappears, and the general health rapidly improves.

The kind of medicine most likely to do good in the treatment of Thrush is that which is most applicable to the constitutional condition of the patient.   Mild alteratives are, in the majority of cases, all which are found necessary ; an open and regular state of the bowels being of the greatest importance.

As an application, borax and honey in the proportion of about a drachm of the former to four of the latter, seems to have obtained the preference to all others, and it is generally the only thing had recourse to locally in the management of the disease.   Preparations of soda, indeed, have been long known as useful applications in subduing irritation of the membrane of the mouth and fauces, whether arising from excoriation or other causes.*

An idea generally prevails among nurses, that the eruption of Thrush extends through the whole alimentary canal, and hence they are accustomed to look on any irritation or redness about the anus as a sign of the termination of the disease ; it being under these circumstances supposed to have finished its travels !   There seems to be no reason to suppose, however, that it extends beyond the mouth and fauces in ordinary instances ; the irritation about the anus being much more satisfactorily explained by reference to the irritation produced by diseased secretions.

An inconvenience of some importance is often sustained by the mother or nurse in suckling infants, during the progress of the disease, from the excoriation of the nipple before alluded to. The cause of this excoriation is generally understood to be the application of the diseased secretion of the tongue and lips of the infant to the part. It is probable that such affection might be entirely prevented, if the existence of Thrush in the infant could be sufficiently soon

---

* A solution of chlorate of soda has lately been introduced to the notice of the profession by Dr. Darling, as possessed of extraordinary powers, not only in subduing irritation in the membrane of the mouth and fauces in ordinary cases, but of restraining or so modifying the action of the salivary glands when under the influence of mercury, as to render the situation of the patient much more comfortable than has been hitherto usual under such circumstances.

discovered : as the simple application of a little warm water to the nipple, after each time that the child has been allowed to suck, would effectually dislodge the cause of the irritation.

The part of the nipple most frequently the seat of this disorder is its base : around which a ring of abraded and extremely tender cutis is generally discovered, sometimes assuming the character of a deep fissure, somewhat resembling those of psoriasis. Extreme tenderness and pain are usually complained of, and if the abraded surface be minutely examined, it is often found covered by coagulable lymph. From the pressure of the dress, or covering of the part, when the affection has existed for a few days, and when a crack or fissure is produced, the sides of such fissure are brought into contact, and are partly glued together in the course of a few hours ; and in such case, the pain, when the child is again applied, is extremely severe, the recently-formed adhesion being forcibly to rn through by its efforts in extending the nipple. This affection is so well understood to arise in the way described, as to render Dr. Bateman's advice in the treatment of Thrush, namely, to change the nurse, extremely difficult to be followed, even if an opportunity should occur of making such arrangement satisfactorily in other respects ; and in the majority of cases, it will be found by far the better plan to wean the child at once. It will be obvious that, even on the account of the latter, this plan ought to be preferred, because the effort of swallowing properly-prepared spoon victuals would be much less painful to it than the attrition and muscular exertion of the affected parts in sucking. If sucking be continued also, it is impossible to have recourse to any applications likely to remove the pain and irritation of the affected nipple with safety, though the disease readily heals if a state of quietude of the part be enjoined, and applications of a sedative kind be had recourse to. Weak spirituous lotions, with a small quantity of ext. litharg. seem entitled to preference on some points ; but the red precipitate ointment, very much diluted, is also used with great advantage.

The Aphtha Adultorem, as it has been termed by Dr. Bateman, is a different disease from the foregoing in many respects, though perhaps attributable, in a great number of cases, to the same causes. It is a distinctly vesicular disease, usually appearing on some of the same parts as the Thrush of infants, the edges of the tongue and the fauces being its most common situations. If the vesicles are observed before the cuticle is ruptured, the fluid they contain is generally found more or less coloured with sanguineous discharge from the denuded cutis, greatly resembling that of the vesicles described in the chapters on Purpura, Pompholyx, &c. ; but when they become broken, the collapsed cuticle exhibits a whitened appearance, and adheres to the affected surface, thus exhibiting some resemblance to Thrush. The diseased surface is exceedingly tender and irritable, and superficial sloughs, to which the cuticle becomes attached, are formed, which do not readily separate. A viscid, offensive discharge, which the patient has much difficulty in getting rid

of, takes place, and seems to excite a great deal of nausea an dvo-
miting, and further exhaustion of strength.

This affection is always found to have originated in low and debilitated states of system. The depressing effects of previous
severe fevers, and in the lower classes of lying-in women, tedious
labours, deficient nourishment, close and unhealthy apartments, are
its usual concomitants ; it is, indeed, both as regards its etiology
and pathology, nearly allied to purpura and the vesicles of scurvy ;
being often in the class of persons mentioned accompanied with the
appearances of the latter diseases, such as petechiæ and blue-co-
loured vesications, on other parts of the body : bleeding from the
gums and inability to masticate, from the pain attending it, are also
often found in the train.

The constitutional condition of the patient always requires the
utmost attention in this disease. The local affection being properly
understood to indicate a state of alarming debility, tonics and
stimulants will be always necessary ; and if the febrile symptoms
afford no objection to the removal of the patient into a purer atmosphere, this step will be attended with great advantage.   It is exceedingly difficult in many cases to give, by any plan of treatment,
that energy and tone to the constitution which is necessary to bring
on a healthy state of the diseased spots, either on the tongue or
other parts.   Weeks will, in some cases, pass away under the diligent employment of the plan of treatment alluded to, without a beneficial change in such respects, and local applications seem to have
little or no effect.   Under these circumstances, there does not seem
to be sufficient energy of the constitution for the slightest natural
attempt at reparation ; and the white sloughy sore, found, perhaps,
on the hands or feet, appears to undergo the process of exsiccation,
neither going forwards to, nor receding further from, the healthy
state, and no appearance of fluid secretion upon it occurring till
death takes place.

Dr. Bateman has applied the term Aphtha Aginosa to a disease
of which he gives the following account. " A species of sore throat,
which is not unfrequently observed during damp and cold autum-
nal seasons, especially in women and children.   It is preceded
by slight febrile symptoms, which seldom continue many days.
On the second or third day, roughness and soreness are perceived
in the throat, which, on inspection, is found to be tumid, especially
the tonsils, uvula, and lower part of the velum pendulum, and considerably inflamed, but of a purplish-red colour.   The same colour
extends along the sides of the tongue, which is covered in the middle
with a thin white crust, through which the elongated and inflamed
papillæ protrude their red points.   Small whitish specks form on
these parts, which usually remain distinct, and heal in a few days,
but occasionally coalesce, and produce patches of superficial ulceration.   The complaint is sometimes continued three weeks or a
month by successive appearances of the Aphthæ, but without any
constitutional disturbance.

"This disease appears to arise from the influence of cold and moisture, unwholesome diet, and acrid effluvia taken into the lungs. In the latter mode it is produced in persons who attend on patients affected with confluent small-pox, scarlatina anginosa, or other malignant fevers.

"Although there is no clear evidence of its propagation by contagion, it is frequently seen to attack several children in the same family, about the same time, or in very quick succession. There appears to be no danger in this affection, and medicine does not materially abbreviate its duration. A light diet, with diluent drinks, and gentle laxatives, where there is a disposition to inactivity in the bowels, constitute the only treatment required for its cure. Leeches and blisters seem to be rather detrimental than advantageous; and cinchona, with mineral acids, to be useless until the decline of the disorder, when they contribute to restore the strength.*

CHAPTER III.

OF DISEASES ANALOGOUS TO PURPURA OR SCURVY, PEMPHIGUS AND POMPHOLYX.

The existence of Pemphigus, or, in other words, of fever of a specific character, accompanied by an eruption of vesicles containing a colourless, or light yellow fluid, has been doubted by the best authorities; nor have the researches of Willan, Bateman, and others, led to any other disclosure, than that such cutaneous affections have now and then occurred in low fevers. To the authorities quoted by these authors in elucidating this matter, may be added that of Rhases, who has described it as occurring under similar circumstances only. According to the latter, it shows itself in small bladders resembling burns, preceded by redness and itching, and terminates in ulcers covered by a dark brown crust. The name, however, has been handed down to us from the time of Sauvages; and within these few years several cases so denominated have been recorded in periodical and other publications: but these may be seen, on attentive perusal, to be merely cases of Pompholyx, attended accidentally by febrile symptoms.

The character of the latter, under which the eruption has appeared, has been in some instances simple at first, and terminating in that of Typhus.† Drs. Porter,‡ Dickson,§ and Bateman,‖ appear

---

* Infantile scurvy would supply all the wants of nomenclature, as regards this disease.—S. P.
† Mr. Frogley's case, vol. xxxi. Med. and Phys. Journ.
§ Edinb. Med. and Surg. Journ. vol. xv. ‡ Ibid. vol. ix.
‖ Synopsis of Cutaneous Diseases.

to agree in the opinion as to the absence of any direct connexion between the cutaneous affection and the fever attending it; and any allusion to the question of recent date, which I have been able to discover elsewhere, seems to afford evidence of its correctness.

If the non-existence of the disease termed Pemphigus be admitted, some modification of the description of Pompholyx by Dr. Willan must necessarily be instituted, as it appears from this author that the latter consists of an "eruption of bullæ, without any inflammation around them, and without fever."

In a very large proportion of the cases which have been recorded of this disease, some acceleration of pulse or other marks of general irritation have preceded it, while it often appears after protracted illness, and is then to be looked upon as a mark of exhaustion.

When the appearance of the vesicles of Pompholyx is not preceded by severe indisposition, it is ushered in by languor, lassitude, and feelings of general debility. It usually first attracts the notice of the patient in the form of a small vesicle about the size of a pea, which in twenty-four hours becomes as large as a walnut, at which time it is commonly broken with the smallest degree of violence. Two or three of these may perhaps be seen at a time on different parts of the body, but their most frequent situation is on the lower extremities. Here and there a small red speck, evidently formed by the rupture of a minute vessel on the surface of the cutis, is discovered intermixed with these, which is to be considered the incipient state of the vesicle. The extravasation of blood is sometimes sufficiently great to give the latter a bluish colour, precisely resembling the vesicles of scurvy described in a preceding page, when, its contents being let out, they assume a resemblance to bloody water. In other cases the fluid consists merely of yellowish serum.*

It does not appear to be confined to any particular class of individuals, but is, notwithstanding, much more frequently seen in young people of delicate constitutions, and accustomed to sedentary habits, than in others. If the disease is seen and properly treated at the commencement, it soon disappears; but if this has not been the case, "the bullæ continue to arise in succession on different parts of the body, and even reappear on the parts first affected, in some cases for several weeks, so that the whole number of bullæ is very great, and when the excoriations are thus multiplied, a slight febrile paroxysm occurs every night, and the patient suffers much from the irritation, and from want of sleep." The character of the excoriation formed by the rupture of the vesicle depends entirely on the degree of constitutional derangement. Sometimes a new and sound skin covers the part in a day or two; in others, it is followed by a superficial sore, surrounded by a red border; while, when it occurs after fevers of debility, it becomes white, sloughy, tedious, and painful.

* Both these appearances are represented in the plate facing the title page, to which the explanation will be found in a subsequent page.

A state of the system or of internal organs seriously interrupting the process of chylification, and in which consequently the nutritive properties of the blood are much reduced, is as favourable to the production of Pompholyx, as of Scurvy and Purpura ; and the analogy of the former to the latter is proved by the badly nourished state of the vessels of the cutis existing in it,* the slightest pressure being generally enough to break them down sufficiently to produce a bruise of the part.

These observations refer generally to Pompholyx as it usually makes its appearance ; but there are three instances recently recorded of its having shown itself in the form of an epidemic. In two of these the disease has been termed Pemphigus ; but, for the reasons detailed, it has been thought proper to notice them here. In truth, the question as to the existence of the fever termed Pemphigus seems of so little practical importance, as to fully justify this line of proceeding.

The first instance is reported by M. Petiet, in the Journal de Médecin, in 1813. It appears, that out of 294 persons comprising the population of a village, thirty-five cases of the disease occurred. It made its appearance with symptoms of a slight febrile character, and itching in different parts of the body, which continued three days before the vesicles appeared. When the latter were broken, the excoriated cutis exhibited a violet brown colour. In some cases the febrile symptoms approached the character of Typhus ; but recovery took place in all cases under the use of refrigerant saline medicines.

The second, observed by Mr. Daniel, of Weldon, in Northamptonshire, in which neighbourhood it occurred,† approaches in similarity to the preceding. A number of hay-makers " were all severally attacked with this disease in a more or less degree ; some so lightly as not to require medical assistance, while others were so alarmingly affected as to excite apprehensions.

" The symptoms of its commencement were similar to a common attack of fever, attended with burning heat of the skin, which was only relieved by the appearance of the vesicle. In the several cases the blister exhibited a puckered appearance, scarcely any becoming larger than the size of a nut. The progress to vesication was extremely rapid, and in the course of a few hours a thin *excoriating ichor* issued from them, exciting great inflammation, swelling, and uneasiness. The general symptoms were violent sickness, pain, or giddiness in the head, prostration of strength, lassitude and general anxiety, aching pain in the limbs, and frequent rigours."

It seems not improbable, that some vegetable poison, with the

---

* When blisters are formed on the surface of the skin without previous inflammation, or evident acrimony, as the vesicles of some cases of Pemphigus may be looked upon to be, I am much disposed to imagine that such arise either from a defect in the cuticle or atony of the extremities of the cutaneous vessels.— *Jackson's Dermato-Pathologia.*

† Medical Repository, vol. vi. p. 277.

operation of which we are not at present acquainted, in its local application, may have exercised its influence in bringing on the disease in the cases above described, as the subjects of it had been employed away from the other part of the labourers in pulling docks. Such appears indeed to have been their own impressions as conveyed to Mr. Daniell. The irritating properties of the fluid of the vesicles, coupled with the exposure to the sun, under which they must have followed their occupation, would have warranted a suspicion, that the action of heat on the surface may have been the exciting cause ; but the vesicles were alike extended over parts well protected by clothing, and those most exposed.

An emetic, followed by aperients and saline medicines, only, seems to have been necessary to the cure.

In 1816, according to the editors of the London Medical Repository,* Pompholyx appeared as an epidemic at Chelsea, and extended several miles on each side of the river.

Calomel joined with other purgatives, and followed by light tonics, quickly brought about recovery.

The foregoing description of the disease given by Mr. Daniell and M. Petiet corresponds generally with those of Dr. Dickson and Dr. Porter,† varying only in the degree of severity of the symptoms. Dr. Porter observes, that in the case which he has detailed, the tunica conjunctiva was much inflamed, and the exposure to light caused a copious flow of scalding tears. A similar observation is made in the first case detailed by Dr. Dickson ; in two others on record also, similar inflammation has been observed.

Anxiety of mind has, in some cases, appeared to exercise considerable influence among the predisposing causes of Pompholyx‡ as well as of scurvy ;§ and where this has manifestly existed during the treatment, the cure has been much retarded.

The constitutional treatment adopted must, of course, be modified according to the existing symptoms. The character of the fever with which the disease is accompanied will influence it to a certain extent ; but the general principle laid down for the management of purpura, in the use of purgatives, is in all cases perfectly applicable here, whether accompanied by fever, or unattended by marks of great constitutional disorder. The same species of deranged biliary secretion, though not to so great an extent, has been generally noticed where the evacuations have been examined ; but such derangement is removed with considerably less difficulty. When things are in a proper state in this respect, the febrile symptoms become mitigated, and the patient receives much benefit from the use of tonics.

The local treatment should consist of puncturing the vesicle as

---

* Vol. vi.   † Edinb. Med. Surg. Journ., vols. x. xv.

‡ Mr. Mayd's case, vol. ii. Med. Repos. The bladders in this contained a fluid resembling " water in which meat had been washed ;" an appearance noticed in a preceding page.

§ See the case of Charpentier, alluded to in the 13th vol. of the above, the works of our early voyagers, &c.

soon as it has attained such a size as to be in danger of being inadvertently broken, and in protecting the collapsed cuticle from being rubbed off, or disturbed. Where the fluid which it contains is not acrid and irritating, this plan will not be followed by any bad effects; though in such instances as those of Mr. Daniell, which have been alluded to, it may perhaps be proper to have recourse to medicinal applications to the surface of the abraded cutis. The utility of the common warm bath is rather questionable ; but the sulphur vapour bath has been in an instance or two found eminently useful.

The disease termed by Dr. Willan Pemphigus infantilis, and by Dr. Stokes* P. gangrenosus, wears the character of a very aggravated state of Pompholyx, with great debility and low fever. According to the former, it " exhibits irregular oblong vesications, or phlyctænæ of a considerable size, and generally flattened at the top : they are at first small and transparent, but as they enlarge, the fluid contained in them assumes a purplish hue, and finally becomes turbid from a slight admixture of pus. They are also surrounded by an inflamed border of a livid red colour. This eruption sometimes appears in infants two or three days after birth, on the neck and upper part of the breast ; on the abdomen, groin, scrotum, and inner parts of the thighs : it has been known, however, to have occurred so late as ten months after birth. When the fluid is discharged after the vesications break, the ulcerated surface is not disposed to heal, but spreads beyond its original boundary, and becomes extremely painful. As the vesications arise one after another in different places, and are all followed by ulcerations ; the disease continues with little remission for several days, generally till the patient expires under the complicated distress arising from pain, loss of sleep, and violent fever. The children thus affected are often weak and emaciated, with a dry shrivelled skin."†

In Ireland, according to Dr. Stokes, it is very commonly seen among the children of the lower class of the poor, and the part most particularly stated as its seat is the back of the ears, but it is by no means confined to this situation. " It occasionally prevails epidemically, and is then preceded by a livid suffusion, slightly elevated above the surrounding parts. In the progress of the disorder, the ulcers enlarge rapidly, are attended with remarkable fætor, very great discharge and livid edges : and if they are situated behind the ears, they destroy the connexion of the posterior cartilage, with the cranium, spread to the meatus auditorius, to the eyes, (the sight of which seemed, in a few cases, to have been destroyed one or two days before death,) and sometimes to the vertex." Great constitutional irritation is produced soon after the vesicles burst ; the energies of the system rapidly decline ; death takes place about the tenth or twelfth day, and is often preceded by convulsions.

The period at which the disease usually makes its appearance is, according to Dr. Stokes's information, from the third month to the

* Med. and Phys. Journ., vol. xix.
† Willan on Cutaneous Diseases, p. 537.

ninth year ; while Dr. Willan describes it to be limited to the first year. This variation, however, may be perhaps explained by the difference in the habits of the people, and in the quality of their food.

The frequently fatal termination of the disease witnessed by Dr. Stokes under ordinary professional management, determined him to have recourse to the recipe of a female reported to be possessed of a nostrum of great efficacy in its treatment. The preparation in question was a green vegetable ointment composed of a farrago of different plants ; it had been used with much success by the country people, and the inquiry set on foot by Dr. S. induced him to think that it was indebted for its virtues to the *scrophularia nodosa,* or great figwort, which formed a part of its composition. An ointment, therefore, made entirely of the latter, was subsequently employed, and was directed to be made use of, preceded by a poultice of oatmeal and porter ; the latter, to have remained on eight hours before the first application of the ointment. " It should be as highly saturated with the green vegetable matter as possible, and when applied, it should be melted, and suffered to cool to the consistence of honey ; it should be applied with a soft feather, and with the utmost gentleness, to the whole surface of the sore." The internal use of yeast is recommended by Dr. S. in conjunction with the above ; and he concludes his observations by remarking that this plan of treatment is decidedly superior to any constituted of the applications usually had recourse to on common surgical principles.

CHAPTER IV.

ON ECTHYMA AND RUPIA.

FIVE species of Ecthyma, all of which occur under the same states of constitution, have originated with the classification of Dr. Willan, and four of them are beautifully represented in Dr. Bateman's 43d and 44th plates.

The general description is as follows : " An eruption of the inflamed pustules, termed phlyzacia,* usually distinct, and arising at a distance from each other. It is commonly indicative of some state of distress under which the constitution labours, and though it is not attended with actual fever, yet a degree of general irritation or erethism is often present with it."

According to our predecessors, so often quoted, there are four forms of this disease : 1. E. vulgare ; 2. E. infantile ; 3. E. luridum ; 4. E. cachecticum. These varieties are, however, the mere results of different degrees of debility of constitution, or irritability of the

* A pustule, commonly of a large size, raised on a hard circular base of a vivid red colour, and succeeded by a thick, hard, dark-coloured scab.

skin of the individual ; the whole of them are distinctly marked by want of energy in the cutaneous vessels as well as of the system at large. Occasionally the eruption is confined to the trunk, but sometimes extends hence to other parts. It is, however, seldom seen on the face or hands. When occurring in infants, it is usually in conjunction with an erythematous redness of the skin particularly observed about the nates, but which also frequently occupies the interstices of the spots elsewhere.

Alibert, who usually finished his various observations on particular diseases with the copious relation of cases, records several under the head of Lepre crustacée, which appear to resemble in all points what we should here denominate Ecthyma, or Rupia. The subjects of the different cases mentioned by him evidently resemble, as regards the condition of their general health, those which have been described as most liable to these diseases here.

Anxiety of mind, accompanied by great bodily exertion, fatigue, low living, the debilitating effects of previous fever, in short, anything reducing the energies of the constitution beyond a certain extent, is capable of producing it, and it is clearly dependent on a similar state of the vessels of the skin to that giving rise to the formation of petechiæ.

Almost the whole of the cases which I have had an opportunity of observing have occurred in young people; the majority in young men, who, with constitutions originally not of the strongest class, had imprudently indulged in excesses and irregularities to a very great extent, accompanied by privation of rest and other depressing circumstances. Very frequently in such cases it is mistaken for a venereal eruption, and the patient himself is readily made to believe in an opinion which his habits have made so probable. If mercury be had recourse to under these circumstances, the disease is much aggravated, the dried scabs of which it is constituted grow on the part rapidly, becoming of a dark brown colour, exceedingly hard, and of a conical form, representing, in fact, on a minute scale, the characters of the disease termed Rupia.

Measles, scarlatina, and many other diseases followed by debility, occasionally prove exciting causes of Ecthyma, in which case the patient, instead of rapidly regaining his strength, is visited simultaneously with the appearance of the eruption, with that restlessness and hurried pulse before alluded to. This state of things, in a trifling degree, sometimes exists for days after the original febrile affection has disappeared, without attracting much attention ; and if the appetite of the patient is not impaired, eventually goes off, leaving him in good health. Now and then, however, instead of gaining strength, the patient loses flesh, passes restless nights, becomes exceedingly languid, subject to regular attacks of hectic fever ; and exhibits a countenance of much anxiety and distress.

When the disease makes its appearance in consequence of an increased degree of debility brought on by any eruptive fever, as measles, &c., it is usually seen in its very earliest stage about the waist. It exhibits a few reddened and slightly elevated spots, covered with a very thin lamina of cuticle, which readily separates. Some of these have a minute elevation in their centre resembling a vesicle: the latter, however, contains nothing like the serum of the herpetic vesicle but a glutinous fluid, which dries upon the part, and forms with the morbid cuticle an elevated scab of a conical form, the basis of which in a day or two is surrounded by a small inflamed areola. The surface of this areola is soon covered with a lamellated scab, and a scale of considerable magnitude is thus formed. During the continuance of the disease, many of these scabs separate; and if the strength is improved, the part beneath returns to a healthy state. Now and then, however, this process is repeated several times, and successive exfoliations continue to be produced.

The minute examination on which the foregoing description of the spots of Ecthyma is founded, has led me to think that the term pustule is very improperly applied to them.

Taking into consideration the state of system existing in all cases as its predisponent cause, it would appear that an active inflammatory and suppurative action on the surface of the body is, of all events, one least to be expected to constitute the essence of the disease, and, accordingly, the kind of examination alluded to discloses a state of vessels materially differing from such an action. I have little doubt that in the formation of every spot of Ecthyma, lesion of some minute vessel takes place, and that a kind of petechiæ is thus produced. From some cause or other either depending on a minor degree of debility of system to that in which the petechiæ remain quiescent, or on mere additional irritability of skin, attempts to repair the mischief are made by the vessels of the part, and inflammatory action takes place.

A minute vessel thus ruptured has its extravasated contents mixed with the coagulable lymph poured out by the vessel labouring to repair the mischief, and the mixed fluids dry on the spot, forming the minuter scabs. The powers of the constitution being unequal to the carrying this process through in a healthy manner, irregular attempts at suppuration take place, and the bulk of the scab is usually surrounded by, and partly mixed with, a minute portion of ill-formed pus. Even this, however, dries on the part, and increases the bulk of the scab. Many of the minute vessels concerned in this process are in too relaxed a state for the occasion, and their secretion is mixed with, and deeply coloured by, red globules of the blood which escape under the increased action.

The scab itself now becomes a cause of irritation to the surface on which its basis rests; the bloody purulent secretion increases, becomes inspissated, and attached to the mass: the ulcerative process gradually destroys the surrounding cutis, and the margin of the sound skin becomes elevated and thickened.

The scab, by the continuance of this process, is gradually elevated, and rapidly enlarges; a fresh and broader ring of growth being added to its base in every three or four days.

At this period it answers to the description of Rupia. This latter named affection, though described by Dr. Bateman, as originating always in a vesicular form, has been satisfactorily ascertained to be the termination of Ecthyma in a variety of cases which have come under my notice when neglected or improperly treated. In ordinary instances, the state of system on which the latter depends being gradually removed by quietude, tonics, good living, and abstinence from any species of indulgence followed by exhaustion, the scabs do not assume the regular conical form of Rupia.

From the description of Dr. Bateman, it would appear that even when Rupia assumes the form of an inflamed. vesicle at its commencement, the subsequent formation of the hard dark-coloured scab is still accomplished in the manner above described. The same states of constitution and condition of vessels of the surface exist in the two cases, and the exudation of lymph from the debilitated vessels in forming the vesicle appears to excite attempts at inflammation in those adjoining, in the same way as in the commencement of the scab of Ecthyma.

The occasional dark colour of the scab in both cases had appeared to me to be produced by a mixture of blood with the coagulable lymph and other secretion constituting its substance, and I am now enabled by inquiry to state this to be the fact. When the scales of Ecthyma are rubbed off in a rude manner, and the skin is abraded on which they were situated, a little blood escapes, but the same kind of secretion continues mixed with the blood, and the scale is harder, considerably darker coloured, and irregular : the formation of an ulcerated sore rarely follows such an accident.

Ecthyma, where the scabs are very small and much diffused, is sometimes mistaken for itch. I have lately met with a case of this kind where sulphur had been used both internally and externally : the external application, however, did not aggravate the cutaneous affection though it did not improve it. The patient rapidly recovered by the use of tonics and sea-bathing.

With respect to the propriety and value of the latter as a remedy it may be well to add, that in several other cases it has produced a manifestly beneficial effect, and I am induced to think that it may be had recourse to in all with great safety and infinite advantage. It appears to afford that kind of stimulus to the cutaneous vessels which is desirable, as well as to act as a tonic to the system generally. Indeed, it would seem at first sight obviously indicated, if the pathology of the disease, which I have endeavoured to give, be correct.

The enlarged, blackened, and prominent scabs of Rupia are never . seen but in the lower classes of society, where long-protracted disease has produced a general wasting of the body, hectic fever, and other symptoms of the lowest degree of debility. In the cases of this kind answering to the Rupia prominens, which have come under my notice in the St. Giles's Infirmary, the patients have been not unfrequently the subjects of syphilis. In one instance of recent occurrence, the patient had severe pains in his head and limbs, on which account he had lately been considered to require mercury. Under the influence of this medicine he grew worse, and the scabs rapidly enlarged—he discontinued it, and was allowed meat, wine, porter, and tonics, and rapidly recovered. In the debilitated state of this patient, and before the constitution had begun to rally, I was induced to apply a strap of adhesive plaster upon the largest of the scabs, and to employ a little pressure, with a view of hastening its separation and preserving it : but on removing the bandage, the surface of the skin and base of the scab were found attached to each other by a stringy slough, which never separated till the constitution was greatly improved. I had been more successful in a former instance of this kind, and by dissolving the incrustation in warm water, ascertained that its substance was largely made up of blood, which had undergone the process of exsiccation mixed with the other fluid secretions of the sore. When the slough is removed from the surface of the sore, the latter exhibits a glassy inactive surface for some time, and the ulceration is usually found to have extended deep into the cellular membrane of the part. The growth of healthy granulations in such a case will be materially expedited by the use of the nitrous acid lotion : a pledget of lint wetted in this should be kept constantly applied, so as to fill the excavation.

Local applications in a disease so markedly dependent on debility of system are seldom called for or productive of benefit, except in cases similar to that above alluded to, where the constitution is improving, the scab detached, and the sore on which it rested beginning to assume a healthy action.

Generally nothing is to be expected from them beyond the temporary alleviation of uneasiness.

In the constitutional management, it will be first necessary to correct the state of the secretions. These will generally be found considerably vitiated and discoloured.

Gentle purgatives, and an occasional dose of some mild mercurial alterative, will be found necessary ; but a state of rest, relaxation from business, and tonics, are generally also required ; and in the employment of these, a sufficient guide will be obtained, from the consideration of the circumstances of the patient, and the history of the disease.

Rupia is spoken of by Rayer in terms of similar import to those I have employed in the foregoing pages ; and there cannot be a doubt that it is merely one of the Protean forms of scurvy.

Rupia. Atonic Ulcers, and their Incrustations.

" Rupia," say Willan and Bateman, " is characterised by small isolated, flattened bullæ, filled with a serous fluid, which soon becomes opaque, puriform, or sanguinolent, and to which succeed black, thick, and prominent scabs, whose bases conceal ulcers of variable depths. Three varieties are observed : rupia *simplex*, rupia *prominens*, and rupia *escharotica*.

" Rupia *simplex* is commonly evolved on the legs, sometimes on the loins and thighs, and more rarely on the other regions of the body. It is proclaimed by one or more flattened bullæ about the size of a shilling, which at first contain a serous and transparent fluid. This fluid, however, soon becomes turbid and purulent, it then grows consistent, and is finally transformed into scabs of a chocolate colour, thicker in their centres than around their circumferences, their outer layer being continuous with the epidermis which appears detached by the serum or pus in which their edges are bathed. Under these scabs, which are detached naturally within a few days, or are accidentally rubbed off, the true skin is found to be excoriated. This superficial sore, left to itself, either heals up, or is covered with another scab, which falls off at a later period, and this process may be repeated several times successively. After cicatrization is accomplished, the skin retains for a very long time a livid, deep red colour.

" In rupia *prominens* the bullæ are larger, the succeeding scabs thicker, and the consequent ulcers deeper. Each bleb is preceded by a circular red spot, over which the cuticle is slowly raised by a blackish-looking thick fluid, which soon concretes, and gives occasion to the formation of a scab, whose thickness and size increase for some days afterwards. The circumference of this incrustation is surrounded by a reddish border a few lines in breadth, the epidermis of which is raised by a fresh effusion of serous fluid, that forms a new incrustation, and adds to the extent of the one originally produced. The areola also spreads slowly around the base of the scab first formed, which itself increases in breadth and thickness during three or four, and sometimes even during seven or eight days. The scab then appears very broad in comparison with its thickness, and is often very aptly likened to an oyster shell. More commonly, however, the incrustation projects in the same degree as it spreads, becomes conical, and bears the greatest resemblance to the shell of the limpet. The scab of rupia generally adheres firmly, and can only be detached with the assistance of moist and emollient applications. The skin once exposed, appears ulcerated to an extent and depth that vary in every instance. If the part affected remain exposed to the air, a new crust is either formed after a greater or shorter interval, or the ulcerative process extends more deeply, and spreads till it sometimes approaches the size of a crown-piece in breadth : the sore in this case looks pale, and bleeds readily. These ulcers, which are usually characterized as *atonic*, and the cure of which is only obtained with extreme slowness, are always succeeded by cicatrices, subject to break open afresh, the brownish livid hue of which continues for a very long time unchanged.

" Rupia *escharotica* is evolved more especially in cachectic infants, and occasionally in elderly persons, or in adults who have suffered from chronic rheumatism, or constitutional syphilis ; it is **most** usually seen upon the legs, the thighs, the scrotum, the abdomen, the loins, the neck, the upper part of the chest, and but very rarely on any part of the upper extremities. It begins by one or two red and livid spots, over which the cuticle is soon raised by the effusion under it of a serous or sero-sanguinolent fluid. These bullæ go on increasing in an irregular manner ; the serum they contain becomes turbid, and acquires a blackish colour. By and by they give way, and the dermis, left naked, appears ulcerated, softened, or gangrenous in different points ; a bloody and very offensive sanies bathes the surface of the sore, the edges of which are livid and not very painful. In adults I have seen rupia escharotica acquire the dimensions of rupia prominens, and small portions of the skin and cellular substance, stricken with sphacelus, become detached slowly from the surface of the ulcerated parts. In children the bullæ of rupia escharotica do not generally acquire so large a size, but they follow each other in greater numbers ; the succeeding sores become very painful, cause fever and sleeplessness, and may exhaust the patient in the space of two or three weeks. In every case the cicatrization of these sores is a very tardy process, and is always expected long before it happens. Ecthyma is a form of cutaneous disease that is often seen along with rupia *simplex*, but very rarely with rupia *escharotica*. I have seen many cases of rupia complicated with purpura and with chronic rheumatism, as also in individuals labouring under constitutional syphilis.

" Scrophulous children, and the offspring of the poor, of delicate constitutions, or who have been weakened by previous illness, are predisposed to rupia. The disease shows itself particularly during the winter season among such as are insufficiently clothed, badly lodged, and ill fed, more especially after any other form of inflammation of the skin, such as variola, scarlatina, measles, &c. I have seen rupia repeatedly complicated with purpura hemorrhagica. The disease also frequently appears among the aged, though those in the vigour of life are by no means exempt from it.

" The small flattened bullæ of rupia very commonly contain a turbid, serous fluid. They cannot be mistaken for the large transparent and prominent blebs of pemphigus. Besides, the rough, thick, and often prominent scabs of rupia, and its succeeding ulcers, are very different and easily distinguishable from the laminated incrustations and superficial excoriations of pemphigus. Yet the pemphigus *infantilis* in which the skin often appears ulcerated in the centre of the blebs, seems in some sort to form the link of connexion between these two diseases. Rupia differs from ecthyma in its primary form, which is bullous, whilst that of ecthyma is pustular ; the base of the pustules of ecthyma is much inflamed, and the scabs with which they become covered at a later stage of their progress are hard, and, as it were, set, or encased within the subtance of the skin ; the circumference of the bullæ of rupia does not present the same degree of inflammation, and their incrustations are much broader, more prominent, and less adherent than those of ecthyma. It must be allowed, however, that the bullæ of rupia become purulent very quickly, and that occasionally the diagnosis is rendered so much the more difficult, as the two eruptions are met with at one time in the same individual. Nevertheless, the prominent incrustations, and the deep, and often intractable ulcers of rupia are very different from the impacted scabs and slighter sores of ecthyma. It does not seem possible that rupia escharotica can ever be confounded either with anthracion, which is surrounded with a broad erysipelatous base, or with frost-bite of the hands and feet presenting bullæ and gangrenous spots.

" Rupia is never a dangerous disease : the *escharotic* species itself is only serious when the eruption is very abundant. When the disease appears on the legs it is always succeeded by intractable ulcers. The duration of the disease cannot be precisely calculated, but depends greatly on the age of the patient, the number and size of the bullæ, on the consequent sores, the degree in which the general constitution is affected, and the influence which certain concomitant maladies, such as scrophula, and chronic affections of the lungs and alimentary canal, may exert on its progress.

" The treatment of rupia is of a general and local nature. The object proposed by the first is to bring about a modification of the constitution acting faultily in a greater or less degree. The milk of a good nurse for infants at the breast, exhausted by hunger, and in misery, through want of proper care ; a nutritious meat diet— beef or mutton—and generous wine mixed with water, for children and adults of lax fibre and a scrofulous habit, a regimen adapted to any form of concomitant disease, if the general health has suffered,— such, in a general way, ought to be held the objects of primary and highest importance.

"The local treatment may be as follows :

"The bullæ of rupia simplex are to be opened if they contain serum, and the parts covered with a soft rag and some lint, the dressings being kept in their places by proper bandages.

"After the fall of the scabs in rupia simplex and rupia prominens, the ulcerated skin is to be bathed with decoction of althea if it be painful; but if the inflammation appear indolent and below the pitch requisite for the production of a new epidermis or the formation of a cicatrix, it may be stimulated with a wash of wine and water, or a weak solution of cream of tartar. I have been in the habit of ordering the sores of rupia to be dusted with cream of tartar, and of all the topical applications I have tried, this is the one that seemed to me to answer best.

"Rest, the horizontal posture of the body and limbs, and continued gentle pressure, assist the cicatrization of the ulcers. Sticking plasters may be employed in some cases where the legs are affected, and the blebs are isolated or few in number ; but whenever the round shape of the ulcers is modified, it is proper to change the adhesive straps for a perforated rag and a compress of lint, maintained by a proper bandage. If the adhesive plasters be continued too long, the parts almost always grow livid and fungous, a state that requires the repeated application of escharotics. The best of these, when they do become necessary, is the nitrate of silver, and the use of this is often followed by good effects. In some cases the nitric or muriatic acid, or the acid nitrate of mercury, may be advantageously employed.

"When the eruption extends to several regions of the body it will be necessary to resort to alkaline and sulphureous baths of regulated strength, alternating these with the simple warm-bath.

"To cleanse the skin and get rid of the incrustations, as also to enable us the better to ascertain the state of the excoriations, I make it almost a general rule to order a warm-bath for the patients we receive into our hospitals. For scrofulous subjects a sulphureous-bath is usually substituted, and this is had recourse to again from time to time during the treatment."

CHAPTER V.

ERYTHEMA NODOSUM.

The erythema nodosum is a singular disease, and in its character most intimately allied to those cases of purpura which occur in apparently healthy states of system. It finds its place here on account of its relationship. The same high degree of derangement of the secretions, and disordered state of those organs in which the process of chylification is carried on, has been noticed constantly where the disease has come under my observation.

The inflammation appears "on the fore part of the legs, and is preceded by slight febrile symptoms for several days. It shows itself in large oval patches, the long diameter of which is parallel with the tibia; these slowly rise into hard and painful protuberances, and as regularly soften and subside in the course of eight or nine days, the red colour turning blue, as if the leg had been bruised."

I have seen the erythema nodosum in children, as well as in grown persons, equally distinctly attended with, and dependent on, the disorder mentioned. In a recent instance, two children who had been placed out at nurse were the subjects of it; the febrile irritation had been severe, and the cutaneous disorder had appeared in the second two or three days after it was discovered in the first. In both cases, after the redness began to die away, a tuberculous hardness and thickening of the cutis and cellular membrane continued for many days, evidently produced by sanguineous effusion into the cellular membrane, during the continuance of the inflammatory action. These cases gave rise, at the commencement, to the suspicion of contagion; but the similarity of the manner of feeding the two children, with that of the history of the previous state of their bowels, cleared up the question. They both rapidly recovered under the free use of purgatives of calomel and jalap, but the blue spots were some weeks in disappearing.

In debilitated states of system in the adults it sometimes takes place, accompanied by those deep-seated pains which have been described as appertaining to purpura, and when the cutaneous affection is seated contiguous to the tibia, considerable pain is referred to this bone, which continues till the redness disappears, and then gradually ceases. Under these circumstances it somewhat resembles the syphilitic node, and has been occasionally mistaken for it. A little inquiry into the history of the case, however, if aided by an attentive examination of the parts, will always be sufficient to guard us against such an error.

There is little to be culled, as regards E. nodosum, from other authors; and as regards the other forms of inflammation of the cutis, denominated E. intertrigo—papulatum—tuberculatum—marginatum—fugax, chronic, &c. it seems a waste of time to the reader to bestow any attention on them further than to say, that they are all speedly got rid of by means of a warm-bath and a dose of senna and salts.

SECTION VI.

Fungoid Diseases of the Cutis and Cuticle.

CHAPTER I.

ICHTHYOSIS.

ICHTHYOSIS, or the fish-skin disease, and warts, constitute the more important parts of this section.

Ichthyosis usually makes its first appearance in a form unequivocally showing its origin in a chronic inflammatory action of the vessels producing the cuticle. A morbid thickening, with a dry and harsh state of this covering, is the first circumstance worthy of observation in the disease. The patient rarely complains of any uneasy sensation at first, but as the thickening of the cuticle increases, a distinct sensation of increased heat is felt, and some marks of irritation and redness are observed on the healthy cutis round the margin of the diseased spot. In the course of a short time the diseased cutis rises above the surrounding parts, and its surface begins to exhibit the appearances of minute and innumerable fissures, which after a short time become elongated, and extensive cracks, intersecting each other, and dividing the surface into innumerable portions, each of which, individually considered, exhibits great similarity in structure to the common wart. When the growth of the diseased structure has attained the height at which roughness of its surface and the minute fissures described occur, it assumes a dusky and dark-brown colour, which colour, as the morbid parts continue to grow, gradually approaches blackness. This appearance, however, does not naturally belong to the disease, but is given to it by the entanglement of dirt, from which it is impossible to protect it, on account of its peculiarity of structure, even though frequent ablutions are had recourse to.

The idea of its analogy in formation and appearance to the skin of any kind of fish, from which it derives its name, is evidently erroneous ; the dry, hard cuticle of the elephant, and some other animals of the larger class, whose skins are not plentifully covered by hair, present a much nearer resemblance ; in point of fact, however, it is a formation which is not analogous to any thing that is often seen but the common wart of old standing, and even this resemblance exists only in structure, the connexion of the latter with the cutis being considerably closer, and its separation effected with much more difficulty and pain. The sensation on drawing the finger along the diseased surface is precisely similar to that occurring from this proceeding on the surface of the larger and old standing warts.

The arms and legs are the most common situations of Ichthyosis : its occurrence on the face, as represented in the 18th plate of Dr. Bateman, is, comparatively, exceedingly rare. Whatever may be its situation, its causes seem to be equally obscure. It has not appeared, in any instance which I have witnessed, to be dependent on, or in any way connected with, constitutional derangement, nor has any cause of local irritation been discovered to have existed before the disease appeared in any one instance.

There are several instances of this deformity recorded, where hereditary origin has been distinctly traced. Dr. Girdlestone* and Mr. Martin† have both furnished cases of this kind to periodical publications, where the disease extended over the greater part of the body. In the cases reported by the first of these gentlemen, the father and grandfather of the patient had been subject to it ; and in those by Mr. Martin, a mother and child, it having made its appearance as early as three months after birth in both, and ultimately extended over every part of the body, except the head and neck. Dr. G. observed an usually florid complexion prevailing in the family whose cases form the subject of his report.

When the disease occurs under such circumstances, it is more than probable that medicinal treatment of every kind will fail to produce any alteration, depending as it does on the original formation of the skin. When only local, and of small extent, remedial measures may be taken with a fair prospect of success. In those cases, how-

* Med. and Phys. Journal, vol. viii.
† Vol. ix. of Medico-Chirurgical Transactions.

ever, which I have had an opportunity of observing, no benefit has been derived, either by the use of pitch or arsenic, both of which have been spoken of as occasionally productive of benefit. Neither has any good resulted from the use of ointments. The frequent soaking of the parts in warm water, and gradually picking off the excrescence, as recommended by Willan and Bateman, will scarcely be practicable, even in cases of the most limited extent ; and if effected, does not at all diminish the morbid disposition in the vessels of the part.

I have had many opportunities of submitting this disease to the combined influence of pressure and the cold lotion. These were situated for the most part on the leg, and as no disorder of the system or digestive organs could be traced, internal medicines were not had recourse to. Considering inordinate action of the vessels on the surface to be manifested both by the heat which was present, and the rapid growth of the excrescence, it appeared that the *modus operandi* of these measures was well adapted to the cases. Adhesive straps were applied as tightly as could be borne with comfort over the whole of the diseased part, extending a little above and below it, and these were supported by a bandage. The latter was kept constantly wetted with the lotion, and at the end of four or five days removed. At this period, the excrescence was found liberated from its attachment to the cutis, and came off in large pieces nearly through its whole extent, exposing a white and ill-formed cuticle, which might be scraped off in great quantities without pain. By perseverance in the same plan of treatment, the cuticle gradually assumed a strong and healthy state, and a complete cure was uniformly effected.

In a case of that form of elephantiasis termed the Barbadoes leg, some years since in St. George's Hospital, the cutis, particularly on the superior part, was covered by the same dry, dark-coloured excrescence, which belong to the older standing cases of Ichthyosis. It is very correctly represented in a drawing of the case made by Mr. Gaskoin ; and is, I understand, constantly observed, more or less extensively, in every instance of this disease. The case in question was, I believe, materially benefited by the diligent and scientific application of pressure by means of bandages, under the direction of Mr. Young.

Of what has beeen termed Ichthyosis cornea, it may be unnecessary to say much here. I am not aware that any light has been thrown on the subject of its pathology since the publication of Bateman's Synopsis. Of horny excrescences arising from the skin, there are a great number of instances recorded. Their removal seems to be a simple operation, not followed by any bad consequences.

A long list of references to cases and descriptions has been given by Dr. Bateman,[*] which the curious reader may be gratified in consulting. The following are the results of Rayer's efforts in contributing to the stock of information regarding it.

* Synopsis.

" Ichthyosis is characterized by a morbid development of the papillæ and thickening of the epidermic lamellæ ; these often assume the form of small irregular compartments, which have been compared to the scales of fish.   The affection may be almost quite general, or it may be confined to a single region.

" When Ichthyosis is general, it is always in those places in which the skin is naturally thick and the epidermis rough, as around the great articulations, on the anterior and external parts of the lower extremities, in front of the patella, over the olecranon, &c., that the altered epidermis becomes thickest. Everywhere else the adventitious layer which it forms over the surface of the skin is much thinner ; it is commonly entirely wanting on the prepuce, eyelids, groins, axillæ, &c., in a word, in every place where the skin is soft and of great delicacy.   This morbid development of the cuticle is also very rarely seen on the palms of the hands and soles of the feet, and when it does appear there, it is always to a less extent than in the regions mentioned as its principal seats.

"At the period of birth, *congenital* ichthyosis is usually but little apparent.   Yet in the anatomical collection of Berlin, there is a fœtal monster preserved, the 'whole surface of whose body is covered with a thick layer of epidermis.   The skin is also several lines in thickness : the cuticle presents numerous fissures, and forms a covering like a coat of mail to the body.   This singular case has been particularly described by M. Steinhausen.*

" The skin of those infants, who are by and by to become the subject of ichthyosis, instead of being smooth and soft as usual, appears sallow, dry, and like shagreen.   The affection is proclaimed with characters less equivocal during the course of the two first months of existence.   The cuticle in some places becomes rough, sallow, of a greyish colour, and to the touch conveys a sensation similar to what is communicated by the skin of the aged in many cases.   An alteration of the epidermis to this extent may continue during a whole lifetime without going farther ; or it may increase with the progress of years until it becomes excessive.

" Ichthyosis may occur several months after birth, with more decided characters.   After having passed by different intermediate degrees of thickening, the epidermis at length appears divided into small irregular compartments, the appearance of which is certainly more analogous to that presented by the legs of fowls than to that of the scales of serpents, although Alibert has designated the variety of Ichthyosis under review by the title of *ichthyose nacrée serpentine.*

" When Ichthyosis shows itself with characters still more decided, it appears upon the extremities, especially in the sense of their extension, under the guise of a thick epidermic layer, which has been likened by some pathologists to the bark of certain trees.   In this, as in the two first varieties, the epidermis appears composed of numbers of small compartments, very irregular in their shapes,

* Steinhausen.   De Singulari Epidermidis Deformatione :   Berlin.—Gazette Médicale, 1831, tom. ii. p. 10.

not imbricated, not more than from two to three lines in diameter, generally broad in proportion as they are thin, and of a greyish or sallow hue. In some rare cases they are shining and in some sort opalescent ; more frequently they are of a deep brown colour ; the surface is then so rough that the hand passed over it experiences a sensation similar to that which is felt when a file, or shagreen, or the skin of certain fishes is handled (*Ichth. nacrée cyprine*, Alib.) These squamæ may be detached without causing any pain, if we except the largest, which always seem to adhere more strongly, and which when removed occasion at least an unpleasant sensation. In every case, when they are detached, whether by friction or in any other manner, they are speedily reproduced with the same characters as before.

" There is a fourth variety of Ichthyosis, extremely rare, but very remarkable. Individuals have been seen whose skins were covered with numerous small and prominent appendices, which could not be taken away without pain, or the subsequent exudation of a reddish or sanguinolent fluid. These appendices are often whitish internally, though black on the surface. One of the most remarkable cases of this singular alteration of the skin occurred in a man, a native of Suffolk, who exhibited himself about the year 1710, and was known under the name of the *porcupine man.* The whole surface of this individual's body, except the face, the palms of the hands, and the soles of the feet, was covered with small excrescences in the form of prickles. These appendages were of a reddish brown colour, and so hard and elastic, that they rustled and made a noise when the hand was passed over the surface. They had appeared two months after birth and fell off every winter to reappear with the summer. The man was in other respects in very good health ; he had six children, all of whom were covered with excrescences like himself ; the hand of one of these children has been figured by Edwards in his *Gleanings,** the hand of the father may be found represented in the 59th volume of the Philosophical Transactions.

" This race of *porcupine men* has been mentioned by several writers, the accounts being taken from the well known family of the name of Lambert. The whole of the males of this family have the body covered with spines. Two brothers, members of it, were examined by Geoffroy St. Hilaire, one aged twenty-two, the other fourteen. The body of the elder, except the head and palms and soles, was entirely covered with spines ; the younger was naked in several places, particularly over the breast ; but a number of brown spots showed that with age he would probably become as completely covered as his brother. The spines on the back of the hand were extremely thick, and might have been compared, as far as the diameter was concerned, to the quills of the porcupine ; those which surrounded the nipple bore a greater resemblance to squamæ ;

* Gleanings of Natural History. London, vol. i. 1758 ; ii. 1760 ; iii. 1764, 4to. pl. 212.

they formed long narrow laminæ, very numerous, closely set and implanted vertically into the skin. This thickening of the epidermis and hair was the effect of a morbid disposition, which was transmitted hereditarily, but only from father to son, the daughters not being affected. Five generations could be reckoned which had been affected in the manner described.*

" Local and accidental Ichthyosis constitutes a fifth variety, very distinct from the preceding species, which in its mode of development bears the greatest analogy to that of corns ; of this nature is the ichthyosis evolved on the anterior and lower part of the thighs in shoemakers, in those places upon which the last rests whilst they are going on with their work, and especially when they are driving nails into the soles of shoes ; of this nature is the scabrous formation which occurs on the elbows of paper-stainers, on the surface of the outer ancle in tailors, &c. I have also observed a morbid development of the papillæ of the tongue, in a man, in all respects in very good health, which I regarded as precisely similar in its nature to local ichthyosis of the skin.

"It is not uncommon for the skin in subjects affected with Ichthyosis, to cast off the adventitious epidermic formations, characteristic of the different varieties of the general disease, during the summer season ; these are, however, constantly reproduced on the approach of autumn. This kind of desquamation has also been observed at other seasons. The skin, divested of its squamæ, shows no signs of inflammation : whether the epidermis have been removed by the influence of the seasons, or by the action of the vapour bath, or any other external application, its colour is natural, only the shallow furrows which occur on its surface are more remarkable than usual. The cutaneous exhalation and follicular secretion appear to be entirely suppressed, or at least are inappreciable.

"Ichthyosis is not accompanied with pruritus, nor with any other morbid sensation. Neither does it seem to have any kind of unfavourable influence on the general constitution ; I have met with several individuals labouring under the first and third varieties of the disease, who enjoyed the most perfect health and were very robust. In these individuals it is probable that the pulmonary exhalation and urinary secretion supplied the place of the defective cutaneous perspiration ; which, however, is sometimes seen to be very copious in the palms of the hands and soles of the feet.

"I have tried the effect of maceration upon portions of the skin of individuals who had laboured under ichthyosis. The small compartments of which the epidermic layer consists, and which give the malady its principal external characters, are readily detached under the form of a greyish or blackish membrane, impregnated with pigmentary matter in the porcupine species, little or not at all coloured in the other varieties of the affection. These small compartments do not overlap each other like the scales of a fish ; the

* Bulletins de Sciences par la Société Philomathique, No. 67, p. 146, an. 11 de la République.

title ichthyosis taken in its literal signification would lead to an erroneous anatomical idea. Tilesius made a few experiments on the nature of the thick black superficial epidermic layer, which was detached in squamæ from the bodies of the brothers Lambert. Buniva has since assured us that the squamæ were nothing more than gelatine, become hard and solid from its combination with a certain quantity of phosphate and carbonate of lime. M. Delvaux has discovered that it also contains a little carbonate of iron and traces of silica; consequently that the squamæ of ichthyosis supplied the same chemical principles as the nails, the hair, and the epidermic productions generally. I have myself demonstrated experimentally that this substance possessed physically and chemically the same properties as the epidermis. Dr. M. Good,* who designates it improperly by the title of *incrustation*, supposes it to be formed by cutaneous secretions containing an excess of calcareous matter. Under the first epidermic layer in ichthyosis, which is commonly coloured by pigmentary matter, a second is found of a dirty white or greyish hue.

"The lines or furrows which the corion presents on its outer surface are much more decided in ichthyosis than in the standard condition. The papillary eminences, which are also much more remarkable than on the healthy skin, are sometimes extremely large; it is indeed to the hypertrophy of these that Telesius ascribes the production of the epidermic spines in the porcupine men. I have ascertained the occurrence of this hypertrophy in the four first varieties of ichthyosis. This excess of development of the epidermic layers recalls an analogous circumstance already observed in a great number of simple cutaneous warty productions.

"Tilesius informs us, that in the brothers Lambert, the cutaneous follicles were obstructed, and full of a thick substance. These organs were but little apparent, and in many places imperceptible, in the great majority of the subjects affected with ichthyosis, whom I have examined. The hair and hair bulbs were found remarkably enlarged in a particular case, the history of which is given by Dr. Martin. To conclude, the corion has always appeared to me to be thicker, harder, and more dense than it is in the natural state.

"In the small number of cases in which opportunities have been found for examining the bodies of individuals affected with ichthyosis, who have died of some accidental intercurrent disease, a variety of organic changes, differing both in kind and situation, have been observed, which appeared in nowise connected either with the development or the existence of this affection of the skin.

"General ichthyosis is by no means a very rare disease, at least in France. I have myself seen above forty cases of the kind. It is known to be transmitted through several successive generations. The history of the Lambert family, which has been given by Geoffroy-St.-Hilaire, Tilesius, and Buniva, affords a remarkable instance

* Study of Medicine, vol. iv. p. 591.

of the affection being only transmitted to the males. Ichthyosis is seldom developed accidentally long after the period of birth. The whole of the male children of the same father and mother, who were themselves free from such an affection, have been seen labouring under ichthyosis. Such was the case of the brothers Brayer, born in the département du Cantal. One of them, John, who became a patient at the hôpital de la Charité, in 1827, assured me that his brother, thirty-seven years of age, was like himself affected with ichthyosis, although neither his father nor his mother had ever been the subjects of a similar affection, of which his three sisters also presented no indication.

"Some pathologists have ascribed the development of ichthyosis to moral affections of the mother during pregnancy; this cause, however, is very problematical. I was once consulted on account of three little boys affected with congenital ichthyosis, born of healthy and well-formed parents; and the mother assured me that she had never been better than during her three pregnancies, in none of which had she suffered anything like mental disquietude or alarm. Neither climate, nor mode of life, nor temperature appear to exercise a marked influence on the production of this disease. It does not appear to be endemical in Haïti, in Paraguay, nor among the inhabitants of the sea-coasts or banks of rivers, who live much upon fish, as it has been stated to be on insufficient authority.

" Women are known to be much more rarely affected with ichthyosis than men.

" Ichthyosis bears but a remote and slight resemblance to the squamous inflammations. Willan and Bateman, and several other pathologists certainly did wrong in uniting these affections into a single group. Ichthyosis almost invariably makes its appearance in the course of the first months of existence, and continues during the whole of after-life. It is not accompanied either with morbid heat, or pruritus, or, in a word, with any other symptom observed in inflammation of the skin.*

In lepra, psoriasis, and pityriasis, the production of squamæ is constantly preceded by redness of the skin, which can be readily made apparent by divesting the integument of the epidermic scales and furfuræ deposited on its surface. In confluent and inveterate lichens, the skin may become rough, brownish, and covered with an infinity of minute scales somewhat similar in appearance to those of slight and partial ichthyosis : but this state is accompanied with pruritus of the most insupportable description, and is preceded by an eruption of papulæ. The simultaneous existence, or the ulterior evolution of papulæ on some neighbouring point of the skin already covered with furfuræ, will always suffice to dissipate any doubt which might be entertained of the true nature of these obscure cases. Local ichthyosis is no less distinct from the squamous and furfura-

---

* This is an assertion which cannot be received as a fact, as regards the disease seen in England. There is always—*always* irritation present, showing itself in some form or other.—S. P.

ceous states which the skin presents around old ulcers, or after chronic and long standing eczemas.

" We know that from the second to the fifth, and sometimes the tenth day after birth, the skin of the new-born infant throws off its epidermis. [Surely Messrs. Rayer and Billard are drawing too largely on our credulity!] This *epidermic exfoliation*, which has been particularly studied by Billard,* can never be confounded with ichthyosis; for, independently of other circumstances, it does not continue longer than a limited period. The skin of the aged is also occasionally affected with an epidermic exfoliation, which differs from the slightest known form of ichthyosis in the absence of all thickening of the cuticle about the knees and elbows, a condition the existence of which is invariable in the latter affection.

"It were useless here to particularise the numerous characters which distinguish ichthyosis from all kinds of horny productions of the skin, and from pelagra; these having been either already mentioned, or being about to be noticed. But I think it necessary to solicit attention to a mistake which must be readily fallen into, seeing that two excellent observers have committed it. Bateman has given a figure in his atlas under the title of *ichthyosis of the face*,† of a case which is certainly one of *ceruminous or sebaceous deposit* from diseased action of the follicles; and Dr. A. T. Thomson has detailed a case of the same description under the same erroneous title !‡ [These are mere assumptions on the part of our author. The individuals named were not likely to have made any such mistake; in point of fact, it is Mr. Rayer who has committed the blunder.] In this affection of the follicles, however, which *I was the first to notice* particularly, the part of the integument affected becomes at first, as it were, unctious or oily; the secretion of the sebaceous follicles then increases; the fluid thrown out upon the surface acquires additional consistency, and finally forms a kind of *squamous crust or layer* of greater or smaller extent. Soft at first, and adhering but slightly, it by and by acquires hardness, and then cannot be removed without occasioning very considerable pain. The skin under this sebaceous deposit is of a vivid red; the orifices of the follicles appear dilated, and sometimes distended with concrete sebaceous matter.

" *Congenital* ichthyosis often disappears for a time in consequence of acute inflammation of the skin. *Local* or *accidental* ichthyosis is often successfully treated by means of flying blisters, or topical stimulants. In two cases in which this chronic alteration of the skin was limited to the legs, Mr. Plumbe succeeded in curing the affection by strapping the affected parts tightly with adhesive plaster, and applying a long roller kept constantly moistened with cold water. The straps were removed every fourth or fifth day,

---

* Traité des Maladies des Enfans nouveau-nés et à la mamelle, 8vo. Paris, 1823, page 32.

† Delineations of Cutaneous Diseases, 4to. London, 1817, pl. xviii.

‡ Bateman, Synopsis, 7th edit. 8vo. London, 1829, p. 8. These cases are quoted in a note to sect. 1187 of this work.—*Tr.*

and brought away with them the accidental epidermic lamina with which they were in contact. The skin under this treatment gradually recovered its natural texture and appearance.[*]

" Patients have been recommended to pick off the squamæ with their nails whilst they were seated in a warm bath, or to get rid of them by rubbing the surface with a flannel cloth or coarse towel after coming out of a simple or sulphureous tepid water bath. (Bateman.) I have always found that though the hardened cuticle could be readily removed in this way, it was constantly reproduced after the lapse of a few days. Dr. Thomson has made the same remark.

" In *general* ichthyosis emollient applications continued for a long time, gentle frictions, mucilaginous and soothing fomentations, tepid baths frequently repeated, or alternated with the watery vapour or the alkaline bath, used with such discretion as to cause no disturbance in the performance of the principal functions, are usefully employed in clearing the skin from the squamæ that cover it, and in keeping it in a state as near as possible to that which characterises the healthy integument."[†]

Of Warts.

Warts appear to be produced by disordered action in a congeries of vessels on the surface of the cutis, commonly originating from the irritation of substances which insinuate themselves into, and perhaps sometimes through the substance of the cuticle. Their structure and formation are almost too well known to require des-

[*] " Plumbe on the Diseases of the Skin, p. 334."

[†] The author whom I have quoted, is, as usual, profuse in his references; videlicet—

Panaroli Iatralogismorum, seu medicinalium observationum pentecostæ quinque, &c.; Romæ, 1652, 4to. Pentecost, v. obs. 9.

Van der Wiel. Obs. rarior. cent. 2 et 2, 8vo. Leidæ, cent. 2, Obs. 35.

M. Donati. De historiâ medicâ mirabili opus, &c. lib. 1 and 3, Mantuæ, 4to.; Schenck. Obs. med. rarior. p. 699.—Vater. Programma de cuticula pueri xv. annorum cutis rhinocerotis aut corticis arboris instar incrassata. Vitemb. 1732.

Willan. On Cutaneous Diseases, 4to. art. Ichthyosis.

Alibert. Dermatoses, 8vo. art. Ichthyose.

Janin de Saint-Just. Journ. compl. des sc. Méd. t. v. page 220.

Ansiaux. Bulletin des sc. Méd. de Férussac. t. xv. p. 289.

Chiappa. Revue Méd. 1829, mars. p. 385.

Tilesius. Ausführlich Beschreibung und Abbildung der heyden sogenanten Stachelschweinmenschen aus der bekannten englischen Familie Lambert, fol. Altenbourg, 1802.

M. Buniva. Particularités les plus remarquables de deux corn-écailleux nommés Jean et Richard Lambert, obsérves à Turin en février et mars de l'an 1809. fig. Mém. de l'Acad. imp. de sciences, lettres et beaux-arts de Turin, 4to. Several additional observations on the same family may be found in the Lon. Med. and Surg. Journ. Sept. 1831.

Martin (P. L.) Medical and Chirurg. Transactions, vol. ix. part 1, p. 153.

Follet. Diss. sur l'Ichthyose cornée. Paris, 1815, p. 280.

Joulhia. Diss. sur l'Ichthyose nacrée. Paris, 1819.

The reader will hardly be able to follow him, be his zeal what it may.

cription ; they are found to be, however, evidently resembling the cuticle, when they are carefully examined, and may be considered an irregalar fungous production of this covering. It is said by some surgical writers, that they are readily and easily destroyed by stimulants or caustics. This is true as regards those occurring on parts covered by an extremely delicate cuticle, as the glans penis, &c., a very slight degree of excitement only being necessary to their destruction in these situations; but when they occur on the hands, the most expeditious mode is as follows :—

A small portion of the Emplas. Lyttæ is to be laid on the crown of the wart, and retained in this situation by adhesive plaster ; in a day or two the substance of the wart is in a soft and moistened state, and a little ring of vesication appears round its base. In this condition the greater part of the wart may be picked or sliced off ; and if this be done, so as to bring it down to a level with the surrounding skin, the application of a little caustic will generally be all that is necessary.

Venereal warts, as they have been termed by English authors (the végétations syphilitiques of Alibert and Rayer,) yield with less trouble to milder escharotics ; a fact which is satisfactorily referrible to the inferior solidity of structure consequent on their rapid growth.*

CONCLUDING REMARKS.

AMONG the affections which do not come under the heads of the foregoing sections, are the varieties of Erythema and Roseola. With respect to the former of these, though the variations in figures of the inflamed surface have given foundation for distinctive designations in the classification of Dr. Willan, which I have already alluded to as respects Roseola, all which has been written is to be found in the following extract from Rayer :—

" ROSEOLA. *Rash, Measles—false Eruption—anomalous, rosy efflorescence—erysipelatous, Rubeola, Roseola.*

" In the first edition of this work, I followed Willan in describing, under the name of Roseola, several eruptions, acute in their nature, not contagious, transitory, and characterised by red spots, variously figured, slightly, or not at all, prominent, and usually preceded or accompanied by febrile symptoms.

---

* Rayer, in his last edition, states that there are two forms of warty excrescences : 1st, the wart with which we are all well acquainted ; an isolated round excrescence : 2d, " *Warty bands*" *on the surface, spreading to a great extent !* I must confess I have never seen this form of the affection. It would be vain to say more on so trifling a subject. After mentioning many vegetable and other escharotics, he declares nitric acid to be the best application ; and there is no doubt that it is adequate to the removal of all cases of the disorder.

"Farther experience has since satisfied me that it was impossible to distinguish several species of roseola from erythema ; moreover, I at one time instituted serious inquiries with a view to ascertain whether another variety, the spots of which greatly resemble those of measles (roseola *infantilis,*) and which forms the principle type of this group, was not itself a mere modification or a variety of measles *without catarrh.* But my facts are not sufficiently decisive, nor is my mind yet sufficiently made up, to allow me to destroy the group formed by Willan, under the title of roseola. At all events, the following may be taken as the characters of the various eruptions he has characterised by his name, the existence of which is not to be disputed, were they ever so variously designated, or ever so differently arranged in different nosological schemes.

"1. Roseola *æstiva.*—This variety, which is sometimes preceded by slight fever, appears first on the arms, the face, and the neck ; within the space of a day or two it spreads to the rest of the body, and causes tingling and severe itching. It appears under the form of small distinct patches, larger, paler, and more irregular than those of measles, separated by numerous intervals where the skin preserves its natural colour. Of a lively red at first, they soon acquire the deep tinge that is proper to them. The pharynx exhibits the same hue, and the patient feels a sort of roughness and dryness when he swallows. The eruption still continues bright on the second day ; immediately after which it begins to decline, slight patches of a dull red continue to the fourth day, and disappear entirely on the fifth along with the constitutional disturbance.

"At times this efflorescence, limited to certain parts of the face and neck, and the upper part of the breast and shoulders, shows itself under the form of very slightly elevated patches, which cause violent itching, but without the sense of prickling that accompanies urticaria. The disease lasts at most a week. The eruption occasionally appears and disappears again and again, without any perceptible cause, or in consequence of violent moral affections, or after the ingestion of spiced food and heating liquors. The recession of the efflorescence is usually attended by derangement of the functions of the stomach, by headache, a state of languor and of lassitude, which the recurrence of the eruption causes immediately to cease.

"This variety of eruption usually occurs during the summer in females of irritable constitution ; it is sometimes connected with the intestinal affections of the season ; it seems to form a kind of a middle state between erythema and urticaria, and requires to be treated by moderate abstinence, acidulated drinks, (when these do not disagree,) and occasionally by gentle laxatives.

"2. Roseola *autumnalis.* This variety attacks children in the autumn, and shows itself under the shape of distinct circular or oval spots, of a dusky red colour, which gradually increase in extent till they have attained the size of a shilling, or something less. They appear principally on the arms, and sometimes end by desquamation ;

this eruption is not accompanied by any great amount of uneasy feelings or of pruritus. It is evidently a variety of erythema.

"3. Roseola *annulata.* This species is sometimes accompanied by febrile symptoms; its duration then is brief; in other cases, there is no disturbance among the functions generally, and the eruption continues during an undetermined period. It appears on almost every part of the body under the form of rosy rings, the central areas of which are of the natural colour of the skin. These rings are not at first more than a line or two in diameter; they enlarge gradually and are sometimes at last an inch and a half in circumference. In the morning the efflorescence is always less vivid. When chronic it has a sallow and discoloured appearance; it revives towards evening or during the night, and causes a sensation of heat, itchiness, and tingling in the skin. If it vanishes or fades, the stomach becomes disordered, languor, vertigo, and pains of the limbs are complained of, symptoms which are generally assuaged by the tepid bath. When the eruption becomes chronic it should be treated by sea-bathing and the use of the mineral acids.

"The description of this variety must be ultimately blended with that of the *erythema annulatum.*

"4. Roseola *infantilis.*—In this variety the spots have smaller intervals of sound skin between them than we observe in Roseola *æstiva.* When the eruption is general, if the appearance of the disease be the only element considered in laying down the diagnosis, it is very apt to be confounded with *common measles.*[*] This variety of roseola attacks children when teething, or it supervenes in the course of their intestinal and febrile affections. Sometimes it only exists for a single night; or it comes and goes successively during several days, accompanied by some disturbance of the principal functions. It may also occur in succession on different parts of the body.

"Whether this variety be a modification of rubeola, or be independent of that disease, it ought to be considered as the type of the group *roseola.*

"5. *Roseola variolosa.*[†]—This exanthema sometimes occurs before the eruption of the natural or inoculated small-pox, preceding the former more rarely than the latter, in which, indeed, it is calculated to appear about once in fifteen cases, in the course of the second day of the eruptive fever, which corresponds with the ninth or tenth day after the inoculation. The efflorescence is first perceived on the arms, the breast, and face, and on the following day it extends to the trunk and extremities. Its long, irregular, and diffused patches, leave numerous intervals between them. More

---

[*] "Underwood is of opinion that this error has been often committed. (On the Diseases of Children, vol. i. p. 87.)"

[†] Dezoteux and Valentin have described the roseola *variolosa* under the title of *eruption anomale rosacé* ('Tr. hist. et prat. de l'Inoculat. 8vo. p. 238).—Baron Dimsdale has given many cases of roseola in consequence of inoculated variola (vide Present method of inoculating for the Small-pox. 8vo. Lond.)

unfrequently this variety of roseola is characterised by an almost generally diffused efflorescence, slightly prominent in some points. It lasts about three days; on the second and third, the variolous pustules may be distinguished amid the roseolar efflorescence by their roundness and prominence, their hardness, and the whiteness of their summits; as soon as they appear the roseola declines. This variety of roseola has been regarded, by several inoculators, as announcing an eruption of distinct small-pox. My observations on natural small-pox, like those of Walker,* lead me to think precisely the contrary.

"The eruption of roseola variolosa is, with difficulty, repelled by exposure to cold, or by cold drinks. By the earlier writers on small-pox it was mistaken for measles; a circumstance that led them to conclude that measles was sometimes turned into small-pox.

"Roseola *variolosa* may be very readily connected with erythema.

"6. Roseola *vaccina*.†—This efflorescence is observed to occur in several children from the ninth to the tenth day after the insertion of the vaccine poison. It comes out as small confluent patches, and sometimes diffused like those of the variolous roseola, and appears about the period when the areola is formed around the vaccine vesicle, from whence it extends irregularly over the entire surface of the body; it is not, however, so general an occurrence as the efflorescence that follows the inoculation of small-pox. It is commonly accompanied by acceleration of the pulse, and great anxiety.

"7. Roseola *miliaris*.—Willan tells us that this variety often accompanies an eruption of miliary vesicles with febrile symptoms. —I have not met with it in my practice.

"8. Roseola *febrilis*.—In continued and typhoid fevers Bateman has observed an efflorescence making its appearance, resembling roseola æstiva, or measles. He saw this roseola occur three times at the close of slight fever. In two of these patients the eruption only lasted two or three days; in the third it appeared on the ninth day of the fever, after a deep sleep and a gentle sweating fit. The spots, of a bright rose-red colour, of an oval shape, slightly prominent, and smooth on their surface, thrown out on the arms and the chest, were more particularly numerous on the inner sides of the arms. The eruption was attended by no pruritus, nor any other sensation. The febrile symptoms all subsided on the same day, and the patient no longer kept his bed. Next day the efflorescence had spread; the spots had become broad and confluent, and their colour, which was weakened, especially towards the centre, had acquired a purple tinge, whilst the margins still continued red and slightly elevated. On the third day, the whole of the patches had a tendency to become livid, and on the fourth, there was scarcely a trace of them remaining, or any symptom of febrile excitement to be observed.

---

* Inquiry into the Small-pox, 8vo. Edinb. 1790. chap. 8.
† Pearson. Observ. &c. in Lond. Philos. Magaz. Jan. 1809.

" This variety of roseola might be described under the head of erythema, with perfect propriety.

" 9. Roseola *rheumatica.*—A roseolar form of efflorescence is sometimes connected with attacks of gout and acute rheumatism. Bateman attended an individual of a gouty habit, in whom a roseola, accompanied by violent fever, extreme languor, complete anorexia, and constipation, appeared upon the lower extremities, the forehead and scalp for the space of a week. The efflorescence ended on the seventh day in desquamation, and in the middle of the night the articulations of the right foot were attacked by gouty inflammation. I have seen roseolar patches occur towards the end of a rheumatic seizure. Case 19. Dr. Schönlein* has also described this variety under the title of *pelliosis rheumatica,* and assigned it the following characters : pains of the articulations and extremities, of greater or less severity, remitting, changing their place, increasing from the influence of cold, and becoming easier from the warmth of the bed ; shivering fits, followed by more or less marked febrile reaction, with quickening of the pulse and increased heat of skin, which is dry.

" Slight gastric symptoms with loss of appetite, clammy state and bitterness of mouth, and furred tongue, are the precursors of the disease. Twenty-four or forty-eight hours afterwards, and often later, a particular eruption makes its appearance, always commencing on the legs, and sometimes going no farther, but most generally coming out on the arms and shoulders at the same time : it rarely appears on the trunk, and never on the face. This eruption consists of small isolated spots, of a round shape, of the size of a millet-seed or small lentil, rarely prominent, and of a deep, or violet, sometimes blackish-red colour. The number of these spots is very variable ; most generally they are neither so numerous nor so closely set as the vesicles of miliaria, or the spots of measles. The fever ceases, and the rheumatic pains either leave the patient or diminish greatly in their severity from the moment this eruption appears. Under the influence of appropriate regimen and treatment, the little spots, whose numbers may be increased by successive crops, grow pale, and the disease ends by a furfuraceous desquamation. If the course of this exanthemata is interrupted, whether by the influence of exposure to cold and moisture, or by the application of discutients, the spots disappear immediately, the rheumatic pains return in increased severity, the joints swell, motion becomes intolerable, and fever is again set up.

" This disease observed at Würzburg, where rheumatic affections are almost endemic and frequently fatal by being complicated with miliaria, attacks adults and males more frequently than females. M. Fuchs says that it is during winter and spring, when the atmosphere is cold and loaded with moisture, that he has most frequently seen this eruption.

* Fuchs, sur le Pelliosis rheumatica (Bullet. des Sc. Med. de Feruss. tom. xviii. p. 274.)

" The plan of treatment pursued at the hospital of Würzburg, consisted in the administration of tartrate of antimony when there were symptoms of gastric derangement ; in the exhibition of the vinum colchici when the rheumatic pains were severe, and the use of diaphoretics, such as the acetate of ammonia and compound powder of ipecacuanha to favour the eruption. Tepid and soothing drinks alone were permitted ; the regimen was simple and anti-phlogistic.

" Petzold,* Nicholson, and Hemming,† have also met with arthritic roseola.   And, to conclude, Dr. Cock‡ has particularly described an ' epidemic eruptive rheumatic fever' which he saw in the West Indies.

" 10. Roscola *cholerica*.§   I saw this variety during the epidemic prevalence of cholera at Paris, in 1832.   After the period of reaction, there occurred in some patients, especially in women, an eruption which most generally appeared on the hands and arms, and then extended to the neck, the breast, the belly, and the upper and lower extremities.   At its commencement it was characterised by patches for the most part of an irregularly circular shape, of a bright' red colour, elevated above the surface, and but slightly itchy.   Very numerous on the hands, arms, and chest, they were less so on various other parts ; in some places they were crowded together, tended to confluence, and had an appearance very analogous to the efflorescence of slight scarlet fever ; in other places the aspect of the eruption was rather like that of measles, and in others even more like that of urticaria.

" I have seen this inflammation complicated with an inflammatory affection of the fauces or tonsils, and its disappearance followed by an aggravation of the general symptoms, and sometimes even by death.   On the chest the spots, occasionally, became confluent, and gave rise to patches as broad as the hand, raised above the general level and pretty well defined.   The eruption then acquired a dirty pink or rose colour.   About the sixth or seventh day the epidermis cracked and was thrown off in large flakes, on almost all the places where the eruption had existed.

Historical Notices and particular Cases of the Disease.

" I have already pointed out the principal treatises upon roscola

* Petzold. Obs. Med. Chir. No. 9.—Nicholson. Lond. Med. Gaz., vol. iii. p. 546.
† Hemming. Beytrage zur Prakt. Arzneykunde. II B.
‡ Cock. Obs. on the Epid. Eruptive Rheumatic Fever of the West Indies, (Edinb. Med. and Surg. Journ., t. xxxiii. p. 43.)
§ Duplay. Memoire sur la roséole consécutive au choléra (Gaz. de santé, in-4. p. 583. Paris, 1832).—Babington. Cutaneous Eruption in Cholera in Lond. Med. Gaz. vol. x. p. 678.—Lepecq-de-la-Cloture avait observé cette eruption à la suite du choléra sporadique (Collect. d'obs. sur les mal. et les constit. epidemiques, p. 1005).

in speaking of the varieties of this efflorescence. Orlov,[*] Seiler,[†] Heim,[‡] and Stromeyer[§] have devoted themselves to pointing out the characters that distinguish it from measles and scarlatina. It is not so easy to establish a line of demarcation between roseola and erythema. No one, indeed, can confound papular erythema without fever, confined to the hands, and its prominent and distinctly circumscribed patches, with the more or less diffused, not prominent red spots of roseola, spread over almost the whole surface of the body, and appearing after a febrile attack analogous to that of eruptive fevers generally. But if a certain number of particular cases of roseola and of erythema be compared, various points will be discovered at which these two diseases meet and run into each other. Let but erythema become somewhat more general than wont, and its spots appear a little more prominent, or the patches of roseola, from some unusual violence of the accompanying inflammation, become particularly prominent, and the appearances of these two exanthemata are the same. So, the eruptions that accompanied cholera, on their first appearance presented several characters that approximated them to the erythema *papulatum ;* these, however, were soon lost, for others held pathognomonic of roseola. Lastly, roseola *annularis* and roseola *autumnalis,* which appear as chronic affections, ought evidently to be classed with erythema.

Sydenham[‖] thought that *roseola* was a variety of rubeola, or measles ; others have imagined that there subsisted between these two affections the same relation as we trace between variola *spuria* and variola *vera ;* whilst a third party[¶] have endeavoured to prove that roseola typified a particular morbid condition distinct from all the other exanthemata.

" Case 1.—*Articular rheumatism, erythematous or roseolous eruption ; muco-enteritis and bronchitis.* Marie Hautefeuille was seized on the 7th of January, 1833, after a hard day's work, with shivering fits, which proved the prelude to a general rheumatic attack. She was treated by bleeding and the tincture of colchicum. On the 26th and 27th a *roseolar eruption* appeared on the fore parts of the breast and abdomen, and over the lumbar region, of which four or five round and well defined patches did not disappear completely with pressure. Upon the abdomen there is a reddish festooned or wavy zone, which extends obliquely from the left hypochondrium to the vicinity of the spinous process of the ilium ; the lower edge of this zone is sharp and well defined, the superior is gradually lost in the healthy skin. The festoons of the lower edge are of unequal breadth. The patient neither complains

---

* Orlov (A. J.) Progr. de rubeol. et morbil. discrimine. 1785, in-4.
† Seiler. Diss. de morbil. inter et rubeol. different. verâ. in-4. Witteb. 1805.
‡ Heim. Journal von Hufeland, 1812.
§ Stromeyer. De rubeol. et morbil. discrimine. in-4. Goetting, 1816.
‖ Opera Med. sec. v. cap. 1.
¶ Hoffman. Opera, t. ii.—Burserius. Institution. vol. 1. Selle, Pyretologia. p. 171.

of heat, itchiness, nor pain in the seat of the eruption, circumstances that sufficiently distinguish it from urticaria.

" On the 28th, the rheumatic and general febrile symptoms recur with new intensity ; the eruption has gone from the breast, and faded on the abdomen. 29th.—Profuse perspiration, which gave relief ; the eruption disappeared from the abdomen. In the night of the 2d of February, the patient wandered, moaned, seemed drowsy, and complained of severe pain in the wrists ; constipation during the last three days,—(*blisters to the fore-arms.*) From this time forward, the rheumatic pains and fever went on declining : and the patient after another relapse of her rheumatism and an attack of diarrhœa, left the hospital on the 25th, recovered.

" Case 2.—*Cholera, Roseola following reaction ; recovery.*— Bougal, aged 43 years, had suffered from cholera for eight days, from the cold or collapsed state of which he had hardly recovered when he was admitted into La Charite. Reaction to a greater extent was excited by a few spoonfuls of Malaga wine, and sinapisms to the extremities : and within a day or two symptoms of greater vigour were manifest.

" On the 21st an eruption appeared over the whole body, but especially on the belly and on the limbs. It is disposed in rather broad patches of an irregularly round form, and a deepish-red colour, not raised above the general level of the skin, and presenting all the characters of roseola ; in some places the efflorescence is uniform, and resembles a little that of scarlatina ; on the breast, however, it is exceedingly like measles : in other respects the patient's state is satisfactory, the eruption is attended by no disturbance of the general functions, and by the 27th, when the patient was quite convalescent, it had vanished without inducing desquamation.

" Many instances of this kind were observed in the wards of La Charité. Two of Dr. Lherminier's patients and three of Dr. Rullier's were affected during the period of reaction in a very similar manner.

" Case 3.—*Roseola, (rubeola spuria?)*—Whilst dressing G. L., aged four years, on the 19th of June, 1825, the nurse observed his face covered with small red spots, without elevation of the skin, and very like measles in colour. Not only were these spots evident at the period of my visit the same day, but others were distinguished on the breast and arms. The child had no fever, had slept well through the preceding night, and was playing as usual ; yet the tongue was foul towards the base, and much dotted ; the appetite too had fallen off for several days, and the bowels were constipated. Pressure on no part of the abdomen caused pain ; the throat was not inflamed, and there was no cough or preternatural flow of tears. Two years previously I had attended this boy and his two brothers for well-marked measles, a circumstance which, considered along with the absence of several of the characteristic symptoms of measles, made me look on this slight eruption as nothing more than roseola, or the false measles of authors—(*gum water, and soup for diet.*) The child slept quietly ; by the 21st the efflorescence was pale ; and two days after it was no longer visible ; a day or two more of perseverance in the same regimen restored the appetite to its usual condition."

It may be fearlessly stated, that it is of no independent importance (except as regards the last, or E. nodosum,) and for the most part merely entitled to notice as symptomatic of derangement of the digestive organs. This admission indeed is distinctly made in the observations appended to the description of each of the variations alluded to both by Willan and Bateman.

The symptomatic inflammation of the skin would be much better understood in the form of a general description. Such description would be divested of that multiplicity of terms which has been so conducive to confusion and intricacy in the works of the authors last mentioned, and it may be effected in a very few words, so as to answer all practical purposes.

The affection in question, whatever may be the form (with the above exception) which it may assume, whether consisting of red patches of an irregular form and short duration ; whether with a smooth, shining surface, or having papulæ distributed on it ; whether with an irregular and defined margin, without elevation, or in the form of slightly elevated tubercles ; requires no external applications, nor can be with propriety interfered with, except through the medium of the constitution.

On the subject of roseola, what has been written may be summed up in a few words. Like simple erythema, it is generally found dependent on some disorder of system, and is to be considered as an effort of nature to mitigate or prevent more important internal disorder.

By roseola is meant " a rose-coloured efflorescence, variously figured, without wheals or papulæ, not contagious, sometimes accompanied by sensations of tingling and itching." It is distributed on the face, neck, and upper extremities, when only partial, rather more frequently than on other parts, but sometimes spreads over the greater portion of the cutaneous surface. If symptoms of febrile irritation, as is sometimes the case, usher in the efflorescence, such symptoms are mitigated by it, and eventually die away as the redness subsides. The rose-colour, from which it takes its name, is not that in which it appears at first, but seems rather to be the result of diminished inflammation : as the form in which it is at first observed is of a much brighter red. Sometimes, like erythema, the redness is diffused ; at others it assumes the forms of rings and spots ; while in others again, irregular lines of a darker colour have their interstices filled up by a lighter shade of red. These variations have been represented in the plates of Dr. Willan, under distinguishing appellations.

Rose-coloured rashes seem to be peculiar to no age or sex, but are, on the whole, more frequent among infants and children than others. Preternatural irritability of skin, with a similar state of constitution, lead to its frequent appearance, in some from the most trifling exciting causes.

The colouring of the plates of Dr. Willan and Dr. Bateman is, in some copies which I have seen, so deficient, as to present no variation of shade between this affection and scarlatina; but the distinction between the two diseases is very considerable, and essentially depends on this point. A due consideration of the co-existent symptoms will enable us to clear up any doubts which may arise.

As in erythema, no local treatment is generally found necessary; but in such case as that inserted by Dr. Willan, communicated by Dr. Currie, of Chester,* where the disease returned annually, and continued from March to October, it would become a matter of necessary consideration.

When the rash appears in connexion with the variolous and vaccine disease, it has been named by Dr. Willan accordingly. It is generally supposed, under such circumstances, that the more important disorder is rendered milder by it.

Venereal eruptions, or those diseases of the skin which are produced by the poison of syphilis, have been recently pronounced to assume no other form than that of scales; the advocates of this doctrine having discovered that the papular or pustular eruptions heretofore appearing under circumstances of suspicion, and characterized as syphilitic, only make their appearance in connexion with, or subsequent to, sores not entitled to such character, or to simple gonorrhœa. Mr. Carmichael, whose opportunities of observation have been of no ordinary kind, has supported this opinion with facts and remarks; which, from the immense importance of the question, ought to be duly weighed and examined by every medical practitioner who has the interest of the profession and of society at heart.

According to Mr. C., the true syphilitic chancre or bubo is followed, as a secondary symptom, by lepra or psoriasis. The line of distinction between these two affections, even when dependent on other causes, is sufficiently indistinct, and when occurring from syphilis, probably still more so; hence the propriety of the general designation proposed of "*scaly venereal disease*" is obvious. The following description, from the pages of Dr. Willan, is that of the diseases in question. "Circular patches, which resemble those of the Lepra nigricans in size and colour, but which are not incrusted. The dryness and harshness of the skin, so remarkable in the Lepra vulgaris and alphoides, do not occur in the venereal lepra; its patches, when somewhat advanced, being as soft and pliable as other parts of the skin. It is, however, proper to observe that every patch originates from a small, hard, reddish protuberance.

* Willan on Cutaneous Diseases, page 438.

As this gradually dilates, the increase of its circumference is not attended with an increasing elevation at the centre : on the contrary, the sides of the patch are sometimes raised, and the central part of it appears a flat surface covered with thin white scales.   The patches are generally distinct, and at a distance from each other : I have seldom seen any of them exceeding the size of a shilling ; yet it is probable they might acquire a greater magnitude, if the progress of the disease were not early arrested by the use of mercury.   When the constitution is under the full influence of this remedy, the sides of the patch shrink and become paler ; the centre is also depressed, but the desquamation proceeds slowly ; and the disease cannot be removed without a perseverance in the course for six or eight weeks. A circular red spot usually appears for some time in the place of every declining patch, and a minute shallow depression, like a cicatrix, is left at the centre ; but no permanent discolouration of the skin remains, as in some other cases.   The leprous form of the syphilis takes place, like other venereal eruptions, at very different periods after infection in different cases.   If no medicines were employed, it would at length terminate in ulcerated blotches.''

The description of the syphilitic psoriasis, page 153 of the same author, is as follows :   " An eruption which very much resembles the Psoriasis guttata, differing from it only by a slighter degree of scaliness, and by the livid red, or dark rose-colour of the patches. The patches vary in their extent, from the size of a silver penny, to that of a section of a pea, but they are not circular.   They rise at first very little, if at all, above the cuticle : as soon, however, as the scales appear on them, they become sensibly elevated, and sometimes the edge or circumference of the patch is higher than the little scales in its centre.   This eruption is usually seen upon the forehead and breast, between the shoulders, or in the inside of the forearms, in the groins, about the inside of the thighs, and upon the skin covering the lower part of the abdomen.''

Mr. Carmichael considers the above description of the " scaly venereal disease,'' or that cutaneous affection exclusively produced by the syphilitic virus, as correct in all its parts, and that all papular or pustular eruptions are connected with other primary diseases not venereal.   Thus, to a simple primary ulcer, excoriation of the glans, and gonorrhœa, he attributes the eruptions of pimples, which are attended by fever, and terminate in desquamation.   When the ulcer has elevated edges, though it be not indurated, phlyzacious pustules occur, preceded by fever, and terminating in ulcers covered with thick crusts, which heal from their margin, and when the disease is on the wane, terminate in red and scaly blotches.   When the primary ulcer has worn a decidedly phagedenic character, the cutaneous disease is at first tubercular, the tubercles being sometimes intermixed with spots approaching to pustules.   Fever also accompanies this form of disease : many of the tubercles or pustules terminate in ulcers covered with a thick crust, which, if undisturbed, grows on the part, and assumes a conical form, its basis being surrounded with a thick phagedenic edge, producing precisely the appearances of what has been named rupia.

There have been several dissentient voices raised against Mr. Carmichael's opinions as above detailed, regarding both the exclusive claim of the scaly eruption to syphilitic origin, and the directness of the connexion of the papular and pustular diseases with the primary sores and affections which he has described. On the first of these points the weight of evidence is decidedly in favour of his doctrine ; but some doubts may, I think, be reasonably entertained, whether a cachectic state of system may not be the sole cause of some of the eruptions he has described as connected with non-syphilitic primary disease.

The phlyzacious pustule, more particularly, seems liable to this suspicion, inasmuch as it is the form which cutaneous disease often assumes where nothing but reduced energy of system exists.

The treatment of these affections is included in the general management of other symptoms of the constitutional disorder which do not form the proper objects for consideration here.

In the chapters on Porrigo, I have omitted the consideration of an affection so designated by Willan and Bateman, from the utter absence of any thing like disease of the cutis in any part of its progress. The affection to which I allude is distinguished by the authors in question by the name of Porrigo decalvans.

No vestige of disease is ever discoverable in this affection in the cutis of the head. The hair gradually falls off without any obvious cause, leaving spots, which exhibit a pale and shining appearance. These are perfectly divested of hair ; and if the surrounding parts, on which the latter is growing, be shaved, they are discovered to form distinct indentations.

Pathologically speaking, the dropping off of the hair is nothing more in this affection than the result of a particular organic structure ceasing to perform its office. In the preliminary remarks to this treatise I have endeavoured to describe the structure in question, the office of which is to secrete and nourish the hair. The falling off of the hair is the consequence of the wasting away of this structure ; and though the causes of the latter are not easily ascertained when occurring in young people in a good state of health, they are sufficiently obvious in opposite states of system, and in old age, the phenomena occurring over the whole scalp in states of great exhaustion and debility from disease in youth where general reduction of fat has been temporarily produced, and in old age, from the general removal of this substance from superficial to internal parts, a change which is much favoured by the diminished energy of circulation which comes on at this period of life.

The similarity of figure which the affected spots assume to the contagious ringworm, often leads to the confounding the affection with the latter disease. It may, however, be distinguished by the absence of redness or irritation, and by the indented appearance described.

Dr. Bateman appears to have suspected the identity of the two affections, and Dr. Willan had seen a case or two in a school where the P. scutulata prevailed very generally ; but it is evident from the preceding description that the only point in which similarity exists is of little or no importance.

When the hair falls off in patches, in the manner described, in children and young people, it often grows again in time as strong as ever : but I do not believe this can be brought about, or at all expedited by any artificial means. I have seen two instances in adults, where the whole of the hair dropped off in a few weeks from the scalp, eyebrows, eyelids, &c., while the individuals were suffering from great grief and ill-health. In one of these an endless variety of stimulant applications had been tried, with the hope of making the hair grow, but without the slightest effect. Indeed, I should be inclined to think, that if any effect at all was produced by this plan, it would be rather that of retarding than expediting the object.

Although I have expressed in a former page a desire to confine the subjects of this work to the consideration of cutaneous diseases, most familiar, and the knowledge of which may be most valuable to English practitioners, I hope to be excused a little addendum, which is not much out of place, and perhaps by some may be considered valuable.

I extract the articles from the work of the author to whom I have been already so much indebted, with the fullest acknowledgments. It is more than probable that some of the diseases described are imaginary. If they are not, the only feeling which a medical practitioner would entertain is that of gratitude that they have not visited this country.

I put the subject-matter to which I allude in the form of an Appendix, disclaiming all pretensions to practical and personal knowledge of it, except as regards those very familiar domestic acquaintances of which my readers will see the names and natural history detailed.

APPENDIX.

Horny productions of the Cutis.

The anomalous horny productions, often conoidal and prominent (*horns*), sometimes flattened (*horny laminæ*), and of various sizes, occasionally observed on the surface of the skin, are formed of a substance very similar to that of the nails and epidermis.

Horny productions are most commonly developed on the head, and on those districts of skin which are most plentifully supplied with sebaceous follicles. The greater number of horny productions result from an affection of one of these follicles. Sir Astley Cooper has given engravings of two cases of these appendages, developed in the cavity of a distended follicle. The simultaneous development of follicular tumours and of these horny appendages has even been observed.

Horny productions, secreted on the inner surface of the follicle, soft at first, soon become hard and tough; they then rise above the level of the skin, and grow both in length and in thickness; they have occasionally been seen several inches long.

During the early periods of their formation, and when they are still of small size, these horny productions are surrounded by a membrane, which makes them appear encysted. By and by this membrane only envelopes the base of the formation. Horny productions do not extend more deeply than the follicles, within the interior of which they appear as it were to be set. They are consequently always moveable, and participate in the motions which the skin receives from the subcutaneous muscles. The cyst, or kind of cyst, in the cavity of which they are implanted, is sometimes the seat of chronic inflammation, which occasionally ends in ulceration.

Horny productions are also frequently evolved on parts attacked with chronic inflammation. M. Julius Cloquet met with a large horny production on the forehead, which made its appearance after a burn. Professor Dubois had an old woman under his care in the Hôspice de Perfectionnement, upon whose forehead there was a conoidal horn six inches and a half high by about seven inches in diameter at the base. There is a drawing of this patient in the collection of the Faculté de Médecine. A contusion or wound of the skin had preceded the growth of this horn. The patient complained of habitual headache, the intensity of which was constantly on the increase. The apex of the horn was solid; its base was of a clearer colour, and much inferior consistency. Circular striæ indicated the successive deposits of the matter of which it consisted, and formed inequalities similar to those which are observed on the horns of certain ruminating animals. The cuticle was arranged around the base of the horn, as it is around the root of the nails; it extended beyond the corion to a greater extent however, namely, by several lines. Portions of the horny production thrown upon lighted coals burned with a smell similar to that which horn diffuses when burned. This production had pushed the integuments of the forehead aside, and especially downwards, to such an extent that the eyelids were habitually closed. The head of the woman diffused a fetid odour.

In a patient of the name of Aumont, who died in the Hôpital de la Charité of a disease of the heart, I observed a squamous and pearly-looking production developed on the cicatrice of a gunshot-wound of the leg, which he had received in 1806, when in the military service. The cicatrice was covered with scattered pearly squamæ, bearing a strong resemblance to the scales of a carp, which they also equalled in size. Detached by means of warm water, or the tepid bath, these squamæ were soon reproduced. Examined after death, the corion presented an arrangement similar to that of the skin, which covers the legs and feet of fowls.

Horny productions may appear on every region of the body. Of seventy-one cases of the kind in which M. Villeneuve was consulted, thirty-seven occurred in women, thirty-six in men, and three in infants. In nine of these cases the horns were situated on the head, in fourteen on the forehead, and in twelve on the thigh: in the other cases they were situated three times on the temples, five times on the nose, twice on the cheek, once on the jaw, four times on the chest, four times on the back, three times on the penis and glans, four times on the ischium, twice on the knee, twice on the ham, once on the leg, twice on the foot, and once on the hand. They have also been seen on the back of the hand, and above the ear.

The cases of horny degeneration of the skin of different parts of the body, and of monstrosity of the integument, observed by Malpighi, Ash, Locke, and Musæus, appear to form a separate group, and to depend on a kind of general disposition, different from the morbid processes, of a purely local character, which eliminate horny productions.

Besides the horny productions developed in the cavities of the sebaceous follicles, similar formations are occasionally seen succeeding a variety of wart. Rose Davène, aged sixty-four, of healthy constitution, and the mother of seven children, consulted me at the Bureau central des Hôpitaux on the 31st of August, on account of a yellowish-coloured horny production, more than two inches in length, the base of which was fixed upon an inflamed patch of the skin, covering the inner surface of the thigh. In the situation which this production now occupied, two small spots (boutons) had appeared about six years ago, which had suppurated. On the seat of these a wart made its appearance, and by and by became covered with a horny top, which she had removed with a ligature; but it soon grew again, and was now of the length mentioned. It was only troublesome in walking; the glands of the groin were healthy. I recommended it to be removed, but do not know whether this was done or not.

Voigtel, Conradi, J. F. Meckel, Otto, and others, have published many interesting facts connected with the history of *manifold horny productions*. In the collection of the Ecole de Médecine at Paris, the hands and feet of an old woman, presented by Beclard, are preserved, which are covered with horny lamellæ of different sizes. The backs of these parts are covered with horny productions of less length than those of the palms and soles. From the latter arise five or six excrescences, as thick as the finger, and from eight to ten inches in length.

The causes of horny productions are extremely obscure; they appear most generally to follow some chronic irritation of the papillæ.

The form, colour, and especially the consistence and structure of these productions, and the smell they exhale in burning, distinguish them sufficiently from the hard, dry, and pyramidal-shaped incrustations, which occasionally cover syphilitic, scrofulous, and cancerous ulcers. These horns are still more distinct from fungous tumours of the dura mater, from nodes of the bones, &c. with which they have been said occasionally to have been confounded. Bony appendages or spiculæ of the femur, humerus, &c., similar to those mentioned by Cabrolius, Vicq d'Azyr, and others, can never be mistaken for horny productions of the skin, even though they projected to a considerable distance after piercing the skin.

I only know of one case which goes to prove that the spontaneous fall of horny productions may be followed by a permanent cure.

When it is held necessary to remove these horny productions, on account of the inconvenience or deformity they occasion, the knife is always to be preferred to every form of caustic application. After including the base of the formation within a circular or elliptical incision, it is necessary to dissect away, or to destroy by means of escharotics, the follicular pouch, the warty excrescence, &c. from which they spring. When this precaution is neglected, or when these productions are simply cut across at their bases or removed with a ligature, they are very apt to grow again.

Little has yet been done in trying to subdue the particular disposition manifested by certain individuals to have horny growths developed upon the surface of their bodies. Fabricius Hildanus tells us, that a young female, having made use of evacuants, emmenagogues, and the sulphureo-aluminous thermal waters of Neuham, was cured for a time of these horny productions with which her skin was covered. The combined action of simple tepid, of alkaline, and of vapour baths would probably prove advantageous, if these appendages were found to adhere but slightly to the skin.

Corns.

Corns are accidental circumscribed epidermic indurations, of a round shape and hard consistence, which are commonly evolved on the upper surface, or on the lateral aspects of the toes, occasionally also on the soles of the feet, especially towards the anterior extremities of the metatarsal bones. Corns compress and irritate, and sometimes inflame and even pierce the subjacent skin; they have also been known to cause inflammation in the joints situated underneath them.

Causes.—The injurious pressure of tight, short, or ill-made shoes immediately upon the skin, or of the toes upon one another, are the most frequent causes of corns.

Corns are generally shaped like the head of a nail, the cuticle of which they consist is so much thickened, that layer after layer of it may be removed with a knife. In the middle of the yellowish hardened epidermis a small point of a white colour may be perceived, which penetrates more deeply than the other parts of the induration. The slightest pressure over this white point occasions severe pain. This nucleus is sometimes surrounded by a slight ecchymosis, situated in the corion.

Corns of the lateral aspects of the toes are usually situated under the projections formed by the articular heads of the phalanges, upon which of course pressure acts more powerfully as well as more continually. They are almost always moist; their centre is depressed, and presents a slight cavity of a greyish tint, that forms a marked contrast with the pearly white which the constant moisture of the feet gives to the thickened cuticular rim that surrounds the corn.

Bunions, callosities, and *indurations,* are names given to thickenings of the cuticle covering the palms of the hands, soles of the feet, and joints of the toes. They do not differ from corns save in being without the central white cone observed in these last, which from penetrating deeply has procured them the title of *clavi pedum.* Pressmen, in printing offices, are liable to indurations of the nature of bunions on certain parts of the palms of their hands, and to painful chaps, which are produced by the alkaline leys used for cleaning type. Hard cuticular indurations frequently occur around the heels, on the inner sides of the great toes, on the inferior surface of all the others, and especially over the digital extremity of the first metatarsal bone.

When pieces of integument, affected with corns, are macerated for some time, the thickened epidermic layers are seen to have depressed and altered the corion beneath them. The central nucleus, firmer and more horny than any other part, is almost always distinct.

Corns may be prevented by wearing easy shoes. These are essential to persons who by their avocations are required to walk a great deal. These individuals ought also to rub their toes with a little tallow, as well as to grease the ends of their stockings, and those points which are in contact with the projecting parts of the foot.

The acute pain produced by corns may be appeased for some time at least by cutting away the exuberant parts of these productions with a sharp knife. The eye of the corn may sometimes be picked out with the nails after soaking the foot in warm water, or softening the induration by the application of a soft cataplasm, or a piece of diachylon plaster, &c. Two or three operations of this kind, at the distance of a fortnight or three weeks from one another, occasionally suffice to remove the evil entirely. A blunt needle, fixed into a handle, is an excellent instrument for working out these central nuclei. The part should afterwards be anointed with a little mutton suet and covered with a bit of soap or diachylon plaster.

Plasters of soap, of mucilage, of gum ammoniacum, of galbanum, various unguents, the leaves of the house-leek, the pith of the rush used by coopers, cotton, fine linen, &c. applied around the toes, will all be found useful in treating corns of the feet, provided the sufferer at the same time reforms his boots and shoes in the essential point of size. It is not without great advantage either, that the central nuclei of corns are preserved from pressure, by being covered with a piece of thickish soft leather spread on one side with diachylon and pierced in its centre with a hole of a size corresponding to that of the nucleus to be defended. Sir A. Carlisle proposes the above means of dressing, continued during about from six weeks, as a simple and very effectual plan of cure.

Corns have also been destroyed by being rubbed with caustic potash, nitrate of silver, muriate of antimony, nitric acid, &c. These agents are attended with many inconveniences in inexperienced hands, and excision appears to be preferable.

Of Hypertrophy of the Corion.

Hypertrophy of the corion or dermis sometimes shows itself externally in the shape of small lenticular tubercles, the colour and consistence of which are very nearly the same as those of the skin that surrounds them.* These tubercles are scarcely ever seen among children,† but they are not rare among adults and individuals of mature years. They occur most commonly on the upper lip, and on the alæ nasi. They are occasionally congenital, and are then frequently surmounted by one or several strong and bristly hairs. They never terminate either in resolution or suppuration, and when they increase in size it is only in the slowest and most gradual manner.

It often happens as one of the consequences of obesity, that the skin in the vicinity of old ulcers, and as an effect of elephantiasis *Arabica*, that the integuments covering the parts affected experience a real hypertrophy, and acquire a mamillated or tuberculated appearance.

---

* So, at least, says M. Rayer.   † It would be strange if they were!

Hypertrophy of the corion and subcutaneous cellular substance may be limited to a single region of the body. This species of structural change occurs more frequently in the nose than in any other part. The affection takes place very slowly and gradually, and most commonly without any evident cause. It generally attacks individuals of mature years. Women appear to be its subjects very rarely. It occurs under three principal forms : in one case the lobes of both alæ nasi are affected with hypertrophy simultaneously, accompanied with a notable development of the vascular rete of the skin, which assumes a vinous red colour ; in another instance one or more small tumours of the form and dimensions of the indurations of rosacea, appear upon the alæ nasi ; in a third instance these two forms of hypertrophy occur combined.

Hypertrophy of the alæ nasi extends by degrees towards the root of the organ. The small tumours or indurations that make their appearance on their surface may continue long stationary, and never exceed a small hazel-nut in size, or they may attain dimensions much more considerable. Not only do they then deform the features, but they impede the free entrance of the air into the nostrils, and even of food into the mouth. The tumour has a knotty uneven surface externally, and is of a deep red or purple colour, traversed by an infinity of fine tortuous vessels. The various nodules composing the general hypertrophy are often separated from each other by deep fissures. When the affection is of very long standing, the integument of the cheeks sometimes acquires a colour and an appearance similar to that of the enlargement of the nose. The sebaceous follicles are very much developed, and the secretion from them sensibly increased. As the disease advances, the enlargement extends in length, hanging down upon the upper lip and over the mouth in one or several lobulated masses.* In the aged, the hypertrophied nose occasionally ulcerates superficially in one or several points of its surface.

The hypertrophy now described, although accompanied with a considerable morbid increase in the vascular rete of the skin, and subcutaneous cellular membrane, differs in point of structure from the sanguineous erectile tumour. When wounded with a cutting instrument, these enlargements bleed profusely indeed, but their lamellar, hard, and dense tissue, which is one of the principal elements in their composition, differs entirely from the open, spongy substance which constitutes the erectile tumour. Both in its external appearance and organization, this alteration of the skin of the nose has still less resemblance to cancer.

Topical blood-letting is an efficient means of checking the progress of this affection.†

---

* Here again is an instance confirmatory of a preceding remark.
† The removal of the exuberant portion of the organ is still more effectual.

Organised Living Creatures infesting the Body of Man, and feeding on his Skin.

Many animals may be developed, or may exist accidentally on the skin of the human subject. Some grow, live, and generate on the surface of the integuments, such as the *pediculus humani corporis*, the *pediculus capitis*, the *pediculus pubis*, and the *pulex irritans ;* others penetrate the epidermis, such as the *acarus scabiei* and the *pulex penetrans.* Other insects deposited within the substance of the human skin, in the state of eggs, are there hatched, live in the form of larvæ, and emerge at length in the shape of winged insects, as the *œstrus*, so common in sheep, oxen, and horses ; finally, one species of the entozoa, the filiaria medinensis, is occasionally met with developed beneath the skin.*

Pediculi, or Lice.

The existence and development of a very great number of pediculi or lice, in a particular region, or over the whole surface of the human body, was formely common.

Pediculi are parasitic insects, the flattened bodies of which are covered with a skin which is hard at the edges and transparent in the centre ; they have a small distinct oval or triangular shaped head,

---

* It has been supposed possible for other animals to exist in the human skin, and Etmuller has been quoted as the authority for this assertion : he informs us, in fact, that he has seen a peculiar disease in new-born children, produced by *small worms* lodged underneath the skin, which occasioned violent itching and irritation, only to be subdued by the expulsion of these animals. According to his account, these pretended worms, which medical men have named *cirones*, or *comedones*, are of a coal-black colour; they have two antennæ, and a tail terminating in a tuft of hair. But in the present day, the observations of Etmuller, and those which have been more recently published upon the same subject by M. Bassignot (*Histoire de la maladie connue sous le nom de crinons qui attaque les nouveau-nés à Seyne, en Provence;* Mémoires de la Société Royale de Médecine, 1776), are generally allowed to be inaccurate. Sebaceous deposits within the follicles of the skin have been mistaken for worms. The *furia infernalis* of Linnæus seems also to be an imaginary worm. The characters which have been assigned to it by this celebrated naturalist, are, in great part, applicable to the *gordias* and *filiaria*.

The larvæ of the genus *musca*, and of several other genera, may be accidentally developed in the meatus auditorius of children who are neglected, on the surface of ulcers, &c. Other insects sometimes inflame the skin by their stings. The *bug (cimex lectularius)*, by means of his trunk, sucks out the blood and throws an acrid fluid of a peculiar nature into the wound.† The bite of this insect is followed by the development of a papular or tubercular elevation of a yellowish red colour. The gnat (*culex pepiens*) produces still more painful stings, followed by small hard tumours, also of a yellowish red, accompanied with heat and violent itchiness. The *rouget*, or *mower's mite (acarus autumnalis*, Linn.), fastens itself to the skin, and produces the most insupportable itching, which is soon followed by small yellowish inflamed tubercles. These insects are destroyed by bathing the skin with pure alcohol or strong vinegar. Finally, other insects, bees, wasps, hornets, spiders, ants, &c. by puncturing the skin, occasion a greater or less degree of pain and irritation.

† Surely this is an unfounded assertion.

supplied in the fore part with a fleshy excrescence, enclosing a small sucker which appears to be simple ; they have two short filiform antennæ, with five joints, and two small round eyes. The corslet, almost square, is narrow in front. They have six short thick feet, all of equal length, each consisting of a haunch of two pieces, of a thigh, a cylindrical leg, and a strong scaly, conical, arched hook at the extremity. The abdomen is round, oval, or oblong, lobulated or incised, and showing eight rings on the sides. It is provided with sixteen sensible stigmata, and with a scaly point at the posterior extremity in the male.

Swammerdam not having been able to discover male organs in any of the pediculi he dissected, and having as constantly met with an ovary, concluded that these insects were hermaphrodites. Leuwenhoeck afterwards succeeded in distinguishing the male from the female, and gave an exact representation of the organs which characterise the male. According to him, the males have a bent stilette which they carry in the abdomen, and with which they can pierce the skin ; he even conceives that the great itchiness which these creatures occasion proceeds from the prick of this sting, the introduction of the sucker into the skin, according to him, scarcely producing any sensation. De Geer mentions his having observed a similar sting at the extremity of the abdomen in several pediculi. According to De Geer, the end of the abdomen is rounded off in the males, whilst in the females, which have no sting or piercer, it is notched.

Pediculi are oviparous, and the females, after the intercourse which fecundates them, deposit their eggs, known by the name of nits, upon the hair and clothes. The young are not long of emerging from the eggs ; they change their skin several times, and after these changes are in a state to generate. To ascertain the periods of propagation, and the length of time required for the growth of these insects, Leuwenhoeck took two females and placed them in a black silk stocking which he wore night and day. At the end of six days each of them, without decreasing in size, had deposited fifty eggs ; at the end of four and twenty days the young ones had produced others in such numbers, that in the course of two months these two females might have seen eighteen thousand of their descendants !

The three species of pediculi observed on the human body are known by the name of pediculus *humani capitis* (De Geer), pediculus *humani corporis* (De Geer), pediculus *pubis* (Linnæus). They all live upon blood, sucked through their trunk, which is not seen unless when in action.

Pediculus capitis.—The body is of a greyish brown, the lobes of the abdomen are rounded. Linnæus considers the *pediculus capitis* as a variety of the *pediculus corporis*, from which it differs in the skin being harder and darker, and the corslet and abdomen being edged on each side by a blackish brown streak. M. Latreille also thinks that they may be considered as a single species. The *pediculus capitis* lives on the head, and, according to Willan, does not quit the hairy scalp spontaneously.

Pediculi capitis are transmitted from one individual to another. Want of cleanliness and diseases of the hairy scalp do not produce them.* If they are often observed among the children of the poor whose heads are not kept clean, and in those who have long fair hair ; if persons who are negligent in removing the scurf which is formed by perspiration and the use of powder, or who are affected with chronic inflammation of the hairy scalp, favus, &c., are often troubled with these insects, and if they are frequently observed among convalescent from a variety of diseases, it is merely because the want of cleanliness insures their propagation, and renders their destruction more difficult, and because certain conditions are more favourable to their existence and increase than others. Some false notions, current among the lower classes, are very favourable to the production of *pediculi ;* they imagine that persons affected with these insects are generally otherwise healthy, that they suck away the *bad blood ;* and, finally, that the existence of a certain number of pediculi on the hairy scalp proves a kind of drain which must not be suppressed without the greatest care.

The existence of pediculi *capitis* is announced by a greater or smaller amount of itchiness. When these insects are numerous, persons affected with them are constantly scratching the head ; in children, the pruritus which follows the first itchiness is sometimes accompanied by loss of sleep, and great nervous irritability. Pediculi multiply to a disgusting degree under the scabs of favus, and in the neighbourhood of the ichorous exudation of eczema of the hairy scalp. But although very numerous, they never cause marasmus, and still less death.

Pediculi capitis may always be destroyed by combing the hair frequently, or by shaving the head, when the hair is covered with nits. The same end is also as fully accomplished by washing the head with an alkaline solution, in which some of the seeds of staphysagria are infused. Oil of lavender, or a decoction of the smaller centauria, has been likewise recommended as an application to the hairy scalp infected with pediculi, or powdered parsley seeds may be sprinkled among the hair with the effect of destroying them ; finally, the head may be lightly rubbed with a small quantity of weakened mercurial ointment. This last means has been said to have produced very serious effects in several children, such as coma, a state of great debility followed by convulsions, &c. I have used it repeatedly, always with success, and never with any ill consequences.

Pediculus corporis.—The body white, broad, and flat, without spots, and the eyes black ; the notches or lobes of the abdomen shorter and less distinct than in the *pediculus capitis.* This species **exists** on those parts of the body which are habitually covered, on

* This is surely a great mistake?

the trunk and extremities, rarely on the head. Its nits are agglo-merated, and generally deposited in the folds of the linen, and other parts of the clothes, of those who live amidst filth, particularly of such as wear flannel, and to whom a change of linen is a rarity. These insects often multiply in a disgusting manner among prison-ers, galley slaves, sailors, and the aged who live in poverty and wretchedness.

The name of *phthiriasis*, or *morbus pedicularis*, has been ap-plied to the development of this species in great numbers. The morbus pedicularis is always the consequence of the successive and multiplied reproduction of one or more of these insects, accidentally contracted.

This insect is found on the surface of the skin of the extremities, and particularly of the chest and axillæ, in the body linen, and on the clothes generally ; the skin is not altered unless the *pediculi* are extremely numerous, and the individuals have been long affected with them. In this case small papular, conical, and reddish eleva-tions, and still more frequently, tubercular spots, and accidental pustules, are frequently seen on the surface. Scratches and excori-ations are also frequently observed. But various other concomi-tant lesions may exist accidentally at the same time, such as prurigo, ecchymosis, &c. Such is the *morbus pedicularis* divested of the hypothetical and unfounded details with which its history is mixed up. As to the spontaneous generation of these insects, Aristotle, Theophrastus, and Avicenna have all admitted it ; attributing the occurrence to a bad state of the body, to heat and putrefaction of the blood, &c. ; but this was at a period when the prodigious fecundity of these animals was not known. Some modern writers have, nevertheless, adopted this old opinion, and have quoted the follow-ing cases in support of it :—1st. An innumerable quantity of pedi-culi are sometimes observed to be developed on the head of a young infant without any appearance of eggs on the hairy scalp, and without either the mother or nurse being affected with pediculi. 2d. M. Mouronval assures us that several patients affected with prurigo *pedicularis*, have come repeatedly to the hôpital St. Louis to beg for advice and relief; simple baths were first administered to cleanse the skin, after which the patients were clothed in fresh linen, and put into a perfectly clean bed ; but, notwithstanding this, the shirts of these individuals were found covered a few moments after with small pediculi, which could only have proceeded from the skin. 3d. In this strange disease (phthiriase), says Lieutaud, lice appear not only in prodigious quantities on the outside, but they also gen-erate under the integuments, and even under the pericranium ; and what is still more surprising, is, that some have even been found on opening dead bodies, which after having perforated the cranium and the two envelopes of the brain, were actually lodged in the *substance of this organ itself.** In opposition to these various

* Lieutaud, and our late friend M. Chabaud, must have been members of the same family ?

assertions, it may be alleged that Lieutaud's cases never occurred ; that the fact quoted by M. Mouronval will only be conclusive after we are assured that there existed neither pediculi upon the body, nor nits upon the hairs when the persons left the bath, which was not ascertained ; and, finally, that M. Bremser's remark cannot be considered as decisive, unless it be proved that the infant had not contracted nits or pediculi in its contact with other persons, and that its clothes had not accidentally become infested, circumstances which would require such minute examination as it would be exceedingly difficult to carry into execution. Nevertheless, I must confess that at the end of any serious illness in children, I have often seen the head become covered almost suddenly with a great quantity of pediculi, and when those about them were not in any way affected in the same manner.*

Some writers assure us, that they have seen small tumours or little elevations on the skin, full of these animals. Forrest, or Forestus, says that his father had seen a case of this kind, and he himself mentions two others. Rust states, that he was once called to a consultation in behalf of a male child, thirteen years of age, labouring under a large tumour on the head, which had resisted every variety of remedy tried for its dispersion. This tumour was very much elevated above the surface, flabby without any fluctuation, and presented no trace either of present or former inflammation, or lesion of the integuments of the cranium. The patient, who seemed cachectic, only complained of an insupportable itchiness in the inside of the tumour. This tumour had appeared after the termination of a nervous fever, and in the space of eight months had increased considerably. An incision was now made into it, when it was found to contain an immense quantity of small white pediculi. It contained nothing else, and the patient soon got well. In these later times, Dr. Heberden has quoted a similar fact from Edward Wilmot. Bernard Valentine gives an account of a man, forty years of age, who experienced the most insupportable itchiness over the whole body, the skin of which was full of small tubercles. Some of these small tumours were opened, and found to contain neither blood, serum, nor pus, but such an immense quantity of pediculi of different sizes, that the patient was nearly dying of fright. Can we reasonably suppose that had these tumours been carefully examined, they would have been found to consist of dilated cutaneous follicles, into which *pediculi* had penetrated ?

The development of *pediculi corporis* has been spoken of as a very serious disease. Some modern writers have related after the ancient traditions, that Herod, Sylla, Ennius, Philip II. of Spain, &c. died of the *morbus pedicularis*. An examination of the viscera of these illustrious men would probably have led to a very different conclusion. Still it is possible, that a great number of these insects infesting infants or aged persons already labouring

---

* English practitioners NEVER see such things.

under some other disease, might occasion such insupportable itchiness and loss of sleep, as would materially aggravate the evils
under which they suffered. Others have thought that the spontaneous development of pediculi might sometimes be salutary. M.
Fournier in the *Dictionnaire des Sciences Medicales*, quotes the case
of an old man affected with rheumatic gout of the right side, in whom
a great number of *pediculi corporis* were developed, although there
had been no neglect of cleanliness chargeable on the patient ; the
pains ceased on this occurrence, and returned after the disappearance
of the *pediculi.*

Pediculi corporis are easily destroyed by sulphureous water-baths,
sulphureo-alkaline frictions, suphureous fumigations, or baths of the
bichloride of mercury. An unguent composed of three parts of
sulphuret of mercury, one part of hydrochlorate of ammonia, and
thirty-two of hog's lard, is also successfully used for this purpose.
The clothes should be fumigated with the vapours of sulphur or of
mercury.

A variety of other compounds have been recommended, into
which the seeds of the staphysagria, cocculus indicus, tobacco, various salts, and mercurial oxydes enter as constituents. The effects
of some of these medicaments should be carefully watched. Frictions with tobacco ointment, or an ointment containing the active
principle of tobacco, have sometimes occasioned convulsions and
vomitings, and mercurial inunction may produce salivation, disorder of the bowels, and other symptoms more or less serious.

Writers who have believed in the spontaneous generation of pediculi, with a view to destroy the unknown cause which gives rise
to the development of these insects, have recommended bleeding,
bitters, purgatives, anti-scorbutics, pills of the protochloride of
mercury, and a variety of other remedies, which may be injurious
or useful, according to the nature of the diseases with which the
individuals are affected upon whom these *pediculi corporis* appear.

Pediculus pubis, Linnæus, Fabricius, Geoffroy.—This species is
rather smaller than the preceding ones, its body is rounder, flatter,
and more compressed, its corslet is very short and almost blends
with the abdomen, which presents posteriorly two projections in
the shape of horns ; the feet are curved underneath. These insects
remain fixed in the same situation and attach themselves very firmly
to the skin, above the level of which they scarcely rise. They are
found at the roots of the hairs of the genital organs, of the beard,
the eyebrows,* the eyelids, and the axillæ ; they also occasionally
propagate on the trunk and extremities, when these are pretty
thickly covered with hair, but it is very remarkable that they never
fix themselves on the hairy scalp. The sting of this species, which
is very sharp, has led some naturalists to denominate it *pediculus*

---

* Here again is an instance showing how different the habits of the lower
class of Frenchmen are from that of the English,—Crab Lice, in the hair of the
eyebrows !

ferox. It is known in France under the name of *morpion*, and in England under the title of *crab-louse.* *Pediculi pubis* occasion an insupportable itchiness. When they are numerous, the skin appears sprinkled over with small red spots, resembling drops of blood, which are said to be produced by the excrement of these insects. Persons affected with them often detach them from the skin with their nails, and papular elevations frequently appear on the points which they occupied. This species propagates like the preceding ones, and increases with great rapidity. Frictions with mercurial ointment over the parts where the *pediculi pubis* appear, generally succeed in destroying them, without there being any occasion to shave off the hair to which the nits of these insects are attached.

Powdering the parts infested with calomel also destroys them. Baths of the bichloride of mercury, sulphureous water-baths, and sulphureous fumigations are more expensive and less efficacious means of getting rid of these insects.

Before concluding this article, I think it necessary to observe, that symptoms similar to those produced by *pediculi* may be occasioned by *ascarides*, an insect very closely allied to the *ixodes*, but capable, according to M. Bory Saint Vincent, of forming a new class, characterised by a small sucker, accompanied with two feelers consisting of four joints. M. Bory Saint Vincent has observed these insects upon a woman about forty years of age, who, after having experienced violent itchiness over the whole body, was very much astonished to see thousands of ascarides on all parts which she had scratched. (*Journ. compl. des Sciences Médicales*, tom. xix. p. 182.) In a case of prurigo *senilis*, Willan also observed an insect, which he says could not be classed either with the genus *pediculus* or *pulex*.

Pulices.   Fleas.[*]

Two species have been observed on the human body : Pulex *irritans*, pulex *penetrans.*

Pulex *irritans*, the common flea, according to Linnæus, is an apterous insect, distinguished by its oval compressed body, covered with a pretty strong skin, and divided into twelve segments, by a small head, very much compressed, rounded superiorly, truncated and ciliated anteriorly, and by two small eyes situated one on either side. Near the origin of the beak, we find a pair of pincers inserted, which are presumed to be antennæ, they are composed of four nearly cylindrical joints : the abdomen is very large ; the extremities are strong, particularly the hind ones, calculated for leaping, and have large legs, and thighs, and tarsi consisting of five joints, the last of which terminates in two elongated hooks. The two fore legs are inserted almost immediately under the head.

* Duméril (C), art. Puce (Dictionnarie des Sciences Naturelles, 8vo. t. xliv., Paris, 1826, pl. 53, n. 3, A).

The bite of a flea occasions as insupportable a sensation as that of a bug. The small ecchymoses which the bites give rise to, differ from petechiæ in presenting a central point, the colour of which is deeper and does not disappear under pressure like the rosy areola which surrounds it. These insects seem to show a particular preference to some persons, and the irritation caused by their bites may occasion serious nervous symptoms in children.

Pulex *penetrans,* of Linnæus,[*] (the penetrating flea, chiko or jigger.) The beak of this species is a third longer than its anterior haunches, which distinguishes it from the preceding one. The chiko is a real scourge to the inhabitants of the West India Islands and South America. This insect has been described by Sloane[†] as occurring in Jamaica, by Maggraf, in Brazil, and by Catesby, in Carolina. The female makes its way under the skin of the inhabitants of these countries, more particularly under the nails of the toes, and towards the heel; and there establishes itself. The itching it occasions is slight at first, and in the commencement nothing is perceptible but a small blackish spot; by degrees, a little tumour, of the size of a pea, appears on the punctured part; it is reddish if the insect be superficially situated, but no change takes place in the colour of the skin if it have penetrated deeply; this small tumour is formed by a kind of bag containing a sanious pus, and a great number of small white, oval, oblong globules, which are the eggs of the insect. Left to itself the tumour bursts spontaneously, and gives rise to an ulcer, over the surface of which the eggs spread and are hatched. These insects are not long of appearing in the surrounding parts, and create in their turn other ulcers, the cure of which is very difficult, and sometimes impossible. In some rare cases, in which the insects insinuate themselves into the skin of the back of the foot, the ulcers, which are the consequence, end in caries of the bones and the loss of the toes. Gangrene has even been known to take place from this cause, and we are farther assured that death may be the result of a long continuance and extensive diffusion of this animal under the skin, negroes being often seen to perish in the colonies from this cause alone. For the rest, those only who are negligent of personal cleanliness are attacked by this insect, which seems to delight in hot, filthy, and ill-ventilated places.

The treatment consists in dislodging the insect; a pin is the implement usually employed to raise the skin, and lay bare the bag, which must be carefully detached without being burst; the only means of destroying the chiko being to get away the bag entire. The membrane only of the bag left behind, is sufficient to produce erythematous inflammation, and ulcers of a very bad description.

[*] Dictionnaire des Sciences Naturelles, pl. 54, fig. 4, 5, *a, a, b.*

[†] Sloane. A voyage to the Islands Madeira, Barbadoes, &c. London, 1707, vol. ii. fol. fig.—Maggraf (G.), Historiæ rerum naturalium Brasiliæ libri octo, Amsterdam, 1648, fcl.—Catesby (Marc.) The natural history of Carolina, &c. London, 1771, fol.—Audouin, Dictionnaire Classique d'Histoire Naturelle, art. Chique.

The negroes are very expert in the operation of extracting these nests of the chiko; so much so that the surgeons of the country are never called upon to practise it. After the extraction, the little wound is dressed with tobacco, either in powder or decoction, or with a decoction of certain acrid plants of the country, with mercurial ointment, calomel, a solution of the nitrate of mercury, &c.

The treatment of the ulcers occasioned by this insect, if they are neglected, becomes exceedingly troublesome, and it is often only after repeated incisions that patients can be freed from the mischiefs attending the continuous presence of these creatures under the skin.

Acarus Scabiei.

M. Raspail, so distinguished for his admirable microscopic researches, has given so complete and faithful an account of this insect, that I cannot do better than copy it here. "This insect, on a cursory view, appears white; with good eyes, a number of reddish brown spots may be perceived covering a part of its body. There is no occasion for a magnifying glass to see it plainly running on a coloured surface. It is about half a millemetre in diameter. A simple lens suffices to enable us to count the feet, distinguish the mouth, and recognise the various details which De Geer has indicated. When the insect crawls, and it is seen through a miscroscope, it appears flattened, and in the transparent parts it presents curved and parallel streaks, which give it the appearance of the scales of a fish. Examined with a glass of the same power, the fore-feet and the head are seen to be susceptible of being concealed under the body, by being curled underneath it, in the same manner as the fore-legs and head of a turtle are drawn under its shell. The conformation of the dorsal surface of this insect is favourable to this action, as it overhangs the whole of the body, and advances like a shield or roof over the fore-feet and head. The posterior portion of the body of this animal, examined particularly, presents eight hairs of unequal lengths, the shortest being situated towards the anus. Four of these hairs belong to the four posterior feet, and the four others are inserted two on each side of the anus, on as many small tubercles, which are not distinguishable unless forced out by pressure with the point of a needle. On the disc of this dorsal surface a series or system of shining points is perceived. On looking at the animal in profile, the large white spot of the centre is found to consist of a protuberance considerably elevated; the anterior and the posterior surface are alike raised, and each of the small curves of the back is surmounted by a stiff transparent hair. The four rows of points which descend toward the anus, and towards the head, present the longest hairs. The contour of the body presents lobes of different forms according to the motions of the animal and the positions which it takes. As for the streaks which we have already mentioned, they cover the whole surface of the body. It was an error to imagine that they were only simple folds of the skin; they form a great cellular net-work, the cells of which are linear and hollow, and the interstices, which I entitle vascular, are in relief. This net-work resists sharp instruments powerfully; so powerfully, indeed, that it would be difficult for the observer, with the best inclination in the world, to transfix the insect with the point of a needle, when he was extracting it; and it is even very difficult with the assistance of the lens and appropriate dissecting instruments, to seize and crush or divide it; it slides and bounds under the instrument, and the stiff hairs which cover its back do not a little serve to increase the difficulty of anatomizing it. Not only is the body then found to possess considerable toughness, but the feet and oval appendage, which, being rendered diaphanous by the refraction of the light, appeared to be of an extremely delicate texture, are discovered to be scaly and horny, and do not yield under the instrument.

Such are the general and particular aspects of this insect when it crawls and presents its back to the eye of the observer. But when it is turned over, and the lower surface of its body is examined, its organization is found so much more complicated as to require a more particular study. The two pairs of fore-legs and the head are then clearly seen to be implanted in so many sheaths, into which, however, it is impossible for them to recede. These sheaths form a sort of plastrum of a singular appearance.

The head is inserted within a notch or angle, the summit of which is prolonged upon the thorax in a line of a golden red. The head is extremely simple in its form, of a purplish colour, and curving downwards to where it ends in a sucker, which does not appear to me to be formed of any apparent system of mandibular forceps. Examined in acetic acid, two transparent vesicles appear on each side of the head, which may be naturally presumed to be the two eyes; on the nucha two pairs of points are seen, each surmounted with a hair. These hairs when they pass the curved head, appear to be of unequal length, because two are inserted on the posterior pair, and two on the anterior pair of points.

The edges of the sheath of the two feet nearest the head extend in two reddish lines, and approach each other at the level of the line which comes from the tail of the creature. The edges of the sheaths of the other two legs, unite in the form of red lines at the convexity of those lines of which we have spoken, a distribution resembling in some sort a fan. The legs are composed of four joints and an oblique basilar piece which appears like a triangle, with its hypothenuse turned towards the posterior part of the body. Each of these articulations is covered with hairs, of which only those on the sides are visible. The last joint is covered with very short prickles, and armed beneath with a stiff hair, which is terminated by a flexible cavity, capable of producing a vacuum, like the soft glutinous pads of certain animals much higher in the scale, such as the tree-frog; these pads enable the insect to fix itself in any position. The joints are not very distinct, and can only be counted after long examination. These five anterior members are half covered, as I said before, by the projection of the dorsal surface of the body.

On the belly two pairs of other organs are seen, which De Geer considered as four hairs enlarged at their bases and attached to the belly. These four supposed hairs are the four hind legs, which, although shorter than the fore-legs, possess the same essential organization, only they are without the apparatus, or accompaniment adapted for progression, which I shall designate under the name of *ambulacrum.* Except this trifling circumstance all that is observed in the fore-feet is found in them. 1st. The reddish line which borders the sheath, the opening of the sheath, the hypothenuse, and the four articulations. Here the *ambulacra* are replaced by very long hairs. Upon the whole, the hairs of the hind feet which are nearest the head, are more developed than those of the two most remote. When seen with the lens, this arrangement of the feet looks very like De Geer's figure, and the red line bordering the sheath appears to be a hair which swells into a red vesicle in the region of the foot, and is produced in the form of a white hair at the top of the vesicle.

The anus sometimes appears projecting and sometimes hidden, but to see it very plainly it is only necessary to let the insect dry, when the dermis, on account of its hardness, retains its shape, the abdomen shrinks, and the direction of the anus is plainly seen through the transparent dermis. This animal externally is as white as snow, except the feet and the sucker ; but if seen by refracted light it appears yellow, like all the white tissues of animals. This happens from the decomposition of the rays of light which pass through the organic substance, a decomposition in virtue of which the least refrangible rays, such as the yellow, alone reach the eye.

Although the hairs of the fore-legs and head of this insect are directed forwards, it is easy to see that having the power of bending its feet and sucker underneath, these hairs can prove no obstacle to its progress through the skin. But the structure which facilitates this operation, is the presence of the hard papillæ of the back, which being directed backwards offers such resistance in that direction, as prevents the creature from receding ; above all, it is assisted by the hard scaly case of the trunk, which forms a covering like the shell of the tortoise. I think I have already remarked that the ventral is streaked in the same manner as the dorsal surface.''

M. Gras has described with much care the furrows or covered ways (cuniculi) already indicated by Bonomo, Casal, and Adams, and at the extremity of which the *acarus scabiei* is generally found. In fact, if the vesicles of scabies in the hands and in some cases in the feet are examined with care, it will be seen that several of them present on their summits or on some part of their surface a small blackish spot ; this sometimes extends in the form of a semi-circle, and is found to be situated on a small whitish point. On

other vesicles a blackish or whitish and sometimes a sinuous punctuated line is seen departing from this point, and traversing almost
the whole surface of the vesicle ; in this case when the epidermis,
which is raised by the serum, is removed, and examined with a
lens, the punctuated line is found to be formed in its substance ;
with the assistance of the light of the sun, a small brownish point
will be seen at the extremity of the cuniculus which is farthest removed from the vesicle, and on raising the epidermis in this place,
the acarus can be extracted. It is worthy of remark, that there is
no communication between this *cuniculus* and the cavity of the
vesicle, and that it is always easy to remove the acarus without
causing any discharge of serum.

M. Gras has never seen two *cuniculi* begin in the same vesicle,
although these covered ways frequently intersect each other. Sometimes the cuniculi are not seen near all the vesicles ; in many individuals affected with scabies they are not even to be met with readily
except on the hands; the burrows generally extend from two to
four lines from the vesicles. M. Gras having placed an acarus on
one of his fingers, the insect took twenty days to trace a burrow of
two lines in length ; another insect was only three days in making
a cuniculus of the same extent.

There is no relation between the number of vesicles and burrows.
In some persons affected with scabies, burrows are seen without
vesicles in their neighbourhood ; but at one end of these burrows,
a small point of skin, which has lost its epidermis, and appears
surrounded by a minute edging of this substance detached, is frequently observed ; in other places, even such a trace of a vesicle
does not exist, as occurred to M. Gras, who having placed acari on
different places in the skin, first observed *cuniculi* formed by the
insects, and vesicles only some time afterwards and at a certain
distance from them.

The insect may be extracted by means of a pin, by introducing the
point obliquely into the epidermis, and turning it back ; the acarus
is often thus discovered at once, which, so long as it is not wetted
with the serum of the vesicle, seems to attach itself with great ease
to the point of the pin. This insect is at first motionless, and it is
not until after three or four minutes that it is seen to move its legs
and begin to walk with rapidity. In a temperature of from fifteen
to eighteen degrees, these insects may be kept alive three or four
days after their extraction.

Acari are very seldom found alive in persons affected with scabies after they have been for two or three days under treatment ;
and yet this disease is seldom cured before the tenth, and sometimes
the fifteenth day. The acarus has been sought for in individuals
affected with other diseases of the skin,—pruriga, eczema, lichen, &c.

In the articles on scabies I have mentioned the principal works
relative to the history and particular study of this insect, but I think
it proper again to quote the work of M. Raspail and that of M. Gras,
whose observations I have here copied. Raspail. *Mém. Compara-*

tif sur l'Histoire Naturelle de l'Insect de la Gale, fig. 8vo. Paris,
1834.—Gras. *Recherches sur l' Acarus ou Scarcopte de la Gale
de l'homme*, 8vo. Paris, 1834.

Filiaria Medinensis.   Guinea Worm, &c.

Under the name of *filiaria*, a genus of the *entoza* is indicated,
the principal characters of which I shall here particularize ; a cylindrical filiform and very long body, decreasing in a very slight degree
only towards the extremities, which are blunt ; a very small articular
mouth, terminal, as is the anus also, in all probability ; the male
organ short, somewhat rounded, and situated before the point of the
tail ; the intestinal canal very distinct, and extending the whole
length of the body.   The *filiaria* inhabits the cellular tissue of
animals of all classes.

The most celebrated of all the species of the filiaria has been
observed in the human subject.   It is known to naturalists by the
name of *filiaria medinensis dracunculus* or *Guinea worm*.   The
body of this species of filiaria is of a dirty white, which becomes
yellow in alcohol ; its size, nearly equal in its whole extent, varies
from that of a thick fiddle-string to that of a straw.   The length of
the filiaria medinensis varies between nine and forty-two inches
(Heath), a foot, a cubit, and more (Kæmpfer), three feet and a half,
Rhenish measure (Gründler), more than two ells (Kunsenmüller),
eight to ten feet (Gallandat), and even eight ells (Fermin).   The
head is furnished with a kind of sucker, formed by the enlargement
of the lip surrounding the mouth, the orifice of which is very small.
The tail is terminated by an inflected hook.   The *filiaria medinensis* bears the greatest resemblance to the *filiaria of the monkey*.

According to the recent researches of Jacobson, certain filiaria at
least are composed, not of one single individual, but of an assem-
blage of individuals, living under or within the same skin.   M.
Jacobson received an Arab into his hospital, who had a tumour near
the outer ancle, which was discovered to be occasioned by a dracunculus ; this, after several fruitless attempts, was extracted by the
ordinary process.   A second tumour having appeared on the other
ancle, it was laid open, and the knife having divided a portion of
the worm longitudinally, a purulent-looking matter issued from the
opening, which, on being examined with a microscope, presented a
crowd of small elongated filiform worms, with the head somewhat
enlarged, and a short tail, much thinner than the body.   Upon
extracting the whole of this dracunculus, all its parts were found to
present the same appearance.   That which had been extracted from
the first tumour, was ascertained on examination to exhibit precisely
the same structure.

M. de Blainville presented a portion of the internal substance of
this worm, collected by M. Jacobson, to the Academy of Sciences.
Seen through a microscope, it was almost entirely formed of small
animals, in perfect accordance with the description given by that

talented anatomist. M. de Blainville thinks that it would be interesting to ascertain whether all dracunculi presented the same peculiar organization. (Gazette Méd. 1834, p. 216.)

The history of this entozöon presents a very remarkable peculiarity, for which it is impossible to account satisfactorily at the present day. The inhabitants of the torrid zone are almost alone ever affected with it. The principal observations which have been made on this animal, have been collected in Arabia-Petrea, on the banks of the Persian Gulph, of the Caspian Sea, and of the Ganges, in Upper Egypt, in Abyssinia, in Guinea, &c. I do not believe that it has been developed in the human subject in Europe.

The filiaria medinensis has been most generally observed in the subcutaneous cellular tissue of the human body, particularly of the lower limbs. Of one hundred and eighty-one cases collected by Sir J. M'Gregor, it is calculated that in one hundred and twenty-four of them, this worm was situated in the feet, in thirty-three in the legs, in eleven in the thighs, in two in the scrotum, and in two in the hands. Kæmpfer found it in the hollow of the ham and in the scrotum ; Pere has seen it in the head, the neck, and the trunk ; Bajon assures us that he has twice seen it under the mucous membrane of the ball of the eye, &c. ; Chardin pretends that the filiaria medinensis almost always occurs singly, while Bajon and Bosmann assure us that it is not uncommon to meet with several of them in one patient at the same time.

It seems to be an established fact that this animal is never developed except on the human body. Lœfler, who lived many years in those parts of Africa, where the inhabitants are troubled with it, never heard that it had, on any occasion, been observed in water ; and Hind, who has also carefully examined the water of these countries, never detected either these worms or their eggs in it. The contrary opinion has arisen from this species of filiaria having been taken for a true *gordius*, which is supposed to possess the power of making its way and living underneath the skin.

The great size which the filiariæ, extracted from the human body, sometimes acquire, would lead one to imagine that these animals do not occasion the phlegmon which proclaims their existence, until some weeks or even months after their formation. This inflammation of the cellular tissue soon ends in suppuration ; and upon the spontaneous or artificial opening of the abscess, one or more inches of the filiaria generally protrude.

This animal is extracted by being repeatedly and very gently drawn out. The imperfect extraction of the filiaria is said to be followed by very serious symptoms.

There is much difficulty in France in procuring the filiaria medinensis, to examine its organization. I have seen one very well preserved in the collection of the Jardin du Roi. It is about three and twenty inches long from the head to the tail, and one line in diameter throughout. It is generally flattened, and the two terminal openings are very distinct. M. Henri Pétroz, chief apothecary to the Hôpital de la Charité, possesses another of these animals, which was extracted from the foot of a negro of Guinea. This worm is about twenty-five inches long ; it is yellowish like cat-gut, which probably arises from its having been so long preserved, dried, and rolled on a small piece of wood. One of the extremities, the tail, is hooked ; and near it, under the microscope, a small tubercle is seen, in the centre of which is an opening. The other extremity, examined by a microscope magnifying twenty-five times, appeared unequal, irregular, and jagged. The head had probably been broken or crushed. Finally, M. de Blainville says, (Grundler, *Traduction Française de l' ouvrage de Bremser sur les vers intestinaux*) that he possesses a filiaria medinensis, which was sent to him by M. De-lorme, the author of some very interesting observations on this entozoon (*Journal de Physique, chimie*, &c. August 1818.)

Several writers have confounded the *filiaria medinensis* with the *gordius aquaticus*. This mistake was the more likely to occur from the circumstance of the body of the *gordius* being in the form of a thread, as well as that of the *filiaria*, but the former differs from the latter, in its body presenting the transverse folds of the annelida, to which natural order it belongs, inhabiting like them soft waters, slimy places subject to inundations, &c., whilst the filiaria are true entozoa. In addition to this, all the gordii examined by M. de Blainville, presented the anterior extremity of the body divided in the form of pincers, which is not the case in the filiaria. Besides, in combating the opinion of Joerdens, who thought that the *gordius aquaticus* might make its way under the skin of the human body, M. Bosc has very judiciously remarked, that the organization of ths species of gordius renders it incapable of piercing the integuments, and that it has never been seen in the subcutaneous cellular tissue of the human subject. After having examined some specimens belonging to this species of *gordius*, which my friend Dr. Asselin and I had collected in the ditches of the forest of Meudon, we became satisfied of the accuracy of the observation of M. Bosc. Finally, it is known that Pallas no where met with so great a number of *gordii aquatici*, as in the lake of Waldei : but he could not learn that this worm had ever been known to make its way under the skin of any of the persons who had bathed in this lake.

Historical Notices.—The *dracunculus* was known to the ancients; "quemadmodum, in quodam Arabiæ loco (ut aient) in tibiis hominum dracunculi vocati nascuntur, nervosa natura, colore crassitudineque lumbricis similes." (Galeni Opera fol. class 4. de locis affectis, lib. 6.) Actius has given a very good description of it, and an account of its treatment after Leonides, (*Tetrabl.* iv. *sermo* xi. *in fol.* 1540, *p.* 800. Avicenna, (*Liber quartus ;* de venâ medeni) speaks also of having seen it. Schenck (*Obs. medic. lib.* v. de dracunculis *Æthiopiæ et Indiæ propriis*) has collected a number of facts relative to the history of the filiaria. In these later times, new gonheau ou Veine de Médine et sur l'usage du sublimé corrosif dans cette maladie, (Journ. de Méd. Janvier 1760);

M'Gregor, (Sir James) medical sketches of the expedition to Egypt from India, 8vo. London, 1804 ; Paton cases of Guinea worm, with observations, (Edinb. Med. and Surg. Journ. t. xi.) Scott, (Will.) and Kennedy, (Alexander) *remarks on the dracunculus*, (Edinb. Med. and Surg. Jour. t. xvii. p. 96.) ; Grant, (Robert) extracts from a correspondence on the filiaria *medinensis* (Edinb. Med. and Surg. Journ. t. xxxv. p. 122). Two cases of the extraction of several *dracunculi* performed in France, have been related by M. Brulatout, (*Journ. de chimie médicale*, t. vi. p. 624). Grundler, in his treatise, *de venâ medinensi*, has given an original representation of the filiaria medinensis, which has been republished in several works, and in particular in the *Encyclopédie Méthodique*, t. xxxix. fig. 3.

Œstrus.

Œstri are dipterous insects characterised by the almost complete absence of a mouth, which has caused them to be designated by M. Duméril *astomata*. Their larvæ deposited under the skin of the human body, and more frequently under that of the ox, occasion small circumscribed and painful inflammations.

The species of œstri which live under the skin of animals have been described with much accuracy by Mr. Clark.[*] (*Trans. of the Linnæan Society of London.*) M. Say (Journ. de Philadelphie, t. ii. p. 363.) thinks with Linnæus that there really exists a species of œstrus, the larvæ of which inhabit the human body ; an opinion which is discountenanced by Fabricius, and other modern entomologists, but which M. Say founds on the following observations. "After a very fatiguing march," he writes to M. Brik, "I went to bathe in the Chama, a small torrent which flows into the lake of Maracaïbo. A short time after coming out of the water, I was stung by a small insect in the left leg, over the anterior and superior part of the tibia. I experienced for several days pretty severe itchiness, but no pain in the part, and continued my journey without suffering any other inconvenience than the following: I felt all at once acute pain in the part, which after having recurred several times, ended by becoming continual. On my arrival and during my stay at *Rosario de Cucuta*, I walked with difficulty ; a tumour of considerable size which had the appearance of phlegmon, and in the middle of which there was a small black spot, existed on the tibia. The ordinary applications were used without success, and the tumour continued to inflame. I remained in this state for several days, experiencing at intervals very great pain. On returning to Maracaïbo, I went down the Cottatumba in a boat, without any awning, and was drenched to the skin with a cold rain, which fell during the night; I suffered greatly from this, and was almost continually tormented by the pain, which then became more severe

[*] This sentence awards, very charily indeed, the merit due to Mr. Bracy Clark. *His* history of the insect is perfectly complete.

than usual. During this passage, which lasted twelve hours, I thought it advisable to scarify the part, and had recourse to the ordinary topical applications, but without success; sometimes I fancied that I felt a living body moving under the skin.

"On my return to Maracaïbo, I was hardly able to walk, and was confined to the house. I continued in this state for two weeks without experiencing any diminution of the pain. The tumour having begun to suppurate, it came into my head to cover it at night with a tobacco poultice made with rum instead of water. During the day I powdered it frequently with the ashes of a cigar. Four days after using this remedy, I felt considerable relief, and on the fifth, with a pair of forceps, I drew out a dead larva. In a few days the sore began to heal, and on the tenth day I was perfectly well again, though from time to time I experienced some pain in the place whence the larva had been extracted. This larva had travelled upon the periosteum of the tibia over a space of two inches in extent, and I attribute the acute pain which I experienced from time to time to the irritation of some of the minute nerves distributed to the parts through which the larva made its way."

M. Say thinks that this larva, which was sent to him by M. Brik, belonged to the œstrus genus. It was full and distended ; the posterior half of the body, thicker than the anterior and rather compressed ; the rings of this part were armed with a transverse series of small hard black tubercles, enlarging towards their bases, and terminating at the points in small filiform horns pointing forwards. These series, six in number on the back and sides, were arranged in pairs, and were three in number under the belly. Near the posterior extremity of the body were a number of small tubercles like the former ones, but not forming a regular series. The anterior half of the body was entirely smooth, cylindrical, or rather in the form of a lengthened and truncated cone, and much smaller in diameter than the posterior part : the folds of the posterior part of the body were short, and the cleft which separated them was narrow.

M. Say compares this larva to that of the œstrus in oxen, horses, sheep, and to that of the hemorrhoidal œstrus, several of the characters of which it presents. "There are," says he, "several opinions with respect to this larva ; by some of the Spaniards and creoles, it is called *ouche,* and said to be nothing more than a worm which crawls upon the body from the earth, penetrates the skin, and is there developed ; others maintain that it is produced by the sting of a winged insect called *zancudo* (the name of *zancudo* is used by the Spaniards of South America, to designate different species of *culex,* which others call *husano*). As for me," adds M. Say, "I am of opinion that this larva is produced by a winged insect which deposits its eggs in the skin after having pierced it."

M. de Humboldt has seen in South America, Indians whose abdomens were covered with small tumours produced as he presumed by the larvæ of the œstrus. Finally, Mr. Howship read a paper to the Medical and Chirurgical Society of London, on the 26th Nov. 1832, on the *œstrus of the human body*, in which is contained an account of two new cases, one of an œstrus found in the cellular tissue of the shoulder of a soldier at Surinam ; the other of an œstrus in the cellular tissue of the scrotum of a young man of Santa Anna, in Colombia. (*Gaz. Médic. de Paris*, 1834, p. 71.)

Hypertrophy of different Layers of the Skin.

These diseases, sometimes preceded or accompanied by fever at the commencement, are almost always followed by permanent *intumescence* or enlargement of the parts affected.

Certain *enlargements* of the limbs, scrotum, labia majora, face, &c. usually accompanied by hypertrophy of the skin, distinct from phlegmon, from œdema, and from bloody tumours, have long been and are still described or designated under the name of *Elephantiasis Arabica*, or Arabian Elephantiasis.

Symptoms.—Elephantiasis Arabica generally attacks the lower limbs ; one limb only is most frequently affected, but both may be implicated either at the same time or successively. (Alard, obs. 1. 3.)   Hendy has described a variety of this disease under the name of *Barbadoes leg ;* M. Alibert designates it by that of *lèpre éléphantine tuberculeuse.*

In a great number of cases this enlargement of the lower extremities is announced in an acute manner by a more or less severe pain in the groin and ham, following the course of the vena saphena, and principal trunks of the lymphatic vessels, and next by the appearance of a *red line* or hard, knotty, tense cord, resembling a chain of small subcutaneous tumours, extending from the bend of the groin to the knee or ankle (Hendy, cases 2, 4, 9.), or from the ankle to the groin (case 8) ; or still otherwise by an attack of erysipelas.   In almost all cases, the skin assumes an erythematous hue, and the subcutaneous cellular tissue becomes the seat of considerable tumefaction.   The neighbouring joints are stiff and contracted ; frequently from the commencement, there are long-continued shivering fits, great thirst, uneasiness, restlessness, violent retching, vomiting, occasionally delirium, then intense heat, accompanied with palpitation of the heart, followed by general or partial sweating, and the cessation of the febrile symptoms.   In the course of one or more months these phenomena return in the shape of paroxysms at shorter or longer intervals, which may vary in number from three (Hendy, case 17,) to fourteen in the course of a year (Hendy, case 16,) or may only recur at the end of seven years. (Hendy, case 19.) These fits, the number and duration of which can neither be foreseen nor calculated, are followed by a progressive increase in the size of the limb, which would appear at first to be owing, in a great measure, to the deposition of a certain quantity of serum or coagulable lymph within the cellular tissue.   The limb afterwards becomes hard, and no longer retains the impression of the

31*

finger.   The lymphatic ganglions of the groin and ham, often very much increased in size, are sometimes otherwise apparently healthy and indolent.   In this second stage of the disease, it exists without any further inconvenience than that which the deformed state of the limb necessarily occasions.   It sometimes assumes such extraordinary shapes, and becomes so entirely out of proportion to the other parts, that it is impossible to form an idea of the extent of alteration undergone without having seen some cases, or the drawings of the disease which have been published.   In one case the tumour is full and uniform like a well filled bag or bladder ; in another it is in divisions, as if each successive fit had formed and left its own particular swelling.

After the first attacks, the skin is usually pliant, and does not exhibit any change of colour ; vessels sometimes appear creeping beneath it, and give it a brownish hue ; by degrees, however, it becomes hard, particularly in the neighbourhood of the ankle joint, and is covered with elevations and small veins ; the epidermis then often becomes thickened as in ichthyosis.   Finally, chaps and fissures are sometimes formed on the limb, which now becomes excessively deformed.   The knee-joint occasionally becomes the seat of very obstinate chronic inflammation in these cases.   Such anomalous developments are not always preceded by the symptoms of acute inflammation of the vessels and lymphatic glands, or of the veins of the inner part of the lower limbs, mentioned in almost all the recent cases of the glandular disease of Barbadoes related by Hendy, and in many others, in which it had been of long standing. (Cases 14, 15, 17, 18, 21.)   In fact, these enlargements sometimes occur after ulcers of the legs (Andral), repeated attacks of erysipelas, of lichen *agrius*, or eczema *rubrum*.   The knotty, hard, and tense cord is not then seen as in the woman Berton (Alard, obs. 1). This symptom was not perceived by M. Bouillaud (Archives Générales de Médecine, t. vi. p. 56), in a woman whose lower extremities, enormously enlarged and as hard *as a stone,* had become like those of an elephant.   The intumescence in this case, in fact, followed the obliteration of the crural veins and the vena cava.   I have given the history of a patient in the first edition of this work, in whom the anomalous development of one of the lower extremities coincide with a varicose state of the veins of the thigh.

Anatomical researches.—Despite these enlargements of the lower extremities, the distended skin may retain its natural thickness, and almost its natural colour ; but hypertrophy more frequently takes place, at least in some parts, and there it bears a close resemblance to a fibrinous deposit, or the buffy coat of the blood ; the epidermis covering it is also generally very much thickened. M. Chevalier (Med. and Chirurg. Transactions, vol. xi. p. 63,) found the papillæ of the skin exceedingly enlarged, lengthened, and projecting from the surface of the dermis ; on the points where these papillæ were less developed, the epidermis was thinner ; the corion was so much hypertrophied, that in some places it was half an inch thick, and ·presented the granular appearance which is observed in large quadrupeds. On its inner surface it adhered to the indurated cellular tissue with which it was evidently blended. In other respects it was neither injected nor altered in its colour.

In the body of a woman who, fifteen years previously, had suffered from an ulcer in the right leg, which had increased to an enormous size, and the skin of which was very hard, rough, and of a dark brown colour, and in some places absolutely black, like that of the hand of a negro, M. Andral found the subcutaneous and intermuscular cellular tissue sensibly hypertrophied and hardened, more and more so as it lay nearer the dermis ; this had also increased considerably in thickness, and in several places could not be separated from the indurated cellular tissue, each seeming to be but different degrees of the same organization. The papillary body lying over the corion was greatly developed, evidently distinct from the dermis, and appearing to stand in the same relation to it as the villi do to the intestinal mucous membrane. Situated over the papillary body, again, and between it and the epidermis, there were three very distinct layers : the innermost of the three penetrating between the eminences of the papillary body, receiving no vessels, and consisting of a fibro-cellular tissue (*couche albide profonde* of Gaulthier, *couche epidermique* of Dutrochet) ; the second, situated more externally, composed of extremely delicate blackish filaments, interwoven in the true sense of the word, forming a network which was exactly similar to the coloured rete of the negro ; finally, a third quite close to the epidermis, and, in particular places, forming only a white line similar to the epidermic layer of the papillæ ; but thicker in others, and hardened as though formed by a series of superposed scales ; this was certainly the *couche albide superficielle* of Gaulthier, the *couche cornée* of Dutrochet. (Archives Générales de Médecine, March, 1823.)

I have made similar observations on the structure of hypertrophied skin in the first edition of this work (vol. ii. p. 560.) M. Gaide and I have since repeated these anatomical researches, and he has published the results. (*Cases of the individuals named Allard and Fournier.*) After having incised the skin in the direction of its thickness, the following layers were discovered, reckoning them from the more internal to the more superficial strata. 1st. Small lobules of adipose tissue, connected together by a healthy laminated tissue, forming a subcutaneous layer. 2d. Above this was placed the corion, represented by a transverse band of a pale yellow colour, evidently hypertrophied, the areolæ of which were less distinct than in the natural state ; it was besides loaded with a great quantity of serum, which was easily made to flow out by compressing it between the fingers. From its inner surface it sent off whitish fibrous prolongations, which penetrated some depth into the subcutaneous cellular tissue. 3d. Above the corion a second layer was seen, composed of parallel fibres, running from the outer surface of the corion towards the epidermis. This second layer,

evidently formed by the papillæ, elongated, and of a ruddy violet
colour, was of unequal thickness in several parts, and varied from
two to three lines and a half in' length.   These two first layers of
the skin were rendered distinct one from the other, both by the opposite directions taken by their fibres, and by a transverse line which
resulted from their difference in colour.   Between the parallel
fibres of the papillary layer, small vessels might be distinguished
by the naked eye ; these were, of course, more distinctly perceived
when examined under the magnifying glass.   The superficial surface of this second layer presented small eminences, mostly lenticular, separated from one another by deep furrows, evidently
formed by the most elongated papillæ, whilst the smaller ones,
united in the same line, gave rise to the formation of the wrinkles
of which I have spoken.   By maceration, the papillæ which formed
these elevations became free, and appeared when examined under
water, like the pile of velvet or plush.   Above the papillæ a third
layer exists, distinct from the epidermis which covers it ; it is that
which has been designated under the name of the *lamina albida
seu cornea.*   In detaching this third layer, very delicate filaments
are seen tending towards little whitish bodies, situated, and as it
were, attached to the inner surface of the *lamina albida* (follicles) :
these small bodies variously disposed, either singly and scattered,
united in parallel series or agglomerated in the form of larger or
smaller patches, all or almost all came away with the *lamina albida*
to which they adhered.   Some of these follicles are perfectly round,
others are longer and terminate in a point at one of their extremities
in the form of tears ; others again, still longer, appear cylindrical ;
some present in their centre, or on their outer aspects, a blackish
point, which would seem to be their orifice.   The epidermic layer,
disposed like the preceding, in the form of a membrane, and like it,
transparent, where it is not formed of accumulated squamæ, is also
in contact, on its inner surface, with small follicles similar to the
preceding ones.   United in general in the form of patches, these
were more particularly apparent on the parts which correspond to
the squamæ.   This disposition of the layers is always seen.   From
the inner surface of the epidermis small prolongations are sent off,
which surround the hairs to their bulbous extremities, and are very
distinct from the follicles.

In elephantiasis *Arabica,* the subcutaneous cellular tissue has been
found harder in proportion as it was nearer to the dermis.   The *adipose tissue* has been known to become enlarged in a very extraordinary manner.   I have also found the cellular tissue infiltrated as
it is in dropsies of long standing.   M. Fabre has seen the subcutaneous cellular tissue converted into a thick, hard, almost fibro-carti-
laginous layer, presenting in several places small ossified plates,
adhering so closely to the aponeurosis of the leg, and to the nerves
and vessels which traverse it, that it was impossible to separate
them.   The subaponeurotic and intermuscular cellular tissue participated in the same alterations, but in a less degree.   In a woman who

died in the hôpital de la Charité, in 1820, whose lower limbs were affected with elephantiasis, M. Andral found, under the skin, and in the place of the muscles of this limb which were reduced to some thin discoloured shreds, an enormous mass of hard, condensed cellular tissue, with cavities here and there filled with serum, and partaking, in more than one place, of all the qualities of cartilage. (Précis de Anatomie Pathologique, t. i. p. 277.)

Hendy has found the lympthatic glands hardened, or in a state of suppuration, and much larger than they are in their natural state. The absorbing vessels were dilated, and their coats so much weakened as to be incapable of standing injection. I never observed these large absorbing vessels in any of the cases of elephantiasis *Arabica* which I have dissected. M. Fabre says, that he found it quite impossible to discover these vessels in the midst of degenerated subcutaneous cellular tissue. I have frequently found inguinal glands of much larger size than they are in a healthy state; but in scrofulous subjects, the same morbid development is observed, without dropsy or any morbid increase in the limb having taken place. In the body of Allard, whose case M. Gaide has published, the lymphatic glands were not found to be larger than those of several other bodies which we examined for the sake of comparison on the same day; the glands of the left groin only were of a deep red, whilst those of the right side were of a milky whiteness. The vessels which were distributed to these glands were not larger than they are in the healthy state. In another patient who died of elephantiasis of both of the lower extremities (case of Fournier), the lymphatic vessels situated on the back of the left foot, and inner edge of the great toe, were as small and delicate as in the healthy state. The glands of the popliteal region had undergone no alteration; but from the left groin to the point where the aorta sends off the renal artery, a string of lymphatic glands existed, each of which was almost of the size of an almond; some of the glands of the groin were red, or reddish, others were white, and easily crushed between the fingers; those lying over the femoral artery, all those which extended from thence along the outer side of the iliac vein, and in the front of the psoas muscle, were white, crushing easily between the fingers, and discharging a whitish fluid like pus, or softened cerebriform matter. Besides this string of altered glands, we discovered others in the cavity of the lesser pelvis beneath the common iliac vein, forming, by their union, a sort of glandular subperitoneal layer, which extended over the internal surface of the ischium. The lymphatic vessels which were distributed to the glands of the groin were not larger than in the healthy state; and although the chain of glands of which I have spoken adhered to the iliac vein, this vessel did not appear to be evidently compressed by it. M. Bourgeoise ascertained that the lymphatic glands of the right leg were not larger than in the healthy state; the lymphatic glands of the ham presented no alteration; the inguinal and pelvic glands, not nearly so large or so numerous as those of the opposite side, disposed in a band round the iliac vessels, penetrated the smaller pelvis; in other respects they presented the same anatomical characters as those of the left side.

In the *compte rendu* of M. Allard's work, and in a case since published (*Archives Générales de Médecine*, vol. ii. p. 215 and 372), M. Bouillaud called the attention of the profession to the obliteration or obstruction of veins, as an occurrence, the influence of which he had already pointed out in causing the development of local dropsical affections.   Since the publication of these researches I have had an opportunity of observing the *contraction* of one of the venæ saphenæ, and the *obliteration* of the other, in a case of elephantiasis of both legs. (*Mémoire de M. Gaide ; obs. d'Allard.*) In the left leg the vena saphena, laid bare along its whole extent, appeared in the form of a cylindrical cord of a yellowish-white colour, and not transparent, about a third less in size than the same vein in its natural state ; the cavity of this vein was found almost obliterated at the point of junction between its middle and lower third ; the vessel having been cut across in this place, a central point was distinguished upon each of the cut extremities into which a fine wire of the diameter of that which is usually inserted into silver catheters could be introduced, though not without difficulty ; the calibre of this vessel had become, in some sort, capillary, through an extent of about two inches; its sides being double their usual thickness ; the vein cut across transversely in any point where it was contracted, continued gaping in the same manner as an artery. The femoral vein towards its junction with the vena saphena, contained clots of recent formation ; most of the other veins of this extremity presented no alteration.   The vena saphena of the left leg contained fibrinous clots of old formation, and adhering by their surface to the internal membrane of the vessel; the calibre of this vessel was not contracted, but its sides, like that of the right vena saphena, were thickened, and resembled those of an artery.   I ought to add that in estimating the thickness of the walls of these vessels, the difference which naturally exists between those of the lower extremities and those of the upper was taken into account, and that to prevent all mistake I took an opportunity of comparing the thickness of the particular veins examined in several subjects.

In the case of elephantiasis of the leg, published by M. Fabre, the vena saphena laid bare from one end to the other, could not be traced in the middle part of the leg ; it was only found at the distance of about four finger breadths below this point; a very fine probe introduced into the upper and lower part of the vein led to two shut sacs.   This vein in the remainder of its course was so much contracted, as only to allow the passage of a small silver stilette with great difficulty along it.   Its sides were hypertrophied, and like those of an artery.   The external saphena, except that it was nowhere obliterated, presented the same appearances ; the anterior tibial and fibular veins contained blood.   The posterior tibial vein was obliterated in a part of its course.   No obstacle to the course of the blood existed in the popliteal, femoral, or external iliac veins, &c.

Hendy found the small *arteries* of parts affected with elephantiasis larger than they are in a healthy state. In two cases seen by M. Gaïde, at the Hôpital Saint Antoine, the arteries of the extremities presented no alteration. In a case, the particulars of which are related by M. Fabre, the anterior tibial and fibular arteries were ossified, they contained a little blood; the posterior tibial vein was converted into a cylindrical bony stem, into which the blood no longer penetrated; the femoral and popliteal artery were also equally ossified; similar ossifications were found in the arteries of the other extremity.

In a particular case which Nægele examined, he found the tibial nerve increased in size, presenting on its surface and in its interior round and oval-shaped nodosities forming so many small cysts, which contained a clear fluid, limpid in some places and turbid in others. In three cases of elephantiasis *Arabica* which I have dissected with great care, the nerves presented no alteration. In M. Fabre's case the great sciatic nerve, after having preserved its natural size to the middle of the thigh, afterwards increased continually in its dimensions till it reached the ham, where it was of such magnitude that its several branches were each much larger than the trunk which sent them off. In the thickness of the external popliteal nerve a gelatinous hydatiform mass was found, of a pale red colour and the size of a small almond, pretty firm in its consistence, and having the medullary fibrils of the nerve parted and applied around it. The branches of the external and internal popliteal nerves were themselves so much increased in bulk that the tibial nerve, the cutaneous muscular branch, and the interior tibial were each four times their natural size, and presented several enlargements. These nerves, although very hard, still preserved pretty evident traces of their peculiar organization. M. Ferrus met with a similar disposition of the nerves in the leg of an old woman affected with elephantiasis.

Hendy commonly found the *muscles* softened and blanched. I have also seen them less deeply coloured than in the healthy state; in a patient of M. Fabre's, several of the muscles were increased, others decreased in size; each of them was converted into a fatty substance. They were very hard, and creaked under the scalpel; the solæus muscle presented this degeneration in a greater degree than any of the rest; here and there, a kind of streak of bony matter appeared, which seemed to occupy the spaces between the muscles, and some of which were connected with certain bony excrescences that rose from the surface of the periosteum of the tibia.

In the cases of elephantiasis of the lower limbs, which I have had the opportunity of examining anatomically, the *bones* had undergone no change. But in several patients, and among others, in a woman whom I attended in the Hôpital Saint Antoine, the tibia of the diseased leg was three times the size of that of the opposite side ; in M. Fabre's patient, too, the inter-osseous ligament of the leg only existed for the space of about an inch in the situation where the interior tibial vessels pass through it ; no further vestiges of it were seen : it was replaced by an osseous uneven lamina covered with asperities, and a line in thickness in some places. This bony lamina adhered so intimately to the tibia and fibula, that these two bones, thus soldered together in their whole extent, really formed no more than one. The surfaces of the lower peroneo-tibial articulation, were so completely united that no trace of division between them could be discovered, even after more than three months of maceration. The circumference of the tibia was almost the double of that which it is in its natural state, that of the fibula in the middle was fully more than triple. These bones, thus closely united, were covered with a prodigious number of shorter or longer asperities, bedded in the soft parts, their edges being prolonged in the form of prominent ridges, twisted in various ways, so as to represent in some sort a series of canals crossed by vessels and nerves, which ramified upon their surface. The upper surface of the bones of the foot presented similar ridges to those on the tibia and fibula. The density of the tissue of the tibia was such that it could only be sawed across with very great difficulty ; in colour and compactness of texture it resembled ivory. The bony surfaces of the tibio-tarsal articulation were healthy ; none of the hard or soft parts of the plantar aspect of the foot participated in these alterations. MM. Ferrus and Cruveilhier have observed similar morbid appearances, and M. Larrey, in his description of elephantiasis, speaks of violent pains felt along the course of the bones.

With respect to lesions of the viscera accompanying or coinciding with elephantiasis *Arabica* in the lower extremities, the following is what I observed in the body of the man Fournier : the larynx. the trachea, and bronchi, were natural ; each of the pleuræ contained from eight to ten ounces of serum. The left lung was crepitating and loaded with serum, which flowed out when pressed between the fingers ; there was no trace of sanguineous engorgement in this lung, but on the posterior part of the lung of the right side there was, and the whole mass of this organ felt firmer and resisted pressure more powerfully than that of the left side.

The pericardium was in a healthy state ; the heart, of the natural size, contained some fibrinous clots in the right cavities ; the thoracic aorta was healthy ; the cavity of the abdomen contained a small quantity of transparent serum. The internal surface of the stomach was divided into two portions by a deep line of demarkation, which separated the anterior exactly from the posterior part ; in the former the mucous membrane was healthy ; the latter was, on the contrary, almost without any trace of mucous membrane, and of a dead white ; its surface was covered with bluish vessels which projected from it, and contained a larger or smaller quantity of blood, so as to give this region a blue and red marbled appearance. The mucous membrane terminated abruptly in a sharply-cut edge in the pyloric region, whilst towards the cardia it was continuous with the mucous membrane in a state of health : in some parts of this region the softened mucous membrane had a greyish aspect, which contrasted on the one side with the alteration which I have met just described, and on the other with the healthy portion of the organ. The small and large intestines presented numerous vascular ramifications, separated from each other by parts in which the softened mucous membrane appeared to have lost some portion of its substance. The mesenteric glands presented no particular appearance, the liver was enlarged, and its yellow substance pretty abundant : the finger could with difficulty be pushed into its substance.

The kidneys, larger than in the healthy state, presented a more decided alteration on the left than on the right side ; their whole substance, but particularly the cortical part, was of a morbid yellowish white, very different from the usual colour of these organs. The brain and its membranes had undergone no alteration.

In another case of a woman named Mary Allard, the peritoneum was covered with numerous granulations over almost its whole extent, particularly where it formed the epiploon ; the portion of this membrane covering the intestinal canal was very much injected. The cavity of the lower pelvis was partly filled with a sero-purulent effusion ; some whitish lines appeared on the stomach, over which the mucous membrane was softened, and very sensibly diminished in thickness. A similar alteration had taken place in several parts of the small intestines. In the large intestines, and particularly in the descending colon, small round ulcers were met with, surrounded by the mucous membrane, blanched, and of a dead white. In the situation of the sigmoid flexure of the colon, the sub-peritoneal cellular tissue was loaded with a profusion of purulent matter ; large sinuses existed here, but without communicating with the intestine ; this alteration extended a considerable way into the cellular tissue of the lower pelvis ; the other viscera of the abdomen appeared to be in a healthy state.

Elephantiasis *Arabica* seldom attacks the superior extremities. M. Alard, however, quotes four cases of the disease occurring in the arms. In one (case 7) the hard and permanent swelling of the left arm occurred after the application of a blister. In another (*Op. Cit.* p. 190) the right arm increased to such a size that it weighed two hundred Genoese pounds, forty of which consisted of serum : the swellings of the arm and fore-arm resembled a distended bladder or skin ; the arteries, the veins, and the nerves had undergone no alteration ; but the lymphatic vessels were very much dilated and loaded with lymph. The third is a case of Fabricius Hildanus. The fourth is extracted from Hendy, who relates several others of the same description, in which the disease appeared in a very acute form by a sort of numbness in the shoulder and arm, the enlargement of a gland in the axilla (case 13) or elbow (case 5) and a red line running over the inner part of the arm and fore-arm.

I have myself observed three cases of elephantiasis of the upper extremity, but the progress of the disease in all was chronic. One occurred in a woman who had had the right breast taken off for a cancerous affection of the mammary gland, and in whom the lymphatic glands, become scirrhous, compressed the axillary vein (first edition, vol. ii. p. 630) ; the second case I met with has been published by M. Gaide ; the sub-clavian axillary and brachial veins were here filled with a fibrinous clot of long standing, adhering to the inner membrane of these vessels, and of a yellowish grey in the centre.   In the third, which was detailed by M. Bonnet, of Poitiers, and in which the left fore-arm was the part affected, we found the basilic vein hard, and filled with a coagulum which adhered closely to its inner membrane ; this hard and solid clot was blanched or of a greyish colour, intermixed with red striæ.

The history of elephantiasis occurring in other regions of the body is less complete ; elephantiasis of the scrotum is almost the only case which has been made the subject of correct anatomical research. Elephantiasis of the hairy scalp is very rare.   M. Ricord has given two cases of it in the *Revue Médicale*, vol. ix. p. 13.

Elephantiasis of the face sometimes only attacks it on one side : such was the case with the patient whose history forms the subject of M. Alard's ninth case, and in whom the elephantiasis would seem to have been complicated with eczema of the ear.   Willier's (Alard, case 2) is a remarkable case of elephantiasis of the face : after committing a debauch at table, this patient experienced a violent pain in the left cheek and below the zygomatic arch ; this pain soon extended under the chin ; the sub-maxillary glands enlarged and felt painful ; the face swelled and became erythematous ; and he experienced nausea and slight shivering fits.   At the end of six months, another attack, after which the patient perceived that the face continued puffed ; this attack was followed by several others, and the face became larger and larger.   In similar cases tumefaction may arrive at such a height, that Schenck speaks of a man whose head exceeded that of an ox in size ; the lower part of the face was entirely covered by the nose, which had to be raised to enable this unhappy being to breathe (Obs. Méd. Rar. Nov., &c. lib. i. p. 12). I have only seen one case of elephantiasis of the face, which supervened after repeated attacks of erysipelas.

This disease causes the breasts to increase to such a size, that they are obliged to be supported by bandages passed round the neck. Salmutius (Cent. 2. Obs. 89.) speaks of a woman whose breasts increased to such a size, that they hung down to her knees.   She had, at the same time, glandular tumours as large as the head of a fœtus, under the axillæ.   M. Borel, physician in Castres, also quotes the case of a woman whose breasts became so large that they were obliged to be supported by straps which passed over the shoulders and neck.

M. Alard relates as a case of elephantiasis, that of a lady of Berlin (*Ephem. nat. cur.* 3 anno. 2, p. 71, 1694,) who had an abdominal tumour, the lower part of which reached to the knees.   This tumour, situated under the skin, outside of the cavity of the peritoneum, was formed of a congeries of small pouches, agglomerated and adhering to one another, like the swimming bladders of some large fish. Seven of these cells, adhering very closely together, formed the circumference of this tumour, and an eighth occupied the centre. Each of these cells was itself divided into several small compartments, which enclosed a clear and limpid fluid, like the white of egg, but of greater consistence in some, and in several similar to the boiled white of an egg. The contents of others again, were yellowish, greenish, or reddish. On the peritoneum being opened, no vestige of disease was found in the abdominal cavity. No sensible alteration had taken place in the veins, which were only found to be a little out of place.

M. Delpech also quotes a case of elephantiasis of the walls of the abdomen, observed in a young woman, four-and-twenty years of age, born at Toulouse, who had three conical tumours on the abdomen, adhering to the sides of this cavity, near the hypogastrium and umbilicus, two on the right side, and the third on the left. These tumours were of the structure of those enlargements which constitute *andrum*; that is to say, a cellular tissue, the cells or areolæ of which are of great magnitude, separated from one another by very extensive and semi-opaque laminæ, covered with very much dilated lymphatic vessels, and some few sanguineous vessels, of extreme tenuity, and but little ramified ; the intestines were filled with a serum, half liquid, half solid, and rendered almost opaque by a considerable portion of albumen.

Next to the lower limbs, the scrotum is the part of the body which is most frequently attacked with elephantiasis *Arabica :* this part and the penis often acquire an enormous size when affected with the disease. This alteration has been improperly designated by Larrey under the name of *sarcocèle d'Egypte ;* by Prosper Alpinus under that of *hernia carnosa ;* and by Kæmpfer, under that of *endemic hydrocele of Malabar.* M. Alard relates three cases of the disease from Hendy (Obs. 16), from Gilbert (Obs. 5), and from the Ephemerides curiosorum naturæ (Obs. 178, p. 212). M. Dumeril has seen a remarkable case of the affection, in a man upon whom all the resources which art possessed had been lavished in vain. M. Delpech relates two cases of it, one of which afforded him the opportunity of performing an extraordinary operation. This last patient, who was thirty-five years of age, had been afflicted with his infirmity for ten years. The skin of the scrotum had become excessively hard, thickened, tuberculated, and intersected with deep wrinkles ; the enlargement, pasty at first, had afterwards acquired a greater consistence, and became hard and very heavy. The tumour formed by the scrotum weighed, when it had attained its maximum size, about sixty pounds ; under or within this shapeless mass, the penis and the testicles were buried ; it seemed to be divided into three unequal masses, two lateral and one anterior, where a sort of umbilicus was seen, through which the urine passed. This patient did not experience the erysipelatous affections, accompanied with fever, shiverings, vomitings, &c. observed by Kæmpfer

(*Amœnit. exotic.* fasc. 3, Obs. 8, p. 557), Hendy, and Gilbert, in similar cases.

Several other inquirers have ascertained that the disease is not always accompanied by these phenomena; and in the patient operated upon by M. Delpech, the same alterations of the skin and cellular tissue which are observed in elephantiasis of the limbs, were found; the organs of generation were healthy.

In another individual who was operated upon by Larrey, one of the testicles was found to be healthy, the other smaller than in its natural state (Campagnes, t. ii. obs. 1, p. 122). In a third case, which has been republished by M. Alard, independently of the alteration in the skin and cellular tissue of the scrotum, it was discovered after death that the testicles were *inflamed like the rest of the parts.* The right testicle, after having been stripped of the tunica vaginalis, was not less than a goose's egg in size. It was divided into three compartments: a gelatinous and thick matter infiltrated the upper and lower parts, and the middle was occupied by a substance nearly the size of a walnut, into which the vasa efferentia emptied themselves, without appearing to have undergone much alteration. The tunica albuginea was much thicker than in the healthy state, and contained a pale fluid, lodged in small divisions like those of a lemon. On opening the tunica vaginalis of the left side, two quarts of serous and almost colourless fluid were discharged; the same state of things was found on the opposite side. Upon removing the integuments covering the penis, which were three fingers thick, this organ was found to be of its natural size, or even smaller than it should have been; the corpora cavernosa could not be inflated as they usually may. All the rest of the body was healthy, except the right kidney, the ulceration of which had no doubt caused the patient's death. Finally, according to Hendy, the disease peculiar to Barbadoes may attack the testicles and the inguinal glands at the same time (Case 10).

Elephantiasis *Arabica* may also be complicated with scrotal hernia, of greater or less magnitude, as it was in the case of a man, named Lajoux, of Toulouse, the details of which the Société de Médecine of that city has published in its transactions. The following case, detailed by M. Fabre, is another instance of the same kind occurring in an old man, seventy-three years of age, who was affected with elephantiasis of the scrotum, and with double inguinal hernia. When this patient was fasting, the upper part of the tumour gave a clear sound on being struck, whilst in the lower part the sound was dull; after eating, however, the dulness extended along the whole of the right side of the tumour, and the sound remained clear in the upper part of the left side. It was enough for the patient whilst fasting, to drink a certain quantity of liquid, to cause the sound immediately to become dull on the right side. M. Fabre concluded, and with reason, that the stomach was displaced in this case, and formed a hernia on the right side.

I saw, some years ago, under the care of M. Dupuytren, in the Hôtel Dieu, a woman of the town, in whom elephantiasis *Arabica* was developed in the labia majora, which were of an enormous size. Similar cases have been collected by Gilbert (Alard, case 2), Larrey (Campagnes d'Egypt, t. ii. p. 127), and by Talrich (Delpech, *mém. cit.*).

Elephantiasis *Arabica* may also be developed on the verge of the anus, a case which Bayle was the first to observe.

Upon dissection, this enlargement of the cellular tissue does not present any appearance similar to scirrhus, in whatever part it is incised. No scirrhous induration or cerebriform matter is discovered, but merely a kind of very hard œdema, a cellular tissue full of a colourless fluid, which can be at all events partially squeezed out by strong pressure. The enlargement is hardly ever confined to the verge of the anus ; it generally extends to a greater or less distance into the cellular tissue of the buttocks, where it terminates gradually (Dict. des Sciences Médic. t. iii. p. 609).

M. Alard gives a curious case as one of elephantiasis of *one side of the body*, but it is deficient in essential details (Op. cit. p. 219).

Finally, certain anomalous enlargements of the tongue, of the uvula, of the sub-mucous cellular tissue of the intestines, and of the sub-serous cellular tissue of the epiploons have been assimilated by some writers to Arabian elephantiasis, and will be described hereafter (see Glos-socele).

Causes. Individuals affected with Arabian Elephantiasis may suffer from a variety of acute and chronic diseases, either before or after the occurrence of the intumescence, which sometimes appears after repeated attacks of eczema. Mentzell and Bayle have seen gout coincide with Arabian Elephantiasis ; patients affected with elephantiasis *Græca* (see elephantiasis *Græca*), have sometimes presented, not only œdema of the lower limbs, which is common, but the true, hard, and bulky enlargement of Arabian elephantiasis also. Upon the whole then, several alterations of the veins (varices, phlebitis, contraction, obliteration, &c. &c. &c.), and various forms of inflammation of the skin (erysipelas, eczema, lichen, ulcers), are the diseases which are the most frequently observed before the development of Arabian elephantiasis, or during its course.

Elephantiasis *Arabica* shows itself above all in those parts of the body in which the venous circulation is lower than usual ;—in the lower extremities, and in the scrotum in men. This disease is neither contagious nor hereditary. It is seen in persons of all ages, most frequently in adults, more rarely in the aged, and in children. Chaussier presented to the society of the faculty of medicine of Paris, on the 1st March, 1810, the leg of a stillborn infant, on which two deep constrictions, and a considerable tumefaction of the back of the foot, very similar to elephantiasis, were seen. Out of thirty cases of this disease which I have met with in Paris, more than half had occurred without any appreciable external cause. There is not perhaps a department in France, in which this singular affection has not been observed.—Delpech assures us, that he has seen numerous cases in Roussillon, particularly in the neighbourhood of Elne. According to Casal, it is very common in the Asturias. (Hist. natur. y médic. de el principado de Astur. p. 321, 323.) I am not aware that any other observations have been made either in England or on the Continent, with the view of ascertaining whether particular topographical conditions or other circumstances have really a marked influence on the development of this disease. In the island of Barbadoes its frequency is attributed to the sudden impression of cold, to the extreme coolnees of the nights, and to the currents of air which blow through all the houses. We are assured that this disease is endemical in some regions of the torrid zone, and it is principally seen on the left bank of the Ganges, in Egypt, Nubia, &c., countries which are continually under influences nearly of the same kind.

Diagnosis.—When Arabian elephantiasis is announced by febrile symptoms, accompanied with pains along the course of the veins, vessels, and lymphatic glands of a limb, it presents almost the same characters as certain œdemas observed in puerperal women, in whom the principal veins of the extremities have been found obstructed by fibrinous clots. (Rayer; article Œdema. Diction. de médecine in twenty-one volumes.) When the skin has become uneven or tubercular in Arabian elephantiasis, the alteration which takes place is somewhat similar to that which is seen in elephantiasis *Græca*, but in the latter the inequalities or tubercles, follow spots of a tawny colour, and do not constitute its principal outward character; whilst in Arabian elephantiasis the lesser swellings and tuberculations are accidental, only appear during the last stages of this disease, and are always accompanied by other lesions of the parts under the skin. The point, in brief, which it is of the greatest consequence to ascertain in cases of elephantiasis *Arabica*, is whether the tumefaction of the parts affected, be produced by indurated cellular tissue, impregnated with serum and hypertrophied by adipose tissue, or by an anomalous development of the skin, muscles, and other tissues which enter as component parts into the organization of the limbs or parts affected; whether the vessels and lymphatic glands be inflamed or not, and whether the enlargement be the result or not of some obstruction to the course of the blood, occasioned by compression, dilatation, contraction, or obliteration of one or several of the veins.

According to Dr. Hendy, in some rare cases, elephantiasis *Arabica* has been known to get well spontaneously. A man affected with elephantiasis of the scrotum, after having had several attacks, was awakened one morning by an uncomfortable dampness round the thighs: this proved to be water which had been effused through a crack in the diseased skin. About six ounces of the fluid were collected in a basin. A few months after this the patient had another recurrence of a similar evacuation from the scrotum, after which this part was reduced almost to its natural size. (Hendy, case 22.)

Treatment.—The inflammatory symptoms observed in the first stage of elephantiasis *Arabica*, should be treated by emollient applications, tepid baths and bleeding; the fears which have been entertained against bleeding are unfounded. I have employed it with success in the paroxysms, the length and intensity of which it certainly moderates. In the chronic stages of the disease, bleeding has been followed by momentary relief at least, when the patient has complained of a feeling of painful distension of the parts affected. I have also seen very happy effects, produced by local bleedings, from the groin, the hollow of the ham, the axillæ, &c. The part affected, placed as constantly as possible in a position that facilitates the return of the blood towards the heart, should be covered with emollient cataplasms, or wrapped in flannels steeped in soothing and narcotic decoctions. If the enlargement is developed on one of the lower extremities, the patient should keep his bed for several weeks.

Emetics and purgatives have been administered at this period with various success; I rarely make use of them myself. The antispasmodic effects of the sublimed oxide of zinc in doses of eight grains a day, have been much spoken of. Hendy assure us, that this remedy allays the sickness and uneasiness, which patients experience in the periodical exacerbations of elephantiasis. Several physicians of the island of Barbadoes, struck with the frequency of the vomitings during the paroxysm, have thought it necessary to encourage the sickness, and even to provoke vomiting by the exhibition of emetics. Dr. Hendy objects to this practice.

In women, pregnancy is a very unfavourable circumstance. A young woman, of Havre, having married against my advice, became pregnant three times; after each pregnancy the right leg, which was affected with elephantiasis, became more and more enlarged.

A great number of patients have been cured by compression, either alone or combined with other means. This method was completely successful in the hands of Bayle and M. Alard, in one of their patients who had been affected with elephantiasis for twelve years. A strong man was employed to press the leg of the patient every morning in all directions, during three quarters of an hour or an hour, after which a roller was firmly applied from the toes to the knee. M. Lisfranc has also been very successful in these cases by the judicious combination of scarifications, compression, and local bleedings. I have myself obtained unhoped-for cures by this method. It is more especially applicable in cases of elephantiasis of the limbs, consisting of simple hypertrophy of the cellular tissue without infiltration of serum. If it does not succeed completely when the tumefaction is partly owing to anomalous development of the muscles and bony tissue, it determines, at all events, the absorption of a certain quantity of fat and serum, and this is a result which we are always happy to obtain. Compression alone is found to suffice in the majority of cases, scarifications seem now to be rarely used; when it is thought necessary to have recourse to them, they should be made at such a distance from each other, as shall prevent the inflamed circles which are or may be formed around them from meeting. The inflammation produced in an extremity by twenty or thirty scarifications of from half an inch to an inch in length, is not in general considerable; if it should increase to any degree of intensity, local and general bleedings must be had recourse to in combination with emollient and cold narcotic applications. It will be necessary to wait until the first scarifications are healed before others are made; several practitioners have recommended blisters and issues to the parts affected with elephantiasis. I have tried the effect of flying blisters. The trials which have been made with arsenical preparations upon patients labouring under elephantiasis, but without any evident advantage, should not be repeated.

Patients who have been affected with elephantiasis should continue to wear a laced stocking, or a tight bandage after recovery, particularly when several of the veins of the extremity are in a varicose state.

Harassed and worn out by the enormous weight of the parts affected, many patients have insisted upon amputation as a last resource for an incurable disease. M. Alard assures us that those who have survived such an operation have become affected with elephantiasis in other parts of the body, or that they seldom failed to sink after one or more attacks of an inflammatory affection of the viscera, to which they seemed to become liable. A woman, who from the age of fifteen years, had suffered frequent attacks of the Barbadoes malady, was so much incommoded by the size of the affected limb, that she begged it might be amputated. This was done; but a short time afterwards she had so violent an attack of the disease in the other leg that she sunk under it (Hendy, case 24). A woman, named Mary Pecout, whose case I have related in the first edition of this work, underwent amputation of the right thigh, in the month of March, 1823, after having been affected with Arabian elephantiasis from the age of seven years. In January, 1825, the disease attacked the right arm, and was treated successfully by M. Lisfranc by local bleedings, scarifications, and blisters. To these unsuccessful cases of amputation, Delpech opposes that in which M. Delmas, chief prosector of the faculty of Montpellier, amputated the arm under circumstances exactly similar, and without its being followed by any relapse. A patient, operated upon by M. Larrey for elephantiasis of the scrotum, was in a fair way of recovery when this celebrated surgeon left him to proceed to Alexandria. Authier, operated upon by Delpech, on the 11th of September, 1820, left the hospital of Montpellier in the early part of February, 1821, apparently in good health, but with a slight cough. On his arrival at Perpignan he was pale and completely blanched in appearance; his pulse was extremely small. He died on the 23d of the same month of inflammation of the liver, of the peritoneum of the right hypochondrium, and of the pleura of the same side. M. Talrich operated with success, in 1811, on a young woman affected with elephantiasis of the sexual organs. Nægele has been equally successful in the amputation of a leg in a case of elephantiasis. The question, therefore, as to the propriety or impropriety of amputation is not entirely settled.

Historical Notices.—The first pretty accurate ideas in regard to this disease are to be found in Rhazés (cum Serapio, Averrhoë, edit. G. Franks), 1533, in Haly-Abas, and in Avicenna (*Libri de re medicâ omnes*, in fol. Venetiis, 1564, vol. i. p. 952, *elephantia*), and it is on this account that the disease we are discussing received the name of *elephantiasis Arabica*, or *Arabian elephantiasis*. The disease has been since noticed by Forrestus (*Opera*, lib. xxiv. p. 453), by Mercuriali (*De morbis cutaneis*, lib. ii. cap. v.) and by Kæmpfer (*Amœnit, exot.* fasc. 3. p. 58). Elephantiasis *Arabica* has been observed in Egypt, by Prosper Alpinus (*Medicina methodica*, Lugd. Batav. 1719), and by the medical officers of the French army that invaded Egypt (Larrey, *Relation hist. et chir. de l'expédit. d'Egypte*, in 8vo. Paris, 1812, 1817). J. Hendy (*On the glandular disease of Barbadoes*, translated into French by Alard, in *Mém. de la Société Médicale d'emul.* t. iv. p. 44) studied it among the natives of Barbadoes. M. Alard, who has published a learned essay on this disease (*D' l' inflammation des Vaisseaux absorbans, lymphatiques dermoides et sous cutanés, éléphantiasis des Arabes*, nouv. édition, in 8vo. fig. Paris, 1824), thought that the anomalous developments which characterise it were constantly preceded by inflammation of the lymphatic vessels and glands. The cases of M. Bouillaud (*Observations d'éléphantiasis des Arabes, Archives générales de Médecine*, t. vi. p. 567), and some others published subsequently in the first edition of this work, and detailed under my own eye by M. Gaide, (*Observ. sur l'éléphantiasis des Arabes*, inserted in the *Archives Gén. de Méd.*, t. vii. p. 553,) would lead us to imagine that varicose states, and contractions, and obliterations of the veins, as also eczematous and erysipelatous inflammations, were in many cases not less efficient causes of these intumescences. The researches of M. Fabre may be consulted with advantage (*Observ. de l'éléphantiasis des Arabes.* Revue Médic. Oct. 1830), and a case of Arabian elephantiasis by Martini and Horack, published under the title of *Obs. rarioris degenerationis cutis in cruribus elephantiasis simulantis*, Lipsiæ, in 4to., 1828 ; as also the various cases which have been published on elephantiasis of the hairy scalp (Ricord, *Revue Méd.* t. ix. p. 13), of the arm (Hensler, *Histor. brachii prætumidi*, in Haller, Disputat. chirurg. vol. v. Ludoff, *Casus éléphantiasis in brachio observatæ*, in 4to. Erford, 1703, in 4to). of the scrotum and penis (Talrich, *Revue Méd.* t. i. p. 180. Wadd, *Cases of diseased prepuce*, 4to. London), on the pathological anatomy of the skin in elephantiasis, by M. Andral (*Revue Méd.* t. xiii. p. 224), and on the treatment of this disease, by Lemasson (*Influence salutaire d'un erysipèle sur une éléphantiasis des Arabes. Journ. Hebd.* 2me serie, t. iv. p. 409), on elephantiasis cured by antiphlogistics, (*Revue Méd.* 2me serie, t. iv. p. 489,) and some cases of amputation published by Nægele (*Archives Gén. de Méd.* t. xiii. p. 126), and by Clot (*Gazette des Hôpitaux*, 1833, p. 388), are all interesting, and all deserve to be consulted.

Barbadoes leg.

In the month of February, 1755, a fever prevailed in the island of Barbadoes, characterised by a cold stage of four or five hours' duration, a hot fit, &c., headache, and frequently severe pains in the back. This fever was sometimes ephemeral, and occasionally lasted no more than four or five days ; it, however, much more generally continued longer, and then there supervened inflammation of the leg similar to that which accompanies the fever of elephantiasis, but without swelling of the lymphatic glands, and without any hard corn in the limb. The inflamed part was of a vivid red ; small phlyctenæ arose here and there over its surface, as in erysipelas, and desquamation took place after the cessation of the inflammatory symptoms.

An epidemic of the same description recurred during the month of February, 1757, but marked by several important varieties, which were probably ascribable to the excessive heat of the weather upon this occasion. The fever was now accompanied with pain in the stomach, nausea, cough, and sometimes with delirium and coma. The local affection was exhibited in the feet, legs, or arms of either side, but never of both sides at once, and was distinguished by the same redness and swelling as in elephantiasis ; the swelling, moreover, increased after the fever had ceased. During the next month many persons were no otherwise affected than with a troublesome cough, which ceased as soon as a tumour appeared on the arm or hand. The disease continued with this phasis till the month of June, when it assumed new features : the heat became more considerable, the thirst greater, the pains in the back and limbs much greater than at first, and the tumours or swellings were apt to fall into suppuration, instead of being resolved, as they were through the preceding stages of the disease. (Hendy, James. *A Treatise on the Glandular Disease of Barbadoes*, 8vo., London, 1784.) Hillary, W. *Obs. on the Air and the concomitant Epidem. Diseases in the Island of Barbadoes*, 8vo., London, 1759.)

Andrum and Perical.—Pedarthorace and Endemic Hydrocele.

These are two affections analogous in their nature to Arabian elephantiasis, which prevail endemically on the Coast of Malabar, in the islands of Ceylon and Japan. The one attacks the foot (*perical*), the other the scrotum (*andrum*).

Perical, or febrile foot, is very common among the natives of Cochin. Young persons are attacked more frequently than adults, and these oftener than the aged. A very general opinion is, that the Christians, among whom the disease is very common, bring it from the Coromandel coast. The disease attacks one or other of the lower extremities ; very rarely both at once, and always occurs on the lowest part.

Those affected with it have an attack of phlegmonous inflammation every month, which vanishes after a few days, but leaves a swelling which degenerates so that the extremity becomes triple, quadruple, and more, its former size. It is uneven, œdematous, hard, of a schirrhous appearance, and often the seat of ulcers which discharge a serous looking fluid. The swelling generally extends to the toes, rarely ascends above the calf of the leg, and never affects the knee. It is occasionally also observed in the thigh, which, however, may be simply infiltrated from the scrotum when this is the part affected, as it is in the endemic hydrocele. Although the tumefied part be hard, and have a brownish and deformed appearance, it never becomes gangrenous, and is not dangerous. It is only painful at the epoch of the periodic inflammation, and is troublesome solely from its weight. When the affection becomes inveterate the limb is apt to be covered with a number of small ulcers, which render the infirmity more unbearable.

Andrum or *endemic hydrocele*, begins with an erysipelatous affection of the scrotum, which is renewed every month at the time of the new moon, and leaves a swelling behind it which is caused by the effusion of a quantity of serous fluid, the quantity of which increases from day to day, and distends the part to such a degree that it has at length to be evacuated by punctures, and scarifications. This fluid is either thin and limpid, or viscid ; it is always reddish in colour, and differs in its qualities according to the temperaments of patients. The disease attacks the natives and Europeans : a residence of a few years is enough to render any one subject to an attack. It is incurable in the country ; but it is neither dangerous nor very troublesome. Still it sometimes happens that the testis is implicated and becomes schirrhous. If patients leave the country, the tumefaction abates gradually, unless it be complicated with sarcocele, a disease for which there is no remedy. (Kæmpfer, *Amœnit. exoticæ*, 4to. Lemgo, 1712.)

Senki.

Senki is the name given by the natives of Japan to a disease which is very common among them, so common, indeed, that among ten adults, it is difficult to find one who has not been affected with it at one time or another. Strangers, too, after a short residence in the country, are liable to the disease. It begins with pain in the belly and spasms, particularly in the abdominal muscles, when a sense of suffocation supervenes from the tension which then occurs between the region of the pubes and that of the false ribs and ensiform cartilage. The disease is sometimes fatal. When it abates, swellings are observed to occur here and there, over the whole surface of the body ; among men it often produces an enormous enlargement of the eyebrows ; among females, the labia majora are beset with a congeries of tubercular or fig-like enlargements. Enlargements of these parts in the two sexes are common in Japan, and may supervene without appearing to be consequences of colic (Kæmpfer, *op. cit.* p. 552).

Mouth Canker, Labri-sulcium, or Cheilocace of Ireland.

Boot describes a disease as prevailing in Ireland, and even in England, among children of four or five years old, characterised by a *tumefaction of the lips*, which become hard, and, by projecting from the gums and teeth, give the countenance a hideous and unnatural expression. Occasionally they are divided by a deep chap or fissure into two parts, as it were, from whence there flows a sanious looking fluid, which dries up into a crust. The upper lip alone presents this anomalous enlargement in some cases, and when both are affected, the upper is always so in a much greater degree than the lower. This disease is *mouth canker, labri-sulcium,* or *cheilocace.* It is very generally accompanied with ulcers in the mouth, on the palate, tongue, and gums. The best treatment, we are told, is to purge patients freely with calomel, and infusions of senna, to make them drink an infusion of fumitory, rumex, patientia, and endive ; to apply leeches to the lips and temples, and even to recur to general blood-letting, if they be strong. The lips should be kept moistened from time to time with a decoction of honey-suckle, &c., or with a solution of sulphate of zinc, and then rubbed over with a linament of the acetas plumbi and oil. A blister may be applied to the arm, if the disease be obstinate, and decoction of sarsaparilla, cinchona, &c., prescribed for all drink (Boot) ; Arnaud (Obs. Med. de affectibus omissis, 12mo. Lond. 1649).

Mercurialius and Bonetus (Sepulchret. anat. lib. i. sect. 21, obs. 17,) also speak of this disease, upon the nature and characters of which there still hangs great obscurity. I have frequently observed a *hard and indolent tumefaction of the lips* in children of scrofulous constitution.

Aleppo Pustule.

Symptoms.—The Aleppo pustule or spot is a disease endemical in Aleppo and several towns of Syria, which attacks almost all and sundry once in their lives. It is characterised by one or several pustules of tardy growth, which get well after having ulcerated, but leave ugly cicatrices behind them. It is denominated habbet el seneh (spot or pustule of a year) in Arabian, because a year is required for it to run through the whole of its stages, that is to say, to be evolved, to suppurate, and to cicatrize. It begins in the shape of a small prominence of a lenticular shape, without heat, pruritus, or pain. This increases insensibly to the end of the fourth or fifth month, when it may have acquired from about six lines to several inches in diameter, and be found to project about three lines. At this time, it becomes the seat of acute pain, the severity of which, however, varies according to the place of the affection. Its surface becomes covered with a whitish moist incrustation, which, when very firm, is either detached completely, or only cracks in different places, and allows a quantity of purulent matter, white or of a yellow colour, and inodorous at first, which is formed slowly in the interior of the tuberculation, to escape. When the crust falls off, the surface it covered is found uneven, granulating, and of the colour of raw meat ; the place is surrounded with a red areola. The crust falls off entirely or in pieces at intervals ; when it remains long adherent the discharge becomes thick, of a dark colour, and extremely fetid. The period of suppuration lasts five or six months, and ends in the formation of a dry and adhering crust, which is complete by the end of the year, when the disease has been left to itself, and the patient is in other respects well. The affected part when first exposed is at first of a vivid red, then becomes of a reddish brown, and by and by approaches nearly to the natural colour of the skin.

All parts of the cutaneous surface may become the seat of the Aleppo pustule, but it is most generally developed on the face and extremities. The inhabitants of Aleppo have it more frequently on the face than on any other part. This is the mark which distinguishes them from the rest of the Syrians. Foreigners on the contrary are seldom attacked with it on the face. When the pustule appears on the joints, over bony projections, and on regions which are but indifferently covered with soft parts, it occasions very severe pain. M. Guilhou relates a case of this disease occurring on the scrotum in a Frenchman ; such an event is very rare.

The inhabitants of Aleppo distinguish two species of this disease by the very objectionable names of *male* pustules, and *female* pustules. The former are *single*, whilst the latter occur in numbers together. Round several of the principal pustules, others of smaller size and in greater or smaller numbers are frequently clustered ; these may be so numerous, that the whole body becomes ulcerated by them. M. Guilhou once saw a Frenchman who had seventy-eight principal pustules, each surrounded by smaller ones in such numbers that the eruption at first sight had the appearance of confluent small-pox. The loss of substance which the pustule occasions is sufficient proof that the whole substance of the skin is implicated. This eruption constantly leaves an indelible cicatrice, which is as various in its form as the ulcers which precede it ; it is depressed, with the edges more or less oblique, sometimes rather deep, but in general superficial. The cicatrice is either smooth or rough, rarely brownish, more frequently white.

This disease is endemical to Aleppo and its environs. M. Guilhou assures us that a similar eruption prevails in Bagdad, on the banks of the Tigris, and of the Euphrates, and in all the villages on the road from Bagdad to Aleppo, such as Mossal, Diarbekir, Hedira, and Orfa. At Aleppo foreigners and natives, without distinction of race, sex, temperament, or profession, are affected by it. It rarely attacks infants at the breast ; it is generally at the age of two or three years that it is developed. There is not a single instance of a child born in Aleppo, having reached the age of two years, without having been affected with this disease. The time at which it attacks strangers is variable. It is very rarely that they suffer from the eruption till after a residence of some months. It is also often years before it shows itself ; but it is the general opinion at Aleppo that it is enough to have passed some days in that city for the disease to be developed sooner or later, in whatever country the individual may afterwards happen to reside. Several facts in confirmation of this statement have been adduced. The Aleppo pustule is not contagious ; it only attacks individuals once in the course of their lives.

This disease is never fatal, but the face may be horribly disfigured by it, particularly when it is situated in the neighbourhood of the eyes, mouth, or nose, &c. The ulcers often partially destroy the eyelids, and the alæ of the nose ; they divide the lips, cause gaps in the external ears, and always leave frightful cicatrices behind them. In fact the cicatrices which the Aleppo pustule leaves, are the worst and most dreadful effects of the disease.

The various curative means tried in this disease have seldom appeared to exert any beneficial influence. It would indeed seem advisable to confine all treatment to the use of soothing washes, to great attention to cleanliness, and to keeping the ulcers from exposure to the air ; still the ceratum minii camphoratum, or cerate of Nuremberg, and the actual cautery, appear to have been employed with advantage before the period of suppuration, during the third or fourth month.

Tara of Siberia.

Gmelin, during his travels in Siberia, in 1740, 1741, 1742, 1743, observed a contagious epidemic, which generally prevails during the months of June and July, in the city of Tara, and on the banks of the river Irtisch. This disease shows itself at first by a species of pustule of a pale colour, and hard to the touch, which appears on different parts of the body. In the course of four or five days these pustules attain the *size of the fist*, without changing colour or becoming less hard. The patient then experiences great weakness, with excessive thirst, loss of appetite, drowsiness, vertigo, præcordial anxiety, difficulty of breathing, fetid breath, paleness of the face, excruciating internal pains, inexpressible anguish, and, if a copious sweat does not break out, death is inevitable from the ninth to the eleventh day.

The treatment which is considered to be infallible is always undertaken by a cossack, who kneads the tumour all round until the blood flows from it, or he plunges a needle into it until the patient complains of the pain of the operation. He then applies a cataplasm of chewed tobacco and sal-ammoniac, which is renewed three or four times in the course of twenty-four hours, and in the space of six or seven days the cure is accomplished. No other drink than warm *quaas* or *quass*, a liquor made of leaven or flour fermented in water, is generallly allowed during this interval, though chicken-broth with horse-radish stewed in it, is occasionally prescribed instead. Milk, meat, fish, and dry vegetables are interdicted.

The parenchyma of these tumours is said to be a spongy bluish kind of flesh. Gmelin treated these tumours by opening them and powdering their interior with the red precipitate of mercury, whilst he prescribed calomel internally.

Horses contract this disease ; the tumours are treated in these animals by the application of the actual cautery. (Gmelin, *Travels in Siberia* from 1733 to 1743, Göttingen, 1731-52, 4 vols. 8vo. fig. in German.)

The Mal Rouge de Cayenne.

In this third group, several endemic diseases which are evidently varieties of Greek elephantiasis, [*the tubercular lepra of the middle ages*] and others which evidently bear a greater analogy to elephantiasis than to syphilis, with which they have sometimes been assimilated.

Under the name of the *Mal Rouge de Cayenne*, Bajon and Bergeron have described Greek elephantiasis, or *tubercular lepra*, which is more common in this island than in any other in the French colonies. *See* Dazille (*Obs. sur les maladies des négres.* 8vo. Paris, 1742, t. i. p. 300); Bajon (*Mémoires pour servir à l'histoire de Cayenne et de la Guyane Française*, Paris, 1777, 1778, in 8vo. 2 vols.) ; the *Rapport des commissaires de la société royale de médecine sur le mal rouge de cayenne ou éléphantiasis*, 8vo. Paris, 1786, and Bergeron's treatise (*Mal rouge observé à Cayenne.* Diss. inaug. Paris, 1823).

Radesyge.

Radesyge is a Norwegian term, synonymous with the Latin phrase *morbus atrox*, by which the physicians of the country have designated a disease of the skin which is looked upon by some as a species of syphilis, and by others as a variety of elephantiasis ; the latter opinion, according to the descriptions we have of the symptoms, seems to me the more probable of the two.

Radesyge shows itself in the cold and foggy season, by a feeling of weight over the whole body, lassitude of the limbs, and itchiness of the skin. Patients afflicted with this malady fly every kind of occupation ; they experience a stiffness in the joints, and head-ache in the frontal region, accompanied with a sense of tightness in the chest, and dyspnœa. The face has a pale, livid, leaden colour, followed by plethoric redness, and running at the nose, which renders the passage of the air into the nasal fossæ difficult. The nose becomes red and swelled, the voice hoarse, the uvula elongated, and wandering pains in the limbs supervene, which subside towards morning after a very copious, viscid, and rather fetid perspiration. Some months, or some years afterwards, a dry, whitish, mealy, or furfuraceous eruption forms on the surface of the integuments, the scales of which fall off and are replaced by thicker ones, which render the skin uneven, hard, and rugous. At other times an extensive discharging eruption is thrown out, which occasions most distressing pruritus. Some patients present an eruption of small spots of various colours and the size of flea-bites, at first on the face, and then over the whole body, which appear and recede alternately, particularly under the influence of a damp atmosphere. These *spots are mostly insensible, and may be pricked with a needle without causing the least pain.* When they ulcerate, a viscid matter runs from them, and they soon become covered with scabs or scales ; a serous ichor often flows from them, which inflames and ulcerates the neighbouring parts with which it comes in contact. These eruptions are accompanied or followed by coppery or lead-coloured tubercles, which are evolved, first on the face, and after-wards over the whole body. By degrees the skin of the forehead thickens and wrinkles, the eyelids tumefy, the cheeks swell, and assume a deep red colour, the lips, swelled and stretched out, give an unnatural size to the mouth. The external ear is folded and convoluted, the eyes are surrounded by a red circle, the look be-comes oblique and menacing,—in one word, the face is so hideous, that it inspires universal disgust and horror. The tubercles once formed, present scales, scabs, and ulcers, on their summits. On examining the posterior fauces, *tubercles* are seen on the uvula, tonsils, and velum palati, to which foul ulcers succeed. These ulcers have callous, hard, uneven edges, from which a reddish and fetid matter is discharged, which dries up into reddish or brownish crusts. The skin situated between the ulcers is also often intersected with ridges or crevices, and the *hair with which it is covered falls out.* The violent pains in the limbs decrease, and sometimes cease entirely, as soon as the skin is affected. Arrived at this pitch, the disease still continues to advance. The ulcers, after having destroyed the integuments and soft parts, extend their ravages to the bones : the discharge is now extremely copious and of a most unbearable odour. Flaps of fungous flesh are detached from the bottom of the ulcers, caries attacks the bony palate, vomer and ossa nasi, the voice changes and becomes weak, articulation is difficult, the hair of the head and of all the other parts of the body, eye-brows, &c. falls off; sometimes even the phalanges of the fingers drop. Patients are at the same time said to have a most voracious appetite and unquenchable thirst. These phenomena are regarded as the forerunners of approaching death. This happens when the strength is entirely exhausted by night sweats and colliquative diarrhœa.

As to the treatment of radesyge, much the same course has been recommended as that followed in elephantiasis, or in cases of syphilitic affection, according as it has been held analogous to the first or to the second of these diseases.

Lepra of Holstein. (Spedalsked.)

Doctor Struve assigns the following characters to this disease the face swelled, sallow and shining; loss of the hair, eyebrows and eyelashes; swelling and change of colour in the nose; the tongue tubercular, the lips swelled and hard ; alteration of the voice and difficulty of breathing. Struve looks upon the lepra of Holstein and radesyge as the same, or as mere varieties of the same affection ; he assimilates them to the lepra of the middle ages. Struve, *Ueber die 'aussatzartige Krankheit Holstein's, allgemein daselbst die Marsch-Kranheit genannt,* 8vo. 1820. Extract, *in Edinb. Med. and Surg. Journ.,* vol. xviii. p. 92

Disease of the Crimea, or Lepra of the Cossacks.

Pallas, Gautier, and Von Martius have described, under this name, a disease which prevails extensively among the inhabitants of the Crimea, and which was introduced, as is said, by the Russian troops engaged in the war with Persia.

According to Martius the lepra Taurica, with or without fever, is announced by the appearance of a great number of livid spots or flat indolent tubercles on the face, trunk, and extremities, principally on the radial side of the wrist. At a later period (according to Martius, in the following year) the spots increase in number and size, and become of a blackish or brown colour ; every part of the body except the skin of the hands, and the bends of the joints may become covered with these spots ; at this period the spots are not painful ; the voice is occasionally hoarse, and the patient is restless and depressed. Still later (about the third year) a feeling of itchiness arises in the parts of the skin affected, similar to that occasioned by the bites of ants. The tubercles now became true flattened tumours, some of which are indolent, whilst others are the seat of intolerable itchiness. The shape of the body and face alters ; the face swells, the lymphatic glands tumefy, and at this stage of the disease great weakness of the internal organs and considerable prostration are observed. In the fourth year great pain in the limbs, and particularly in the joints, comes on ; the sleep and appetite, which up to this time were unaffected, fail ; the strength decreases by degree, the spots and tumours assume a reddish brown hue, becomes hard, rough, and covered with squamæ. Scirrhous lumps are seen under the skin of the face and extremities, and under the tongue. In the *fifth year*, the swellings begin to give way, and, particularly on the feet, are followed by ulcers of bad character, from which a fetid sanies is discharged, or which are covered with thick scabs. These ulcers have been seen to involve the loss of all the points of the fingers in succession. Tormented by a sense of insupportable heat in other parts, the patients by scratching give rise to ulcers, even of a more serious nature than the first. The manner in which the natives treat these ulcers, farther, renders the slightest wounds of the greatest consequence. Finally, (sixth year) the cheeks, lips, velum palati and tongue, are corroded by ulcers, which often form on the inside of the nose, and in the trachea, when they occasion great interruption to the breathing. The nails by this time have generally been lost or are much altered. The viscera become more and more diseased, and death at length concludes the patient's misery. Martius's (Henricus,) *de lepra taurica, specimen medico-practicum,* 8vo. Lipsiæ, 1806.

Lepra Anæsthesiaca of India.

Robinson gives the characters of this disease in the following terms :

One or two circumscribed spots or patches of a deeper colour than that of the skin around them, appear on the feet or hands, and sometimes on the trunk and face ; these spots are neither prominent nor depressed, they are shining and wrinkled ; the wrinkles do not run into the surrounding healthy skin. The spots extend slowly until the skin of the legs and arms, and by degrees that of the whole body, when the disease is so extensive, is totally deprived of feeling. No perspiration takes place from the surfaces affected, neither are they itchy, nor painful, and it very seldom happens that they are swollen. In a more advanced stage of the disease the pulse becomes very slow, (fifty to sixty pulsations in a minute,) and soft without being small ; constipation of the bowels follows ; the toes and fingers are benumbed as though with cold, shining, slightly swelled, and somewhat stiff. The patient is indolent, slow in understanding the questions put to him, and seems to be constantly half asleep. The soles of the feet, and palms of the hands present hard and dry cracks; a furfuraceous matter is deposited under the nails, which raises them and occasions the skin around them to ulcerate. The legs and forearms swell; the skin is every where rough and chapped ; at the same time ulcers form on the metacarpal and metatarsal articulations of the fingers, and toes, in the line of flexion, and in the corresponding parts of the articulations of the trunk, without any evident tumefaction or pain ; pieces of skin half an inch in length become gangrenous and fall off, leaving the pale and flaccid muscles bare ; these in their turn mortify, and by and by are also cast off. Different joints may be thus attacked and destroyed in succession, by the slow, but uninterrupted progress of this terrible disease, which renders those who are affected with it, objects of horror to all who approach them. The pains in this affection are not insupportable ; the appetite is unaffected, and patients, horribly mutilated, sometimes live long without appearing to be disgusted with life. They are finally carried off by dysentery and diarrhœa. Robinson assures us, that although *tubercular* elephantiasis sometimes shows itself during the course of elephantiasis *anæthesiaca*, it is not necessarily consequent on it. (*Transact. of the Medic. and Chirurg. Society of London*, vol. x.)

Jewish Leprosy. (Saraat.)

Moses (xiii. and xiv. chap. of Leviticus), in his laws concerning leprosy (Saraat), points out the signs or marks by which the Jewish priests might recognize it.* Dom Calmet has given the following summary of these: the first indication is an outward tumour; the second is a pustule or abscess; the third a whitish or red and shining spot, to which the epithets *white, brilliant*, are often applied. Whoever presented one or more of these marks, was shut up for seven or fourteen days. The certain signs of leprosy were, first, a whitish, reddish, and shining spot; second, the hair pale and red on the same place; third, the part more deeply sunk than the rest of the skin.

A simple white spot was not sufficient to cause a man to be declared a leper; it was necessary that this should grow and increase. When the whole body was white from the head to the feet, it was a *pure leprosy*, and when the flesh was covered with white tumours, and the hair, on the parts where these were seen, had changed colour and become white, it was an *inveterate* leprosy, and rooted in the skin.

If in a cicatrice or in a place which had been burned, a white tumour or a whitish spot was seen, shining or red, more depressed than the surrounding parts, and the hair of which had become fair or pale, it was a mark of a *true leprosy*. Finally, when any place was seen in the head of man or woman, more sunk than the rest, if the spot increased, it was leprosy, whether the hair changed colour or not. On the head of a bald person, spots, either whiter or redder and more shining than the surrounding parts, indicated leprosy.

The Jewish leprosy has been assimilated with several other diseases. Bartholinus, J. Leclerc, and others, associate it with tubercular elephantiasis: Hillary and Adams think that it was nothing more than the frambœsia of Africa; Bateman believes it to correspond with the *leuce* of the Greeks, with the *baras* of the Arabians, and with the third species of *vitiligo* of Celsus; Lorry and several others regard it as a distinct disease; but at the present day, possessing these scanty accounts only, it is evidently impossible to form any thing like a definite and just idea of the nature of this disease.

Historical Notices.—On the Jewish leprosy, consult Mead. (Medica Sacra, London, 1749,) Dom Calmet, (*Diss. ou recherches sur la nature et les effets de la lepre dans la Sainte Bible*, Latin

---

* Moses is the more accurate describer and better author as regards this disease. Certainly some changes have taken place in its features, but there is evidence enough remaining to show that the sacred historian described a disease identical with that we now call by common consent, "Lepra."

et Français, 8vo. 1820, t. iii. p. 19); Oussel, (*Diss. Philologico-medica de leprâ cutic Hebræorum,*) Schilling (*de Leprâ*, p. 63), and Boussille-Chamseru (*Recherches sur le véritable caractère de la lèpre des Hébreux.* Mém. de la Soc. Méd. d'Emulation, t. iii. p. 335.)

Malum Mortuum.

The *malum mortuum* observed in the middle ages, has been described by Theodoric and Joannes de Vigo. " Quædam infirmitas nascitur circa tibias et brachia, quæ *mal morto* appellatur. Sunt enim ulcera livida et sicca modicæ saniei generativa, quandoque fiunt de purâ melancholiâ naturali ; quandoque e melancholiâ cum admissione phlegmatis salsi ; si illud, cognoscitur per nigras pustulas sine pruritu ; si hoc, livescit locus cum pruritu e mordicationibus." (Théodoric, *Chirurg.*, lib. iii. c. 49.) " Malum mortuum est squalida scabies maligna et corrupta in brachiis, coxis et tibiis, faciens pustulas crustosas cum saniositate subtus ad instar lupini. . . . Sumitur per viam contagionis. . . . In signis, curis et causis plurimùm confert cum morbo gallico ; quæ uni conferunt, alteri conferre videntur . . . . Pustulæ sunt aliquantulum extra cutim elevatæ cum colore mori semimaturi . . . *scarificatione profundâ usque ad os parùm aut nihil patiens sentire videtur.* (De Vigo, *Tract. in arte Chirurg.*, etc. c. v. p. 3.)

The *malum mortuum* has been assimilated with gangrenous elephantiasis and elephantiasis anæsthesiaca.

M. Alibert (*note on the genus spiloplaxia*, malum mortuum *of some pathologists*, Revue Médicale, 1829, vol. iv. 169) has given the details of a case of disease which he met with in the course of his practice, and which he believes to be similar to the malum mortuum of the middle ages.

In two thirds of the persons who have been affected the disease appeared to me to get well spontaneously ; others still retain some of the symptoms.   It has not as yet been fatal to any one.   A man named Pierre Françoise Goudey, twenty-eight years of age, was the first who was attacked, it is now twenty-eight months ago. The first symptoms were general weakness and disinclinaton to work, then pains in the limbs, which lasted about two months, afterwards inflammatory and aphthous swellings of the lips and inside of the mouth, for nine months ; at the same time inflammation occurred in the posterior fauces, and an extinction of the voice supervened which lasted for three months ; an inflammatory affection also appeared on the scrotum, which the patient attributed to the rubbing of his drawers of coarse new linen cloth.   No vestiges now remain of the disease, nor of any of its symptoms.   Goudey communicated the affection under which he laboured to his three young children, all of whom suffered from the swelled and aphthous state of the lips ; one only from the inflammatory symptoms of the throat, and the hoarseness.   His wife, with whom he cohabited, was the only individual of the family who did not catch the infection from him, which would seem to indicate that the union of the sexes is a means very little apt to communicate this disease, although it is regarded as a peculiar modification of syphilis. Goudey having been made prisoner, and detained for three days, by a troop of Austrians, at Mountbéliard, at the time of the second invasion, pretends that he contracted the disease by drinking out of the same glass immediately after a soldier of that nation, who, he said, had the disease on the lips. It was not until some time after his return home that Goudey felt the first symptoms of his malady. Elizabeth Goudey, fourteen years of age, assures me that she caught the infection from the children of the above-named Goudey, her relations, from having eaten with them; she experienced the pains in the limbs, swelling of the lips, inflammation of the throat and loss of voice. Her brother, Claude Françoise Goudey, about fifteen years of age, contracted the disease some time after his sister, and experienced the same symptoms, with ophthalmia, in addition, which lasted several months; the eyelids, indeed, are still slightly injected, and the eyes watery. The wife of Jean Baptiste Goudey thinks she caught the disease from Elizabeth Goudey, from her being so frequently in the house, and going often there to her meals. The disease shows itself in her by very intense pains in the limbs. These pains began in the lower limbs, and attacked, successively, the shoulders, elbows, and wrists, and lasted about five months. She had had a general pustular eruption, though the head was the part more particularly affected; the marks of this still exist in the shape of spots of a livid red. Her lips were not inflamed, but she had aphthæ on the tongue, and inflammation of the throat, which, as well as the hoarseness, was still complained of. The husband of this woman took this disease six months after her; he only suffered from pains in the limbs for a fortnight; his throat is in a high state of inflammation, and this is accompanied by an almost complete extinction of voice. The inhabitants of Chavanne are persuaded that this disease is particularly propagated through the medium of the implements used in taking food, which is the more likely to be the case, as it is known that country people make use of these one after another without any attention to cleanliness. The following cases seem to strengthen these opinions . . . . . .

"Since the month of March I have had those individuals under my care who still showed any of the symptoms of the disease; I recommended the use of the warm-bath, and of tonics, and mercurial preparations, particularly the liquor of Van Swieten, internally. I had the satisfaction of finding that this treatment succeeded, and of seeing the disease entirely disappear from Chavanne, without having been communicated to the adjoining districts." (Journal compl. du Dict. des Sc. Méd. t. v. p. 134.)

Disease of the Bay of St. Paul. (Canada)

Between the years 1776 and 1780, a disease, which has been designated under the names of the *Disease of the Bay of Saint Paul, le mal de chicot, le mal des éboulemens*, appeared in Canada, particularly in the bay of St. Paul. According to Doctor Bowman,[*] who was sent by Governor Hamilton to investigate the nature of the disease, it was announced by the appearance of a number of small pustules on the lips, tongue, and inside of the mouth. These pustules, which resembled small aphthæ, advanced rapidly ; and children have been seen whose tongues were almost entirely destroyed by them. The whitish and puriform matter they contain communicates the infection to those who touch it. Patients are tormented with nocturnal pains in the bones, but these generally subside when ulcers appear on the skin and in the interior of the mouth ; cervical, axillary, and inguinal bubos are often met with ; at a more advanced stage, the body becomes covered with pruriginous tetters, which soon disappear. The bones of the nose, palate, cranium, pelvis, thighs, arms, and hands, become affected with nodes and caries : all the functions become greatly disordered ; the senses are disturbed, and patients die a prey to the most acute sufferings. Some individuals, however, are so robust that they stand out against this complication of infirmities for many years, dragging on a most miserable life ; entire limbs have been sometimes known to sphacelate and fall off. This frightful disease spares no one, but it seems to rage with peculiar virulence among children. It is, above all, by the sexual act, that it is communicated or transmitted from one individual to another.

Decoctions of the roots of patientia, of arctium lappa, and sarsaparilla, are the remedies usually employed to arrest its progress. A decoction of a species of fir, or beer, made with a decoction of the branches and bark of the pine' of Canada (Pinus Canadensis) has also been approved. The inhabitants of some parts of Canada, and among others, those of the Bay of Saint Paul, where the disease spread extensively, pretend that it was brought among them by the English. The peculiarity most worthy of remark connected with the history of this disease is, that it rarely attacks the organs of generation, and that it may be contracted without any actual intercourse with individuals affected with it, even without touching them immediately.

Swediaur observes, that Dr. Bowman's description, however imperfect it may be, recalls to his mind the account which the writers of the fifteenth century have given of syphilis. The similarity to Scherlievo is still more striking.

Disease of Fiume, or Scherlievo.

This epidemic disease, the origin of which was attributed to four sailors, who were supposed to have brought it from Turkey, appeared in 1800, in the districts Scherlievo, Gronemico, Fiume, &c. It was supposed by others, again, to have been imported in 1790,

---

[*] These details are from Swediaur, who seems to have mistaken the name of Bowman for that of Beaumont, and that of Hamilton for that of Haldiman. (Adams, *Obs. on Morbid Poisons*, p. 194.)

from Kukulianova, by a peasant named Kumsut. A short time after his return, his father and mother were affected by it, and afterwards propagated it in Scherlievo, &c. The disease spread with so much rapidity in 1801 in the provinces of Buccari, Fiume, Viccodal, and Fuccini, that in a population of from fourteen to fifteen thousand individuals, it was calculated that more than four thousand five hundred were affected with it. MM. Percy 'and Laurent assure us that a commission of physicians, appointed in September, 1801, found more than thirteen thousand persons affected with this disease, out of a population of thirty-eight thousand. It re-appeared in 1808 and 1809, raging especially in Scherlievo, where it seemed to be kept up by the filth of the lower orders of the people, whose damp cabins are shared with their domestic animals.

This disease usually commences with lassitude of the limbs, and pains in the bones of the arms, thighs, and spine, which increase during the night ; the voice soon becomes hoarse, and deglutition difficult ; the face is flushed, the velum palati, the uvula, the tonsils, and sometimes the pharynx and larynx are red. Soon after, a species of aphthæ burst, and discharge an ichor, which erodes the neighbouring parts ; small ulcers are formed, which unite and create a sore of various dimensions, but always of a round shape, of an ashy colour, and with hard, raised, and dark red edges. These ulcers, which are in some cases evolved with great rapidity, cover the uvula, the tonsils, the velum palati, and the surface of the cheeks and lips ; caries affects the bones of the nose, when pus, of unbearable fetidity, is discharged ; the voice changes more and more, and is at last entirely lost. The exostoses which had appeared from the beginning, occasionally but rarely shrink and vanish, along with the pains which accompany them, as soon as a pustular eruption is evolved upon the skin. Dr. Lambini, however, relates four cases, which prove that the pains in the bones became more violent, notwithstanding the treatment employed, and lasted throughout the whole course of the disease.

When Scherlievo commences by a pustular eruption, M. Boué says that it is announced by violent itchiness, which lessens as the eruption comes to an end. The pustules are of a coppery colour, round, and of various extent. They most frequently appear on the forehead and hairy scalp, but they are also seen on the inner surface of the thighs, legs, and arms, and round the anus and genital organs. An acrid ichor sometimes flows from them, which inflames the skin ; at other times this discharge dries and forms scabs ; the disease often remains stationary in this state. After the scabs have fallen off, the skin retains marks of a coppery hue, which it is difficult to remove.

Scherlievo has been known to begin with various sized blotches of a coppery colour, in the centre of which ulcers are seen, from which a matter is poured out, that by drying forms scabs similar to those which cover the pustules. These blotches are generally surrounded by an areola of a coppery hue, and give the patient a most

hideous aspect.   It is related as a fact worthy of notice that the
genital organs in women are more frequently the seat of this disease
than those of men.   Doctor Cambieri, among the immense number
of cases which came under his notice, only found one of gonorrhœa
which came on after the desiccation of the pustules of the skin, and
which disappeared as soon as the eruption was restored.   As for
the ulcers which so frequently erode the scrotum, they always
appear as secondary to the general infection.   '

The transmission of Scherlievo is seldom the consequence of
sexual intercourse, but is the effect of simple intermediate contact :
the clothes, table utensils, such as glasses, spoons, forks, napkins,
&c., and an atmosphere charged with the breath of those infected,
are all sufficient to sow the seeds of it.   Children have been known
to bring the disease with them into the world, or to have it communicated by the nurses who suckled them.   It hardly ever appears by bubos in the groins, or enlargements of any of the other
lymphatic glands.

When this disease appears in the form of pustules, spots, or
ulcers in the mouth, it yields readily to antivenereal remedies.   The
prognosis is more unfavourable when patients have been weakened
by fruitless treatment, or by previous complaints ; when the ulcers
have reached, and occasioned caries of the bones, or when the
patients are debauched, indulging in intemperance and neglect of
personal cleanliness.   The treatment of Scherlievo does not in any
particular differ from that of syphilis.   I have been assured that the
bi-chloride of mercury given in the syrup of Cuisinier (comp. of
senna and sarsaparilla) has always proved the most effectual means
of subduing it ; and that, when caries had attacked the bones, the
treatment might be concluded with advantage by ten or twelve
mercurial frictions.   Opium combined with mercury is employed
with complete success against the pains in the bones.   The protochloride of mercury, mixed in the cerate with which the ulcerated
pustules are dressed, and the solution of corrosive sublimate diluted,
used as a gargle or wash to the ulcers of the mouth, always expedite
the cure.

Messrs. Percy and Laurent have proposed the establishment of
a Lazaretto and the disinfection of the dwellings and clothes of
the poorer classes by chemical agency, as a means of getting rid of
the disease of Scherlievo entirely (Percy and Laurent *Dict. des Sc.
Méd.* art. Mal. de Fiume).   But if, as all seems to authorise us in
believing, Scherlievo be only syphilis under another name, this
recommendation would require to be modified.

Facaldine.

Under this name is designated a variety of syphilis, which is
said to have been introduced, in 1786, into the village of Facaldo, in
the province of Bellune, bordering on the Tyrol, by a female mendi-
cant labouring under *venereal itch,* and ulcers and warts **on the**

vulva. The following characters are assigned to Facaldine : Scabious eruption of a syphilitic nature, which attacks adults and children, ulcers in the throat and nasal fossæ, destruction of the nose, serpiginous ulcers, which erode the skin in various directions ; gummy tumours are seldom seen ; pains in the bones occur very rarely, and exostosis hardly ever. In adults blenorrhagic discharges take place from, and ulcers occur on the genital organs ; bubos and several species of excrescences are also frequent occurrences. Cure by means of mercurial medicines.

Marcolini has related a case, and given a representation of Facaldine, which he considers a variety of Scherlievo. Marcolini, Memorie Medico-Chirurgiche, Milano, 1829, p. 18. See besides Zucchinelli, Ann. Univers. di Medicina. Milano.

Morbus Brunno-Gallicus. (Moravia.)

An epidemic appeared in Brünn, in 1578, which, in the space of two or three months, attacked forty persons in the city, and almost a hundred in the suburbs ; a considerable number of the country people were also affected. This disease presented symptoms similar to those of syphilis. The disease was generally supposed to have been propagated by baths, and the practice of cupping, which is in common use among the inhabitants.

Thomas Jordan, the historian of this epidemic, thus describes its characters : "Interim insuetâ quâdam ignaviâ, seu torpore gravati : pigri, segnes, inertes ad consueta munia obeunda, animo quoque abjecto, tristes vultu, cùm nec mens neque manus et pedes officium facerent, veluti umbræ, non homines, passim oberrantes conspiciebantur. Nativus faciei color in pallidum : vigor ipse oculorum in torvum, circulo fusco sicut mulieribus menstruatis deformem, subitò immutatus : frons exporrectior in caperatam et nubilam degeneraverat. Manifestis tum se prodidit indiciis. Vestigia cucurbitularum turgescentia, extemplò ardor invasit immensus et immedicabilis, quem fœdi abscessûs et ulcera excepere putrida, sanie taboque fluitantia ; circumcirca pustulæ, palmum quoque latæ, achoribus floridæ, quibus dehiscentibus acu aut medicamine discissis, profluxit pituita tenuis, serosa, marcida, sanieque mucosa : aliis etiam acris et erodens : tum caro cucurbitæ ambitu circumsepta, corrosa, putrescens, tetrum ut è telephiis ac phagedænicis ulceribus invexit fœtorem. Ubi admiratione dignum initio, quòd è tot affixis cucurbitis, cùm alii decem plus minùs, tres quoque tantum nonnulli apponi jussissent, una duntaxat, aut ad summum duæ (socrui Laurentiæ sartoris è quindecim, tres) ex iis omnibus in fœdam transiverint vomicam. Nonnullis universum corpus pustulis conspersum, facies informis, triste supercilium, horrendus vultus, dorsum, pectus, abdomen, pedes, loci à summo ad imum porriginosâ scabie, crustaceis ulceribus supra cutem paulum elevatis, duorum cruciatorum nummum, vel unguis pollicis amplitudine, ambitu rubente, candidâ superficie (ut tinea quam barbari vocant) polluta, et deturpata, cernere erat. Manabant hæc quoque pingui liquore, mucore lento, qui non pus, sed saniem refe'rret luridam. Imò, scabie sublatâ et sanatâ, maculæ atræ, diversæque ab impetiginibus, aut vitiliginibus, plumbe, et fusci coloris, remanserunt. Progressu morbi, in capite calli concrevere, qui summo cum dolore et ejullatu rupti vel dissecti, melleum quippiam, resinosum et tenax, ceu ex conferis arsoribus laticem extillare videmus, lentum inquam et viscidum, marescentis pituitæ argumentum, exudabant. Abscessus hi sordidi verèque cacoëthes, postquam magnâ difficultate expurgati, et non minore carne rursum productâ, coaluissent, novum prorupit symptoma. Universi corporis actus, brachia, scapulæ, cubitus, humeri, suræ, tibiæ, pedes imi, puncturis quibusdam quasi aculeis intensissimè vellebantur, ac si ferris descinderentur, aut forcipibus ignitis laniarentur (sic ægri sensum doloris exprimebant) potissimùm ubi tibia maximè excarnis, lacertorum non fulta thoris, à solo periosteo vestitur. . . . . Nulla tamen quies, perpetua vociferatio, lachrymæ, gemitus, indesinenter torquentibus doloribus, nocte potissimùm, cum fessa membra sopore dulci reficiuntur, illis noctes pervigiles, ob cruciatuum vehementiam," &c.

Various remedies were tried for this disease : but the following method seems to have been generally successful. After having bled the plethoric patients, and exhibited some purgative medicine, decoctions of guaiacum, turbith mineral in pills, and the expressed juices of wild endive and fumitory were administered, whilst the ulcers were dressed with mercurial ointment.

Amboyna Pustule, or Amboyna Pox.

J. Bontius (Medicina Indorum, 4to. Lugd. Batav. 1718 ; De tophis, gummatis ac exulcerationibus endemiis in insulâ Amboynâ ac Moluccis præcipuè quas nostratas *Amboynse pocken vocant*,) gives the following description of a disease, endemic in the island of Amboyna ; —

"Endemius, seu popularis quidam morbus in Amboynâ, ac Moluccis insulis præcipuè oritur, qui symptomatis suis admodum similis est morbo venereo. Sed in his inter se differunt quod hic sine congressu venereo quoque nasci solet. Erumpunt in facie, brachiis, ac cruribus tophi, seu tumores, duri primùm ac scirrhosi ac tam crebri per universum corpus, quam clavi ac verrucæ oriuntur in manibus ac pedibus in patriâ ; si verò eos ulcerari contingat, materiam lentam, ac gummosam à se reddunt, attamen tam acrem, ac mordacem, ut profunda, ac cava ulcera inde oriantur, cum labiis callosis, ac inversis ; fœdum ac deforme malum, et cum lue venereâ conveniens, nisi quòd hic tanti dolores non adsint, nec caries in ossibus tam facilè oriatur, nisi per curantis incuriam. Hic affectus originem trahit, primùm ex peculiari cœli et soli istius genio ; tum ex aere, vaporibus salsis, è mari undique ascendentibus infecto ; cibis præterea crassis, ac melancholicis et pituitosis, ut sunt pisces marini, quorum hìc magna captura est, quibus incolæ assiduo escuntur, quod reliquæ annonæ sit satis indiga regio. Magnum etiam momentum huic malo adfert usus placentarum, quas, vice panis, per totum istum tractum edunt, ac ab incolis *sago* vocatur, et est è corticibus arborum excussa farina. Ad hæc confert potus importunus liquoris cujusdam saguër vocati, qui fermè eodem modo ex arbore elicitur, quo è palmâ indicâ seu coquos arbore liquor iste, quem incolæ Towac, Lusitani Vinho de Palmâ vocant. Hic immoderatè sumptus non secus ac vinum et cervisia inebriat ; caput ac nervos infestat, hinc etiam in his insulis crebrior est ista paralyseos species, quam *Beriberii* supra diximus appellari. Quantum ad curam attinet, ea, si recens sit hoc malum, non admodum difficilis est. Sin inveteratum, jam molestior est curatio. Porrò iisdem fermè remediis cedit, quibus lues venerea, obstructiones lienis, leuco-phleg-matia, ac ipse hydrops, et cæteri chronici ac rebelles morbi. Decocta hìc itaque parentur è Chinæ radice, salsparilla, Guajaco et corticibus ejusdem, quibus incoquantur anagalidis aquaticæ, seu beccabungæ, *m. ij.* Post peccans materia vehementioribus catharticis educenda est : nam levia hìc non possunt. Talia sunt extract. Guttæ cambodja, elaterium ; et si his non cedit, ad chymica, et mineralia deveniendum est : ut sunt mercurius vitæ, seu butyrum antimonii, turbith minerale, tum mercurius præcipitatus albus. Unguenta quoque mercurialia secundum artem parata hìc externè adhibenda sunt."

This disease very much resembles Scherlievo, which, it would seem, ought to be assimilated with syphilis.

Sibbens.

This disease has been observed in Scotland, particularly in the counties of Ayr, Galloway, and Dumfries.

According to Gilchrist, *sibbens* shows itself under several forms. Sometimes inflammation of the velum palati and surrounding parts takes place, and is accompanied with a kind of white eschar, or a superficial ulcer of a bright red colour. At the same time aphthæ or small white spots or eschars often occur on the velum palati or insides of the cheeks. Small elevations, of a pearly or milky colour, also usually appear on the commissures of the lips. Often, too, a very small excrescence or *fleshy growth* is developed, resembling a raspberry, and which becomes covered with a scab. This growth is almost a certain indication of the disease, even when the sore throat does not exist. Dr. Trotter, who has also described this affection of the mouth, compares its appearance to that of toasted cheese.

Another form of this disease is that of destructive ulceration, which may cause the entire loss of the velum palati ; and the death from inanition, of infants at the breast, deglutition becoming impossible.

Sibbens appears in other cases *on the skin,* and under different aspects. Sometimes the whole surface of the body is spotted and clouded with a coppery and dusky red blush. At other times clusters of pustules appear, over which several successive desquamations of the epidermis take place. *Scabby eruptions* of the hairy scalp, forehead, inner sides of the thighs, &c. are accompanied with little hard lumps in the thickness of the skin, and a feeling of unpleasant itchiness. At other times, a species of tumour, similar to furuncles, are seen on the arms, shoulders, face, legs, and feet, which give rise to ulcers that perforate the whole thickness of the skin, and lay bare the muscles, which they sometimes also corrode. Adams is disposed to believe that these ulcers are the result of the immediate contact of the virulent matter proper to this disease.

Finally, the soft and spongy *raspberry tumours*, which have been mentioned (whence the name of *sibbens* or *sivvens*, which is derived from sibbens *frambœsia*), are the last symptoms of the disease ; they do not seem to occur in all places alike, for several other forms of the disease are observed, several indeed, which Gilchrist himself had never seen.

According to the same observer, the bones are not affected in this disease ; Bell, on the contrary, speaks of *nodes* and *caries*.

Sibbens is rarely communicated by sexual intercourse ; the alterations which are sometimes seen in the genital organs take place consecutively. The disease is more frequently transmitted by nursing, and the common use of the same utensils, the use of the same *pipe* for instance.

The silence of contemporary writers on this disease leads me to imagine that it is now extinct.

The treatment adopted for sibbens, bears the greatest resemblance to that employed in syphilis.

Pian de Nerac.

(Departement de Lot et Garonne).

Raulin has described an epidemic disease which seemed to be analogous to Scherlievo and Facaldine, under the objectionable name of *Pian de Nerac :*—

"At the end of the month of June, 1752, a singular epidemic disease appeared at Nerac ; it was a species of lepra or frambœsia (*pian*), similar to that which affects the negroes in the gulph of Mexico. It spreads among children at the breast ; those affected by it begin to fall off : by degrees pustules appear on the face, mouth, neck, buttocks, and thighs. Nurses contract this eruption on the breasts, and it afterwards appears over the whole body. The pustules are generally round, hard, and rather callous ; from some of them a yellowish ichor is discharged ; others are covered with a pulverulent crust ; these pustules covering the body, become confluent, and appear to form only a single incrustation ; they degenerate into deep ulcers which lay bare the bones, and occasion death ; towards the end of December it was calculated that more than forty infants had already been affected with this disease. The treatment that was most successful was the use of an ointment made with one ounce of pure mercury, rubbed till the globules had disappeared in Venice turpentine, one ounce of lard, and one scruple of camphor, well mixed together. Mercurial frictions were tried upon several women, but mercury without camphor was often ineffectual. Children were cured in a fortnight, but it was necessary to continue the treatment for several days afterwards. The origin and cause of this disease were entirely unknown."

Frambœsia, or Yaws. (Pian, French.)

The identity of *yaws* and *pian* seem to be demonstrated, although some differences appear in the exposition of the symptoms of the two diseases, given by the English observers who have studied *yaws* in Guinea and in Jamaica, and by the French physicians who have observed *pian* in Saint Domingo, Guadaloupe, Brazil, Jamaica, &c ; these differences, however, seem at the most but to characterise two varieties of the same affection.

Yaws begin by a state of languor and weakness, with pains in the joints, and in most cases with fever, which runs highest among children. Before the eruption, the skin is often covered with a white dust (Thomson), as though it had been powdered. Some days afterwards, spots similar to flea-bites or small papulæ appear on the skin, particularly on the forehead (Thomson). These elevations increase in size during from six to ten days ; at the end of this time, a scab is formed on their top, from under the edge of which a crude purulent matter is discharged.

The size of the pustules still increases, and they become covered with irregular scabs not very adherent. Many of these pustules acquire the size of a shilling. On raising the scab, an ulcer of a bad or gangrenous description is discovered. It does not assume the fungous appearance which it ultimately acquires, at any definite period of the disease : this sometimes occurs a month after the breaking out of the eruption, and sometimes three months afterwards ; the development of these fungous growths seems to depend very much on the constitution of patients, taking place earlier in those who are lusty and well fed. A second eruption sometimes makes its appearance preceded by fever, the progress of which is the same as of that which has gone before it, so that the eruption may occasionally be seen in different states in the same individual. The elevations are broader and more numerous on the face, groins, axillæ, verge of the anus, and labia majora, than on any other part of the body. New eruptions take place as soon as the first begin to dry off, so that after these successive eruptions, the number of pustules existing together is sometimes very considerable. If the writers who have studied this disease are to be credited, there is always one pustule larger and more raised than the others, which it is more difficult to cure, and which is designated by the name of the *mother yaw.*

This affection is said to be sometimes accompanied with nocturnal pains, and swellings of the bones, ulcers in the pharynx, &c. In this last case the disease becomes very distressing ; the ulcers of the throat resemble toasted cheese ; they never become fungous, and a great part of the palate is very apt to be destroyed by them.

The fungus of framboesia in patients of a good constitution is red, like a piece of flesh ; in delicate or diseased subjects, it is white like a piece of cauliflower ; it bleeds on the slightest touch ; in the latter case it is less raised than in the former.   After having continued some time in this state, the fungus gradually diminishes in dimensions and in height, shrinking and generally disappearing without leaving any scar, except in places where the inflammation has been very great.   (Thomson.)   The cicatrices that then result resemble those of cow-pox, only they are larger and more superficial.

In the successive eruptions of yaws, there is often one pustule which does not heal like the rest ; left to itself it is apt to produce caries of the neighbouring bones.

Yaws appear under a different aspect in cachectic persons : the pustules are smaller, the eruption which is less copious is successive ; the fungous growths which usually characterise the disease, either do not exist all, or are very small and *watery*.

The length of time during which this disease continues is uncertain ; in some patients it lasts six months, in others a year ; in general its term may be stated at about eight months.   Fever, and the symptoms generally, are mostly very well marked in weakly and ill-fed children.

We are informed that framboesia usually occurs among the ill-fed negroes, whose skin is continually irritated by a burning sun, by the bites of insects, and by the rancid oils with which they are in the habit of anointing their bodies.   This disease is contagious, is transmitted by the union of the sexes, by suckling, or by the application of the matter of the pustules or cutaneous ulcers, to any part of the excoriated skin ; perhaps it is also transmitted in still other ways.

It is difficult to determine the time of *incubation*.   Thomson relates that a certain number of negroes in good health were sent with their children to a sugar farm in a mountainous district, and the healthy children having eaten and drunk with those at the farm who were affected with yaws, three of the former were, seven weeks afterwards, attacked with fever and pains, followed by a general eruption ; the others were not affected until three weeks later ; at the end of eight months all were cured.   Thomson inoculated a child in five places with the matter of an ulcer from which the scab had been removed.   Three of the punctures healed ; the two others looked like simple scratches during three weeks, when they formed into small ulcers, which spread until they had assumed a gangrenous appearance with jagged edges, &c.   Seven weeks afterwards papulæ appeared on the forehead, and by-and-bye extended over the whole body ; the *fungus* now formed ; the patient had an abundant eruption which lasted nine months. The two ulcers which followed the punctures never became fungous, but they left deep scars. Thomson relates that variolous matter having been taken from a little negress affected with yaws, the child, who was inoculated with this pus, had the small-pox in a very mild form, and was afterwards attacked with yaws. The blood of a negro covered with yaws, was inoculated upon four children, in five different places, without producing the disease.

The usual progress of cow-pox, small-pox, and varicella, is not impeded by the existence of yaws.

The matter of yaws has not been found to occasion any eruption in rabits or dogs, even after repeated inoculations.

The disease can only be communicated once to the human subject. A woman, however, who suckles a child affected with yaws may have the breast excoriated; death has even been known to follow this circumstance ; phagedenic ulceration of the breasts, which extended to the axillæ and was accompanied with great irritation, took place and proved fatal.

If a person affected with an ulcer of considerable size, contracts yaws, it may happen that this ulcer continues, and presents the granulated aspect characteristic of yaws without the eruption appearing. If this ulcer be healed, an eruption similar to that which occurs in the usual course of the malady is developed ; left to itself the ulcer gets into a progressively worse and worse condition.

Thomson says, that an old Scotch physician who had long been familiar with yaws in Jamaica, was struck on his return to Scotland with the identity of sibbens and this disease in most of their phenomena.

Yaws is regarded by some writers as a modification of syphilis, by others as a peculiar disease of the skin.

According to Thomson the usual practice in the island of Jamaica is to leave the disease to the efforts of nature. Good food is recommended, and moderate work : sulphur, a decoction of the sudorific woods, and antimonial preparations are exhibited in the cases of children with success. A great many diseases are generally attributed to the *dregs* of yaws. Thomson thinks that the number of these has been greatly exaggerated.

Finally, Thomson is convinced that though mercurial preparations may cause the symptoms to disappear in the course of a month, they generally re-appear at a later period, and with greater violence than before. Hunter had already declared against the use of mercury in Frambœsia.

The remaing affections alluded to by Rayer, are,—
1. Acrodynia. (Page 1176.)
2. A disease in Melada, an island in the Gulf of Venice.
3. Carate. A disease peculiar to the countries bordering the Cordilleras.

4. Pinta. A discolouration of the skin, formed of blue spots, the brief details of which are extremely curious.

5. The Ignis Sacer of the middle ages.

6. Sudor Anglicus, of which this author observes that the description by W. John Kaye, or Caius, is the best extant, which he forthwith gives in the original Latin of that learned individual.

7: Epinyctis. The description and history of which consists of a quotation from Hippocrates and Celsus.

8. The Mentagra of the Latins (a quotation from Pliny).

9. The Waren of Westphalia (a quotation from Henry à Bra !) &c. &c.

The grease and itch of animals, with an article on glanders, culled from the English journals, conclude M. Rayer's work. It would be unjust to omit to say, that much amusement and information may be obtained by the curious reader in the perusal of these articles, for they bear the character of research and inquiry which belongs to every part of the work.

THE END.

[*Philadelphia, July,* 1841.]

THE

SELECT MEDICAL LIBRARY

(*NEW SERIES*)

AND

𝖅𝖚𝖑𝖑𝖊𝖙𝖎𝖓 𝖔𝖋 𝕸𝖊𝖉𝖎𝖈𝖆𝖑 𝕾𝖈𝖎𝖊𝖓𝖈𝖊.

EDITED BY

JOHN BELL, M.D.

LECTURER ON MATERIA MEDICA AND THERAPEUTICS, FELLOW OF THE COLLEGE
OF PHYSICIANS OF PHILADELPHIA, MEMBER OF THE AMERICAN PHILOSO-
PHICAL SOCIETY, CORRESPONDING SECRETARY OF THE MEDICAL
COLLEGE OF PHILADELPHIA, ETC., ETC.

A PERIODICAL publication so decidedly and obviously beneficial, to the profession for whose use it is intended, as the *Select Medical Library, and Bulletin of Medical Science*, needs little to be said of it in the way of explanation and eulogy.

No argument is required to convince the members of a liberal profession, of he necessity, as a point of honour not less than of policy, of constantly adding to their store of knowledge, from the contributions of genius and experience in different ages and countries. The proposed plan will be found eminently conducive to this desirable object. By it, both the past and the present will be made to minister to the future; and a seasonable variety offered for every taste and exigency in medical reading. To the physician living in distant States, frequent intelligence of all medical improvements, discoveries, and writings, and reprints of the best works, can be sent, through the medium of the *Bulletin* and *Library*, with nearly as much readiness, and at the same cost, as to his professional brother living in any of the great cities on the seabord. The distant reader will receive this work by mail as he would his newspaper.

The *Bulletin* will be a record of the discoveries and positive improvements in medicine and surgery, and their kindred departments—made with brevity and yet with clearness, and divested of speculative comments and discussions. It will also contain a notice of new works in medicine, at home and abroad.

In the *Library*, the editor will introduce such notes and annotations to the works published under his supervision as he may deem to be necessary.

In the *Eclectic* Journal*, the history of cases which have a definite bearing and application, summaries of opinions and practice, criticisms brief and pertinent, have found a place. The circumstances which exert an influence over the health, both of individuals and of communities, receive a due share of notice; it being as much the duty of the physician to foresee, and by timely warning to prevent, as, after its infliction, to cure disease. In the performance of this task, the more salient points have been presented,—the actual wants of the profession indicated, and all proper and practicable ameliorations, both scientific and ethical, suggested.

---

* The *first Series*, comprising four years, or forty-eight monthly numbers, was issued under the title of SELECT MEDICAL LIBRARY and *Eclectic Journal of Medicine*, at $10 per annum, a few sets of which are yet at the Publishers' disposal.

13

STANDARD WORKS

PUBLISHED IN

𝕿𝖍𝖊 𝕾𝖊𝖑𝖊𝖈𝖙 𝕸𝖊𝖉𝖎𝖈𝖆𝖑 𝕷𝖎𝖇𝖗𝖆𝖗𝖞

From NOVEMBER, 1836, *to* OCTOBER, 1837.

1. LECTURES on the MORBID ANA-
TOMY, NATURE, and TREAT-
MENT of ACUTE and CHRONIC
DISEASES. By the late John Arm-
strong, M.D.; Author of "Practical
Illustrations of Typhous and Scarlet
Fever," &c. Edited by Joseph Rix,
Member of the Royal College of Sur-
geons.

The *British and Foreign Medical Review*
says of this work:

"We admire, in almost every page, the precise and cautious practical directions; the striking allusions to instructive cases; the urgent
recommendations of the pupil to be careful, to
be diligent in observation, to avoid hurry and
heedlessness, to be attentive to the poor. Nothing
can be more excellent than the rules laid down
for all the parts of the delicate management of
fever patients: nothing more judicious than the
general instructions arising out of the lecturer's
perfect knowledge of mankind. . . . . . His prudent admonitions respecting the employment of
some of the heroic remedies, as mercury, arsenic,
and colchicum, attest his powers of observation
and his practical merits." "The pious office of
preserving and publishing his Lectures has been
performed by Mr. Rix, with singular ability."

2. OBSERVATIONS on the PRINCI-
PAL MEDICAL INSTITUTIONS
and PRACTICE of FRANCE, ITA-
LY and GERMANY: with Notices of
the Universities, and Cases from Hos-
pital Practice: With an Appendix on
ANIMAL MAGNETISM and HO-
MŒOPATHY. By Edwin Lee,
Member of the Royal College of Sur-
geons, &c.

"Mr. Lee has judiciously selected some clinical cases, illustrating the practice pursued at the
different hospitals, and he has wound up the volume with an amusing account of animal magnetism and homœopathy—those precious effusions of German idealty, for which we refer to
the work itself."—*Medico-Chirurg. Rev.*

3. A THERAPEUTIC ARRANGE-
MENT and SYLLABUS of MATE-
RIA MEDICA. By James Johnstone,
M.D., Fellow of the College of Physi-
cians, and Physician to the General
Hospital, Birmingham.

"This book cannot but be particularly useful
to those who intend to lecture or write upon the
Materia Medica; as well as to the students for
whose particular use it is prepared."—*Brit. and
For. Med Rev.*

4. A TREATISE ON TETANUS, being
the ESSAY for which the Jacksonian
Prize for the year 1834, was awarded,

by the Royal College of Surgeons in
London. By Thomas Blizard Curl-
ing, Assistant Surgeon to the London
Hospital, &c.

"This book should be in the library of every
surgeon and physician. It is a valuable work of
reference. It does not pretend to originality, for
originality on such a subject was not wanted.
But a compendium of facts *was* wanted, and such
a compendium is this volume. We cannot part
from Mr. Curling without thanking him for the
information we have received in reading his
work, and for the matter it has enabled us to
offer to our readers."—*Medico-Chir. Rev.*

5. PRACTICAL OBSERVATIONS on
DISEASES of the HEART, LUNGS,
STOMACH, LIVER, &c., OCCA-
SIONED BY SPINAL IRRITA-
TION: AND ON THE NERVOUS
SYSTEM IN GENERAL, AS A
SOURCE OF ORGANIC DISEASE.
Illustrated by Cases. By John Marshall,
M.D.

6. A TREATISE ON INSANITY AND
OTHER DISEASES AFFECTING
THE MIND. By James Cowles Prichard, M.D. F.R.S. Corresponding Member of the Institute of France, &c.

"The author is entitled to great respect for his
opinions, not only because he is well known as
a man of extensive erudition, but also on account of his practical acquaintance with the
subject on which he writes. The work, we may
safely say, is the best, as well as the latest, on
mental derangement, in the English language."
—*Medico Chir. Rev.*

7. BOUILLAUD ON ACUTE ARTI-
CULAR RHEUMATISM IN GEN-
ERAL. *Translated from the French, for
this Library,* by James Kitchen, M.D.,
Philada.

8. A PRACTICAL TREATISE ON
THE PRINCIPAL DISEASES OF
THE LUNGS. Considered especially in relation to the particular Tissues affected, illustrating
the different kinds of Cough. By
G. Hume Weatherhead, M.D., Member
of the Royal College of Physicians,
Lecturer on the Principles and Practice
of Medicine, and on Materia Medica
and Therapeutics, &c. &c.

EPIDEMICS of the MIDDLE AGES.
From the German of I. F. C. Hecker,
M.D., &c. &c. Translated by R. G.
Babington, M.D. F.R.S.—

9. No. I.—THE BLACK DEATH IN THE 14th CENTURY.

" Hecker's account of the ' Black Death,' which ravaged so large a portion of the globe in the fourteenth century, may be mentioned as a work worthy of our notice, both as containing many interesting details of this tremendous pestilence, and as exhibiting a curious specimen of medical hypothesis."—*Cyclopedia of Practical Medicine— History of Medicine by Dr. Bostock.*

10. No. II.—THE DANCING MANIA.

" Medical History has long been in need of the chapter which this book supplies; and the deficiency could not have been remedied at a better season. On the whole, this volume ought to be popular; to the profession it must prove highly acceptable, as conveying so much information, touching an important subject which had almost been suffered to be buried in oblivion, and we think that to Dr. Babington especial thanks are due for having naturalised so interesting a production. The style of the translation, we may add, is free from foreign idioms: it reads like an English original."—*Lond. Med. Gaz.*

11. LECTURES on Subjects connected with CLINICAL MEDICINE. By P. M. Latham, M.D. Fellow of the Royal College of Physicians, and Physician to St. Bartholomew's Hospital.

" We strongly recommend them [Latham's Lectures] to our readers; particularly to pupils attending the practice of our hospitals."—*Lond Med. Gaz.*

12. ELEMENTS OF SURGERY, in THREE PARTS. By Robert Liston, Fellow of the Royal College of Surgeons in London and Edinburgh, Surgeon to the Royal Infirmary, Senior Surgeon to the Royal Dispensatory for the City and County of Edinburgh, Professor of Surgery in the London University, &c.

" In the present work, an endeavour has been made, in the first place, to lay down, correctly and concisely, the general principles which ought to guide the Practitioner in the management of constitutional disturbance, however occasioned.

" The observations introduced to illustrate the doctrines inculcated, are given as briefly as is consistent with an accurate detail of symptoms and results. The descriptions of particular diseases have been sketched and finished from nature; and, it is hoped, with such fidelity, that their resemblance will be readily recognized.

" Such modes of operating are described, as have been repeatedly and successfully performed by the Author."

JOURNAL DEPARTMENT.

IN THE *ECLECTIC JOURNAL OF MEDICINE*, Vol. I., *or First Year,*

HAVE BEEN PUBLISHED,

ORIGINAL ARTICLES on Retrospection in Medicine, Clinical Medicine, Medical Education, Medical College of Philadelphia, Phrenology, &c. Numerous articles on Physiology, Pathology, Therapeutics, Midwifery, Surgery, and Hygiene, embracing new views of disease and modes of practice. *Reviews* of new Works, &c. &c.

In the Library, Second Year,

FROM NOVEMBER, 1837, TO OCTOBER, 1838,

HAVE BEEN PUBLISHED,

1. A PRACTICAL TREATISE ON DISEASES OF THE SKIN, arranged with a view to their Constitutional Causes and Local Character, &c. By SAMUEL PLUMBE, late Senior Surgeon to the Royal Metropolitan Infirmary for Children, &c. Illustrated with Splendid Coloured Copperplate and Lithographic Engravings.

" This work, which has long been a standard on Cutaneous Diseases, is, in this new edition, brought down by the Author to the present state of our knowledge; a clear compendium is presented of the recent discoveries of Chevallier, Breschet, and Vauzème. The illustrations of cutaneous disease are happily exhibited, and are essential accompaniments. It is an able, instructive, and elaborate production, and indis pensable to the medical man."—*Annals of Medicine, January,* 1837.

PLUMBE *on Diseases of the Skin.*—" This excellent Treatise upon an order of diseases, the pa

thology of which is, in general, as obscure as the treatment is empirical, has just been republished in the Select Medical Library, edited by Dr. John Bell, of this city. We hail with pleasure the appearance of any new work calculated to elucidate the intricate and ill-understood subject of skin-diseases. The late Dr. Mackintosh, in his Practice of Physic, recommends it as the ' best pathological and practical treatise on this class of diseases, which is to be found in any language.' "—*Phil. Med. Exam., Jan.* 17, 1838.

" This work is one of the most excellent on the Diseases of the Skin, in the English language."—*West. Jour. of Med. and Phys. Sciences, Jan.* 1838.

2. THE MEDICAL PROPERTIES of the NATURAL ORDER RANUNCULACEÆ, &c. &c. By A. TURNBULL, M.D.

3. THE GUMS; with late Discoveries on their Structure, Growth, Connections, Diseases, and Sympathies. By GEORGE

15

WAITE, Member of the London Royal College of Physicians.

4. A PRACTICAL TREATISE ON MIDWIFERY; Containing the Results of Sixteen Thousand Six Hundred and Fifty-four Births, occurring in the Dublin Lying-in Hospital. By ROBERT COLLINS, M.D., Late Master of the Institution.

"Several reprints of great value have already appeared in the Library—among others, Prichard on Insanity, Curling on Tetanus, Latham's Clinical Lectures, &c. The Number for the present month commences Collins's Treatise on Midwifery, a work rich in statistical details."— *Phil. Med. Exam., Jan. 17, 1838.*

"The seventeenth Number of Dr. Bell's Select Medical Library contains the conclusion of 'A Practical Treatise on Midwifery, by Robert Collins, M.D., Physician of the Dublin Lying-in Hospital.'—The author of this work has employed the numerical method of M. Louis; and by accurate tables of classification, enables his readers to perceive, at a glance, the consequences of the diversified conditions, in which he saw his patients. A vast amount of information is thus obtained, which is invaluable to those who duly appreciate precision in the examination of cases."—*Balt. Chron.*

5. A PRACTICAL TREATISE ON THE MANAGEMENT AND DISEASES OF CHILDREN. By RICHARD T. EVANSON, M.D., Professor of Medicine,—and HENRY MAUNSELL, M.D. Professor of Midwifery,—in the College of Surgeons in Ireland.

"The authors of the work before us, have had the advantage of investigating the subject of Infantile Diseases, conjointly in a public institution—an advantage which no private medical man, however extensive his practice, could probably have. The observations being made conjointly too, offer a greater guarantee of correctness and authenticity, than if they emanated from a single source, however respectable. From their acquaintance, also, with foreign works, they have been able to bring up the Anatomy, Physiology, Pathology, and even Therapeutics, to a far higher level than is to be found in any previous work in the English language.

"The second chapter embraces the Management and Physical Education of Children. This chapter ought to be printed in gold letters, and hung up in the nursery of every family. It would save many lives, and prevent much suffering."—*Medico-Chirurg. Rev.*

"We know of no work to which, on the whole, so little can be objected in matter or manner. It is an elegant and practical compendium of Infantile Diseases; a safe guide in the Management of Children; and completely fulfils the purposes proposed."—*British Annals of Medicine,* No. VIII.

6. THE SURGEON'S PRACTICAL GUIDE IN DRESSING, and in the Methodic APPLICATION of BANDAGES. Illustrated by ONE HUNDRED ENGRAVINGS. By THOMAS CUTLER, M.D. late Staff Surgeon in the Belgian Army.

"CUTLER on Bandages, with one hundred illustrative Engravings, will be invaluable to the great majority of the profession, throughout this country. But few have had the opportunity, which a large hospital only affords, of becoming

acquainted with the best mode of applying apparatus, in cases of wounds, fractures, dislocations, &c. The plates and descriptions of this work, give this important information."—*Balt. Chron.*

7. ON THE INFLUENCE OF PHYSICAL AGENTS ON LIFE. By W. F. EDWARDS, M.D., F.R.S., etc. Translated from the French, by Drs. Hodgkin and Fisher. To which are added, some Observations on ELECTRICITY, and Notes to the work.

"This is a work of standard authority in Medicine; and, in a physiological point of view, is pre-eminently the most valuable publication of the present century; the experimental investigation instituted by the author, having done much towards solving many problems hitherto but partially understood. The work was originally presented in parts to the Royal Academy of Science of Paris, and so highly did they estimate the labours of the author, and so fully appreciate the services by him thus rendered to science and to humanity, that they awarded him, though a foreigner, the prize founded for the promotion of experimental physiology.

"His researches relate to what are denominated the Physical Agents, viz: Temperature, as modified in degree and duration; Electricity; Air, as regards quantity, motion or rest, density or rarity; Water, as a liquid and in a state of vapour; and Light; and his object is to show the effects produced on the human system by these agents which surround and are incessantly exercising an influence upon us.

"It is hardly necessary for us to say, that the design has been executed in a masterly manner, and that the profession is under deep obligations to Dr. Edwards, for so satisfactorily performing his task, and furnishing it with such a body of facts, and such a vast number of experiments, in illustration and confirmation of his views."—*Prov. Jour.*

8. PROF. HORNER'S NECROLOGICAL NOTICE OF DR. P. S. PHYSICK; Delivered before the American Philos. Society, May 4, 1838.

9. ESSAYS ON PHYSIOLOGY AND HYGIENE; viz:

I. REID'S EXPERIMENTAL INVESTIGATION INTO THE FUNCTIONS OF THE EIGHTH PAIR OF NERVES.

II. EHRENBERG'S MICROSCOPICAL OBSERVATIONS ON THE BRAIN AND NERVES (WITH NUMEROUS ENGRAVINGS).

III. ON THE COMBINATION OF MOTOR AND SENSITIVE NERVOUS ACTIVITY; by Professor STROMEYER, Hanover.

IV. VEGETABLE PHYSIOLOGY.

V. EXPERIMENTS ON THE BRAIN, SPINAL MARROW, AND NERVES. By Prof. MAYER, of Bonn (WITH WOODCUTS).

16

VI. PUBLIC HYGIENE.

VII. PROGRESS OF THE ANATOMY AND PHYSIOLOGY OF THE NERVOUS SYSTEM, DURING 1836, By Professor MULLER.

VIII. VITAL STATISTICS.

10. CURIOSITIES OF MEDICAL EXPERIENCE. BY J. G. MILLINGEN, Surgeon to the Forces, Member of the Medical Society of the Ancient Faculty of Paris, etc., etc.

" *Curiosities of Medical Experience.* By J. G. MILLINGEN, Surgeon to the Forces, etc. The Author or Compiler derived the idea which prompted him to write this work from D'Israeli's 'Curiosities of Literature;' and, in our view, he has made a book equally curious in its way with that one. The heads of his chapters are numerous and varied; and all his subjects are treated in an agreeable and comprehensible style to the general reader. The drift of the Author, too, is decidedly useful. We shall endeavour to give some extracts from this work." —*Nat. Gaz.*

11. MEDICAL CLINIC; or, Reports of Medical Cases: By G. ANDRAL, Professor of the Faculty of Medicine of Paris, etc. Condensed and Translated, with Observations extracted from the Writings of the most distinguished Medical Authors: By D. SPILLAN, M.D., etc., etc.; containing *Diseases of the Encephalon, &c.*, with Extracts from OLLIVIER's Work on *Diseases of the Spinal Cord* and its Membranes.

12. AN ESSAY ON DEW, and several Appearances connected with it; by WILLIAM CHARLES WELLS, M.D F.R.S., etc.

JOURNAL DEPARTMENT.

IN THE *ECLECTIC JOURNAL OF MEDICINE*, VOL. II., *or Second Year*,

HAVE BEEN PUBLISHED,

ORIGINAL ARTICLES on Animal Magnetism, Laryngeal Phthisis, Elephantiasis, the Use and Abuse of the Pessary, Dislocation of the Elbow Joint, Lithotripsy, Pneumonia Typhoides, Excision of the Neck of the Uterus, the Plague of Athens, translated from the Greek of Thucydides; the Application of Turpentine in Tetanus, Medical Schools and Professorships, the Use of the Balsam of Copaiba in Diseases of the Mucous Membrane of the Intestinal Canal, on an Improved Auriscope—with engraving; the Solar Speculum—with engraving, &c.; *Digests* and *Reviews* of several new Works. Numerous Articles on Physiology, Chemistry, Pathology, Therapeutics, Midwifery, Surgery, and Hygiene, with an Index and Title-page,—forming a handsome volume of near *five hundred* closely printed pages; which, with the L I B R A R Y, amounts to near *THREE THOUSAND* pages for TEN DOLLARS; containing as much matter, (and that selected from the best authors,) as in ordinary medical works would occupy five thousand pages, or twelve volumes.

In the Library, Third Year,

COMMENCING NOVEMBER, 1838, AND ENDING OCTOBER, 1839,

HAVE BEEN PUBLISHED,

1. LECTURES ON THE PHYSIOLOGY AND DISEASES OF THE CHEST; including the Principles of Physical and General Diagnosis. Delivered during the Spring Sessions of 1836 and 1837, at the Anatomical School, Kinnerton Street, near St. George's Hospital. BY CHARLES J. B. WILLIAMS, M.D., F.R.S. Professor of the Principles and Practice of Medicine in University College, London. With Engravings.

2. ESSAY UPON THE QUESTION, IS MEDICAL SCIENCE FAVORABLE TO SCEPTICISM? By JAMES W. DALE, M.D., of Newcastle, Delaware.

17

3. LECTURES ON THE PRINCIPLES OF SURGERY. By John Hunter, F.R.S. With Notes by James F. Palmer, Senior Surgeon to the St. George's and St. James's Dispensaries, &c. &c. With Plates.

"We have perused these lectures with no ordinary feelings of satisfaction. They embody an immense amount of important facts, directed with no common skill to the illustration and improvement of medical science generally, and of the surgical department in particular. Indeed we have no hesitation in saying, that, whatever be the position of the reader in the profession, he will not relinquish the perusal of these lectures without the consciousness of having usefully employed the time which he may have bestowed upon them. For they constitute, in the fullest sense of the term, a philosophical disquisition on the science of Surgery; and hence, embracing the great principles on which the whole art of healing rests, their interest will be felt by all who regard Medicine as a true branch of science, and who delight to witness the gradual development of principles in the right interpretation of the phenomena of nature."

"We cannot bring our notice of the present volume to a close without offering our testimony to the admirable manner in which the editor and annotator has fulfilled his part of the undertaking. The advancements and improvements that have been effected, up to our own day, not only in practical surgery, but in all the collateral departments, are constantly brought before the reader's attention in clear and concise terms."—*Brit. & For. Med. Rev.*

"The surgical lectures alone were sufficient to fix us long in our chair, and our pains were amply compensated by the perusal of the very words in which Hunter had instructed his class."

"The rescuing of these lectures from that oblivion which they must needs have fallen into in private hands, alone constitutes the editor of Hunter's works a benefactor to the student and the scholar."—*Medical Gazette.*

4. ON DENGUE; ITS HISTORY, PATHOLOGY, AND TREATMENT. By S. Henry Dickson, M.D., Professor of the Institutes and Practice of Medicine in the Medical College of S.C.

5. OUTLINES OF GENERAL PATHOLOGY. By George Freckleton, M.D., Fellow of the Royal College of Physicians.

6. URINARY DISEASES and their TREATMENT. By Robert Willis, M.D., Physician to the Royal Infirmary for Children, &c. &c.

"We do not know that a more competent author than Dr. Willis could have been found to undertake the task; possessing, as it is evident from his work that he does possess, an accurate acquaintance with the subject in all its details, considerable personal experience in the diseases of which he treats, capacity for lucid arrangement, and a style of communication commendable in every respect."

"Our notice of Dr. Willis's work most here terminate. It is one which we have read and trust again to read with profit. The history of discovery is successfully given; cases curious and important; illustrative of the various subjects have been selected from many new sources, as well as detailed from the author's own experi-

ence, chemical analyses, not too elaborate, have been afforded, which will be most convenient to those who wish to investigate the qualities of the urine in disease; the importance of attending to this secretion in order to a proper understanding of disease is strongly insisted upon; in short, a book has been composed, which was much required, and which we can conscientiously and confidently recommend as likely to be useful to all classes of practitioners.—*Brit. & For. Med. Rev.*

7. LECTURES on BLOOD-LETTING. By Henry Clutterbuck, M.D.

8. THE LIFE OF JOHN HUNTER, F.R.S. By Drewry Ottley.

"In the summing up of Mr. Hunter's character, Mr. Ottley exhibits equal judgment and candour."—*Brit. & For. Med. Rev.*

9. HUNTER'S TREATISE ON THE VENEREAL DISEASE. With Notes by Dr. Babington. With Plates.

"Under the hands of Mr. Babington, who has performed his task as editor in a very exemplary manner, the work has assumed quite a new value, and may now be as advantageously placed in the library of the student as in that of the experienced surgeon.—*Brit. & For. Med. Rev.*

"The notes, in illustration of the text, contain a summary of our present knowledge on the subject; the manner in which these notes are constructed is at once clever and perspicuous; and the modes of treatment prescribed, spring from a right apprehension of the disease. We would recommend to the reader the note on the primary venereal sore; the note itself is an essay in every word of which we fully concur."—*Med. Gazette.*

10. A TREATISE ON THE TEETH. By John Hunter. With Notes by Thomas Bell, F.R.S. With Plates.

"The treatise on the teeth is edited by Mr. Bell, a gentleman accomplished in his art. Mr. Bell has studied his subject with the greatest minuteness and care; and in appropriate notes at the first of the page corrects the author with the air of a gentleman, and the accuracy of a man of science. The matter contained in these short notes forms an ample scholium to the text; and without aiming at the slightest display of learning, they at the same time exhibit a ready knowledge on every point, and an extensive information both of comparative anatomy and pathology.—*Med. Gazette.*

11. MEDICAL AND TOPOGRAPHICAL OBSERVATIONS upon the MEDITERRANEAN and upon PORTUGAL, SPAIN, AND OTHER COUNTRIES. By G. R. B. Horner, M.D., Surgeon U. S. Navy, and Honorary Member of the Philadelphia Medical Society. With Engravings.

"An uncommonly interesting book is presented to those who have any disposition to know the things medical in Portugal, Spain, and other countries," and "will doubtless be read, also, with marked satisfaction by all who have a taste for travels."—*Bost. Med. and Surg. Jour.*

12. LECTURES ON THE BLOOD, and on the CHANGES which it UNDERGOES DURING DISEASE. By F. Magendie, M.D.

18

JOURNAL DEPARTMENT.

IN THE *ECLECTIC JOURNAL OF MEDICINE*, Vols. III., IV., or *Third* and *Fourth Years*, 1838–40, HAVE BEEN PUBLISHED,

ORIGINAL REVIEWS and BIBLIOGRAPHICAL NOTICES, viz., of Hosack's Lectures on the Theory and Practice of Physic, Walker on Intermarriage, the Works of John Hunter, Introductory Lectures, Granville on Counter-Irritation, Gallup's Outlines of the Institutes of Medicine, Bouvier on Club Foot, Harris's Dental Surgery, Vimont on Human and Comparative Physiology, &c., &c.: *also*, Selected *Reviews* of Lonsdale and Burke on Fractures, Foissac on the Influence of Climate, Lecanu and Denis on the Chemistry of the Blood, Gondret &c. on Counter-Irritation, &c., &c.: *Papers* on Club Foot, Yellow Fever, Pathology of the Ovaria, Extirpation of the Parotid Gland, Endermic Medicine, Simple Ulceration of the Stomach, Artificial Digestion, Diseases of the Kidneys, Diseases of the Spine, Irritable Bladder, Fibres of the Spinal Marrow, Experiments on the Blood, Galvanic Experiments on a Dead Body, &c., &c., and numerous other articles on *Therapeutics, Pathology, Surgery, and Midwifery.*

*** *Each Volume of the JOURNAL contains above 500 pages of closely printed matter.*

In the Library, Fourth Year,

COMMENCING NOVEMBER 1839, AND ENDING OCTOBER 1840,

HAVE BEEN PUBLISHED,

1. MEDICAL NOTES AND REFLECTIONS. By HENRY HOLLAND, M.D., F.R.S., Fellow of the Royal College of Physicians, and Physician Extraordinary to the Queen.

2. CLINICAL REMARKS ON SOME CASES OF LIVER ABSCESS PRESENTING EXTERNALLY. By JOHN G. MALCOLMSON, M.D. Surgeon Hon. E. I. C. Service, Fellow of the Royal Asiatic Society, and the Geological Society, London.

3. HISTORICAL NOTICES ON THE OCCURRENCE OF INFLAMMATORY AFFECTIONS OF THE INTERNAL ORGANS AFTER EXTERNAL INJURIES AND SURGICAL OPERATIONS. By WILLIAM THOMPSON, M.D., &c. &c.

4. A EXPERIMENTAL INVESTIGATION INTO THE FUNCTIONS OF THE EIGHTH PAIR OF NERVES. By John Reid, M.D., &c.

5. TREATISE ON THE BLOOD, INFLAMMATION, AND GUN-SHOT WOUNDS. By JOHN HUNTER, F.R.S. With Notes, by JAMES F. PALMER, Senior Surgeon to the St. George's and St. James's Dispensary, &c., &c.

6. A PRACTICAL TREATISE ON VENEREAL DISORDERS, AND MORE ESPECIALLY ON THE HISTORY AND TREATMENT OF CHANCRE. BY PHILIPPE RICORD, M.D., Surgeon to the Venereal Hospital at Paris.

7. A TREATISE ON INFLAMMATION. By JAMES MACARTNEY, F.R.S., F.L.S., &c., &c. Member of the Royal College, of Surgeons, London, &c., &c.

8. AMUSSAT'S LECTURES ON THE RETENTION OF URINE, CAUSED BY STRICTURES OF THE URETHRA, and on the Diseases of the Prostate, translated from the French by James P. Jervey, M.D.

9. OBSERVATIONS ON CERTAIN PARTS OF THE ANIMAL ŒCONOMY, Inclusive of several papers from the Philosophical Transactions, &c. By JOHN HUNTER, F.R.S., &c., &c. With Notes by RICHARD OWEN, F.R.S.

"One distinctive feature of the present edition of Hunter's works has been already mentioned, viz: in the addition of illustrative notes, which are not thrown in at hazard, but are written by men who are already eminent for their skill and attainments on the particular subjects which they have thus illustrated. By this means, whilst we have the views entire of John Hunter in the text, we are enabled by reference to the accompanying notes, to see wherein the author is borne out by the positive knowledge of the present day, or to what extent his views require modification and correction. The names of the gentlemen who have in this manner assisted Mr. Palmer, are guarantees of the successful performance of their task."

10. HINTS ON THE MEDICAL EXAMINATION OF RECRUITS FOR THE ARMY; and on the Discharge of Soldiers from the Service on Surgeon's Certificate : Adapted to the Service of the United States. By THOMAS HENDERSON, M. D., Assistant Surgeon U. S. Army, &c., &c.

11. AN ESSAY ON HYSTERIA; being an analysis of its irregular and aggravated forms; including Hysterical Hemorrhage

and Hysterical Ischuria. With numerous Illustrations and Curious Cases. By THOMAS LAYCOCK, House Surgeon to the York County Hospital.

12. A TREATISE ON THE CAUSES AND CONSEQUENCES OF HABITUAL CONSTIPATON. By JOHN BURNE, M.D., Fellow of the Royal College of Physicians, Physician to the Westminster Hospital, &c. &c.

" For some interesting cases illustrative of this work, the author is indebted to Dr. Williams, Dr. Stroud, Dr. Callaway. Mr. Morgan, Mr. Taunton, Dr. Roots, Sir Astley Cooper, Sir Benjamin Brodie, Mr. Tupper, Mr. Butler, Dr. Paris, Mr. Dendy, Dr. Hen. U. Thomson," &c.—*Preface.*

13. A TREATISE ON MENTAL DISEASES. By M. ESQUIROL.

☞ The Works published in either year, as enumerated above, with the ECLECTIC JOURNAL, can be obtained bound in 6 vols. for $13; or, the whole Series, 24 volumes, bound in uniform style, for $52.

TERMS OF SUBSCRIPTION

TO

THE SELECT MEDICAL LIBRARY,

AND

BULLETIN OF MEDICAL SCIENCE.

PUBLISHED QUARTERLY,

IN JANUARY, APRIL, JULY, AND OCTOBER.

EACH Number of the Library will consist of one or more approved works on some branch of Medicine, including, of course, Surgery and Obstetrics.

Every work in the Library will be completed in the number in which it is begun, unless the subject naturally admits of division ; and hence the size of the numbers will vary. It will be done up in a strong paper cover, and each work labelled on the back; thus obviating the *immediate necessity* of binding.

Subscribers will receive fourteen hundred pages of closely printed matter of Library in the year.

To each number of the Library will be appended a BULLETIN of MEDICAL SCIENCE.

FIVE DOLLARS *per annum*, in advance ; and in no single instance, out of the chief cities, will this rule be departed from.

Subscribers who wish to receive the Library direct from us must remit the amount of their subscription ; as none of our agents are authorized to receive money on our account ; nor will we hold ourselves responsible for any defalcations on their part.

BARRINGTON & HASWELL,
PUBLISHERS, 293 *Market Street, Philadelphia.*

The attention of the Medical Faculty is respectfully invited to this Periodical. In the fifth year, which is now in course of successful publication, have been published, in the January number, A PRACTICAL DICTIONARY OF MATERIA MEDICA, *including the Composition, Preparation and Uses of Medicines ; and a large number of Extemporaneous Formulæ : together with important Toxicological Observations ; on the Basis of Brande's Dictionary of Materia Medica and Practical Pharmacy;* by JOHN BELL, M.D., Lecturer on Materia Medica and Therapeutics, &c. &c. ; and in the April number, *Schill's Outlines of Pathological Semeiology, translated by D. Spillan, M.D.,* &c. &c. ; with *Aretæus on the Causes and Signs of Acute Disease :* — containing also, in the BULLETIN, Notices of New Works, Medical Schools, and other valuable professional information.

20

www.ingramcontent.com/pod-product-compliance
Lightning Source LLC
Chambersburg PA
CBHW031349290326
41932CB00044B/782